MW00814618

The Periodic Table of Elements

Edited by Paul F. Kisak

Contents

Chapter 1

Periodic table

This article is about the table used in chemistry. For other uses, see Periodic table (disambiguation).

Group→	1	2	3	4	5	6	7	8	9	10	11	12	13	14	15	16	17	18
↓Period																		
1	1 H																	2 He
2	3 Li	4 Be											5 B	6 C	7 N	8 O	9 F	10 Ne
3	11 Na	12 Mg											13 Al	14 Si	15 P	16 S	17 Cl	18 Ar
4	19 K	20 Ca	21 Sc	22 Ti	23 V	24 Cr	25 Mn	26 Fe	27 Co	28 Ni	29 Cu	30 Zn	31 Ga	32 Ge	33 As	34 Se	35 Br	36 Kr
5	37 Rb	38 Sr	39 Y	40 Zr	41 Nb	42 Mo	43 Tc	44 Ru	45 Rh	46 Pd	47 Ag	48 Cd	49 In	50 Sn	51 Sb	52 Te	53 I	54 Xe
6	55 Cs	56 Ba	*	72 Hf	73 Ta	74 W	75 Re	76 Os	77 Ir	78 Pt	79 Au	80 Hg	81 Tl	82 Pb	83 Bi	84 Po	85 At	86 Rn
7	87 Fr	88 Ra	**	104 Rf	105 Db	106 Sg	107 Bh	108 Hs	109 Mt	110 Ds	111 Rg	112 Cn	113 Uut	114 Fl	115 Uup	116 Lv	117 Uus	118 Uuo

*	57 La	58 Ce	59 Pr	60 Nd	61 Pm	62 Sm	63 Eu	64 Gd	65 Tb	66 Dy	67 Ho	68 Er	69 Tm	70 Yb	71 Lu
**	89 Ac	90 Th	91 Pa	92 U	93 Np	94 Pu	95 Am	96 Cm	97 Bk	98 Cf	99 Es	100 Fm	101 Md	102 No	103 Lr

Standard form of the periodic table (color legend below)

The **periodic table** is a tabular arrangement of the chemical elements, ordered by their atomic number (number of protons in the nucleus), electron configurations, and recurring chemical properties. The table also shows four rectangular blocks: s-, p- d- and f-block. In general, within one row (period) the elements are metals on the lefthand side, and non-metals on the righthand side.

The rows of the table are called periods; the columns are called groups. Six groups (columns) have names as well as numbers: for example, group 17 elements are the halogens; and group 18, the noble gases. The periodic table can be used to derive relationships between the properties of the elements, and predict the properties of new elements yet to be discovered or synthesized. The periodic table provides a useful framework for analyzing chemical behavior, and is widely used in chemistry and other sciences.

Although precursors exist, Dmitri Mendeleev is generally credited with the publication, in 1869, of the first widely recognized periodic table. He developed his table to illustrate periodic trends in the properties of the then-known elements.

Mendeleev also predicted some properties of then-unknown elements that would be expected to fill gaps in this table. Most of his predictions were proved correct when the elements in question were subsequently discovered. Mendeleev's periodic table has since been expanded and refined with the discovery or synthesis of further new elements and the development of new theoretical models to explain chemical behavior.

All elements from atomic numbers 1 (hydrogen) to 118 (ununoctium) have been discovered or reportedly synthesized, with elements 113, 115, 117, and 118 having yet to be confirmed. The first 94 elements exist naturally, although some are found only in trace amounts and were synthesized in laboratories before being found in nature.[n 1] Elements with atomic numbers from 95 to 118 have only been synthesized in laboratories. It has been shown that elements 95 to 100 once occurred in nature but currently do not.[1] Synthesis of elements having higher atomic numbers is being pursued. Numerous synthetic radionuclides of naturally occurring elements have also been produced in laboratories.

1.1 Overview

For large cell versions, see Periodic table (large cells).

Some presentations include an element zero (i.e. a substance composed purely of neutrons), although this is uncommon. See, for example. Philip Stewart's Chemical Galaxy. Each chemical element has a unique atomic number representing the number of protons in its nucleus. Most elements have differing numbers of neutrons among different atoms, with these variants being referred to as isotopes. For example, carbon has three naturally occurring isotopes: all of its atoms have six protons and most have six neutrons as well, but about one per cent have seven neutrons, and a very small fraction have eight neutrons. Isotopes are never separated in the periodic table; they are always grouped together under a single element. Elements with no stable isotopes have the atomic masses of their most stable isotopes, where such masses are shown, listed in parentheses.[2]

In the standard periodic table, the elements are listed in order of increasing atomic number (the number of protons in the nucleus of an atom). A new row (*period*) is started when a new electron shell has its first electron. Columns (*groups*) are determined by the electron configuration of the atom; elements with the same number of electrons in a particular subshell fall into the same columns (e.g. oxygen and selenium are in the same column because they both have four electrons in the outermost p-subshell). Elements with similar chemical properties generally fall into the same group in the periodic table, although in the f-block, and to some respect in the d-block, the elements in the same period tend to have similar properties, as well. Thus, it is relatively easy to predict the chemical properties of an element if one knows the properties of the elements around it.[3]

As of 2014, the periodic table has 114 confirmed elements, comprising elements 1 (hydrogen) to 112 (copernicium), 114 (flerovium) and 116 (livermorium). Elements 113, 115, 117 and 118 have reportedly been synthesised in laboratories, but none of these claims have been officially confirmed by the International Union of Pure and Applied Chemistry (IUPAC), nor are they named. As such these elements are currently identified by their atomic number (e.g., "element 113"), or by their provisional systematic name ("ununtrium", symbol "Uut").[4]

A total of 94 elements occur naturally; the remaining 20 elements, from americium to copernicium, and flerovium and livermorium, occur only when synthesised in laboratories. Of the 94 elements that occur naturally, 84 are primordial. The other 10 naturally occurring elements occur only in decay chains of primordial elements.[1] No element heavier than einsteinium (element 99) has ever been observed in macroscopic quantities in its pure form, nor has astatine (element 85); francium (element 87) has been only photographed in the form of light emitted from microscopic quantities (300,000 atoms).[5]

1.1.1 Layout variants

In the most common graphic presentation of the periodic table, the main table has 18 columns and the lanthanides and the actinides are shown as two additional rows below the main body of the table,[6] with two placeholders shown in the main table, between barium and hafnium, and radium and rutherfordium, respectively. These placeholders can be asterisk-like markers, or a contracted range description of elements ("57–71"). This convention is entirely a matter of formatting practicality. The same table structure can be shown in a 32-column format, with the lanthanides and actinides in the main table's row 6 and 7.

However, based on the chemical and physical properties of elements, many alternative table *structures* have been constructed.

1.2 Grouping methods

1.2.1 Groups

Main article: Group (periodic table)

A *group* or *family* is a vertical column in the periodic table. Groups usually have more significant periodic trends than periods and blocks, explained below. Modern quantum mechanical theories of atomic structure explain group trends by proposing that elements within the same group generally have the same electron configurations in their valence shell.[7] Consequently, elements in the same group tend to have a shared chemistry and exhibit a clear trend in properties with increasing atomic number.[8] However, in some parts of the periodic table, such as the d-block and the f-block, horizontal similarities can be as important as, or more pronounced than, vertical similarities.[9][10][11]

Under an international naming convention, the groups are numbered numerically from 1 to 18 from the leftmost column (the alkali metals) to the rightmost column (the noble gases).[12] Previously, they were known by roman numerals. In America, the roman numerals were followed by either an "A" if the group was in the s- or p-block, or a "B" if the group was in the d-block. The roman numerals used correspond to the last digit of today's naming convention (e.g. the group 4 elements were group IVB, and the group 14 elements was group IVA). In Europe, the lettering was similar, except that "A" was used if the group was before group 10, and "B" was used for groups including and after group 10. In addition, groups 8, 9 and 10 used to be treated as one triple-sized group, known collectively in both notations as group VIII. In 1988, the new IUPAC naming system was put into use, and the old group names were deprecated.[13]

Some of these groups have been given trivial (unsystematic) names, as seen in the table below, although some are rarely used. Groups 3–10 have no trivial names and are referred to simply by their group numbers or by the name of the first member of their group (such as 'the scandium group' for Group 3), since they display fewer similarities and/or vertical trends.[12]

Elements in the same group tend to show patterns in atomic radius, ionization energy, and electronegativity. From top to bottom in a group, the atomic radii of the elements increase. Since there are more filled energy levels, valence electrons are found farther from the nucleus. From the top, each successive element has a lower ionization energy because it is easier to remove an electron since the atoms are less tightly bound. Similarly, a group has a top to bottom decrease in electronegativity due to an increasing distance between valence electrons and the nucleus.[14] There are exceptions to these trends, however, an example of which occurs in group 11 where electronegativity increases farther down the group.[15]

1.2.2 Periods

Main article: Period (periodic table)

A *period* is a horizontal row in the periodic table. Although groups generally have more significant periodic trends, there are regions where horizontal trends are more significant than vertical group trends, such as the f-block, where the lanthanides and actinides form two substantial horizontal series of elements.[16]

Elements in the same period show trends in atomic radius, ionization energy, electron affinity, and electronegativity. Moving left to right across a period, atomic radius usually decreases. This occurs because each successive element has an added proton and electron, which causes the electron to be drawn closer to the nucleus.[17] This decrease in atomic radius also causes the ionization energy to increase when moving from left to right across a period. The more tightly bound an element is, the more energy is required to remove an electron. Electronegativity increases in the same manner as ionization energy because of the pull exerted on the electrons by the nucleus.[14] Electron affinity also shows a slight trend across a period. Metals (left side of a period) generally have a lower electron affinity than nonmetals (right side of a period), with the exception of the noble gases.[18]

1.2.3 Blocks

Main article: Block (periodic table)
 Specific regions of the periodic table can be referred to as *blocks* in recognition of the sequence in which the electron

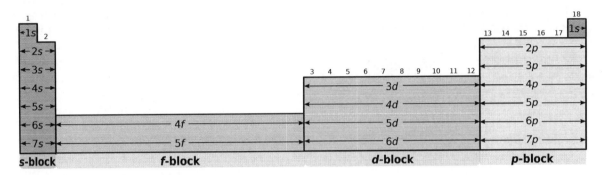

Left to right: s-, f-, d-, p-block in the periodic table

shells of the elements are filled. Each block is named according to the subshell in which the "last" electron notionally resides.[19][n 2] The s-block comprises the first two groups (alkali metals and alkaline earth metals) as well as hydrogen and helium. The p-block comprises the last six groups, which are groups 13 to 18 in IUPAC (3A to 8A in American) and contains, among other elements, all of the metalloids. The d-block comprises groups 3 to 12 (or 3B to 2B in American group numbering) and contains all of the transition metals. The f-block, often offset below the rest of the periodic table, has no group numbers and comprises lanthanides and actinides.[20]

1.2.4 Metals, metalloids and nonmetals

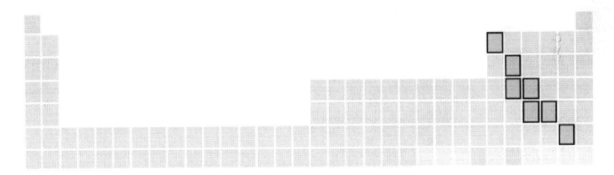

Metals, metalloids, nonmetals, and elements with unknown chemical properties in the periodic table. Sources disagree on the classification of some of these elements.

According to their shared physical and chemical properties, the elements can be classified into the major categories of metals, metalloids and nonmetals. Metals are generally shiny, highly conducting solids that form alloys with one another and salt-like ionic compounds with nonmetals (other than the noble gases). The majority of nonmetals are coloured or colourless insulating gases; nonmetals that form compounds with other nonmetals feature covalent bonding. In between metals and nonmetals are metalloids, which have intermediate or mixed properties.[21]

Metal and nonmetals can be further classified into subcategories that show a gradation from metallic to non-metallic properties, when going left to right in the rows. The metals are subdivided into the highly reactive alkali metals, through the less reactive alkaline earth metals, lanthanides and actinides, via the archetypal transition metals, and ending in the physically and chemically weak post-transition metals. The nonmetals are simply subdivided into the polyatomic nonmetals, which, being nearest to the metalloids, show some incipient metallic character; the diatomic nonmetals, which are essentially nonmetallic; and the monatomic noble gases, which are nonmetallic and almost completely inert. Specialized groupings such as the refractory metals and the noble metals, which are subsets (in this example) of the transition metals, are also known[22] and occasionally denoted.[23]

Placing the elements into categories and subcategories based on shared properties is imperfect. There is a spectrum of properties within each category and it is not hard to find overlaps at the boundaries, as is the case with most classification schemes.[24] Beryllium, for example, is classified as an alkaline earth metal although its amphoteric chemistry and tendency to mostly form covalent compounds are both attributes of a chemically weak or post transition metal. Radon is classified as a nonmetal and a noble gas yet has some cationic chemistry that is more characteristic of a metal. Other classification schemes are possible such as the division of the elements into mineralogical occurrence categories, or crystalline structures. Categorising the elements in this fashion dates back to at least 1869 when Hinrichs[25] wrote that simple boundary lines could be drawn on the periodic table to show elements having like properties, such as the metals and the nonmetals, or the gaseous elements.

1.3 Periodic trends

Main article: Periodic trends

1.3.1 Electron configuration

Main article: Electronic configuration
 The electron configuration or organisation of electrons orbiting neutral atoms shows a recurring pattern or periodicity. The electrons occupy a series of electron shells (numbered shell 1, shell 2, and so on). Each shell consists of one or more subshells (named s, p, d, f and g). As atomic number increases, electrons progressively fill these shells and subshells more or less according to the Madelung rule or energy ordering rule, as shown in the diagram. The electron configuration for neon, for example, is $1s^2 2s^2 2p^6$. With an atomic number of ten, neon has two electrons in the first shell, and eight electrons in the second shell—two in the s subshell and six in the p subshell. In periodic table terms, the first time an electron occupies a new shell corresponds to the start of each new period, these positions being occupied by hydrogen and the alkali metals.[26][27]

Since the properties of an element are mostly determined by its electron configuration, the properties of the elements likewise show recurring patterns or periodic behaviour, some examples of which are shown in the diagrams below for atomic radii, ionization energy and electron affinity. It is this periodicity of properties, manifestations of which were noticed well before the underlying theory was developed, that led to the establishment of the periodic law (the properties of the elements recur at varying intervals) and the formulation of the first periodic tables.[26][27]

1.3.2 Atomic radii

Main article: Atomic radius
 Atomic radii vary in a predictable and explainable manner across the periodic table. For instance, the radii generally decrease along each period of the table, from the alkali metals to the noble gases; and increase down each group. The radius increases sharply between the noble gas at the end of each period and the alkali metal at the beginning of the next period. These trends of the atomic radii (and of various other chemical and physical properties of the elements) can be explained by the electron shell theory of the atom; they provided important evidence for the development and confirmation of quantum theory.[28]

The electrons in the 4f-subshell, which is progressively filled from cerium (element 58) to ytterbium (element 70), are not particularly effective at shielding the increasing nuclear charge from the sub-shells further out. The elements immediately following the lanthanides have atomic radii that are smaller than would be expected and that are almost identical to the atomic radii of the elements immediately above them.[29] Hence hafnium has virtually the same atomic radius (and

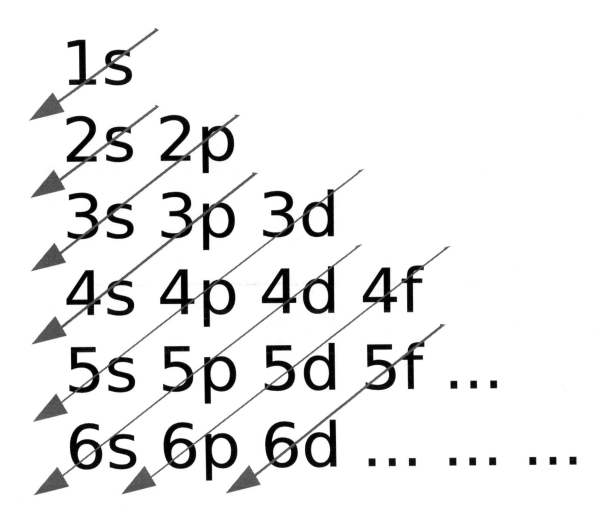

Approximate order in which shells and subshells are arranged by increasing energy according to the Madelung rule

chemistry) as zirconium, and tantalum has an atomic radius similar to niobium, and so forth. This is known as the lanthanide contraction. The effect of the lanthanide contraction is noticeable up to platinum (element 78), after which it is masked by a relativistic effect known as the inert pair effect.[30] The d-block contraction, which is a similar effect between the d-block and p-block, is less pronounced than the lanthanide contraction but arises from a similar cause.[29]

1.3.3   Ionization energy

Main article: Ionization energy

The first ionization energy is the energy it takes to remove one electron from an atom, the second ionization energy is the energy it takes to remove a second electron from the atom, and so on. For a given atom, successive ionization energies increase with the degree of ionization. For magnesium as an example, the first ionization energy is 738 kJ/mol and the second is 1450 kJ/mol. Electrons in the closer orbitals experience greater forces of electrostatic attraction; thus, their removal requires increasingly more energy. Ionization energy becomes greater up and to the right of the periodic table.[30]

Large jumps in the successive molar ionization energies occur when removing an electron from a noble gas (complete electron shell) configuration. For magnesium again, the first two molar ionization energies of magnesium given above correspond to removing the two 3s electrons, and the third ionization energy is a much larger 7730 kJ/mol, for the removal of a 2p electron from the very stable neon-like configuration of Mg^{2+}. Similar jumps occur in the ionization energies of

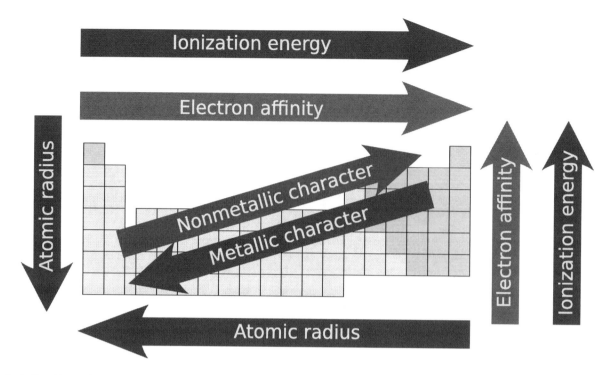

Periodic table trends (arrows direct an increase)

other third-row atoms.[30]

1.3.4 Electronegativity

Main article: Electronegativity

Electronegativity is the tendency of an atom to attract electrons.[31] An atom's electronegativity is affected by both its atomic number and the distance between the valence electrons and the nucleus. The higher its electronegativity, the more an element attracts electrons. It was first proposed by Linus Pauling in 1932.[32] In general, electronegativity increases on passing from left to right along a period, and decreases on descending a group. Hence, fluorine is the most electronegative of the elements,[n 4] while caesium is the least, at least of those elements for which substantial data is available.[15]

There are some exceptions to this general rule. Gallium and germanium have higher electronegativities than aluminium and silicon respectively because of the d-block contraction. Elements of the fourth period immediately after the first row of the transition metals have unusually small atomic radii because the 3d-electrons are not effective at shielding the increased nuclear charge, and smaller atomic size correlates with higher electronegativity.[15] The anomalously high electronegativity of lead, particularly when compared to thallium and bismuth, appears to be an artifact of data selection (and data availability)—methods of calculation other than the Pauling method show the normal periodic trends for these elements.[33]

1.3.5 Electron affinity

Main article: Electron affinity

The electron affinity of an atom is the amount of energy released when an electron is added to a neutral atom to form a negative ion. Although electron affinity varies greatly, some patterns emerge. Generally, nonmetals have more positive electron affinity values than metals. Chlorine most strongly attracts an extra electron. The electron affinities of the noble gases have not been measured conclusively, so they may or may not have slightly negative values.[36]

Electron affinity generally increases across a period. This is caused by the filling of the valence shell of the atom; a group 17 atom releases more energy than a group 1 atom on gaining an electron because it obtains a filled valence shell and is

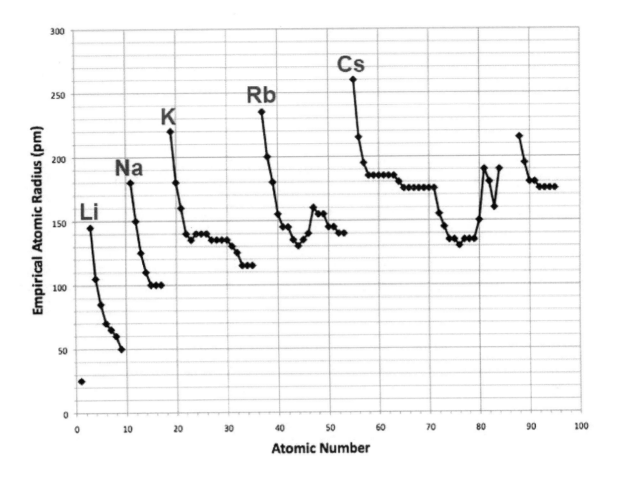

Atomic number plotted against atomic radius[In 3]

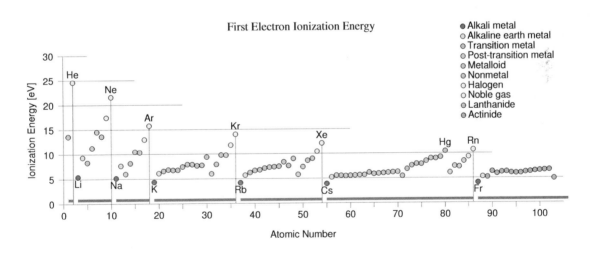

Ionization energy: each period begins at a minimum for the alkali metals, and ends at a maximum for the noble gases

therefore more stable.[36]

A trend of decreasing electron affinity going down groups would be expected. The additional electron will be entering an orbital farther away from the nucleus. As such this electron would be less attracted to the nucleus and would release less energy when added. However, in going down a group, around one-third of elements are anomalous, with heavier elements

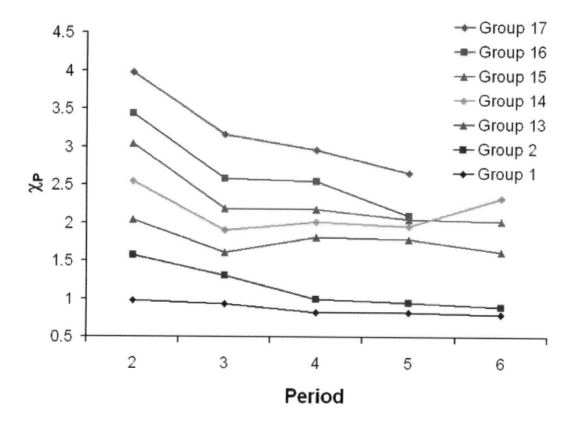

Graph showing increasing electronegativity with growing number of selected groups

having higher electron affinities than their next lighter congenors. Largely, this is due to the poor shielding by d and f electrons. A uniform decrease in electron affinity only applies to group 1 atoms.[37]

1.3.6 Metallic character

The lower the values of ionization energy, electronegativity and electron affinity, the more metallic character the element has. Conversely, nonmetallic character increases with higher values of these properties.[38] Given the periodic trends of these three properties, metallic character tends to decrease going across a period (or row) and, with some irregularities (mostly) due to poor screening of the nucleus by d and f electrons, and relativistic effects,[39] tends to increase going down a group (or column or family). Thus, the most metallic elements (such as caesium and francium) are found at the bottom left of traditional periodic tables and the most nonmetallic elements (oxygen, fluorine, chlorine) at the top right. The combination of horizontal and vertical trends in metallic character explains the stair-shaped dividing line between metals and nonmetals found on some periodic tables, and the practice of sometimes categorizing several elements adjacent to that line, or elements adjacent to those elements, as metalloids.[40][41]

1.4 History

Main article: History of the periodic table

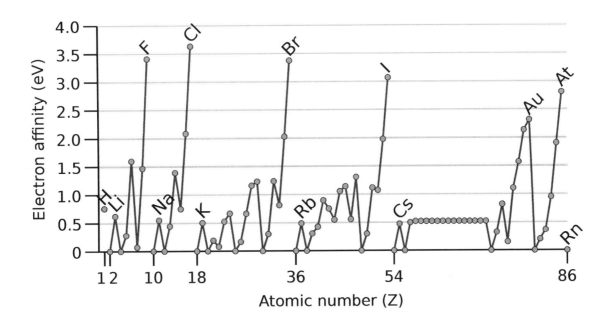

Dependence of electron affinity on atomic number.[34] *Values generally increase across each period, culminating with the halogens before decreasing precipitously with the noble gases. Examples of localized peaks seen in hydrogen, the alkali metals and the group 11 elements are caused by a tendency to complete the s-shell (with the 6s shell of gold being further stabilized by relativistic effects and the presence of a filled 4f sub shell). Examples of localized troughs seen in the alkaline earth metals, and nitrogen, phosphorus, manganese and rhenium are caused by filled s-shells, or half-filled p- or d-shells.*[35]

1.4.1   First systemization attempts

In 1789, Antoine Lavoisier published a list of 33 chemical elements, grouping them into gases, metals, nonmetals, and earths.[42] Chemists spent the following century searching for a more precise classification scheme. In 1829, Johann Wolfgang Döbereiner observed that many of the elements could be grouped into triads based on their chemical properties. Lithium, sodium, and potassium, for example, were grouped together in a triad as soft, reactive metals. Döbereiner also observed that, when arranged by atomic weight, the second member of each triad was roughly the average of the first and the third;[43] this became known as the Law of Triads.[44] German chemist Leopold Gmelin worked with this system, and by 1843 he had identified ten triads, three groups of four, and one group of five. Jean-Baptiste Dumas published work in 1857 describing relationships between various groups of metals. Although various chemists were able to identify relationships between small groups of elements, they had yet to build one scheme that encompassed them all.[43]

In 1858, German chemist August Kekulé observed that carbon often has four other atoms bonded to it. Methane, for example, has one carbon atom and four hydrogen atoms. This concept eventually became known as valency; different elements bond with different numbers of atoms.[45]

In 1862, Alexandre-Emile Béguyer de Chancourtois, a French geologist, published an early form of periodic table, which he called the telluric helix or screw. He was the first person to notice the periodicity of the elements. With the elements arranged in a spiral on a cylinder by order of increasing atomic weight, de Chancourtois showed that elements with similar properties seemed to occur at regular intervals. His chart included some ions and compounds in addition to elements. His paper also used geological rather than chemical terms and did not include a diagram; as a result, it received little attention until the work of Dmitri Mendeleev.[46]

In 1864, Julius Lothar Meyer, a German chemist, published a table with 44 elements arranged by valency. The table showed that elements with similar properties often shared the same valency.[47] Concurrently, William Odling (an English chemist) published an arrangement of 57 elements, ordered on the basis of their atomic weights. With some irregularities and gaps, he noticed what appeared to be a periodicity of atomic weights among the elements and that this accorded with 'their usually received groupings.'[48] Odling alluded to the idea of a periodic law but did not pursue it.[49] He subsequently proposed (in 1870) a valence-based classification of the elements.[50]

1 H																	2 He
3 Li	4 Be											5 B	6 C	7 N	8 O	9 F	10 Ne
11 Na	12 Mg											13 Al	14 Si	15 P	16 S	17 Cl	18 Ar
19 K	20 Ca	21 Sc	22 Ti	23 V	24 Cr	25 Mn	26 Fe	27 Co	28 Ni	29 Cu	30 Zn	31 Ga	32 Ge	33 As	34 Se	35 Br	36 Kr
37 Rb	38 Sr	39 Y	40 Zr	41 Nb	42 Mo	43 Tc	44 Ru	45 Rh	46 Pd	47 Ag	48 Cd	49 In	50 Sn	51 Sb	52 Te	53 I	54 Xe
55 Cs	56 Ba	57 -71	72 Hf	73 Ta	74 W	75 Re	76 Os	77 Ir	78 Pt	79 Au	80 Hg	81 Tl	82 Pb	83 Bi	84 Po	85 At	86 Rn
87 Fr	88 Ra	89 -103	104 Rf	105 Db	106 Sg	107 Bh	108 Hs	109 Mt	110 Ds	111 Rg	112 Cn	113 Uut	114 Fl	115 Uup	116 Lv	117 Uus	118 Uuo

57 La	58 Ce	59 Pr	60 Nd	61 Pm	62 Sm	63 Eu	64 Gd	65 Tb	66 Dy	67 Ho	68 Er	69 Tm	70 Yb	71 Lu
89 Ac	90 Th	91 Pa	92 U	93 Np	94 Pu	95 Am	96 Cm	97 Bk	98 Cf	99 Es	100 Fm	101 Md	102 No	103 Lr

■ Known in antiquity
▨ also known when (akw) Levoisier published his list of elements (1789)
□ akw Mendeleev published his periodic table (1869)
□ akw Deming published his periodic table (1923)

▨ akw Seaborg published his periodic table (1945)
□ also known (ak) up to 2000
▨ ak to 2012

The discovery of the elements mapped to significant periodic table development dates (pre-, per- and post-)

No.		No.		No.		No.		No.		No.		No.		No.	
H	1	F	8	Cl	15	Co & Ni	22	Br	29	Pd	36	I	42	Pt & Ir	50
Li	2	Na	9	K	16	Cu	23	Rb	30	Ag	37	Cs	44	Os	51
G	3	Mg	10	Ca	17	Zn	24	Sr	31	Cd	38	Ba & V	45	Hg	52
Bo	4	Al	11	Cr	19	Y	25	Ce & La	33	U	40	Ta	46	Tl	53
C	5	Si	12	Ti	18	In	26	Zr	32	Sn	39	W	47	Pb	54
N	6	P	13	Mn	20	As	27	Di & Mo	34	Sb	41	Nb	48	Bi	55
O	7	S	14	Fe	21	Se	28	Ro & Ru	35	Te	43	Au	49	Th	56

Newlands's periodic table, as presented to the Chemical Society in 1866, and based on the law of octaves

English chemist John Newlands produced a series of papers from 1863 to 1866 noting that when the elements were listed in order of increasing atomic weight, similar physical and chemical properties recurred at intervals of eight; he likened such periodicity to the octaves of music.[51][52] This so termed Law of Octaves, however, was ridiculed by Newlands' contemporaries, and the Chemical Society refused to publish his work.[53] Newlands was nonetheless able to draft a table of the elements and used it to predict the existence of missing elements, such as germanium.[54] The Chemical Society only acknowledged the significance of his discoveries five years after they credited Mendeleev.[55]

In 1867, Gustavus Hinrichs, a Danish born academic chemist based in America, published a spiral periodic system based on atomic spectra and weights, and chemical similarities. His work was regarded as idiosyncratic, ostentatious and labyrinthine and this may have militated against its recognition and acceptance.[56][57]

1.4.2 Mendeleev's table

Dmitri Mendeleev

Russian chemistry professor Dmitri Mendeleev and German chemist Julius Lothar Meyer independently published their periodic tables in 1869 and 1870, respectively.[58] Mendeleev's table was his first published version; that of Meyer was an expanded version of his (Meyer's) table of 1864.[59] They both constructed their tables by listing the elements in rows or

ОПЫТЪ СИСТЕМЫ ЭЛЕМЕНТОВЪ,
ОСНОВАННОЙ НА ИХЪ АТОМНОМЪ ВѢСѢ И ХИМИЧЕСКОМЪ СХОДСТВѢ.

			Ti=50	Zr=90	?=180.
			V=51	Nb=94	Ta=182.
			Cr=52	Mo=96	W=186.
			Mn=55	Rh=104,4	Pt=197,1.
			Fe=56	Ru=104,4	Ir=198.
		Ni=Co=59		Pd=106,6	Os=199.
H=1			Cu=63,4	Ag=108	Hg=200.
	Be= 9,4	Mg=24	Zn=65,2	Cd=112	
	B=11	Al=27,3	?=68	Ur=116	Au=197?
	C=12	Si=28	?=70	Sn=118	
	N=14	P=31	As=75	Sb=122	Bi=210?
	O=16	S=32	Se=79,4	Te=128?	
	F=19	Cl=35,5	Br=80	I=127	
Li=7	Na=23	K=39	Rb=85,4	Cs=133	Tl=204.
		Ca=40	Sr=87,6	Ba=137	Pb=207.
		?=45	Ce=92		
		?Er=56	La=94		
		?Yt=60	Di=95		
		?In=75,6	Th=118?		

Д. Менделѣевъ

A version of Mendeleev's 1869 periodic table: An experiment on a system of elements. Based on their atomic weights and chemical similarities. *This early arrangement presents the periods vertically, and the groups horizontally.*

columns in order of atomic weight and starting a new row or column when the characteristics of the elements began to repeat.[60]

The recognition and acceptance afforded to Mendeleev's table came from two decisions he made. The first was to leave gaps in the table when it seemed that the corresponding element had not yet been discovered.[61] Mendeleev was not the first chemist to do so, but he was the first to be recognized as using the trends in his periodic table to predict the properties of those missing elements, such as gallium and germanium.[62] The second decision was to occasionally ignore the order suggested by the atomic weights and switch adjacent elements, such as tellurium and iodine, to better classify them into chemical families. Later in 1913, Henry Moseley determined experimental values of the nuclear charge or atomic number of each element, and showed that Mendeleev's ordering actually corresponds to the order of increasing atomic number.[63]

The significance of atomic numbers to the organization of the periodic table was not appreciated until the existence and properties of protons and neutrons became understood. Mendeleev's periodic tables used atomic weight instead of atomic number to organize the elements, information determinable to fair precision in his time. Atomic weight worked well enough in most cases to (as noted) give a presentation that was able to predict the properties of missing elements more accurately than any other method then known. Substitution of atomic numbers, once understood, gave a definitive, integer-based sequence for the elements, and Moseley predicted that the only missing elements (in 1913) between aluminum (Z=13) and gold (Z=79) (in 1913) were Z = 43, 61, 72 and 75, which were all later discovered. The sequence of atomic numbers is still used today even as new synthetic elements are being produced and studied.[64]

1.4.3   Second version and further development

Reihen	Gruppe I. — R^2O	Gruppe II. — RO	Gruppe III. — R^2O^3	Gruppe IV. RH^4 RO^2	Gruppe V. RH^3 R^2O^5	Gruppe VI. RH^2 RO^3	Gruppe VII. RH R^2O^7	Gruppe VIII. — RO^4
1	H=1							
2	Li=7	Be=9.4	B=11	C=12	N=14	O=16	F=19	
3	Na=23	Mg=24	Al=27.3	Si=28	P=31	S=32	Cl=35.5	
4	K=39	Ca=40	—=44	Ti=48	V=51	Cr=52	Mn=55	Fe=56, Co=59, Ni=59, Cu=63.
5	(Cu=63)	Zn=65	—=68	—=72	As=75	Se=78	Br=80	
6	Rb=85	Sr=87	?Yt=88	Zr=90	Nb=94	Mo=96	—=100	Ru=104, Rh=104, Pd=106, Ag=108.
7	(Ag=108)	Cd=112	In=113	Sn=118	Sb=122	Te=125	J=127	
8	Cs=133	Ba=137	?Di=138	?Ce=140	—	—	—	— — — —
9	(—)	—	—	—	—	—	—	
10	—	—	?Er=178	?La=180	Ta=182	W=184	—	Os=195, Ir=197, Pt=198, Au=199.
11	(Au=199)	Hg=200	Tl=204	Pb=207	Bi=208	—	—	
12	—	—	—	Th=231	—	U=240	—	— — — —

Mendeleev's 1871 periodic table with eight groups of elements. Dashes represented elements unknown in 1871.

In 1871, Mendeleev published his periodic table in a new form, with groups of similar elements arranged in columns rather than in rows, and those columns numbered I to VIII corresponding with the element's oxidation state. He also gave detailed predictions for the properties of elements he had earlier noted were missing, but should exist.[65] These gaps were subsequently filled as chemists discovered additional naturally occurring elements.[66] It is often stated that the last naturally occurring element to be discovered was francium (referred to by Mendeleev as *eka-caesium*) in 1939.[67] However, plutonium, produced synthetically in 1940, was identified in trace quantities as a naturally occurring primordial element in 1971.[68][n 5]

The popular[69] periodic table layout, also known as the common or standard form (as shown at various other points in this article), is attributable to Horace Groves Deming. In 1923, Deming, an American chemist, published short (Mendeleev style) and medium (18-column) form periodic tables.[70][n 6] Merck and Company prepared a handout form of Deming's 18-column medium table, in 1928, which was widely circulated in American schools. By the 1930s Deming's table was appearing in handbooks and encyclopaedias of chemistry. It was also distributed for many years by the Sargent-Welch Scientific Company.[71][72][73]

With the development of modern quantum mechanical theories of electron configurations within atoms, it became apparent that each period (row) in the table corresponded to the filling of a quantum shell of electrons. Larger atoms have more electron sub-shells, so later tables have required progressively longer periods.[74]

Period	Series	I a	I b	II a	II b	III a	III b	IV a	IV b	V a	V b	VI a	VI b	VII a	VII b	VIII a	VIII	VIII	VIII b
1	I	1 H																	2 He
2	II	3 Li		4 Be		5 B		6 C		7 N		8 O		9 F					10 Ne
3	III	11 Na		12 Mg		13 Al		14 Si		15 P		16 S		17 Cl					18 Ar
4	IV	19 K		20 Ca			21 Sc		22 Ti		23 V		24 Cr		25 Mn	26 Fe	27 Co	28 Ni	
4	V		29 Cu		30 Zn	31 Ga		32 Ge		33 As		34 Se		35 Br					36 Kr
5	VI	37 Rb		38 Sr			39 Y		40 Zr		41 Nb		42 Mo		43 Tc	44 Ru	45 Rh	46 Pd	
5	VII		47 Ag		48 Cd	49 In		50 Sn		51 Sb		52 Te		53 I					54 Xe
6	VIII	55 Cs		56 Ba			57–71		72 Hf		73 Ta		74 W		75 Re	76 Os	77 Ir	78 Pt	
6	IX		79 Au		80 Hg	81 Tl		82 Pb		83 Bi		84 Po		85 At					86 Rn
7	X	87 Fr		88 Ra			89–103		104 Rf		105 Db		106 Sg		107 Bh	108 Hs	109 Mt	110 Ds	
7	XI		111 Rg		112 Cn	113 Uut		114 Fl		115 Uup		116 Lv		117 Uus					118 Uuo

	I	II	III	IV	V	VI	VII	VIII
Higher oxides	R_2O	RO	R_2O_3	RO_2	R_2O_5	RO_3	R_2O_7	RO_4
Volatile hydrogen compounds			$[(RH_3)_x]$	RH_4	RH_3	RH_2	RH	

57 La	58 Ce	59 Pr	60 Nd	61 Pm	62 Sm	63 Eu	64 Gd	65 Tb	66 Dy	67 Ho	68 Er	69 Tm	70 Yb	71 Lu
89 Ac	90 Th	91 Pa	92 U	93 Np	94 Pu	95 Am	96 Cm	97 Bk	98 Cf	99 Es	100 Fm	101 Md	102 No	103 Lr

Eight-column form of periodic table, updated with all elements discovered to 2015

In 1945, Glenn Seaborg, an American scientist, made the suggestion that the actinide elements, like the lanthanides, were filling an f sub-level. Before this time the actinides were thought to be forming a fourth d-block row. Seaborg's colleagues advised him not to publish such a radical suggestion as it would most likely ruin his career. As Seaborg considered he did not then have a career to bring into disrepute, he published anyway. Seaborg's suggestion was found to be correct and he subsequently went on to win the 1951 Nobel Prize in chemistry for his work in synthesizing actinide elements.[75][76][n 7]

Although minute quantities of some transuranic elements occur naturally,[1] they were all first discovered in laboratories. Their production has expanded the periodic table significantly, the first of these being neptunium, synthesized in 1939.[77] Because many of the transuranic elements are highly unstable and decay quickly, they are challenging to detect and characterize when produced. There have been controversies concerning the acceptance of competing discovery claims for some elements, requiring independent review to determine which party has priority, and hence naming rights. The most recently accepted and named elements are flerovium (element 114) and livermorium (element 116), both named on 31 May 2012.[78] In 2010, a joint Russia–US collaboration at Dubna, Moscow Oblast, Russia, claimed to have synthesized six atoms of ununseptium (element 117), making it the most recently claimed discovery.[79]

1.5 Alternative structures

Main article: Alternative periodic tables

There are many periodic tables with structures other than that of the standard form. Within 100 years of the appearance of Mendeleev's table in 1869 it has been estimated that around 700 different periodic table versions were published.[80] As well as numerous rectangular variations, other periodic table formats have included, for example,[n 8] circular, cubic,

Glenn T. Seaborg who, in 1945, suggested a new periodic table showing the actinides as belonging to a second f-block series

cylindrical, edificial (building-like), helical, lemniscate, octagonal prismatic, pyramidal, separated, spherical, spiral, and triangular forms. Such alternatives are often developed to highlight or emphasize chemical or physical properties of the elements that are not as apparent in traditional periodic tables.[80]

A popular[81] alternative structure is that of Theodor Benfey (1960). The elements are arranged in a continuous spiral,

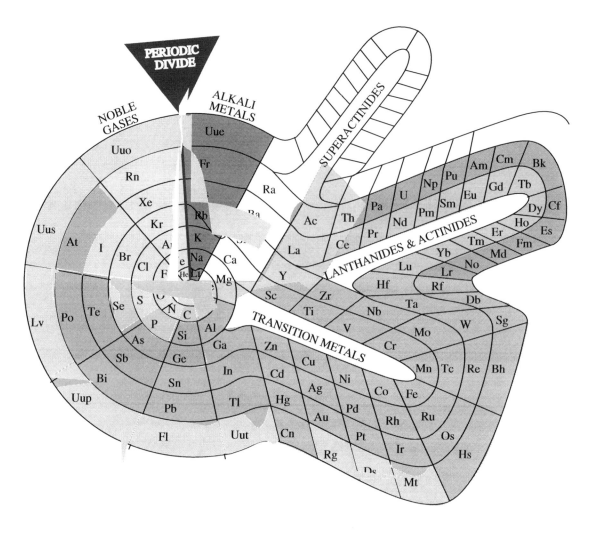

Theodor Benfey's spiral periodic table

with hydrogen at the center and the transition metals, lanthanides, and actinides occupying peninsulas.[82]

Most periodic tables are two-dimensional;[1] however, three-dimensional tables are known to as far back as at least 1862 (pre-dating Mendeleev's two-dimensional table of 1869). More recent examples include Courtines' Periodic Classification (1925),[83] Wringley's Lamina System (1949),[84] Giguère's Periodic helix (1965)[85][n 9] and Dufour's Periodic Tree (1996).[86] Going one better, Stowe's Physicist's Periodic Table (1989)[87] has been described as being four-dimensional (having three spatial dimensions and one colour dimension).[88]

The various forms of periodic tables can be thought of as lying on a chemistry–physics continuum.[89] Towards the chemistry end of the continuum can be found, as an example, Rayner-Canham's 'unruly'[90] Inorganic Chemist's Periodic Table (2002),[91] which emphasizes trends and patterns, and unusual chemical relationships and properties. Near the physics end of the continuum is Janet's Left-Step Periodic Table (1928). This has a structure that shows a closer connection to the order of electron-shell filling and, by association, quantum mechanics.[92] Somewhere in the middle of the continuum is the ubiquitous common or standard form of periodic table. This is regarded as better expressing empirical trends in physical state, electrical and thermal conductivity, and oxidation numbers, and other properties easily inferred from traditional techniques of the chemical laboratory.[93]

1.6 Open questions and controversies

1.6.1 Elements with unknown chemical properties

Although all elements up to ununoctium have been discovered, of the elements above hassium (element 108), only copernicium (element 112) and flerovium (element 114) have known chemical properties. The other elements may behave differently from what would be predicted by extrapolation, due to relativistic effects; for example, flerovium has been predicted to possibly exhibit some noble-gas-like properties, even though it is currently placed in the carbon group.[94] More recent experiments have suggested, however, that flerovium behaves chemically like lead, as expected from its periodic table position.[95]

1.6.2 Further periodic table extensions

Main article: Extended periodic table

It is unclear whether new elements will continue the pattern of the current periodic table as period 8, or require further adaptations or adjustments. Seaborg expected the eighth period to follow the previously established pattern exactly, so that it would include a two-element s-block for elements 119 and 120, a new g-block for the next 18 elements, and 30 additional elements continuing the current f-, d-, and p-blocks.[97] More recently, physicists such as Pekka Pyykkö have theorized that these additional elements do not follow the Madelung rule, which predicts how electron shells are filled and thus affects the appearance of the present periodic table.[98]

1.6.3 Element with the highest possible atomic number

The number of possible elements is not known. A very early suggestion made by Elliot Adams in 1911, and based on the arrangement of elements in each horizontal periodic table row, was that elements of atomic weight greater than 256± (which would equate to between elements 99 and 100 in modern-day terms) did not exist.[99] A higher—more recent—estimate is that the periodic table may end soon after the island of stability,[100] which is expected to center around element 126, as the extension of the periodic and nuclides tables is restricted by proton and neutron drip lines.[101] Other predictions of an end to the periodic table include at element 128 by John Emsley,[1] at element 137 by Richard Feynman,[102] and at element 155 by Albert Khazan.[1][n 10]

Bohr model

The Bohr model exhibits difficulty for atoms with atomic number greater than 137, as any element with an atomic number greater than 137 would require 1s electrons to be traveling faster than c, the speed of light.[103] Hence the non-relativistic Bohr model is inaccurate when applied to such an element.

Relativistic Dirac equation

The relativistic Dirac equation has problems for elements with more than 137 protons. For such elements, the wave function of the Dirac ground state is oscillatory rather than bound, and there is no gap between the positive and negative energy spectra, as in the Klein paradox.[104] More accurate calculations taking into account the effects of the finite size of the nucleus indicate that the binding energy first exceeds the limit for elements with more than 173 protons. For heavier elements, if the innermost orbital (1s) is not filled, the electric field of the nucleus will pull an electron out of the vacuum, resulting in the spontaneous emission of a positron;[105] however, this does not happen if the innermost orbital is filled, so that element 173 is not necessarily the end of the periodic table.[106]

1.6.4 Placement of hydrogen and helium

Simply following electron configurations, hydrogen (electronic configuration 1s^1) and helium (1s^2) should be placed in groups 1 and 2, above lithium ([He]2s^1) and beryllium ([He]2s^2).[19] However, such placing is rarely used outside of the context of electron configurations: When the noble gases (then called "inert gases") were first discovered around 1900, they were known as "group 0," reflecting no chemical reactivity of these elements known at that point, and helium was placed on the top that group, as it did share the extreme chemical inertness seen throughout the group. As the group changed its formal number, many authors continued to assign helium directly above neon, in the group 18; one of the examples of such placing is the current IUPAC table.[107]

Hydrogen's chemical properties are not very close to those of the alkali metals, which occupy the group 1, and on that basis hydrogen is sometimes placed elsewhere: one of the most common alternatives is in group 17; one of the factors behind it is the strictly univalent predominantly non-metallic chemistry of hydrogen, and that of fluorine (the element placed on the top of the group 17) is strictly univalent and non-metallic. Sometimes, to show how hydrogen has properties both corresponding to those of the alkali metals and the halogens, it may be shown in two columns simultaneously.[108] Another suggestion is above carbon in group 14: placed that way, it fits well into the trend of increasing trends of ionization potential values and electron affinity values, and is not too stray from the electronegativity trend.[109] Finally, hydrogen is sometimes placed separately from any group; this is based on how general properties of hydrogen differ from that of any group: unlike hydrogen, the other group 1 elements show extremely metallic behavior; the group 17 elements commonly form salts (hence the term "halogen"); elements of any other group show some multivalent chemistry. The other period 1 element, helium, is sometimes placed separately from any group as well.[110] The property that distinguishes helium from the rest of the noble gases (even though the extraordinary inertness of helium is extremely close to that of neon and argon[111]) is that in its closed electron shell, helium has only two electrons in the outermost electron orbital, while the rest of the noble gases have eight.

1.6.5 Groups included in the transition metals

The definition of a transition metal, as given by IUPAC, is an element whose atom has an incomplete d sub-shell, or which can give rise to cations with an incomplete d sub-shell.[112] By this definition all of the elements in groups 3–11 are transition metals. The IUPAC definition therefore excludes group 12, comprising zinc, cadmium and mercury, from the transition metals category.

Some chemists treat the categories "d-block elements" and "transition metals" interchangeably, thereby including groups 3–12 among the transition metals. In this instance the group 12 elements are treated as a special case of transition metal in which the d electrons are not ordinarily involved in chemical bonding. The recent discovery that mercury can use its d electrons in the formation of mercury(IV) fluoride (HgF_4) has prompted some commentators to suggest that mercury can be regarded as a transition metal.[113] Other commentators, such as Jensen,[114] have argued that the formation of a compound like HgF_4 can occur only under highly abnormal conditions. As such, mercury could not be regarded as a transition metal by any reasonable interpretation of the ordinary meaning of the term.[114]

Still other chemists further exclude the group 3 elements from the definition of a transition metal. They do so on the basis that the group 3 elements do not form any ions having a partially occupied d shell and do not therefore exhibit any properties characteristic of transition metal chemistry.[115] In this case, only groups 4–11 are regarded as transition metals.

1.6.6 Period 6 and 7 elements in group 3

Although scandium and yttrium are always the first two elements in group 3 the identity of the next two elements is not settled. They are either lanthanum and actinium; or lutetium and lawrencium. There are strong chemical and physical arguments supporting the latter arrangement[116][117] but not all authors have been convinced.[118] Most working chemists are not aware there is any controversy.[119]

Lanthanum and actinium are traditionally depicted as the remaining group 3 members.[120][121] It has been suggested that this layout originated in the 1940s, with the appearance of periodic tables relying on the electron configurations of the elements and the notion of the differentiating electron. The configurations of caesium, barium and lanthanum are $[Xe]6s^1$, $[Xe]6s^2$ and $[Xe]5d^16s^2$. Lanthanum thus has a 5d differentiating electron and this establishes "it in group 3 as the first member of the d-block for period 6."[122] A consistent set of electron configurations is then seen in group 3: scandium $[Ar]3d^14s^2$, yttrium $[Kr]4d^15s^2$ and lanthanum $[Xe]5d^16s^2$. Still in period 6, ytterbium was assigned an electron configuration of $[Xe]4f^{13}5d^16s^2$ and lutetium $[Xe]4f^{14}5d^16s^2$, "resulting in a 4f differentiating electron for lutetium and firmly establishing it as the last member of the f-block for period 6."[122]

In other tables, lutetium and lawrencium are the remaining group 3 members.[123] It has been known since the early 20th century that, "yttrium and (to a lesser degree) scandium are closer in their chemical properties to lutetium and the other heavy rare earths [i.e. lanthanides] than they are to lanthanum."[122] Accordingly, lutetium rather than lanthanum was assigned to group 3 by some chemists in the 1920s and 30s. Later spectroscopic work found that the electron configuration of ytterbium was in fact $[Xe]4f^{14}6s^2$. This meant that ytterbium and lutetium—the latter with $[Xe]4f^{14}5d^16s^2$—both had 14 f electrons, "resulting in a d rather than an f differentiating electron" for lutetium and making it an "equally valid candidate" with $[Xe]5d^16s^2$ lanthanum, for the group 3 periodic table position below yttrium.[122] Several physicists in the 1950s and 60s opted for lutetium, in light of a comparison of several of its physical properties with those of lanthanum.[122] This arrangement, in which lanthanum is the first member of the f-block, is disputed by some authors since lanthanum lacks any f electrons. However, it has been argued that this is not valid concern given other periodic table anomalies— thorium, for example, has no f electrons yet is part of the f-block.[124] As for lawrencium, its electron configuration was confirmed in 2015 as $[Rn]5f^{14}7s^27p^1$. Such a configuration represents another periodic table anomaly, regardless of whether lawrencium is located in the f-block or the d-block, as the only potentially applicable p-block position has been reserved for ununtrium with its predicted electron configuration of $[Rn]5f^{14}6d^{10}7s^27p^1$.[125]

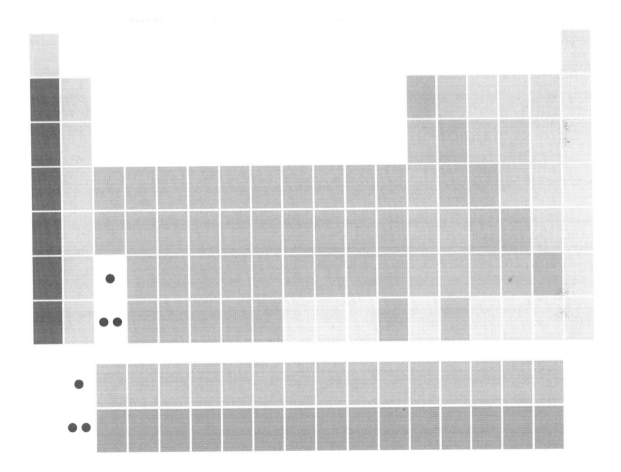

Some tables, including the table on the IUPAC site,[126][n 11] place footnote markers in the two positions below scandium and yttrium, and show both lanthanum and lutetium, and actinium and lawrencium as being part of, respectively, the lanthanide series and the actinide series of elements. This arrangement emphasizes similarities in the chemistry of the 15 lanthanide elements (La–Lu) over electron configuration arguments. The actinides are more diverse in their behavior. Most early members show some similarities to transition metals; actinium and the later members are more like lanthanides.[127]

1.6.7 Optimal form

The many different forms of periodic table have prompted the question of whether there is an optimal or definitive form of periodic table. The answer to this question is thought to depend on whether the chemical periodicity seen to occur among the elements has an underlying truth, effectively hard-wired into the universe, or if any such periodicity is instead the product of subjective human interpretation, contingent upon the circumstances, beliefs and predilections of human observers. An objective basis for chemical periodicity would settle the questions about the location of hydrogen and helium, and the composition of group 3. Such an underlying truth, if it exists, is thought to have not yet been discovered. In its absence, the many different forms of periodic table can be regarded as variations on the theme of chemical periodicity, each of which explores and emphasizes different aspects, properties, perspectives and relationships of and among the elements.[n 12] The ubiquity of the standard or medium-long periodic table is thought to be a result of this layout having a good balance of features in terms of ease of construction and size, and its depiction of atomic order and periodic trends.[49][128]

1.7 See also

- Abundance of the chemical elements

- Atomic electron configuration table

- Element collecting

- List of elements

- List of periodic table-related articles

- Table of nuclides

- The Mystery of Matter: Search for the Elements (PBS film)

- Timeline of chemical element discoveries

1.8 Notes

[1] The elements discovered initially by synthesis and later in nature are technetium (Z=43), promethium (61), astatine (85), neptunium (93), and plutonium (94).

[2] There is an inconsistency and some irregularities in this convention. Thus, helium is shown in the p-block but is actually an s-block element, and (for example) the d-subshell in the d-block is actually filled by the time group 11 is reached, rather than group 12.

[3] The noble gases, astatine, francium, and all elements heavier than americium were left out as there is no data for them.

[4] While fluorine is the most electronegative of the elements under the Pauling scale, neon is the most electronegative element under other scales, such as the Allen scale.

[5] John Emsley, in his book, *Nature's Building Blocks,* writes that americium, curium, berkelium and californium (elements 95–98) can occur naturally as trace amounts in uranium ores by neutron capture and beta decay. This assertion appears to lack independent substantiation. See: Emsley J. (2011). *Nature's Building Blocks: An A-Z Guide to the Elements* (New ed.). New York, NY: Oxford University Press, p. 109.

[6] An antecedent of Deming's 18-column table may be seen in Adams' 16-column Periodic Table of 1911. Adams omits the rare earths and the 'radioactive elements' (i.e. the actinides) from the main body of his table and instead shows them as being 'careted in only to save space' (rare earths between Ba and eka-Yt; radioactive elements between eka-Te and eka-I). See: Elliot Q. A. (1911). "A modification of the periodic table". *Journal of the American Chemical Society.* **33**(5): 684–688 (687).

[7] A second extra-long periodic table row, to accommodate known and undiscovered elements with an atomic weight greater than bismuth (thorium, protactinium and uranium, for example), had been postulated as far back as 1892. Most investigators, however, considered that these elements were analogues of the third series transition elements, hafnium, tantalum and tungsten. The existence of a second inner transition series, in the form of the actinides, was not accepted until similarities with the electron structures of the lanthanides had been established. See: van Spronsen, J. W. (1969). *The periodic system of chemical elements.* Amsterdam: Elsevier. p. 315–316, ISBN 0-444-40776-6.

[8] See *The Internet database of periodic tables* for depictions of these kinds of variants.

[9] The animated depiction of Giguère's periodic table that is widely available on the internet (including from here) is erroneous, as it does not include hydrogen and helium. Giguère included hydrogen, above lithium, and helium, above beryllium. See: Giguère P.A. (1966). "The "new look" for the periodic system". *Chemistry in Canada* **18** (12): 36–39 (see p. 37).

[10] Karol (2002, p. 63) contends that gravitational effects would become significant when atomic numbers become astronomically large, thereby overcoming other super-massive nuclei instability phenomena, and that neutron stars (with atomic numbers on the order of 10^{21}) can arguably be regarded as representing the heaviest known elements in the universe. See: Karol P. J. (2002). "The Mendeleev–Seaborg periodic table: Through Z = 1138 and beyond". *Journal of Chemical Education* **79** (1): 60–63.

[11] Although this form of the table is sometimes referred to as the "approved" or "official" IUPAC periodic table, "IUPAC has not approved any specific form of the periodic table..." See: Leigh, G. J. (January–February 2009). "Periodic Tables and IUPAC". *Chemistry International* **31** (1).

[12] Scerri, one of the foremost authorities on the history of the periodic table (Sella 2013), favoured the concept of an optimal form of periodic table but has recently changed his mind and now supports the value of a plurality of periodic tables. See: Sella A. (2013). 'An elementary history lesson'. *New Scientist.* 2929, 13 August: 51, accessed 4 September 2013; and Scerri, E. (2013). 'Is there an optimal periodic table and other bigger questions in the philosophy of science.'. 9 August, accessed 4 September 2013.

1.9   References

[1] Emsley, John (2011). *Nature's Building Blocks: An A-Z Guide to the Elements* (New ed.). New York, NY: Oxford University Press. ISBN 978-0-19-960563-7.

[2] Greenwood, pp. 24–27

[3] Gray, p. 6

[4] Koppenol, W. H. (2002). "Naming of New Elements (IUPAC Recommendations 2002)" (PDF). *Pure and Applied Chemistry* **74** (5): 787–791. doi:10.1351/pac200274050787.

[5] Silva, Robert J. (2006). "Fermium, Mendelevium, Nobelium and Lawrencium". In Morss; Edelstein, Norman M.; Fuger, Jean. *The Chemistry of the Actinide and Transactinide Elements* (3rd ed.). Dordrecht, The Netherlands: Springer Science+Business Media. ISBN 1-4020-3555-1.

[6] Gray, p. 11

[7] Scerri 2007, p. 24

[8] Messler, R. W. (2010). *The essence of materials for engineers.* Sudbury, MA: Jones & Bartlett Publishers. p. 32. ISBN 0-7637-7833-8.

[9] Bagnall, K. W. (1967). "Recent advances in actinide and lanthanide chemistry". In Fields, P.R.; Moeller, T. *Advances in chemistry, Lanthanide/Actinide chemistry.* Advances in Chemistry **71**. American Chemical Society. pp. 1–12. doi:10.1021/ba-1967-0071. ISBN 0-8412-0072-6.

[10] Day, M. C., Jr.; Selbin, J. (1969). *Theoretical inorganic chemistry* (2nd ed.). New York: Nostrand-Rienhold Book Corporation. p. 103. ISBN 0-7637-7833-8.

[11] Holman, J.; Hill, G. C. (2000). *Chemistry in context* (5th ed.). Walton-on-Thames: Nelson Thornes. p. 40. ISBN 0-17-448276-0.

[12] Leigh, G. J. (1990). *Nomenclature of Inorganic Chemistry: Recommendations 1990.* Blackwell Science. ISBN 0-632-02494-1.

[13] Fluck, E. (1988). "New Notations in the Periodic Table" (PDF). *Pure Appl. Chem.* (IUPAC) **60** (3): 431–436. doi:10.1351/pac1. Retrieved 24 March 2012.

[14] Moore, p. 111

[15] Greenwood, p. 30

[16] Stoker, Stephen H. (2007). *General, organic, and biological chemistry.* New York: Houghton Mifflin. p. 68. ISBN 978-0-618-73063-6. OCLC 52445586.

[17] Mascetta, Joseph (2003). *Chemistry The Easy Way* (4th ed.). New York: Hauppauge. p. 50. ISBN 978-0-7641-1978-1. OCLC 52047235.

[18] Kotz, John; Treichel, Paul; Townsend, John (2009). *Chemistry and Chemical Reactivity, Volume 2* (7th ed.). Belmont: Thomson Brooks/Cole. p. 324. ISBN 978-0-495-38712-1. OCLC 220756597.

[19] Gray, p. 12

[20] Jones, Chris (2002). *d- and f-block chemistry.* New York: J. Wiley & Sons. p. 2. ISBN 978-0-471-22476-1. OCLC 300468713.

[21] Silberberg, M. S. (2006). *Chemistry: The molecular nature of matter and change* (4th ed.). New York: McGraw-Hill. p. 536. ISBN 0-07-111658-3.

[22] Manson, S. S.; Halford, G. R. (2006). *Fatigue and durability of structural materials.* Materials Park, Ohio: ASM International. p. 376. ISBN 0-87170-825-6.

[23] Bullinger, Hans-Jörg (2009). *Technology guide: Principles, applications, trends.* Berlin: Springer-Verlag. p. 8. ISBN 978-3-540-88545-0.

[24] Jones, B. W. (2010). *Pluto: Sentinel of the outer solar system.* Cambridge: Cambridge University Press. pp. 169–71. ISBN 978-0-521-19436-5.

[25] Hinrichs, G. D. (1869). "On the classification and the atomic weights of the so-called chemical elements, with particular reference to Stas's determinations". *Proceedings of the American Association for the Advancement of Science* **18** (5): 112–124.

[26] Myers, R. (2003). *The basics of chemistry.* Westport, CT: Greenwood Publishing Group. pp. 61–67. ISBN 0-313-31664-3.

[27] Chang, Raymond (2002). *Chemistry* (7 ed.). New York: McGraw-Hill. pp. 289–310; 340–42. ISBN 0-07-112072-6.

[28] Greenwood, p. 27

[29] Jolly, W. L. (1991). *Modern Inorganic Chemistry* (2nd ed.). McGraw-Hill. p. 22. ISBN 978-0-07-112651-9.

[30] Greenwood, p. 28

[31] IUPAC, *Compendium of Chemical Terminology,* 2nd ed. (the "Gold Book") (1997). Online corrected version: (2006–) "Electronegativity".

[32] Pauling, L. (1932). "The Nature of the Chemical Bond. IV. The Energy of Single Bonds and the Relative Electronegativity of Atoms". *Journal of the American Chemical Society* **54** (9): 3570–3582. doi:10.1021/ja01348a011.

[33] Allred, A. L. (1960). "Electronegativity values from thermochemical data". *Journal of Inorganic and Nuclear Chemistry* (Northwestern University) **17** (3–4): 215–221. doi:10.1016/0022-1902(61)80142-5. Retrieved 11 June 2012.

[34] Huheey, Keiter & Keiter, p. 42

[35] Siekierski, Slawomir; Burgess, John (2002). *Concise chemistry of the elements.* Chichester: Horwood Publishing. pp. 35–36. ISBN 1-898563-71-3.

[36] Chang, pp. 307–309

[37] Huheey, Keiter & Keiter, pp. 42, 880–81

[38] Yoder, C. H.; Suydam, F. H.; Snavely, F. A. (1975). *Chemistry* (2nd ed.). Harcourt Brace Jovanovich. p. 58. ISBN 0-15-506465-7.

[39] Huheey, Keiter & Keiter, pp. 880–85

[40] Sacks, O (2009). *Uncle Tungsten: Memories of a chemical boyhood.* New York: Alfred A. Knopf. pp. 191, 194. ISBN 0-375-70404-3.

[41] Gray, p. 9

[42] Siegfried, Robert (2002). *From elements to atoms a history of chemical composition.* Philadelphia, Pennsylvania: Library of Congress Cataloging-in-Publication Data. p. 92. ISBN 0-87169-924-9.

[43] Ball, p. 100

[44] Horvitz, Leslie (2002). *Eureka!: Scientific Breakthroughs That Changed The World.* New York: John Wiley. p. 43. ISBN 978-0-471-23341-1. OCLC 50766822.

[45] van Spronsen, J. W. (1969). *The periodic system of chemical elements.* Amsterdam: Elsevier. p. 19. ISBN 0-444-40776-6.

[46] "Alexandre-Emile Bélguier de Chancourtois (1820-1886)" (in French). Annales des Mines history page. Retrieved 18 September 2014.

[47] Venable, pp. 85–86; 97

[48] Odling, W. (2002). "On the proportional numbers of the elements". *Quarterly Journal of Science* **1**: 642–648 (643).

[49] Scerri, Eric R. (2011). *The periodic table: A very short introduction.* Oxford: Oxford University Press. ISBN 978-0-19-958249-5.

[50] Kaji, M. (2004). "Discovery of the periodic law: Mendeleev and other researchers on element classification in the 1860s". In Rouvray, D. H.; King, R. Bruce. *The periodic table: Into the 21st Century.* Research Studies Press. pp. 91–122 (95). ISBN 0-86380-292-3.

[51] Newlands, John A. R. (20 August 1864). "On Relations Among the Equivalents". *Chemical News* **10**: 94–95.

[52] Newlands, John A. R. (18 August 1865). "On the Law of Octaves". *Chemical News* **12**: 83.

[53] Bryson, Bill (2004). *A Short History of Nearly Everything.* Black Swan. pp. 141–142. ISBN 978-0-552-15174-0.

[54] Scerri 2007, p. 306

[55] Brock, W. H.; Knight, D. M. (1965). "The Atomic Debates: 'Memorable and Interesting Evenings in the Life of the Chemical Society'". *Isis* (The University of Chicago Press) **56** (1): 5–25. doi:10.1086/349922.

[56] Scerri 2007, pp. 87, 92

[57] Kauffman, George B. (March 1969). "American forerunners of the periodic law". *Journal of Chemical Education* **46** (3): 128–135 (132). Bibcode:1969JChEd..46..128K. doi:10.1021/ed046p128.

[58] Mendelejew, Dimitri (1869). "Über die Beziehungen der Eigenschaften zu den Atomgewichten der Elemente". *Zeitschrift für Chemie* (in German): 405–406.

[59] Venable, pp. 96–97; 100–102

[60] Ball, pp. 100–102

[61] Pullman, Bernard (1998). *The Atom in the History of Human Thought.* Translated by Axel Reisinger. Oxford University Press. p. 227. ISBN 0-19-515040-6.

[62] Ball, p. 105

[63] Atkins, P. W. (1995). *The Periodic Kingdom.* HarperCollins Publishers, Inc. p. 87. ISBN 0-465-07265-8.

[64] Samanta, C.; Chowdhury, P. Roy; Basu, D.N. (2007). "Predictions of alpha decay half lives of heavy and superheavy elements". *Nucl. Phys. A* **789**: 142–154. arXiv:nucl-th/0703086. Bibcode:2007NuPhA.789..142S. doi:10.1016/j.nuclphysa.2007.04.001.

[65] Scerri 2007, p. 112

[66] Kaji, Masanori (2002). "D.I. Mendeleev's Concept of Chemical Elements and the Principle of Chemistry" (PDF). *Bull. Hist. Chem.* (Tokyo Institute of Technology) **27** (1): 4–16. Retrieved 11 June 2012.

[67] Adloff, Jean-Pierre; Kaufman, George B. (25 September 2005). "Francium (Atomic Number 87), the Last Discovered Natural Element". The Chemical Educator. Retrieved 26 March 2007.

[68] Hoffman, D. C.; Lawrence, F. O.; Mewherter, J. L.; Rourke, F. M. (1971). "Detection of Plutonium-244 in Nature". *Nature* **234** (5325): 132–134. Bibcode:1971Natur.234..132H. doi:10.1038/234132a0.

[69] Gray, p. 12

[70] Deming, Horace G (1923). *General chemistry: An elementary survey.* New York: J. Wiley & Sons. pp. 160, 165.

[71] Abraham, M; Coshow, D; Fix, W. *Periodicity:A source book module, version 1.0* (PDF). New York: Chemsource, Inc. p. 3.

[72] Emsley, J (7 March 1985). "Mendeleyev's dream table". *New Scientist*: 32–36(36).

[73] Fluck, E (1988). "New notations in the period table". *Pure & Applied Chemistry* **60** (3): 431–436(432). doi:10.1351/pac1988.

[74] Ball, p. 111

[75] Scerri 2007, pp. 270–71

[76] Masterton, William L.; Hurley, Cecile N.; Neth, Edward J. *Chemistry: Principles and reactions* (7th ed.). Belmont, CA: Brooks/Cole Cengage Learning. p. 173. ISBN 1-111-42710-0.

[77] Ball, p. 123

[78] Barber, Robert C.; Karol, Paul J; Nakahara, Hiromichi; Vardaci, Emanuele; Vogt, Erich W. (2011). "Discovery of the elements with atomic numbers greater than or equal to 113 (IUPAC Technical Report)". *Pure Appl. Chem.* **83** (7): 1485. doi:10.1351/PAC-REP-10-05-01.

[79] Эксперимент по синтезу 117-го элемента получает продолжение[Experiment on sythesis of the 117th element is to be continued] (in Russian). JINR. 2012.

[80] Scerri 2007, p. 20

[81] Emsely, J; Sharp, R (21 June 2010). "The periodic table: Top of the charts". *The Independent.*

[82] Seaborg, Glenn (1964). "Plutonium: The Ornery Element". *Chemistry* **37** (6): 14.

[83] Mark R. Leach. "1925 Courtines' Periodic Classification". Retrieved 16 October 2012.

[84] Mark R. Leach. "1949 Wringley's Lamina System". Retrieved 16 October 2012.

[85] Mazurs, E.G. (1974). *Graphical Representations of the Periodic System During One Hundred Years.* Alabama: University of Alabama Press. p. 111. ISBN 978-0-8173-3200-6.

[86] Mark R. Leach. "1996 Dufour's Periodic Tree". Retrieved 16 October 2012.

[87] Mark R. Leach. "1989 Physicist's Periodic Table by Timothy Stowe". Retrieved 16 October 2012.

[88] Bradley, David (20 July 2011). "At last, a definitive periodic table?". *ChemViews Magazine.* doi:10.1002/chemv.201000107.

[89] Scerri 2007, pp. 285–86

[90] Scerri 2007, p. 285

[91] Mark R. Leach. "2002 Inorganic Chemist's Periodic Table". Retrieved 16 October 2012.

[92] Scerri, Eric (2008). "The role of triads in the evolution of the periodic table: Past and present". *Journal of Chemical Education* **85** (4): 585–89 (see p.589). Bibcode:2008JChEd..85..585S. doi:10.1021/ed085p585.

[93] Bent, H. A.; Weinhold, F (2007). "Supporting information: News from the periodic table: An introduction to "Periodicity symbols, tables, and models for higher-order valency and donor–acceptor kinships"". *Journal of Chemical Education* **84** (7): 3–4. doi:10.1021/ed084p1145.

[94] Schändel, Matthias (2003). *The Chemistry of Superheavy Elements*. Dordrecht: Kluwer Academic Publishers. p. 277. ISBN 1-4020-1250-0.

[95] Scerri 2011, pp. 142–143

[96] Fricke, B.; Greiner, W.; Waber, J. T. (1971). "The continuation of the periodic table up to Z = 172. The chemistry of superheavy elements". *Theoretica chimica acta* (Springer-Verlag) **21** (3): 235–260. doi:10.1007/BF01172015. Retrieved 28 November 2012.

[97] Frazier, K. (1978). "Superheavy Elements". *Science News* **113** (15): 236–238. doi:10.2307/3963006. JSTOR 3963006.

[98] Pyykkö, Pekka (2011). "A suggested periodic table up to Z ≤ 172, based on Dirac–Fock calculations on atoms and ions". *Physical Chemistry Chemical Physics* **13** (1): 161–168. Bibcode:2011PCCP...13..161P. doi:10.1039/c0cp01575j. PMID 20967377.

[99] Elliot, Q. A. (1911). "A modification of the periodic table". *Journal of the American Chemical Society* **33** (5): 684–688 (688). doi:10.1021/ja02218a004.

[100] Glenn Seaborg (c. 2006). "transuranium element (chemical element)". Encyclopædia Britannica. Retrieved 16 March 2010.

[101] Cwiok, S.; Heenen, P.-H.; Nazarewicz, W. (2005). "Shape coexistence and triaxiality in the superheavy nuclei". *Nature* **433** (7027): 705–9. Bibcode:2005Natur.433..705C. doi:10.1038/nature03336. PMID 15716943.

[102] Column: The crucible Ball, Philip in Chemistry World, Royal Society of Chemistry, Nov. 2010

[103] Eisberg, R.; Resnick, R. (1985). *Quantum Physics of Atoms, Molecules, Solids, Nuclei and Particles*. Wiley.

[104] Bjorken, J. D.; Drell, S. D. (1964). *Relativistic Quantum Mechanics*. McGraw-Hill.

[105] Greiner, W.; Schramm, S. (2008). "American Journal of Physics" **76**. p. 509., and references therein.

[106] Ball, Philip (November 2010). "Would Element 137 Really Spell the End of the Periodic Table? Philip Ball Examines the Evidence". Royal Society of Chemistry. Retrieved 30 September 2012.

[107] IUPAC (2013-05-01). "IUPAC Periodic Table of the Elements" (PDF). *iupac.org*. IUPAC. Retrieved 2015-09-20.

[108] Seaborg, Glenn Theodore (1945). "The chemical and radioactive properties of the heavy elements". *Chemical English Newspaper* **23** (23): 2190–2193.

[109] Cronyn, Marshall W. (August 2003). "The Proper Place for Hydrogen in the Periodic Table". *Journal of Chemical Education* **80** (8): 947–951. Bibcode:2003JChEd..80..947C. doi:10.1021/ed080p947.

[110] Greenwood, throughout the book

[111] Lewars, Errol G. (2008-12-05). *Modeling Marvels: Computational Anticipation of Novel Molecules*. Springer Science & Business Media. p. 69–71. ISBN 9781402069734.

[112] IUPAC, *Compendium of Chemical Terminology*, 2nd ed. (the "Gold Book") (1997). Online corrected version: (2006–) "transition element".

[113] Xuefang Wang; Lester Andrews; Sebastian Riedel; Martin Kaupp (2007). "Mercury Is a Transition Metal: The First Experimental Evidence for HgF_4". *Angew. Chem. Int. Ed.* **46** (44): 8371–8375. doi:10.1002/anie.200703710. PMID 17899620.

[114] William B. Jensen(2008). "Is Mercury Now a Transition Element?". *J. Chem. Educ.* **85**(9):1182–1183. Bibcode:2008JChEd..85. doi:10.1021/ed085p1182.

[115] Rayner-Canham, G; Overton, T. *Descriptive inorganic chemistry* (4th ed.). New York: W H Freeman. pp. 484–485. ISBN 0-7167-8963-9.

[116] Thyssen, P.; Binnemanns, K. (2011). "1: Accommodation of the rare earths in the periodic table: A historical analysis". In Gschneidner Jr., K. A.; Büzli, J-C. J.; Pecharsky, V. K. *Handbook on the Physics and Chemistry of Rare Earths* **41**. Amsterdam: Elsevier. pp. 80–81. ISBN 978-0-444-53590-0.

[117] Keeler, J.; Wothers, P. (2014). *Chemical Structure and Reactivity: An Integrated Approach*. Oxford: Oxford University. p. 259. ISBN 978-0-19-9604135.

[118] Scerri, E. (2012). "Mendeleev's Periodic Table Is Finally Completed and What To Do about Group 3?". *Chemistry International* **34** (4).

[119] Castelvecchi, Davide (8 April 2015). "Exotic atom struggles to find its place in the periodic table". *Nature News*. Retrieved 20 Sep 2015.

[120] Emsley, J. (2011). *Nature's Building Blocks* (new ed.). Oxford: Oxford University. p. 651. ISBN 978-0-19-960563-7.

[121] See, for example: "Periodic Table". Royal Society of Chemistry. Retrieved 20 Sep 2015.

[122] William B. Jensen (1982). "The Positions of Lanthanum (Actinium) and Lutetium (Lawrencium) in the Periodic Table". *J. Chem. Educ.* **59** (8): 634–636. doi:10.1021/ed059p634.

[123] See, for example: Brown, T. L.; LeMay Jr., H. E.; Bursten, B. E.; Murphy, C. J. (2009). *Chemistry: The Central Science* (11th ed.). Upper Saddle River, New Jersey: Pearson Education. p. endpapers. ISBN 0-13-235848-4.

[124] Scerri, E (2015). "Five ideas in chemical education that must die - part five". *educationinchemistryblog*. Royal Society of Chemistry. Retrieved Sep 19, 2015. It is high time that the idea of group 3 consisting of Sc, Y, La and Ac is abandoned

[125] Jensen, W. B. (2015). "Some Comments on the Position of Lawrencium in the Periodic Table" (PDF). Retrieved 20 Sep 2015.

[126] "Periodic Table of the Elements". International Union of Pure and Applied Chemistry. Retrieved 3 April 2010.

[127] Owen, S. M. (1991). *A Guide to Modern Inorganic Chemistry*. Harlow, Essex: Longman Scientific & Technical. p. 190. ISBN 0-58-206439-2.

[128] Francl, Michelle (May 2009). "Table manners" (PDF). *Nature Chemistry* **1** (2):97–98. Bibcode:2009NatCh...1...97F. doi:10.1038/. PMID 21378810.

1.10 Bibliography

- Ball, Philip (2002). *The Ingredients: A Guided Tour of the Elements*. Oxford: Oxford University Press. ISBN 0-19-284100-9.

- Chang, Raymond (2002). *Chemistry* (7th ed.). New York: McGraw-Hill Higher Education. ISBN 978-0-19-284100-1.

- Gray, Theodore (2009). *The Elements: A Visual Exploration of Every Known Atom in the Universe*. New York: Black Dog & Leventhal Publishers. ISBN 978-1-57912-814-2.

- Greenwood, Norman N.; Earnshaw, Alan (1984). *Chemistry of the Elements*. Oxford: Pergamon Press. ISBN 0-08-022057-6.

- Huheey, JE; Keiter, EA; Keiter, RL. *Principles of structure and reactivity* (4th ed.). New York: Harper Collins College Publishers. ISBN 0-06-042995-X.

- Moore, John (2003). *Chemistry For Dummies*. New York: Wiley Publications. p. 111. ISBN 978-0-7645-5430-8. OCLC 51168057.

- Scerri, Eric (2007). *The periodic table: Its story and its significance*. Oxford: Oxford University Press. ISBN 0-19-530573-6.

- Scerri, Eric R. (2011). *The periodic table: A very short introduction*. Oxford: Oxford University Press. ISBN 978-0-19-958249-5.

- Venable, F P (1896). *The Development of the Periodic Law*. Easton PA: Chemical Publishing Company.

1.11   External links

- M. Dayah. "Dynamic Periodic Table". Retrieved 14 May 2012.

- Brady Haran. "The Periodic Table of Videos". University of Nottingham. Retrieved 14 May 2012.

- Mark Winter. "WebElements: the periodic table on the web". University of Sheffield. Retrieved 14 May 2012.

- Mark R. Leach. "The INTERNET Database of Periodic Tables". Retrieved 14 May 2012.

Chapter 2

Atomic number

See also: List of elements by atomic number
In chemistry and physics, the **atomic number** of a chemical element (also known as its **proton number**) is the number of

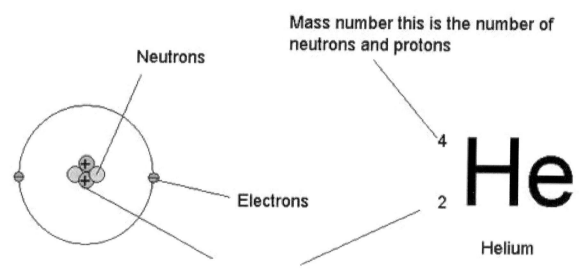

Mass number this is the number of neutrons and protons

Neutrons

4

$_2$ He

Electrons

Helium

Protons:
This number lets us know how many protons there are. In a neutral atom this is also the same as the number of electrons.

An explanation of the superscripts and subscripts seen in atomic number notation. Atomic number is the number of protons, and therefore also the total positive charge, in the atomic nucleus.

protons found in the nucleus of an atom of that element, and therefore identical to the charge number of the nucleus. It is conventionally represented by the symbol Z. The atomic number uniquely identifies a chemical element. In an uncharged atom, the atomic number is also equal to the number of electrons.

The atomic number, Z, should not be confused with the mass number, A, which is the number of nucleons, the total number of protons and neutrons in the nucleus of an atom. The number of neutrons, N, is known as the neutron number of the atom; thus, $A = Z + N$ (these quantities are always whole numbers). Since protons and neutrons have approximately the same mass (and the mass of the electrons is negligible for many purposes) and the mass defect of nucleon binding is always small compared to the nucleon mass, the atomic mass of any atom, when expressed in unified atomic mass units

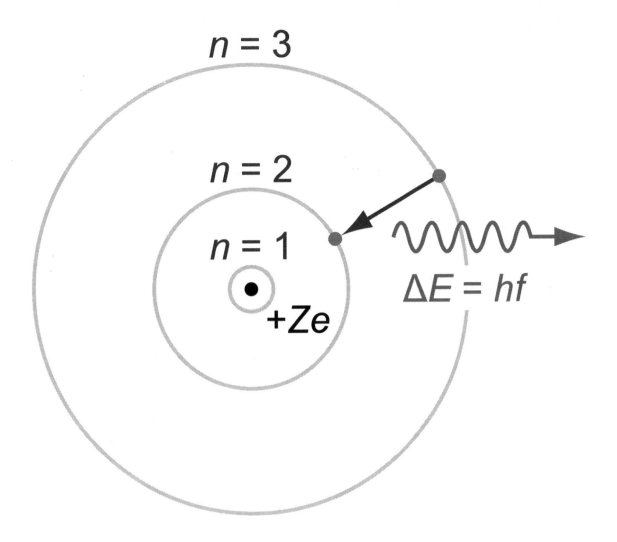

*The **Rutherford–Bohr model** of the hydrogen atom (Z = 1) or a hydrogen-like ion (Z > 1). In this model it is an essential feature that the photon energy (or frequency) of the electromagnetic radiation emitted (shown) when an electron jumps from one orbital to another, be proportional to the mathematical square of atomic charge (Z^2). Experimental measurement by Henry Moseley of this radiation for many elements (from Z = 13 to 92) showed the results as predicted by Bohr. Both the concept of atomic number and the Bohr model were thereby given scientific credence.*

(making a quantity called the "relative isotopic mass"), is roughly (to within 1%) equal to the whole number A.

Atoms with the same atomic number Z but different neutron numbers N, and hence different atomic masses, are known as isotopes. A little more than three-quarters of naturally occurring elements exist as a mixture of isotopes (see monoisotopic elements), and the average isotopic mass of an isotopic mixture for an element (called the relative atomic mass) in a defined environment on Earth, determines the element's standard atomic weight. Historically, it was these atomic weights of elements (in comparison to hydrogen) that were the quantities measurable by chemists in the 19th century.

The conventional symbol Z comes from the German word **Zahl** meaning number/numeral/figure, which, prior to the modern synthesis of ideas from chemistry and physics, merely denoted an element's numerical place in the periodic table, whose order is approximately, but not completely, consistent with the order of the elements by atomic weights. Only after 1915, with the suggestion and evidence that this Z number was also the nuclear charge and a physical characteristic of atoms, did the word *Atomzahl* (and its English equivalent *atomic number*) come into common use in this context.

2.1 History

2.1.1 The periodic table and a natural number for each element

Loosely speaking, the existence or construction of a periodic table of elements creates an ordering of the elements, and so they can be numbered in order.

Dmitri Mendeleev claimed that he arranged his first periodic tables in order of atomic weight ("Atomgewicht").[1] However, in consideration of the elements' observed chemical properties, he changed the order slightly and placed tellurium (atomic weight 127.6) ahead of iodine (atomic weight 126.9).[1][2] This placement is consistent with the modern practice of ordering the elements by proton number, Z, but that number was not known or suspected at the time.

A simple numbering based on periodic table position was never entirely satisfactory, however. Besides the case of iodine and tellurium, later several other pairs of elements (such as argon and potassium, cobalt and nickel) were known to have nearly identical or reversed atomic weights, thus requiring their placement in the periodic table to be determined by their chemical properties. However the gradual identification of more and more chemically similar lanthanide elements, whose atomic number was not obvious, led to inconsistency and uncertainty in the periodic numbering of elements at least from lutetium (element 71) onwards (hafnium was not known at this time).

2.1.2 The Rutherford-Bohr model and van den Broek

In 1911, Ernest Rutherford gave a model of the atom in which a central core held most of the atom's mass and a positive charge which, in units of the electron's charge, was to be approximately equal to half of the atom's atomic weight, expressed in numbers of hydrogen atoms. This central charge would thus be approximately half the atomic weight (though it was almost 25% different from the atomic number of gold ($Z = 79$, $A = 197$), the single element from which Rutherford made his guess). Nevertheless, in spite of Rutherford's estimation that gold had a central charge of about 100 (but was element $Z = 79$ on the periodic table), a month after Rutherford's paper appeared, Antonius van den Broek first formally suggested that the central charge and number of electrons in an atom was *exactly* equal to its place in the periodic table (also known as element number, atomic number, and symbolized Z). This proved eventually to be the case.

2.1.3 Moseley's 1913 experiment

The experimental position improved dramatically after research by Henry Moseley in 1913.[3] Moseley, after discussions with Bohr who was at the same lab (and who had used Van den Broek's hypothesis in his Bohr model of the atom), decided to test Van den Broek's and Bohr's hypothesis directly, by seeing if spectral lines emitted from excited atoms fitted the Bohr theory's postulation that the frequency of the spectral lines be proportional to the square of Z.

To do this, Moseley measured the wavelengths of the innermost photon transitions (K and L lines) produced by the elements from aluminum ($Z = 13$) to gold ($Z = 79$) used as a series of movable anodic targets inside an x-ray tube.[4] The square root of the frequency of these photons (x-rays) increased from one target to the next in an arithmetic progression. This led to the conclusion (Moseley's law) that the atomic number does closely correspond (with an offset of one unit for K-lines, in Moseley's work) to the calculated electric charge of the nucleus, i.e. the element number Z. Among other things, Moseley demonstrated that the lanthanide series (from lanthanum to lutetium inclusive) must have 15 members—no fewer and no more—which was far from obvious from the chemistry at that time.

2.1.4 The proton and the idea of nuclear electrons

In 1915 the reason for nuclear charge being quantized in units of Z, which were now recognized to be the same as the element number, was not understood. An old idea called Prout's hypothesis had postulated that the elements were all made of residues (or "protyles") of the lightest element hydrogen, which in the Bohr-Rutherford model had a single electron and a nuclear charge of one. However, as early as 1907 Rutherford and Thomas Royds had shown that alpha particles, which had a charge of +2, were the nuclei of helium atoms, which had a mass four times that of hydrogen, not two times.

If Prout's hypothesis were true, something had to be neutralizing some of the charge of the hydrogen nuclei present in the nuclei of heavier atoms.

In 1917 Rutherford succeeded in generating hydrogen nuclei from a nuclear reaction between alpha particles and nitrogen gas, and believed he had proven Prout's law. He called the new heavy nuclear particles protons in 1920 (alternate names being proutons and protyles). It had been immediately apparent from the work of Moseley that the nuclei of heavy atoms have more than twice as much mass as would be expected from their being made of hydrogen nuclei, and thus there was required a hypothesis for the neutralization of the extra protons presumed present in all heavy nuclei. A helium nucleus was presumed to be composed of four protons plus two "nuclear electrons" (electrons bound inside the nucleus) to cancel two of the charges. At the other end of the periodic table, a nucleus of gold with a mass 197 times that of hydrogen, was thought to contain 118 nuclear electrons in the nucleus to give it a residual charge of $+79$, consistent with its atomic number.

2.1.5   The discovery of the neutron makes Z the proton number

All consideration of nuclear electrons ended with James Chadwick's discovery of the neutron in 1932. An atom of gold now was seen as containing 118 neutrons rather than 118 nuclear electrons, and its positive charge now was realized to come entirely from a content of 79 protons. After 1932, therefore, an element's atomic number Z was also realized to be identical to the proton number of its nuclei.

2.2   The symbol of Z

The conventional symbol Z possibly comes from the German word *Atomzahl* (atomic number).[5] However, prior to 1915, the word *Zahl* (simply *number*) was used for an element's assigned number in the periodic table.

2.3   Chemical properties

Each element has a specific set of chemical properties as a consequence of the number of electrons present in the neutral atom, which is Z (the atomic number). The configuration of these electrons follows from the principles of quantum mechanics. The number of electrons in each element's electron shells, particularly the outermost valence shell, is the primary factor in determining its chemical bonding behavior. Hence, it is the atomic number alone that determines the chemical properties of an element; and it is for this reason that an element can be defined as consisting of *any* mixture of atoms with a given atomic number.

2.4   New elements

The quest for new elements is usually described using atomic numbers. As of 2010, elements with atomic numbers 1 to 118 have been observed. Synthesis of new elements is accomplished by bombarding target atoms of heavy elements with ions, such that the sum of the atomic numbers of the target and ion elements equals the atomic number of the element being created. In general, the half-life becomes shorter as atomic number increases, though an "island of stability" may exist for undiscovered isotopes with certain numbers of protons and neutrons.

2.5   See also

- History of the periodic table
- Effective atomic number
- Atomic theory

- Prout's hypothesis

2.6 References

[1] The Periodic Table of Elements, American Institute of Physics

[2] The Development of the Periodic Table, Chemsoc

[3] Ordering the Elements in the Periodic Table, Royal Chemical Society

[4] Moseley's paper with illustrations

[5] Origin of symbol Z

Russian chemist Dmitri Mendeleev created a periodic table of the elements that ordered them numerically by atomic weight, yet occasionally used chemical properties in contradiction to weight.

Niels Bohr's 1913 Bohr model of the atom required van den Broek's atomic number of nuclear charges, and Bohr believed that Moseley's work contributed greatly to the acceptance of the model.

Henry Moseley helped develop the concept of atomic number by showing experimentally (1913) that Van den Broek's 1911 hypothesis combined with the Bohr model nearly correctly predicted atomic X-ray emissions.

Chapter 3

Electron configuration

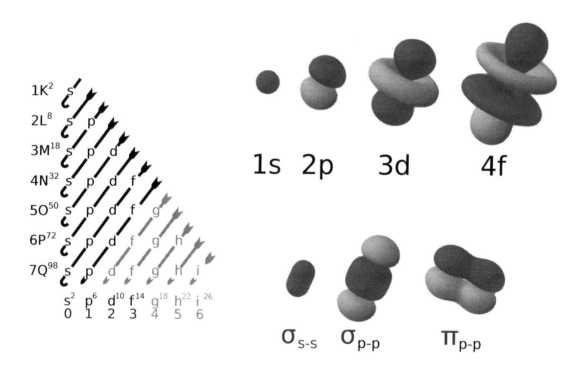

$$1s_2^2\,2s_4^2\,2p_{10}^6\,3s_{12}^2\,3p_{18}^6\,4s_{20}^2\,3d_{30}^{10}\,4p_{36}^6\,5s_{38}^2\,4d_{48}^{10}\,5p_{54}^6\,6s_{56}^2\,4f_{70}^{14}\,5d_{80}^{10}\,6p_{86}^6\,7s_{88}^2\,5f_{102}^{14}\,6d_{112}^{10}\,7p_{118}^6$$

Electron atomic and molecular orbitals

In atomic physics and quantum chemistry, the **electron configuration** is the distribution of electrons of an atom or molecule (or other physical structure) in atomic or molecular orbitals.[1] For example, the electron configuration of the neon atom is $1s^2\,2s^2\,2p^6$.

Electronic configurations describe electrons as each moving independently in an orbital, in an average field created by all other orbitals. Mathematically, configurations are described by Slater determinants or configuration state functions.

According to the laws of quantum mechanics, for systems with only one electron, an energy is associated with each electron configuration and, upon certain conditions, electrons are able to move from one configuration to another by the emission or absorption of a quantum of energy, in the form of a photon.

Knowledge of the electron configuration of different atoms is useful in understanding the structure of the periodic table

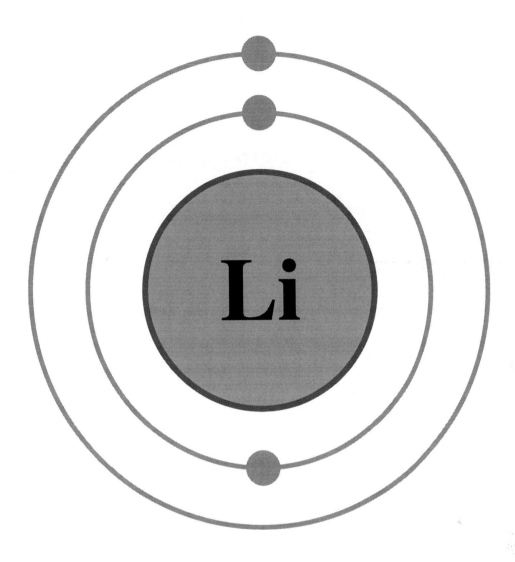

A Bohr diagram of lithium

of elements. The concept is also useful for describing the chemical bonds that hold atoms together. In bulk materials, this same idea helps explain the peculiar properties of lasers and semiconductors.

3.1   Shells and subshells

See also: Electron shell

Electron configuration was first conceived of under the Bohr model of the atom, and it is still common to speak of shells and subshells despite the advances in understanding of the quantum-mechanical nature of electrons.

An electron shell is the set of allowed states that share the same principal quantum number, n (the number before the letter in the orbital label), that electrons may occupy. An atom's nth electron shell can accommodate $2n^2$ electrons, *e.g.*

the first shell can accommodate 2 electrons, the second shell 8 electrons, and the third shell 18 electrons. The factor of two arises because the allowed states are doubled due to electron spin—each atomic orbital admits up to two otherwise identical electrons with opposite spin, one with a spin +1/2 (usually denoted by an up-arrow) and one with a spin −1/2 (with a down-arrow).

A subshell is the set of states defined by a common azimuthal quantum number, ℓ, within a shell. The values $\ell = 0, 1, 2, 3$ correspond to the s, p, d, and f labels, respectively. The maximum number of electrons that can be placed in a subshell is given by $2(2\ell + 1)$. This gives two electrons in an s subshell, six electrons in a p subshell, ten electrons in a d subshell and fourteen electrons in an f subshell.

The numbers of electrons that can occupy each shell and each subshell arise from the equations of quantum mechanics,[2] in particular the Pauli exclusion principle, which states that no two electrons in the same atom can have the same values of the four quantum numbers.[3]

3.2   Notation

See also: Atomic orbital

Physicists and chemists use a standard notation to indicate the electron configurations of atoms and molecules. For atoms, the notation consists of a sequence of atomic orbital labels (e.g. for phosphorus the sequence 1s, 2s, 2p, 3s, 3p) with the number of electrons assigned to each orbital (or set of orbitals sharing the same label) placed as a superscript. For example, hydrogen has one electron in the s-orbital of the first shell, so its configuration is written $1s^1$. Lithium has two electrons in the 1s-subshell and one in the (higher-energy) 2s-subshell, so its configuration is written $1s^2\,2s^1$ (pronounced "one-s-two, two-s-one"). Phosphorus (atomic number 15) is as follows: $1s^2\,2s^2\,2p^6\,3s^2\,3p^3$.

For atoms with many electrons, this notation can become lengthy and so an abbreviated notation is used, since all but the last few subshells are identical to those of one or another of the noble gases. Phosphorus, for instance, differs from neon ($1s^2\,2s^2\,2p^6$) only by the presence of a third shell. Thus, the electron configuration of neon is pulled out, and phosphorus is written as follows: $[\text{Ne}]\,3s^2\,3p^3$. This convention is useful as it is the electrons in the outermost shell that most determine the chemistry of the element.

For a given configuration, the order of writing the orbitals is not completely fixed since only the orbital occupancies have physical significance. For example, the electron configuration of the titanium ground state can be written as either $[\text{Ar}]\,4s^2\,3d^2$ or $[\text{Ar}]\,3d^2\,4s^2$. The first notation follows the order based on the Madelung rule for the configurations of neutral atoms; 4s is filled before 3d in the sequence Ar, K, Ca, Sc, Ti. The second notation groups all orbitals with the same value of n together, corresponding to the "spectroscopic" order of orbital energies that is the reverse of the order in which electrons are removed from a given atom to form positive ions; 3d is filled before 4s in the sequence Ti^{4+}, Ti^{3+}, Ti^{2+}, Ti^+, Ti.

The superscript 1 for a singly occupied orbital is not compulsory. It is quite common to see the letters of the orbital labels (s, p, d, f) written in an italic or slanting typeface, although the International Union of Pure and Applied Chemistry (IUPAC) recommends a normal typeface (as used here). The choice of letters originates from a now-obsolete system of categorizing spectral lines as "sharp", "principal", "diffuse" and "fundamental" (or "fine"), based on their observed fine structure: their modern usage indicates orbitals with an azimuthal quantum number, l, of 0, 1, 2 or 3 respectively. After "f", the sequence continues alphabetically "g", "h", "i"... ($l = 4, 5, 6...$), skipping "j", although orbitals of these types are rarely required.[4][5]

The electron configurations of molecules are written in a similar way, except that molecular orbital labels are used instead of atomic orbital labels (see below).

3.3   Energy — ground state and excited states

The energy associated to an electron is that of its orbital. The energy of a configuration is often approximated as the sum of the energy of each electron, neglecting the electron-electron interactions. The configuration that corresponds to the lowest electronic energy is called the ground state. Any other configuration is an excited state.

As an example, the ground state configuration of the sodium atom is $1s^2 2s^2 2p^6 3s$, as deduced from the Aufbau principle (see below). The first excited state is obtained by promoting a 3s electron to the 3p orbital, to obtain the $1s^2 2s^2 2p^6 3p$ configuration, abbreviated as the 3p level. Atoms can move from one configuration to another by absorbing or emitting energy. In a sodium-vapor lamp for example, sodium atoms are excited to the 3p level by an electrical discharge, and return to the ground state by emitting yellow light of wavelength 589 nm.

Usually, the excitation of valence electrons (such as 3s for sodium) involves energies corresponding to photons of visible or ultraviolet light. The excitation of core electrons is possible, but requires much higher energies, generally corresponding to x-ray photons. This would be the case for example to excite a 2p electron to the 3s level and form the excited $1s^2 2s^2 2p^5 3s^2$ configuration.

The remainder of this article deals only with the ground-state configuration, often referred to as "the" configuration of an atom or molecule.

3.4 History

Niels Bohr (1923) was the first to propose that the periodicity in the properties of the elements might be explained by the electronic structure of the atom.[6] His proposals were based on the then current Bohr model of the atom, in which the electron shells were orbits at a fixed distance from the nucleus. Bohr's original configurations would seem strange to a present-day chemist: sulfur was given as 2.4.4.6 instead of $1s^2 \; 2s^2 \; 2p^6 \; 3s^2 \; 3p^4$ (2.8.6).

The following year, E. C. Stoner incorporated Sommerfeld's third quantum number into the description of electron shells, and correctly predicted the shell structure of sulfur to be 2.8.6.[7] However neither Bohr's system nor Stoner's could correctly describe the changes in atomic spectra in a magnetic field (the Zeeman effect).

Bohr was well aware of this shortcoming (and others), and had written to his friend Wolfgang Pauli to ask for his help in saving quantum theory (the system now known as "old quantum theory"). Pauli realized that the Zeeman effect must be due only to the outermost electrons of the atom, and was able to reproduce Stoner's shell structure, but with the correct structure of subshells, by his inclusion of a fourth quantum number and his exclusion principle (1925):[8]

> *It should be forbidden for more than one electron with the same value of the main quantum number* n *to have the same value for the other three quantum numbers* k *[l], j [m_l] and m [m_s].*

The Schrödinger equation, published in 1926, gave three of the four quantum numbers as a direct consequence of its solution for the hydrogen atom:[2] this solution yields the atomic orbitals that are shown today in textbooks of chemistry (and above). The examination of atomic spectra allowed the electron configurations of atoms to be determined experimentally, and led to an empirical rule (known as Madelung's rule (1936),[9] see below) for the order in which atomic orbitals are filled with electrons.

3.5 Atoms: Aufbau principle and Madelung rule

The Aufbau principle (from the German *Aufbau*, "building up, construction") was an important part of Bohr's original concept of electron configuration. It may be stated as:[10]

> *a maximum of two electrons are put into orbitals in the order of increasing orbital energy: the lowest-energy orbitals are filled before electrons are placed in higher-energy orbitals.*

The principle works very well (for the ground states of the atoms) for the first 18 elements, then decreasingly well for the following 100 elements. The modern form of the Aufbau principle describes an order of orbital energies given by Madelung's rule (or Klechkowski's rule). This rule was first stated by Charles Janet in 1929, rediscovered by Erwin Madelung in 1936,[9] and later given a theoretical justification by V.M. Klechkowski[11]

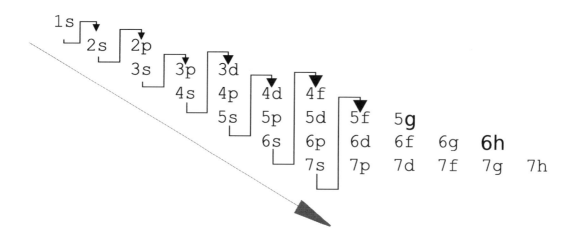

The approximate order of filling of atomic orbitals, following the arrows from 1s to 7p. (After 7p the order includes orbitals outside the range of the diagram, starting with 8s.)

1. Orbitals are filled in the order of increasing $n+l$;

2. Where two orbitals have the same value of $n+l$, they are filled in order of increasing n.

This gives the following order for filling the orbitals:

1s, 2s, 2p, 3s, 3p, 4s, 3d, 4p, 5s, 4d, 5p, 6s, 4f, 5d, 6p, 7s, 5f, 6d, 7p, (8s, 5g, 6f, 7d, 8p, and 9s)

In this list the orbitals in parentheses are not occupied in the ground state of the heaviest atom now known (Uuo, Z = 118).

The Aufbau principle can be applied, in a modified form, to the protons and neutrons in the atomic nucleus, as in the shell model of nuclear physics and nuclear chemistry.

3.5.1 Periodic table

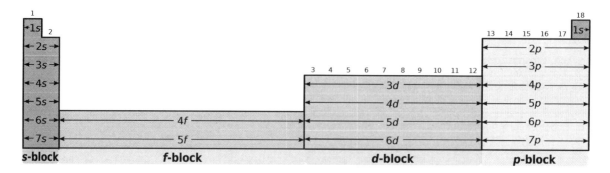

Electron configuration table

The form of the periodic table is closely related to the electron configuration of the atoms of the elements. For example, all the elements of group 2 have an electron configuration of [E] ns^2 (where [E] is an inert gas configuration), and have notable similarities in their chemical properties. In general, the periodicity of the periodic table in terms of periodic table blocks is clearly due to the number of electrons (2, 6, 10, 14...) needed to fill s, p, d, and f subshells.

The outermost electron shell is often referred to as the "valence shell" and (to a first approximation) determines the chemical properties. It should be remembered that the similarities in the chemical properties were remarked on more than a century before the idea of electron configuration.[12] It is not clear how far Madelung's rule *explains* (rather than simply describes) the periodic table,[13] although some properties (such as the common +2 oxidation state in the first row of the transition metals) would obviously be different with a different order of orbital filling.

3.5.2   Shortcomings of the Aufbau principle

The Aufbau principle rests on a fundamental postulate that the order of orbital energies is fixed, both for a given element and between different elements; in both cases this is only approximately true. It considers atomic orbitals as "boxes" of fixed energy into which can be placed two electrons and no more. However, the energy of an electron "in" an atomic orbital depends on the energies of all the other electrons of the atom (or ion, or molecule, etc.). There are no "one-electron solutions" for systems of more than one electron, only a set of many-electron solutions that cannot be calculated exactly[14] (although there are mathematical approximations available, such as the Hartree–Fock method).

The fact that the Aufbau principle is based on an approximation can be seen from the fact that there is an almost-fixed filling order at all, that, within a given shell, the s-orbital is always filled before the p-orbitals. In a hydrogen-like atom, which only has one electron, the s-orbital and the p-orbitals of the same shell have exactly the same energy, to a very good approximation in the absence of external electromagnetic fields. (However, in a real hydrogen atom, the energy levels are slightly split by the magnetic field of the nucleus, and by the quantum electrodynamic effects of the Lamb shift.)

3.5.3   Ionization of the transition metals

The naïve application of the Aufbau principle leads to a well-known paradox (or apparent paradox) in the basic chemistry of the transition metals. Potassium and calcium appear in the periodic table before the transition metals, and have electron configurations [Ar] $4s^1$ and [Ar] $4s^2$ respectively, i.e. the 4s-orbital is filled before the 3d-orbital. This is in line with Madelung's rule, as the 4s-orbital has $n+l = 4$ ($n = 4$, $l = 0$) while the 3d-orbital has $n+l = 5$ ($n = 3$, $l = 2$). After calcium, most neutral atoms in the first series of transition metals (Sc-Zn) have configurations with two 4s electrons, but there are two exceptions. Chromium and copper have electron configurations [Ar] $3d^5\,4s^1$ and [Ar] $3d^{10}\,4s^1$ respectively, i.e. one electron has passed from the 4s-orbital to a 3d-orbital to generate a half-filled or filled subshell. In this case, the usual explanation is that "half-filled or completely filled subshells are particularly stable arrangements of electrons".

The apparent paradox arises when electrons are *removed* from the transition metal atoms to form ions. The first electrons to be ionized come not from the 3d-orbital, as one would expect if it were "higher in energy", but from the 4s-orbital. This interchange of electrons between 4s and 3d is found for all atoms of the first series of transition metals.[15] The configurations of the neutral atoms (K, Ca, Sc, Ti, V, Cr, ...) usually follow the order 1s, 2s, 2p, 3s, 3p, 4s, 3d, ...; however the successive stages of ionization of a given atom (such as Fe^{4+}, Fe^{3+}, Fe^{2+}, Fe^+, Fe) usually follow the order 1s, 2s, 2p, 3s, 3p, 3d, 4s, ...

This phenomenon is only paradoxical if it is assumed that the energy order of atomic orbitals is fixed and unaffected by the nuclear charge or by the presence of electrons in other orbitals. If that were the case, the 3d-orbital would have the same energy as the 3p-orbital, as it does in hydrogen, yet it clearly doesn't. There is no special reason why the Fe^{2+} ion should have the same electron configuration as the chromium atom, given that iron has two more protons in its nucleus than chromium, and that the chemistry of the two species is very different. Melrose and Eric Scerri have analyzed the changes of orbital energy with orbital occupations in terms of the two-electron repulsion integrals of the Hartree-Fock method of atomic structure calculation.[16]

Similar ion-like $3d^x4s^0$ configurations occur in transition metal complexes as described by the simple crystal field theory, even if the metal has oxidation state 0. For example, chromium hexacarbonyl can be described as a chromium atom (not ion) surrounded by six carbon monoxide ligands. The electron configuration of the central chromium atom is described as $3d^6$ with the six electrons filling the three lower-energy d orbitals between the ligands. The other two d orbitals are at higher energy due to the crystal field of the ligands. This picture is consistent with the experimental fact that the complex is diamagnetic, meaning that it has no unpaired electrons. However, in a more accurate description using molecular orbital theory, the d-like orbitals occupied by the six electrons are no longer identical with the d orbitals of the free atom.

3.5.4   Other exceptions to Madelung's rule

There are several more exceptions to Madelung's rule among the heavier elements, and it is more and more difficult to resort to simple explanations, such as the stability of half-filled subshells. It is possible to predict most of the exceptions by Hartree–Fock calculations,[17] which are an approximate method for taking account of the effect of the other electrons on orbital energies. For the heavier elements, it is also necessary to take account of the effects of Special Relativity on the energies of the atomic orbitals, as the inner-shell electrons are moving at speeds approaching the speed of light. In general, these relativistic effects[18] tend to decrease the energy of the s-orbitals in relation to the other atomic orbitals.[19] The table below shows the ground state configuration in terms of orbital occupancy, but it does not show the ground state in terms of the sequence of orbital energies as determined spectroscopically. For example, in the transition metals, the 4s orbital is of a higher energy than the 3d orbitals; and in the lanthanides, the 6s is higher than the 4f and 5d. The ground states can be seen in the Electron configurations of the elements (data page).

The electron-shell configuration of elements beyond rutherfordium has not yet been empirically verified, but they are expected to follow Madelung's rule without exceptions until element 120.[22]

3.6   Electron configuration in molecules

In molecules, the situation becomes more complex, as each molecule has a different orbital structure. The molecular orbitals are labelled according to their symmetry,[23] rather than the atomic orbital labels used for atoms and monatomic ions: hence, the electron configuration of the dioxygen molecule, O_2, is written $1\sigma_g^2\,1\sigma_u^2\,2\sigma_g^2\,2\sigma_u^2\,3\sigma_g^2\,1\pi_u^4\,1\pi_g^2$,[24][25] or equivalently $1\sigma_g^2\,1\sigma_u^2\,2\sigma_g^2\,2\sigma_u^2\,1\pi_u^4\,3\sigma_g^2\,1\pi_g^2$.[1] The term $1\pi_g^2$ represents the two electrons in the two degenerate π^*-orbitals (antibonding). From Hund's rules, these electrons have parallel spins in the ground state, and so dioxygen has a net magnetic moment (it is paramagnetic). The explanation of the paramagnetism of dioxygen was a major success for molecular orbital theory.

The electronic configuration of polyatomic molecules can change without absorption or emission of a photon through vibronic couplings.

3.6.1   Electron configuration in solids

In a solid, the electron states become very numerous. They cease to be discrete, and effectively blend into continuous ranges of possible states (an electron band). The notion of electron configuration ceases to be relevant, and yields to band theory.

3.7   Applications

The most widespread application of electron configurations is in the rationalization of chemical properties, in both inorganic and organic chemistry. In effect, electron configurations, along with some simplified form of molecular orbital theory, have become the modern equivalent of the valence concept, describing the number and type of chemical bonds that an atom can be expected to form.

This approach is taken further in computational chemistry, which typically attempts to make quantitative estimates of chemical properties. For many years, most such calculations relied upon the "linear combination of atomic orbitals" (LCAO) approximation, using an ever larger and more complex basis set of atomic orbitals as the starting point. The last step in such a calculation is the assignment of electrons among the molecular orbitals according to the Aufbau principle. Not all methods in calculational chemistry rely on electron configuration: density functional theory (DFT) is an important example of a method that discards the model.

For atoms or molecules with more than one electron, the motion of electrons are correlated and such a picture is no longer exact. A very large number of electronic configurations are needed to exactly describe any multi-electron system, and no energy can be associated with one single configuration. However, the electronic wave function is usually dominated by a very small number of configurations and therefore the notion of electronic configuration remains essential for multi-electron systems.

A fundamental application of electron configurations is in the interpretation of atomic spectra. In this case, it is necessary to supplement the electron configuration with one or more term symbols, which describe the different energy levels available to an atom. Term symbols can be calculated for any electron configuration, not just the ground-state configuration listed in tables, although not all the energy levels are observed in practice. It is through the analysis of atomic spectra that the ground-state electron configurations of the elements were experimentally determined.

3.8   See also

- Born–Oppenheimer approximation
- Electron configurations of the elements (data page)
- Periodic table (electron configurations)
- Atomic orbital
- Energy level
- Term symbol
- Molecular term symbol
- HOMO/LUMO
- Periodic Table Group
- d electron count
- Extension of the periodic table beyond the seventh period Discusses the limits of the periodic table

3.9   Notes

[1] IUPAC, *Compendium of Chemical Terminology*, 2nd ed. (the "Gold Book") (1997). Online corrected version: (2006–) "configuration (electronic)".

[2] In formal terms, the quantum numbers n, ℓ and $m\ell$ arise from the fact that the solutions to the time-independent Schrödinger equation for hydrogen-like atoms are based on spherical harmonics.

[3] IUPAC, *Compendium of Chemical Terminology*, 2nd ed. (the "Gold Book") (1997). Online corrected version: (2006–) "Pauli exclusion principle".

[4] Weisstein, Eric W. (2007). "Electron Orbital". *wolfram*.

[5] Ebbing, Darrell D.; Gammon, Steven D. (2007-01-12). *General Chemistry*. p. 284. ISBN 978-0-618-73879-3.

[6] Bohr, Niels (1923). "Über die Anwendung der Quantumtheorie auf den Atombau.I".*Zeitschrift für Physik* **13**:117. Bibcode:1923Z. doi:10.1007/BF01328209.

[7] Stoner, E.C. (1924). "The distribution of electrons among atomic levels". *Philosophical Magazine (6th Ser.)* **48** (286): 719–36. doi:10.1080/14786442408634535.

[8] Pauli, Wolfgang (1925). "Über den Einfluss der Geschwindigkeitsabhändigkeit der elektronmasse auf den Zeemaneffekt". *Zeitschrift für Physik* **31**: 373. Bibcode:1925ZPhy...31..373P. doi:10.1007/BF02980592. English translation from Scerri, Eric R. (1991). "The Electron Configuration Model, Quantum Mechanics and Reduction" (PDF). *Br. J. Phil. Sci.* **42** (3): 309–25. doi:10.1093/bjps/42.3.309.

[9] Madelung, Erwin (1936). *Mathematische Hilfsmittel des Physikers*. Berlin: Springer.

[10] IUPAC, *Compendium of Chemical Terminology*, 2nd ed. (the "Gold Book") (1997). Online corrected version: (2006–) "aufbau principle".

[11] Wong, D. Pan (1979). "Theoretical justification of Madelung's rule". *Journal of Chemical Education* **56** (11): 714–18. Bibcode:1979JChEd..56..714W. doi:10.1021/ed056p714.

[12] The similarities in chemical properties and the numerical relationship between the atomic weights of calcium, strontium and barium was first noted by Johann Wolfgang Döbereiner in 1817.

[13] Scerri, Eric R. (1998). "How Good Is the Quantum Mechanical Explanation of the Periodic System?" (PDF). *Journal of Chemical Education* **75** (11): 1384–85. Bibcode:1998JChEd..75.1384S. doi:10.1021/ed075p1384. Ostrovsky, V.N. (2005). "On Recent Discussion Concerning Quantum Justification of the Periodic Table of the Elements". *Foundations of Chemistry* **7** (3): 235–39. doi:10.1007/s10698-005-2141-y.

[14] Electrons are identical particles, a fact that is sometimes referred to as "indistinguishability of electrons". A one-electron solution to a many-electron system would imply that the electrons could be distinguished from one another, and there is strong experimental evidence that they can't be. The exact solution of a many-electron system is a n-body problem with $n \geq 3$ (the nucleus counts as one of the "bodies"): such problems have evaded analytical solution since at least the time of Euler.

[15] There are some cases in the second and third series where the electron remains in an s-orbital.

[16] Melrose, Melvyn P.; Scerri, Eric R. (1996). "Why the 4s Orbital is Occupied before the 3d". *Journal of Chemical Education* **73** (6): 498–503. Bibcode:1996JChEd..73..498M. doi:10.1021/ed073p498.

[17] Meek, Terry L.; Allen, Leland C. (2002). "Configuration irregularities: deviations from the Madelung rule and inversion of orbital energy levels". *Chem. Phys. Lett.* **362** (5–6): 362–64. Bibcode:2002CPL...362..362M. doi:10.1016/S0009-2614(02)00919-3.

[18] IUPAC, *Compendium of Chemical Terminology*, 2nd ed. (the "Gold Book") (1997). Online corrected version: (2006–) "relativistic effects".

[19] Pyykkö, Pekka (1988). "Relativistic effects in structural chemistry". *Chem. Rev.* **88** (3): 563–94. doi:10.1021/cr00085a006.

[20] Miessler, G. L.; Tarr, D. A. (1999). *Inorganic Chemistry* (2nd ed.). Prentice-Hall. p. 38.

[21] Scerri, Eric R. (2007). *The periodic table: its story and its significance*. Oxford University Press. pp. 239–240. ISBN 0-19-530573-6.

[22] Hoffman, Darleane C.; Lee, Diana M.; Pershina, Valeria (2006). "Transactinides and the future elements". In Morss; Edelstein, Norman M.; Fuger, Jean. *The Chemistry of the Actinide and Transactinide Elements* (3rd ed.). Dordrecht, The Netherlands: Springer Science+Business Media. ISBN 1-4020-3555-1.

[23] The labels are written in lowercase to indicate that they correspond to one-electron functions. They are numbered consecutively for each symmetry type (irreducible representation in the character table of the point group for the molecule), starting from the orbital of lowest energy for that type.

[24] Levine I.N. *Quantum Chemistry* (4th ed., Prentice Hall 1991) p.376 ISBN 0-205-12770-3

[25] Miessler G.L. and Tarr D.A. *Inorganic Chemistry* (2nd ed., Prentice Hall 1999) p.118 ISBN 0-13-841891-8

3.10 References

- Jolly, William L. (1991). *Modern Inorganic Chemistry* (2nd ed.). New York: McGraw-Hill. pp. 1–23. ISBN 0-07-112651-1.

- Scerri, Eric (2007). *The Periodic System, Its Story and Its Significance*. New York: Oxford University Press. ISBN 0-19-530573-6.

3.11 External links

- What does an atom look like? Configuration in 3D

- The trouble with the aufbau principle Eric Scerri, *Education in Chemistry*, 7 November 2013

Chapter 4

Chemical property

A **chemical property** is any of a material's properties that becomes evident during, or after, a chemical reaction; that is, any quality that can be established only by changing a substance's chemical identity.[1] Simply speaking, chemical properties cannot be determined just by viewing or touching the substance; the substance's internal structure must be affected greatly for its chemical properties to be investigated. When a substance goes under a chemical reaction, the properties will change drastically, resulting in chemical change.However a catalytic property would also be a chemical property.

Chemical properties can be contrasted with physical properties, which can be discerned without changing the substance's structure. However, for many properties within the scope of physical chemistry, and other disciplines at the boundary between chemistry and physics, the distinction may be a matter of researcher's perspective. Material properties, both physical and chemical, can be viewed as supervenient; i.e., secondary to the underlying reality. Several layers of superveniency are possible.

Chemical properties can be used for building chemical classifications. They can also be useful to identify an unknown substance or to separate or purify it from other substances. Materials science will normally consider the chemical properties of a substance to guide its applications.

4.1   Examples of chemical properties

- Heat of combustion

- Enthalpy of formation

- Toxicity

- Chemical stability in a given environment

- Flammability (The ability to burn)

- Preferred oxidation state(s)

4.2   See also

- Physical property

- Chemical structure

- Material properties

- Biological activity

- Quantitative structure–activity relationship (QSAR)

- Lipinski's Rule of Five, describing molecular properties of drugs

4.3   References

[1] William L. Masterton, Cecile N. Hurley, "Chemistry: Principles and Reactions", 6th edition. Brooks/Cole Cengage Learning, 2009, p.13 (Google books)

Chapter 5

Block (periodic table)

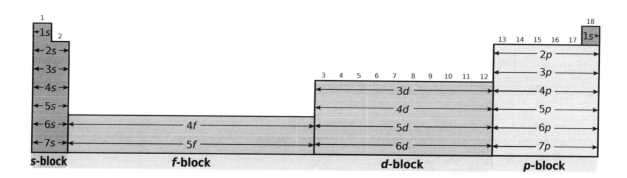

A **block** of the periodic table of elements is a set of adjacent groups. The term appears to have been first used by Charles Janet.[1] The respective highest-energy electrons in each element in a block belong to the same atomic orbital type. Each block is named after its characteristic orbital; thus, the blocks are:

- s-block

- p-block

- d-block

- f-block

- g-block (hypothetical)

The block names (s, p, d, f and g) are derived from the spectroscopic notation for the associated atomic orbitals: **s**harp, **p**rincipal, **d**iffuse and **f**undamental, and then g which follows f in the alphabet.

The following is the order for filling the "subshell" orbitals, according to the Aufbau principle, which also gives the linear order of the "blocks" (as atomic number increases) in the periodic table:

1s, 2s, 2p, 3s, 3p, 4s, 3d, 4p, 5s, 4d, 5p, 6s, 4f, 5d, 6p, 7s, 5f, 6d, 7p, ...

For discussion of the nature of why the energies of the blocks naturally appear in this order in complex atoms, see atomic orbital and electron configuration.

The "periodic" nature of the filling of orbitals, as well as emergence of the **s**, **p**, **d** and **f** "blocks" is more obvious, if this order of filling is given in matrix form, with increasing principal quantum numbers starting the new rows ("periods") in the matrix. Then, each subshell (composed of the first two quantum numbers) is repeated as many times as required for each pair of electrons it may contain. The result is a compressed periodic table, with each entry representing two successive elements:

5.1 Periodic table

There is an approximate correspondence between this nomenclature of blocks, based on electronic configuration, and groupings of elements based on chemical properties. The s-block and p-block together are usually considered as the main group elements, the d-block corresponds to the transition metals, and the f-block are the lanthanides and the actinides. However, not everyone agrees on the exact membership of each set of elements, so that for example the Group 12 elements Zn, Cd and Hg are considered as main group by some scientists and transition metals by others. Groups (columns) in the f-block (between groups 2 and 3) are not numbered.

In periodic tables organized by blocks, helium is placed next to hydrogen, instead of on top of neon as in tables organized by chemical properties. This is because helium is in the s-block, with its outer (and only) electrons in the 1s atomic orbital. In addition to the blocks listed in this table, there is a hypothetical g-block which is not pictured here. The g-block elements can be seen in the expanded extended periodic table. Also, lutetium and lawrencium are placed under scandium and yttrium to reflect their status as d-block elements[2] (although it has been argued that lanthanum and actinium should instead hold these positions, as they have no electrons in the 4f and 5f orbitals, respectively, while lutetium and lawrencium do).[3]

5.2 References

[1] Charles Janet, *La classification hélicoïdale des éléments chimiques*, Beauvais, 1928

[2] Scerri, Eric. "Mendeleev's table finally completed and what to do about group 3".

[3] Lavelle, Laurence. "Lanthanum (La) and Actinium (Ac) Should Remain in the d-Block" (PDF). *lavelle.chem.ucla.edu*. Retrieved 9 November 2014.

Chapter 6

Period (periodic table)

Group→	1	2	3	4	5	6	7	8	9	10	11	12	13	14	15	16	17	18
↓Period																		
1	1 H																	2 He
2	3 Li	4 Be											5 B	6 C	7 N	8 O	9 F	10 Ne
3	11 Na	12 Mg											13 Al	14 Si	15 P	16 S	17 Cl	18 Ar
4	19 K	20 Ca	21 Sc	22 Ti	23 V	24 Cr	25 Mn	26 Fe	27 Co	28 Ni	29 Cu	30 Zn	31 Ga	32 Ge	33 As	34 Se	35 Br	36 Kr
5	37 Rb	38 Sr	39 Y	40 Zr	41 Nb	42 Mo	43 Tc	44 Ru	45 Rh	46 Pd	47 Ag	48 Cd	49 In	50 Sn	51 Sb	52 Te	53 I	54 Xe
6	55 Cs	56 Ba	*	72 Hf	73 Ta	74 W	75 Re	76 Os	77 Ir	78 Pt	79 Au	80 Hg	81 Tl	82 Pb	83 Bi	84 Po	85 At	86 Rn
7	87 Fr	88 Ra	**	104 Rf	105 Db	106 Sg	107 Bh	108 Hs	109 Mt	110 Ds	111 Rg	112 Cn	113 Uut	114 Fl	115 Uup	116 Lv	117 Uus	118 Uuo

*	57 La	58 Ce	59 Pr	60 Nd	61 Pm	62 Sm	63 Eu	64 Gd	65 Tb	66 Dy	67 Ho	68 Er	69 Tm	70 Yb	71 Lu	
**	89 Ac	90 Th	91 Pa	92 U	93 Np	94 Pu	95 Am	96 Cm	97 Bk	98 Cf	99 Es	100 Fm	101 Md	102 No	103 Lr	

In the periodic table of the elements, each numbered row is a period.

In the periodic table of the elements, elements are arranged in a series of rows (or **periods**) so that those with similar properties appear in a column. Elements of the same period have the same number of electron shells; with each group across a period, the elements have one more proton and electron and become less metallic. This arrangement reflects the *periodic* recurrence of similar properties as the atomic number increases. For example, the alkaline metals lie in one group (group 1) and share similar properties, such as high reactivity and the tendency to lose one electron to arrive at a noble-gas electronic configuration. The periodic table of elements has a total of 118 elements.

Modern quantum mechanics explains these periodic trends in properties in terms of electron shells. As atomic number increases, shells fill with electrons in approximately the order shown at right. The filling of each shell corresponds to a row in the table.

In the s-block and p-block of the periodic table, elements within the same period generally do not exhibit trends and similarities in properties (vertical trends down groups are more significant). However in the d-block, trends across periods become significant, and in the f-block elements show a high degree of similarity across periods.

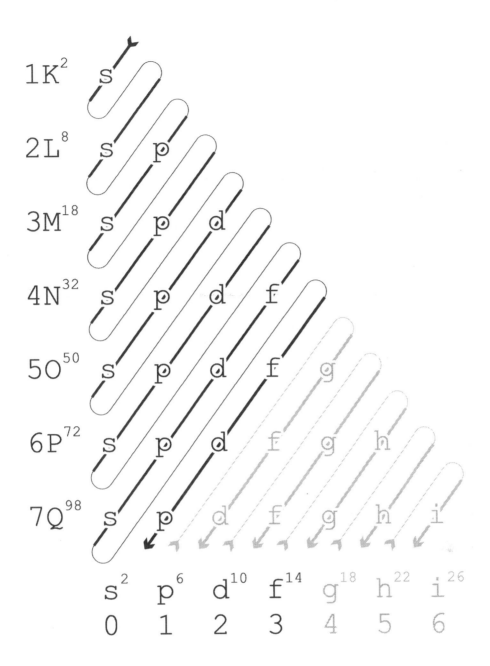

The Madelung energy ordering rule describes the order in which orbitals are arranged by increasing energy according to the Madelung rule. Each diagonal corresponds to a different value of n + l.

6.1 Periods

Seven periods of elements occur naturally on Earth. For period 8, which includes elements which may be synthesized after 2013, see the extended periodic table.

A group in chemistry means a family of objects with similarities like different families. There are 7 periods, going horizontally across the periodic table.

6.1.1 Period 1

The first period contains fewer elements than any other, with only two, hydrogen and helium. They therefore do not follow the octet rule. Chemically, helium behaves as a noble gas, and thus is taken to be part of the group 18 elements. However, in terms of its nuclear structure it belongs to the s block, and is therefore sometimes classified as a group 2 element, or simultaneously both 2 and 18. Hydrogen readily loses and gains an electron, and so behaves chemically as both a group 1 and a group 17 element.

- Hydrogen (H) is the most abundant of the chemical elements, constituting roughly 75% of the universe's elemental mass.[1] Ionized hydrogen is just a proton. Stars in the main sequence are mainly composed of hydrogen in its plasma state. Elemental hydrogen is relatively rare on Earth, and is industrially produced from hydrocarbons such as methane. Hydrogen can form compounds with most elements and is present in water and most organic compounds.[2]

- Helium (He) exists only as a gas except in extreme conditions.[3] It is the second lightest element and is the second most abundant in the universe.[4] Most helium was formed during the Big Bang, but new helium is created through nuclear fusion of hydrogen in stars.[5] On Earth, helium is relatively rare, only occurring as a byproduct of the natural decay of some radioactive elements.[6] Such 'radiogenic' helium is trapped within natural gas in concentrations of up to seven percent by volume.[7]

6.1.2 Period 2

Period 2 elements involve the 2s and 2p orbitals. They include the biologically most essential elements besides hydrogen: carbon, nitrogen, and oxygen.

- Lithium (Li) is the lightest metal and the least dense solid element.[8] In its non-ionized state it is one of the most reactive elements, and so is only ever found naturally in compounds. It is the heaviest primordial element forged in large quantities during the Big Bang.

- Beryllium (Be) has one of the highest melting points of all the light metals. Small amounts of beryllium were synthesised during the Big Bang, although most of it decayed or reacted further within stars to create larger nucleii, like carbon, nitrogen or oxygen. Beryllium is classified by the International Agency for Research on Cancer as a group 1 carcinogen.[9] Between 1% and 15% of people are sensitive to beryllium and may develop an inflammatory reaction in their respiratory system and skin, called chronic beryllium disease.[10]

- Boron (B) does not occur naturally as a free element, but in compounds such as borates. It is an essential plant micronutrient, required for cell wall strength and development, cell division, seed and fruit development, sugar transport and hormone development,[11][12] though high levels are toxic.

- Carbon (C) is the fourth most abundant element in the universe by mass after hydrogen, helium and oxygen[13] and is the second most abundant element in the human body by mass after oxygen,[14] the third most abundant by number of atoms.[15] There are an almost infinite number of compounds that contain carbon due to carbon's ability to form long stable chains of C—C bonds.[16][17] All organic compounds, those essential for life, contain at least one atom of carbon;[16][17] combined with hydrogen, oxygen, nitrogen, sulfur, and phosphorus, carbon is the basis of every important biological compound.[17]

- Nitrogen (N) is found mainly as mostly inert diatomic gas, N_2, which makes up 78% of the Earth's atmosphere. It is an essential component of proteins and therefore of life.

- Oxygen (O) comprising 21% of the atmosphere and is required for respiration by all (or nearly all) animals, as well as being the principal component of water. Oxygen is the third most abundant element in the universe, and oxygen compounds dominate the Earth's crust.

- Fluorine (F) is the most reactive element in its non-ionized state, and so is never found that way in nature.

- Neon (Ne) is a noble gas used in neon lighting.

6.1.3 Period 3

All period three elements occur in nature and have at least one stable isotope. All but the noble gas argon are essential to basic geology and biology.

- Sodium (Na) is an alkali metal. It is present in Earth's oceans in large quantities in the form of sodium chloride (table salt).

- Magnesium (Mg) is an alkaline earth metal. Magnesium ions are found in chlorophyll.

- Aluminium (Al) is a post-transition metal. It is the most abundant metal in the Earth's crust.

- Silicon (Si) is a metalloid. It is a semiconductor, making it the principal component in many integrated circuits. Silicon dioxide is the principal constituent of sand. As Carbon is to Biology, Silicon is to Geology.

- Phosphorus (P) is a nonmetal essential to DNA. It is highly reactive, and as such is never found in nature as a free element.

- Sulfur (S) is a nonmetal. It is found in two amino acids: cysteine and methionine.

- Chlorine (Cl) is a halogen. It is used as a disinfectant, especially in swimming pools.

- Argon (Ar) is a noble gas, making it almost entirely nonreactive. Incandescent lamps are often filled with noble gases such as argon in order to preserve the filaments at high temperatures.

6.1.4 Period 4

From left to right, aqueous solutions of: $Co(NO_3)_2$ (red); $K_2Cr_2O_7$ (orange); K_2CrO_4 (yellow); $NiCl_2$ (green); $CuSO_4$ (blue); $KMnO_4$ (purple).

Period 4 includes the biologically essential elements potassium and calcium, and is the first period in the d-block with the lighter transition metals. These include iron, the heaviest element forged in main-sequence stars and a principal component of the earth, as well as other important metals such as cobalt, nickel, copper, and zinc. Almost all have biological roles.

6.1.5 Period 5

Period 5 contains the heaviest few elements that have biological roles, molybdenum and iodine. (Tungsten, a period 6 element, is the only heavier element that has a biological role.) It includes technetium, the lightest exclusively radioactive element.

6.1.6 Period 6

Period 6 is the first period to include the f-block, with the lanthanides (also known as the rare earth elements), and includes the heaviest stable elements. Many of these heavy metals are toxic and some are radioactive, but platinum and gold are largely inert.

6.1.7 Period 7

All elements of period 7 are radioactive. This period contains the heaviest element which occurs naturally on earth, californium. All of the subsequent elements in the period have been synthesized artificially. Whilst one of these(einsteinium) is now available in macroscopic quantities, most are extremely rare, having only been prepared in microgram amounts or less. Some of the later elements have only ever been identified in laboratories in quantities of a few atoms at a time.

Although the rarity of many of these elements means that experimental results are not very extensive, periodic and group trends in behaviour appear to be less well defined for period 7 than for other periods. Whilst francium and radium do show typical properties of Groups 1 and 2 respectively, the actinides display a much greater variety of behaviour and oxidation states than the lanthanides. These peculiarities of period 7 may be due to a variety of factors, including a large degree of spin-orbit coupling and relativistic effects, ultimately caused by the very high positive electrical charge from their massive atomic nuclei.

6.1.8 Period 8

Main article: Extended periodic table

No element of the eighth period has yet been synthesized. A g-block is predicted. It is not clear if all elements predicted for the eighth period are in fact physically possible. There may therefore be no ninth period.

6.2 References

[1] Palmer, David (November 13, 1997). "Hydrogen in the Universe". NASA. Retrieved 2008-02-05.

[2] "hydrogen". *Encyclopædia Britannica*. 2008.

[3] "Helium: physical properties". WebElements. Retrieved 2008-07-15.

[4] "Helium: geological information". WebElements. Retrieved 2008-07-15.

[5] Cox, Tony (1990-02-03). "Origin of the chemical elements". *New Scientist*. Retrieved 2008-07-15.

[6] "Helium supply deflated: production shortages mean some industries and partygoers must squeak by.". Houston Chronicle. 2006-11-05.

[7] Brown, David (2008-02-02). "Helium a New Target in New Mexico". American Association of Petroleum Geologists. Retrieved 2008-07-15.

[8] Lithium at WebElements.

[9] "IARC Monograph, Volume 58". International Agency for Research on Cancer. 1993. Retrieved 2008-09-18.

[10] Information about chronic beryllium disease.

[11] "Functions of Boron in Plant Nutrition" (PDF). U.S. Borax Inc.

[12] Blevins, Dale G.; Lukaszewski, Krystyna M. (1998). "Functions of Boron in Plant Nutrition". *Annual Review of Plant Physiology and Plant Molecular Biology* **49**: 481–500. doi:10.1146/annurev.arplant.49.1.481. PMID 15012243.

[13] Ten most abundant elements in the universe, taken from *The Top 10 of Everything*, 2006, Russell Ash, page 10. Retrieved October 15, 2008.

[14] Chang, Raymond (2007). *Chemistry, Ninth Edition*. McGraw-Hill. p. 52. ISBN 0-07-110595-6.

[15] Freitas Jr., Robert A. (1999). *Nanomedicine*. Landes Bioscience. Tables 3-1 & 3-2. ISBN 1-57059-680-8.

[16] "Structure and Nomenclature of Hydrocarbons". Purdue University. Retrieved 2008-03-23.

[17] Alberts, Bruce; Alexander Johnson; Julian Lewis; Martin Raff; Keith Roberts; Peter Walter. *Molecular Biology of the Cell*. Garland Science.

Chapter 7

Group (periodic table)

Group →	1	2	3	4	5	6	7	8	9	10	11	12	13	14	15	16	17	18
↓Period																		
1	1 H																	2 He
2	3 Li	4 Be											5 B	6 C	7 N	8 O	9 F	10 Ne
3	11 Na	12 Mg											13 Al	14 Si	15 P	16 S	17 Cl	18 Ar
4	19 K	20 Ca	21 Sc	22 Ti	23 V	24 Cr	25 Mn	26 Fe	27 Co	28 Ni	29 Cu	30 Zn	31 Ga	32 Ge	33 As	34 Se	35 Br	36 Kr
5	37 Rb	38 Sr	39 Y	40 Zr	41 Nb	42 Mo	43 Tc	44 Ru	45 Rh	46 Pd	47 Ag	48 Cd	49 In	50 Sn	51 Sb	52 Te	53 I	54 Xe
6	55 Cs	56 Ba	*	72 Hf	73 Ta	74 W	75 Re	76 Os	77 Ir	78 Pt	79 Au	80 Hg	81 Tl	82 Pb	83 Bi	84 Po	85 At	86 Rn
7	87 Fr	88 Ra	**	104 Rf	105 Db	106 Sg	107 Bh	108 Hs	109 Mt	110 Ds	111 Rg	112 Cn	113 Uut	114 Fl	115 Uup	116 Lv	117 Uus	118 Uuo

*	57 La	58 Ce	59 Pr	60 Nd	61 Pm	62 Sm	63 Eu	64 Gd	65 Tb	66 Dy	67 Ho	68 Er	69 Tm	70 Yb	71 Lu
**	89 Ac	90 Th	91 Pa	92 U	93 Np	94 Pu	95 Am	96 Cm	97 Bk	98 Cf	99 Es	100 Fm	101 Md	102 No	103 Lr

In the periodic table of the elements, each numbered column is a group.

In chemistry, a **group** (also known as a *family*) is a column of elements in the periodic table of the chemical elements. There are 18 numbered groups in the periodic table, but the f-block columns (between groups 2 and 3) are not numbered. The elements in a group have similar physical or chemical characteristic of the outermost electron shells of their atoms (i.e., the same core charge), as most chemical properties are dominated by the orbital location of the outermost electron. There are three systems of group numbering. The modern numbering *group 1* to *group 18* is recommended by the International Union of Pure and Applied Chemistry (IUPAC). It replaces two older naming schemes that were mutually confusing. Also, groups may be identified by their topmost element or have a specific name. For example, group 16 is variously described as *oxygen group* and *chalcogen*.

7.1   CAS and old IUPAC numbering

Two earlier group number systems exist: *CAS* (Chemical Abstracts Service) and *old IUPAC*. Both use numerals (Arabic or Roman) and letters *A* and *B*. Both systems agree on the numbers. The numbers indicate approximately the highest oxidation number of the elements in that group, and so indicate similar chemistry with other elements with the same numeral. The number proceeds in a linearly increasing fashion for the most part, once on the left of the table, and once on the right (see List of oxidation states of the elements), with some irregularities in the transition metals. However, the two systems use the letters differently. For example, potassium (K) has one valence electron. Therefore, it is located in group 1. Calcium (Ca) is in group 2, for it contains two valence electrons.

In the old IUPAC system the letters A and B were designated to the left (A) and right (B) part of the table, while in the CAS system the letters A and B are designated to main group elements (A) and transition elements (B). The old IUPAC system was frequently used in Europe while the CAS is most common in America. The new IUPAC scheme was developed to replace both systems as they confusingly used the same names to mean different things. The new system simply numbers the groups increasingly from left to right on the standard periodic table. The IUPAC proposal was first circulated in 1985 for public comments,[1] and was later included as part of the 1990 edition of the *Nomenclature of Inorganic Chemistry*.[2]

7.2   Group names

In history, several sets of group names have been used:[1][3]

7.3   References

[1] Fluck, E. (1988). "New Notations in the Periodic Table" (PDF). *Pure Appl. Chem.* (IUPAC) **60** (3): 431–436. doi:10.1351/pac198860030431. Retrieved 24 March 2012.

[2] Leigh, G. J. *Nomenclature of Inorganic Chemistry: Recommendations 1990*. Blackwell Science, **1990**. ISBN 0-632-02494-1.

[3] IUPAC (2005). "Nomenclature of inorganic chemistry" (PDF).

7.4   Further reading

- Scerri, E. R. (2007). *The periodic table, its story and its significance*. Oxford University Press. ISBN 978-0-19-530573-9.

Chapter 8

Halogen

This article is about the chemical series. For other uses, see Halogen (disambiguation).

The **halogens** or **halogen elements** (/ˈhælədʒin/) are a group in the periodic table consisting of five chemically related elements: fluorine (F), chlorine (Cl), bromine (Br), iodine (I), and astatine (At). The artificially created element 117 (ununseptium) may also be a halogen. In the modern IUPAC nomenclature, this group is known as **group 17**.

The name 'halogen' means 'salt-producing'. When halogens react with metals they produce a wide range of salts, including calcium fluoride, sodium chloride (common salt), silver bromide and potassium iodide.

The group of halogens is the only periodic table group that contains elements in three of the four main states of matter at standard temperature and pressure. All of the halogens form acids when bonded to hydrogen. Most halogens are typically produced from minerals or salts. The middle halogens, that is chlorine, bromine and iodine, are often used as disinfectants. Organobromides are the most important class of flame retardants. Elemental halogens are lethally to dangerously toxic.

8.1 History

The fluorine mineral fluorospar was known as early as 1529. Early chemists realized that fluorine compounds contain an undiscovered element, but were unable to isolate it. In 1869, George Gore, an English chemist, ran a current of electricity through hydrofluoric acid and discovered fluorine, but he was unable to prove his results at the time. In 1886, Henri Moissan, a chemist in Paris, performed electrolysis on potassium bifluoride dissolved in waterless hydrofluoric acid, and successfully produced fluorine.[1]

Hydrochloric acid was known to alchemists and early chemists. However, elemental chlorine was not produced until 1774, when Carl Wilhelm Scheele heated hydrochloric acid with manganese dioxide. Scheele called the element "dephlogisticated muriatic acid", which is how chlorine was known for 33 years. In 1807, Humphry Davy investigated chlorine and discovered that it is an actual element. Chlorine was used as a poison gas during World War I.[1]

Bromine was discovered in the 1820s by Antoine-Jérôme Balard. Balard discovered bromine by passing chlorine gas through a sample of brine. He originally proposed the name *muride* for the new element, but the French Academy changed the element's name to bromine.[1]

Iodine was discovered by Bernard Courtois, who was using seaweed ash as part of a process for saltpeter manufacture. Courtois typically boiled the seaweed ash with water to generate potassium chloride. However, in 1811, Courtois added sulfuric acid to his process, and found that his process produced purple fumes that condensed into black crystals. Suspecting that these crystals were a new element, Courtois sent samples to other chemists for investigation. Iodine was proven to be a new element by Joseph Gay-Lussac.[1]

In 1931, Fred Allison claimed to have discovered element 85 with a magneto-optical machine, and named the element Alabamine, but was mistaken. In 1937, Jajendralal De claimed to have discovered element 85 in minerals, and called the element dakine, but he was also mistaken. An attempt at discovering element 85 in 1939 by Horia Hulublei and Yvette

Cauchois via spectroscopy was also unsuccessful, as was an attempt in the same year by Walter Minder, who discovered an iodine-like element resulting from beta decay of radium. Element 85, now named astatine, was produced successfully in 1940 by Dale R. Corson, K.R. Mackenzie, and Emilio G. Segrè, who bombarded bismuth with alpha particles.[1]

8.1.1   Etymology

In 1842, the Swedish chemist Baron Jöns Jakob Berzelius proposed the term "halogen" – ἅλς (*háls*), "salt" or "sea", and γεν- (*gen-*), "to produce" – for the four elements (fluorine, chlorine, bromine, and iodine) that produce a sea-salt-like substance when they form a compound with a metal.[2] The word "halogen" had actually first been proposed in 1811 by Johann Salomo Christoph Schweigger as a name for the newly discovered element chlorine, but Davy's proposed term for this element eventually won out, and Schweigger's term was kept at Berzelius' suggestion as the term for the element group that contains chlorine.[3]

Fluorine's name comes from the Latin word *fluere*, meaning "to flow". Chlorine's name comes from the Greek word *chloros*, meaning "greenish-yellow". Bromine's name comes from the Greek word *bromos*, meaning "stench". Iodine's name comes from the Greek word *iodes*, meaning "violet". Astatine's name comes from the Greek word *astatos*, meaning "unstable".[1]

8.2   Characteristics

8.2.1   Chemical

The halogens show trends in chemical bond energy moving from top to bottom of the periodic table column with fluorine deviating slightly. (It follows trend in having the highest bond energy in compounds with other atoms, but it has very weak bonds within the diatomic F_2 molecule.) This means, as you go down the periodic table, the reactivity of the element will decrease because of the increasing size of the atoms [4]

Halogens are highly reactive, and as such can be harmful or lethal to biological organisms in sufficient quantities. This high reactivity is due to the high electronegativity of the atoms due to their high effective nuclear charge. Because the halogens have seven valence electrons in their outermost energy level, they can gain an electron by reacting with atoms of other elements to satisfy the octet rule. Fluorine is one of the most reactive elements, attacking otherwise-inert materials such as glass, and forming compounds with the usually inert noble gases. It is a corrosive and highly toxic gas. The reactivity of fluorine is such that, if used or stored in laboratory glassware, it can react with glass in the presence of small amounts of water to form silicon tetrafluoride (SiF_4). Thus, fluorine must be handled with substances such as Teflon (which is itself an organofluorine compound), extremely dry glass, or metals such as copper or steel, which form a protective layer of fluoride on their surface.

The high reactivity of fluorine allows paradoxically some of the strongest bonds possible, especially to carbon. For example, Teflon is fluorine bonded with carbon and is extremely resistant to thermal and chemical attack and has a high melting point.

Molecules

Diatomic halogen molecules   The halogens form homonuclear diatomic molecules (not proven for astatine). Due to relatively weak intermolecular forces, chlorine and fluorine form part of the group known as "elemental gases".

The elements become less reactive and have higher melting points as the atomic number increases.

Compounds

Hydrogen halides   All of the halogens have been observed to react with hydrogen to form hydrogen halides. For fluorine, chlorine, and bromine, this reaction is in the form of:

$$H_2 + X_2 \rightarrow 2HX$$

However, hydrogen iodide and hydrogen astatide can split back into their constituent elements.[6]

The hydrogen-halogen reactions get gradually less reactive toward the heavier halogens. A fluorine-hydrogen reaction is explosive even when it is dark and cold. A chlorine-hydrogen reaction is also explosive, but only in the presence of light and heat. A bromine-hydrogen reaction is even less explosive; it is explosive only when exposed to flames. Iodine and astatine only partially react with hydrogen, forming equilibria.[6]

All halogens form binary compounds with hydrogen known as the hydrogen halides: hydrogen fluoride (HF), hydrogen chloride (HCl), hydrogen bromide (HBr), hydrogen iodide (HI), and hydrogen astatide (HAt). All of these compounds form acids when mixed with water. Hydrogen fluoride is the only hydrogen halide that forms hydrogen bonds. Hydrochloric acid, hydrobromic acid, hydroiodic acid, and hydroastatic acid are all strong acids, but hydrofluoric acid is a weak acid.[7]

All of the hydrogen halides are irritants. Hydrogen fluoride and hydrogen chloride are highly acidic. Hydrogen fluoride is used as an industrial chemical, and is highly toxic, causing pulmonary edema and damaging cells.[8] Hydrogen chloride is also a dangerous chemical. Breathing in gas with more than fifty parts per million of hydrogen chloride can cause death in humans.[9] Hydrogen bromide is even more toxic and irritating than hydrogen chloride. Breathing in gas with more than thirty parts per million of hydrogen bromide can be lethal to humans.[10] Hydrogen iodide, like other hydrogen halides, is toxic.[11]

Metal halides   Main article: Metal halides

All the halogens are known to react with sodium to form sodium fluoride, sodium chloride, sodium bromide, sodium iodide, and sodium astatide. Heated sodium's reaction with halogens produces bright-orange flames. Sodium's reaction with chlorine is in the form of:

$$2Na + Cl_2 \rightarrow 2NaCl^{[6]}$$

Iron reacts with fluorine, chlorine, and bromine to form Iron(III) halides. These reactions are in the form of:

$$2Fe + 3X_2 \rightarrow 2FeX_3^{[6]}$$

However, when iron reacts with iodine, it forms only iron(II) iodide.

Iron wool can react rapidly with fluorine to form the white compound iron(III) fluoride even in cold temperatures. When chlorine comes into contact with heated iron, they react to form the black iron (III) chloride. However, if the reaction conditions are moist, this reaction will instead result in a reddish-brown product. Iron can also react with bromine to form iron(III) bromide. This compound is reddish-brown in dry conditions. Iron's reaction with bromine is less reactive than its reaction with fluorine or chlorine. Hot iron can also react with iodine, but it forms iron(II) iodide. This compound may be gray, but the reaction is always contaminated with excess iodine, so it is not known for sure. Iron's reaction with iodine is less vigorous than its reaction with the lighter halogens.[6]

Interhalogen compounds   Main article: Interhalogen

Interhalogen compounds are in the form of XY_n where X and Y are halogens and n is one, three, five, or seven. Interhalogen compounds contain at most two different halogens. Large interhalogens, such as ClF_3 can be produced by a reaction of a pure halogen with a smaller interhalogen such as ClF. All interhalogens except IF_7 can be produced by directly combining pure halogens in various conditions.[12]

Interhalogens are typically more reactive than all diatomic halogen molecules except F_2 because interhalogen bonds are weaker. However, the chemical properties of interhalogens are still roughly the same as those of diatomic halogens. Many interhalogens consist of one or more atoms of fluorine bonding to a heavier halogen. Chlorine can bond with up to 3 fluorine atoms, bromine can bond with up to five fluorine atoms, and iodine can bond with up to seven fluorine atoms. Most interhalogen compounds are covalent gases. However, there are some interhalogens that are liquids, such as BrF_3, and many iodine-containing interhalogens are solids.[12]

Organohalogen compounds   Many synthetic organic compounds such as plastic polymers, and a few natural ones, contain halogen atoms; these are known as *halogenated* compounds or organic halides. Chlorine is by far the most abundant of the halogens in seawater, and the only one needed in relatively large amounts (as chloride ions) by humans. For example, chloride ions play a key role in brain function by mediating the action of the inhibitory transmitter GABA and are also used by the body to produce stomach acid. Iodine is needed in trace amounts for the production of thyroid hormones such as thyroxine. On the other hand, neither fluorine nor bromine is believed to be essential for humans. Organohalogens are also synthesized through the nucleophilic abstraction reaction.

Polyhalogenated compounds   Polyhalogenated compounds are industrially created compounds substituted with multiple halogens. Many of them are very toxic and bioaccumulate in humans, and have a very wide application range. They include PCBs, PBDEs, and perfluorinated compounds (PFCs), as well as numerous other compounds.

Reactions

Reactions with water   Fluorine reacts vigorously with water to produce oxygen (O_2) and hydrogen fluoride (HF):[13]

$$2 F_2(g) + 2 H_2O(l) \rightarrow O_2(g) + 4 HF(aq)$$

Chlorine has maximum solubility of ca. 7.1 g Cl_2 per kg of water at ambient temperature (21 °C).[14] Dissolved chlorine reacts to form hydrochloric acid (HCl) and hypochlorous acid, a solution that can be used as a disinfectant or bleach:

$$Cl_2(g) + H_2O(l) \rightarrow HCl(aq) + HClO(aq)$$

Bromine has a solubility of 3.41 g per 100 g of water,[15] but it slowly reacts to form hydrogen bromide (HBr) and hypobromous acid (HBrO):

$$Br_2(g) + H_2O(l) \rightarrow HBr(aq) + HBrO(aq)$$

Iodine, however, is minimally soluble in water (0.03 g/100 g water at 20 °C) and does not react with it.[16] However, iodine will form an aqueous solution in the presence of iodide ion, such as by addition of potassium iodide (KI), because the triiodide ion is formed.

8.2.2   Physical and atomic

The table below is a summary of the key physical and atomic properties of the halogens. Data marked with question marks are either uncertain or are estimations partially based on periodic trends rather than observations.

Isotopes

Fluorine has one stable and naturally occurring isotope, fluorine-19. However, there are trace amounts in nature of the radioactive isotope fluorine-23, which occurs via cluster decay of protactinium-231. A total of eighteen isotopes of fluorine have been discovered, with atomic masses ranging from 14 to 31. Chlorine has two stable and naturally occurring isotopes, chlorine-35 and chlorine-37. However, there are trace amounts in nature of the isotope chlorine-36, which occurs via spallation of argon-36. A total of 24 isotopes of chlorine have been discovered, with atomic masses ranging from 28 to 51.[1]

There are two stable and naturally occurring isotopes of bromine, bromine-79 and bromine-81. A total of 32 isotopes of bromine have been discovered, with atomic masses ranging 67 to 98. There is one stable and naturally occurring isotope of iodine, iodine-127. However, there are trace amounts in nature of the radioactive isotope iodine-129, which occurs via spallation and from the radioactive decay of uranium in ores. Several other radioactive isotopes of iodine have also been created naturally via the decay of uranium. A total of 38 isotopes of iodine have been discovered, with atomic masses ranging from 108 to 145.[1]

There are no stable isotopes of astatine. However, there are three naturally occurring radioactive isotopes of astatine produced via radioactive decay of uranium, neptunium, and plutonium. These isotopes are astatine-215, astatine-217, and astatine-219. A total of 31 isotopes of astatine have been discovered, with atomic masses ranging from 193 to 223.[1]

8.3 Production

Approximately six million metric tons of the fluorine mineral fluorite are produced each year. Four hundred-thousand metric tons of hydrofluoric acid are made each year. Fluorine gas is made from hydrofluoric acid produced as a by-product of phosphoric acid manufacture. Approximately 15,000 metric tons of fluorine gas are made per year.[1]

The mineral halite is the mineral that is most commonly mined for chlorine, but the minerals carnallite and sylvite are also mined for chlorine. Forty million metric tons of chlorine are produced each year by the electrolysis of brine.[1]

Approximately 450,000 metric tons of bromine are produced each year. Fifty percent of all bromine produced is produced in the United States, 35% in Israel, and most of the remainder in China. Historically, bromine was produced by adding sulfuric acid and bleaching powder to natural brine. However, in modern times, bromine is produced by electrolysis, a method invented by Herbert Dow. It is also possible to produce bromine by passing chlorine through seawater and then passing air through the seawater.[1]

In 2003, 22,000 metric tons of iodine were produced. Chile produces 40% of all iodine produced, Japan produces 30%, and smaller amounts are produced in Russia and the United States. Until the 1950s, iodine was extracted from kelp. However, in modern times, iodine is produced in other ways. One way that iodine is produced is by mixing sulfur dioxide with nitrate ores, which contain some iodates. Iodine is also extracted from natural gas fields.[1]

Even though astatine is naturally occurring, it is usually produced by bombarding bismuth with alpha particles.[1]

From left to right: chlorine, bromine, and iodine at room temperature. Chlorine is a gas, bromine is a liquid, and iodine is a solid. Fluorine could not be included in the image due to its high reactivity.

8.4 Applications

Both chlorine and bromine are used as disinfectants for drinking water, swimming pools, fresh wounds, spas, dishes, and surfaces. They kill bacteria and other potentially harmful microorganisms through a process known as sterilization. Their reactivity is also put to use in bleaching. Sodium hypochlorite, which is produced from chlorine, is the active ingredient of most fabric bleaches, and chlorine-derived bleaches are used in the production of some paper products. Chlorine also reacts with sodium to create sodium chloride, which is another name for table salt.

Halogen lamps are a type of incandescent lamp using a tungsten filament in bulbs that have a small amounts of a halogen, such as iodine or bromine added. This enables the production of lamps that are much smaller than non-halogen incandescent lightbulbs at the same wattage. The gas reduces the thinning of the filament and blackening of the inside of the bulb resulting in a bulb that has a much greater life. Halogen lamps glow at a higher temperature (2800 to 3400 Kelvin) with a whiter color than incandescent bulbs. However, this requires bulbs to be manufactured from fused quartz rather than silica glass to reduce breakage.[22]

In drug discovery, the incorporation of halogen atoms into a lead drug candidate results in analogues that are usually more lipophilic and less water-soluble.[23] As a consequence, halogen atoms are used to improve penetration through lipid membranes and tissues. It follows that there is a tendency for some halogenated drugs to accumulate in adipose tissue.

The chemical reactivity of halogen atoms depends on both their point of attachment to the lead and the nature of the halogen. Aromatic halogen groups are far less reactive than aliphatic halogen groups, which can exhibit considerable chemical reactivity. For aliphatic carbon-halogen bonds, the C-F bond is the strongest and usually less chemically reactive than aliphatic C-H bonds. The other aliphatic-halogen bonds are weaker, their reactivity increasing down the periodic table. They are usually more chemically reactive than aliphatic C-H bonds. As a consequence, the most common halogen substitutions are the less reactive aromatic fluorine and chlorine groups.

8.5   Biological role

Fluoride anions are found in ivory, bones, teeth, blood, eggs, urine, and hair of organisms. Fluoride anions in very small amounts are essential for humans. There are 0.5 milligrams of fluorine per liter of human blood. Human bones contain 0.2 to 1.2% fluorine. Human tissue contains approximately 50 parts per billion of fluorine. A typical 70-kilogram human contains 3 to 6 grams of fluorine.[1]

Chloride anions are essential to a large number of species, humans included. The concentration of chlorine in the dry weight of cereals is 10 to 20 parts per million, while in potatoes the concentration of chloride is 0.5%. Plant growth is adversely affected by chloride levels in the soil falling below 2 parts per million. Human blood contains an average of 0.3% chlorine. Human bone typically contains 900 parts per million of chlorine. Human tissue contains approximately 0.2 to 0.5% chlorine. There is a total of 95 grams of chlorine in a typical 70-kilogram human.[1]

Some bromine in the form of the bromide anion is present in all organisms. A biological role for bromine in humans has not been proven, but some organisms contain organobromine compounds. Humans typically consume 1 to 20 milligrams of bromine per day. There are typically 5 parts per million of bromine in human blood, 7 parts per million of bromine in human bones, and 7 parts per million of bromine in human tissue. A typical 70-kilogram human contains 260 milligrams of bromine.[1]

Humans typically consume less than 100 micrograms of iodine per day. Iodine deficiency can cause intellectual disability. Organoiodine compounds occur in humans in some of the glands, especially the thyroid gland, as well as the stomach, epidermis, and immune system. Foods containing iodine include cod, oysters, shrimp, herring, lobsters, sunflower seeds, seaweed, and mushrooms. However, iodine is not known to have a biological role in plants. There are typically 0.06 milligrams per liter of iodine in human blood, 300 parts per billion of iodine in human bones, and 50 to 700 parts per billion of iodine in human tissue. There are 10 to 20 milligrams of iodine in a typical 70-kilogram human.[1]

Astatine has no biological role.[1]

8.6   Toxicity

The halogens tend to decrease in toxicity towards the heavier halogens.[24]

Fluorine gas is extremely toxic; breathing fluorine gas at a concentration of 0.1% for several minutes is lethal. Hydrofluoric acid is also toxic, being able to penetrate skin and cause highly painful burns. In addition, fluoride anions are toxic, but not as toxic as pure fluorine. Fluoride can be lethal in amounts of 5 to 10 grams. Prolonged consumption of fluoride above concentrations of 1.5 mg/L is associated with a risk of dental fluorosis, an aesthetic condition of the teeth.[25] At concentrations above 4 mg/L, there is an increased risk of developing skeletal fluorosis, a condition in which bone fractures become more common due to the hardening of bones. Current recommended levels in water fluoridation, a way to prevent dental caries, range from 0.7 to 1.2 mg/L to avoid the detrimental effects of fluoride while at the same time reaping the benefits.[26] People with levels between normal levels and those required for skeletal fluorosis tend to have symptoms similar to arthritis.[1]

Chlorine gas is highly toxic. Breathing in chlorine at a concentration of 3 parts per million can rapidly cause a toxic reaction. Breathing in chlorine at a concentration of 50 parts per million is highly dangerous. Breathing in chlorine at a concentration of 500 parts per million for a few minutes is lethal. Breathing in chlorine gas is highly painful.[24] Hydrochloric acid is a dangerous chemical.[1]

Pure bromine is somewhat toxic, but less toxic than fluorine and chlorine. One hundred milligrams of bromine are lethal.[1] Bromide anions are also toxic, but less so than bromine. Bromide has a lethal dose of 30 grams.[1]

Iodine is somewhat toxic, being able to irritate the lungs and eyes, with a safety limit of 1 milligram per cubic meter. When taken orally, 3 grams of iodine can be lethal. Iodide anions are mostly nontoxic, but these can also be deadly if ingested in large amounts.[1]

Astatine is very radioactive and thus highly dangerous.[1]

8.7 Superhalogen

Main article: Superatom

Certain aluminium clusters have superatom properties. These aluminium clusters are generated as anions (Al−n with n = 1, 2, 3, ...) in helium gas and reacted with a gas containing iodine. When analyzed by mass spectrometry one main reaction product turns out to be Al13I−.[27] These clusters of 13 aluminium atoms with an extra electron added do not appear to react with oxygen when it is introduced in the same gas stream. Assuming each atom liberates its 3 valence electrons, this means 40 electrons are present, which is one of the magic numbers for sodium and implies that these numbers are a reflection of the noble gases.

Calculations show that the additional electron is located in the aluminium cluster at the location directly opposite from the iodine atom. The cluster must therefore have a higher electron affinity for the electron than iodine and therefore the aluminium cluster is called a superhalogen (i.e., the vertical electron detachment energies of the moieties that make up the negative ions are larger than those of any halogen atom).[28] The cluster component in the Al13I− ion is similar to an iodide ion or a bromide ion. The related Al13I−2 cluster is expected to behave chemically like the triiodide ion.[29][30]

8.8 See also

- Pseudohalogen
- Halogen bond
- Halogen lamp
- Interhalogen

8.9   Notes

[1] The number given in parentheses refers to the measurement uncertainty. This uncertainty applies to the least significant figure(s) of the number prior to the parenthesized value (i.e., counting from rightmost digit to left). For instance, 1.00794(7) stands for 1.00794±0.00007, while 1.00794(72) stands for 1.00794±0.00072.[17]

[2] The average atomic weight of this element changes depending on the source of the chlorine, and the values in brackets are the upper and lower bounds.[18]

[3] The element does not have any stable nuclides, and the value in brackets indicates the mass number of the longest-lived isotope of the element.[18]

8.10   References

[1] Emsley, John (2011). *Nature's Building Blocks*. ISBN 0199605637.

[2] Online Etymology Dictionary halogen.

[3] Snelders, H. A. M. (1971). "J. S. C. Schweigger: His Romanticism and His Crystal Electrical Theory of Matter". *Isis* **62** (3): 328. doi:10.1086/350763. JSTOR 229946.

[4] Page 43, Edexcel International GCSE chemistry revision guide, Curtis 2011

[5] Greenwood & Earnshaw 1998, p. 804.

[6] Jim Clark (2011). "Assorted reactions of the halogens". Retrieved February 27, 2013

[7] Jim Clark (2002). "THE ACIDITY OF THE HYDROGEN HALIDES". Retrieved February 24, 2013

[8] "Facts about hydrogen fluoride". 2005. Retrieved February 2013

[9] "Hydrogen chloride". Retrieved February 24, 2013

[10] "Hydrogen bromide". Retrieved February 24, 2013

[11] "Poison Facts:Low Chemicals: Hydrogen Iodid". Retrieved 2015-04-12.

[12] Saxena, P. B (2007). *Chemistry Of Interhalogen Compounds*. ISBN 9788183562430. Retrieved February 27, 2013.

[13] "The Oxidising Ability of the Group 7 Elements". Chemguide.co.uk. Retrieved 2011-12-29.

[14] "Solubility of chlorine in water". Resistoflex.com. Retrieved 2011-12-29.

[15] "Properties of bromine". bromaid.org.

[16] "Iodine MSDS publisher = Hazard.com". 1998-04-21. Retrieved 2011-12-29.

[17] "Standard Uncertainty and Relative Standard Uncertainty". *CODATA reference*. National Institute of Standards and Technology. Retrieved 26 September 2011.

[18] Wieser, Michael E.; Coplen, Tyler B. (2011). "Atomic weights of the elements 2009 (IUPAC Technical Report)" (PDF). *Pure Appl. Chem.* (IUPAC) **83** (2): 359–396. doi:10.1351/PAC-REP-10-09-14. Retrieved 5 December 2012.

[19] Lide, D. R., ed. (2003). *CRC Handbook of Chemistry and Physics* (84th ed.). Boca Raton, FL: CRC Press.

[20]Slater,J.C. (1964). "Atomic Radii in Crystals".*Journal of Chemical Physics***41**(10):3199–3205.Bibcode:1964JChPh..41.3S. doi:10.1063/1.1725697.

[21] Bonchev, Danail; Kamenska, Verginia (1981). "Predicting the properties of the 113–120 transactinide elements". *The Journal of Physical Chemistry* (ACS Publications) **85** (9): 1177–86. doi:10.1021/j150609a021.

[22] "The Halogen Lamp". *Edison Tech Center*. Edison Steinmetz Center, Schenectady, New York. Retrieved 2014-09-05.

[23] Thomas, G. (2000). *Medicinal Chemistry an Introduction*. John Wiley & Sons, West Sussex, UK. ISBN 978-0-470-02597-0.

[24] Gray, Theodore (2010). *The Elements*. ISBN 9781579128951.

[25] Fawell, J.; Bailey, K.; Chilton, J.; Dahi, E.; Fewtrell, L.; Magara, Y. (2006). "Guidelines and standards". *Fluoride in Drinking-water* (PDF). World Health Organization. pp. 37–9. ISBN 92-4-156319-2.

[26] "CDC Statement on the 2006 National Research Council (NRC) Report on Fluoride in Drinking Water". Centers for Disease Control and Prevention. July 10, 2013. Retrieved August 1, 2013.

[27] Bergeron, D. E.; Castleman, A. Welford; Morisato, Tsuguo; Khanna, Shiv N. (2004). "Formation of $Al_{13}I^-$: Evidence for the Superhalogen Character of Al_{13}". *Science* **304** (5667): 84. Bibcode:2004Sci...304...84B. doi:10.1126/science.1093902. PMID 15066775.

[28] Giri, Santanab; Behera, Swayamprabha; Jena, Puru (2014). "Superhalogens as Building Blocks of Halogen-Free Electrolytes in Lithium-Ion Batteries†". *Angewandte Chemie* **126** (50): 14136. doi:10.1002/ange.201408648.

[29] Ball, Philip (16 April 2005). "A New Kind of Alchemy". *New Scientist*.

[30] Bergeron, D. E.; Roach, P. J.; Castleman, A. W.; Jones, N. O.; Khanna, S. N. (2005). "Al Cluster Superatoms as Halogens in Polyhalides and as Alkaline Earths in Iodide Salts". *Science* **307** (5707): 231. Bibcode:2005Sci...307..231B. doi:10.1126/ .PMID15653497.

8.11 Further reading

- Greenwood, Norman N.; Earnshaw, Alan (1997). *Chemistry of the Elements* (2nd ed.). Butterworth-Heinemann. ISBN 0080379419.

Chapter 9

Noble gas

The **noble gases** make a group of chemical elements with similar properties. Under standard conditions, they are all odorless, colorless, monatomic gases with very low chemical reactivity. The six noble gases that occur naturally are helium (He), neon (Ne), argon (Ar), krypton (Kr), xenon (Xe), and the radioactive radon (Rn).

For the first six periods of the periodic table, the noble gases are exactly the members of **group 18** of the periodic table. It is possible that due to relativistic effects, the group 14 element flerovium exhibits some noble-gas-like properties,[1] instead of the group 18 element ununoctium.[2] Noble gases are typically highly unreactive except when under particular extreme conditions. The inertness of noble gases makes them very suitable in applications where reactions are not wanted. For example: argon is used in lightbulbs to prevent the hot tungsten filament from oxidizing; also, helium is breathed by deep-sea divers to prevent oxygen and nitrogen toxicity.

The properties of the noble gases can be well explained by modern theories of atomic structure: their outer shell of valence electrons is considered to be "full", giving them little tendency to participate in chemical reactions, and it has been possible to prepare only a few hundred noble gas compounds. The melting and boiling points for a given noble gas are close together, differing by less than 10 °C (18 °F); that is, they are liquids over only a small temperature range.

Neon, argon, krypton, and xenon are obtained from air in an air separation unit using the methods of liquefaction of gases and fractional distillation. Helium is sourced from natural gas fields which have high concentrations of helium in the natural gas, using cryogenic gas separation techniques, and radon is usually isolated from the radioactive decay of dissolved radium, thorium, or uranium compounds (since those compounds give off alpha particles). Noble gases have several important applications in industries such as lighting, welding, and space exploration. A helium-oxygen breathing gas is often used by deep-sea divers at depths of seawater over 55 m (180 ft) to keep the diver from experiencing oxygen toxemia, the lethal effect of high-pressure oxygen, and nitrogen narcosis, the distracting narcotic effect of the nitrogen in air beyond this partial-pressure threshold. After the risks caused by the flammability of hydrogen became apparent, it was replaced with helium in blimps and balloons.

9.1   History

Noble gas is translated from the German noun *Edelgas*, first used in 1898 by Hugo Erdmann[3] to indicate their extremely low level of reactivity. The name makes an analogy to the term "noble metals", which also have low reactivity. The noble gases have also been referred to as *inert gases*, but this label is deprecated as many noble gas compounds are now known.[4] *Rare gases* is another term that was used,[5] but this is also inaccurate because argon forms a fairly considerable part (0.94% by volume, 1.3% by mass) of the Earth's atmosphere due to decay of radioactive potassium-40.[6]

Pierre Janssen and Joseph Norman Lockyer discovered a new element on August 18, 1868 while looking at the chromosphere of the Sun, and named it helium after the Greek word for the Sun, ήλιος (*ílios* or *helios*).[7] No chemical analysis was possible at the time, but helium was later found to be a noble gas. Before them, in 1784, the English chemist and physicist Henry Cavendish had discovered that air contains a small proportion of a substance less reactive than nitrogen.[8] A century later, in 1895, Lord Rayleigh discovered that samples of nitrogen from the air were of a different density than

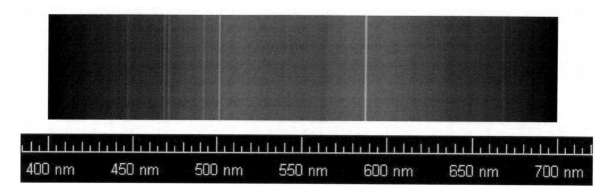

Helium was first detected in the Sun due to its characteristic spectral lines.

nitrogen resulting from chemical reactions. Along with Scottish scientist William Ramsay at University College, London, Lord Rayleigh theorized that the nitrogen extracted from air was mixed with another gas, leading to an experiment that successfully isolated a new element, argon, from the Greek word αργός (*argós*, "inactive").[8] With this discovery, they realized an entire class of gases was missing from the periodic table. During his search for argon, Ramsay also managed to isolate helium for the first time while heating cleveite, a mineral. In 1902, having accepted the evidence for the elements helium and argon, Dmitri Mendeleev included these noble gases as group 0 in his arrangement of the elements, which would later become the periodic table.[9]

Ramsay continued to search for these gases using the method of fractional distillation to separate liquid air into several components. In 1898, he discovered the elements krypton, neon, and xenon, and named them after the Greek words κρυπτός (*kryptós*, "hidden"), νέος (*néos*, "new"), and ξένος (*xénos*, "stranger"), respectively. Radon was first identified in 1898 by Friedrich Ernst Dorn,[10] and was named *radium emanation*, but was not considered a noble gas until 1904 when its characteristics were found to be similar to those of other noble gases.[11] Rayleigh and Ramsay received the 1904 Nobel Prizes in Physics and in Chemistry, respectively, for their discovery of the noble gases;[12][13] in the words of J. E. Cederblom, then president of the Royal Swedish Academy of Sciences, "the discovery of an entirely new group of elements, of which no single representative had been known with any certainty, is something utterly unique in the history of chemistry, being intrinsically an advance in science of peculiar significance".[13]

The discovery of the noble gases aided in the development of a general understanding of atomic structure. In 1895, French chemist Henri Moissan attempted to form a reaction between fluorine, the most electronegative element, and argon, one of the noble gases, but failed. Scientists were unable to prepare compounds of argon until the end of the 20th century, but these attempts helped to develop new theories of atomic structure. Learning from these experiments, Danish physicist Niels Bohr proposed in 1913 that the electrons in atoms are arranged in shells surrounding the nucleus, and that for all noble gases except helium the outermost shell always contains eight electrons.[11] In 1916, Gilbert N. Lewis formulated the *octet rule*, which concluded an octet of electrons in the outer shell was the most stable arrangement for any atom; this arrangement caused them to be unreactive with other elements since they did not require any more electrons to complete their outer shell.[14]

In 1962, Neil Bartlett discovered the first chemical compound of a noble gas, xenon hexafluoroplatinate.[15] Compounds of other noble gases were discovered soon after: in 1962 for radon, radon difluoride,[16] and in 1963 for krypton, krypton difluoride (KrF

2).[17] The first stable compound of argon was reported in 2000 when argon fluorohydride (HArF) was formed at a temperature of 40 K (−233.2 °C; −387.7 °F).[18]

In December 1998, scientists at the Joint Institute for Nuclear Research working in Dubna, Russia bombarded plutonium (Pu) with calcium (Ca) to produce a single atom of element 114,[19] flerovium (Fl).[20] Preliminary chemistry experiments have indicated this element may be the first superheavy element to show abnormal noble-gas-like properties, even though it is a member of group 14 on the periodic table.[21] In October 2006, scientists from the Joint Institute for Nuclear Research and Lawrence Livermore National Laboratory successfully created synthetically ununoctium (Uuo), the seventh element in group 18,[22] by bombarding californium (Cf) with calcium (Ca).[23]

9.2   Physical and atomic properties

The noble gases have weak interatomic force, and consequently have very low melting and boiling points. They are all monatomic gases under standard conditions, including the elements with larger atomic masses than many normally solid elements.[11] Helium has several unique qualities when compared with other elements: its boiling and melting points are lower than those of any other known substance; it is the only element known to exhibit superfluidity; it is the only element that cannot be solidified by cooling under standard conditions—a pressure of 25 standard atmospheres (2,500 kPa; 370 psi) must be applied at a temperature of 0.95 K (−272.200 °C; −457.960 °F) to convert it to a solid.[26] The noble gases up to xenon have multiple stable isotopes. Radon has no stable isotopes; its longest-lived isotope, ^{222}Rn, has a half-life of 3.8 days and decays to form helium and polonium, which ultimately decays to lead.[11] Melting and boiling points generally increase going down the group.

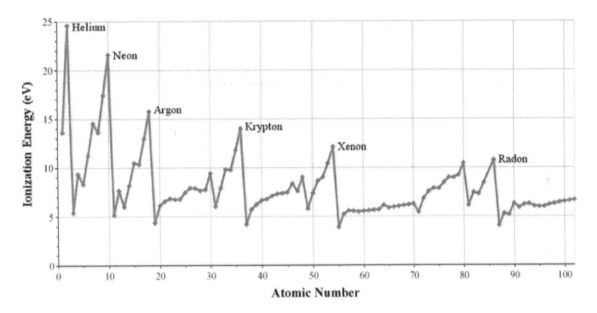

This is a plot of ionization potential versus atomic number. The noble gases, which are labeled, have the largest ionization potential for each period.

The noble gas atoms, like atoms in most groups, increase steadily in atomic radius from one period to the next due to the increasing number of electrons. The size of the atom is related to several properties. For example, the ionization potential decreases with an increasing radius because the valence electrons in the larger noble gases are farther away from the nucleus and are therefore not held as tightly together by the atom. Noble gases have the largest ionization potential among the elements of each period, which reflects the stability of their electron configuration and is related to their relative lack of chemical reactivity.[24] Some of the heavier noble gases, however, have ionization potentials small enough to be comparable to those of other elements and molecules. It was the insight that xenon has an ionization potential similar to that of the oxygen molecule that led Bartlett to attempt oxidizing xenon using platinum hexafluoride, an oxidizing agent known to be strong enough to react with oxygen.[15] Noble gases cannot accept an electron to form stable anions; that is, they have a negative electron affinity.[27]

The macroscopic physical properties of the noble gases are dominated by the weak van der Waals forces between the atoms. The attractive force increases with the size of the atom as a result of the increase in polarizability and the decrease in ionization potential. This results in systematic group trends: as one goes down group 18, the atomic radius, and with it the interatomic forces, increases, resulting in an increasing melting point, boiling point, enthalpy of vaporization, and solubility. The increase in density is due to the increase in atomic mass.[24]

The noble gases are nearly ideal gases under standard conditions, but their deviations from the ideal gas law provided important clues for the study of intermolecular interactions. The Lennard-Jones potential, often used to model intermolecular interactions, was deduced in 1924 by John Lennard-Jones from experimental data on argon before the development of quantum mechanics provided the tools for understanding intermolecular forces from first principles.[28] The theoretical analysis of these interactions became tractable because the noble gases are monatomic and the atoms spherical, which means that the interaction between the atoms is independent of direction, or isotropic.

9.3 Chemical properties

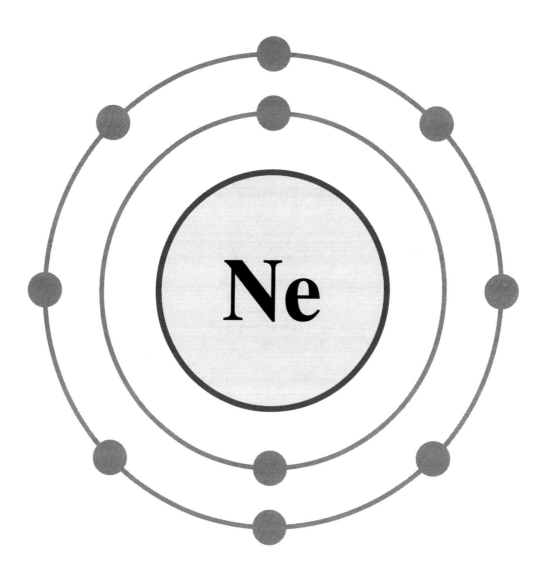

Neon, like all noble gases, has a full valence shell. Noble gases have eight electrons in their outermost shell, except in the case of helium, which has two.

The noble gases are colorless, odorless, tasteless, and nonflammable under standard conditions. They were once labeled *group 0* in the periodic table because it was believed they had a valence of zero, meaning their atoms cannot combine with those of other elements to form compounds. However, it was later discovered some do indeed form compounds, causing this label to fall into disuse.[11]

9.3.1   Configuration

Main article: Noble gas configuration

Like other groups, the members of this family show patterns in its electron configuration, especially the outermost shells resulting in trends in chemical behavior:

The noble gases have full valence electron shells. Valence electrons are the outermost electrons of an atom and are normally the only electrons that participate in chemical bonding. Atoms with full valence electron shells are extremely stable and therefore do not tend to form chemical bonds and have little tendency to gain or lose electrons.[29] However, heavier noble gases such as radon are held less firmly together by electromagnetic force than lighter noble gases such as helium, making it easier to remove outer electrons from heavy noble gases.

As a result of a full shell, the noble gases can be used in conjunction with the electron configuration notation to form the *noble gas notation*. To do this, the nearest noble gas that precedes the element in question is written first, and then the electron configuration is continued from that point forward. For example, the electron notation of phosphorus is $1s^2$ $2s^2$ $2p^6$ $3s^2$ $3p^3$, while the noble gas notation is [Ne] $3s^2$ $3p^3$. This more compact notation makes it easier to identify elements, and is shorter than writing out the full notation of atomic orbitals.[30]

9.3.2   Compounds

Main article: Noble gas compound

The noble gases show extremely low chemical reactivity; consequently, only a few hundred noble gas compounds have been formed. Neutral compounds in which helium and neon are involved in chemical bonds have not been formed (although there is some theoretical evidence for a few helium compounds), while xenon, krypton, and argon have shown only minor reactivity.[31] The reactivity follows the order Ne < He < Ar < Kr < Xe < Rn.

In 1933, Linus Pauling predicted that the heavier noble gases could form compounds with fluorine and oxygen. He predicted the existence of krypton hexafluoride (KrF_6) and xenon hexafluoride (XeF_6), speculated that XeF_8 might exist as an unstable compound, and suggested xenic acid could form perxenate salts.[32][33] These predictions were shown to be generally accurate, except that XeF_8 is now thought to be both thermodynamically and kinetically unstable.[34]

Xenon compounds are the most numerous of the noble gas compounds that have been formed.[35] Most of them have the xenon atom in the oxidation state of +2, +4, +6, or +8 bonded to highly electronegative atoms such as fluorine or oxygen, as in xenon difluoride (XeF_2), xenon tetrafluoride (XeF_4), xenon hexafluoride (XeF_6), xenon tetroxide (XeO_4), and sodium perxenate (Na_4XeO_6). Xenon reacts with fluorine to form numerous xenon fluorides according to the following equations:

$$Xe + F_2 \rightarrow XeF_2$$
$$Xe + 2F_2 \rightarrow XeF_4$$
$$Xe + 3F_2 \rightarrow XeF_6$$

Some of these compounds have found use in chemical synthesis as oxidizing agents; XeF_2, in particular, is commercially available and can be used as a fluorinating agent.[36] As of 2007, about five hundred compounds of xenon bonded to other elements have been identified, including organoxenon compounds (containing xenon bonded to carbon), and xenon bonded to nitrogen, chlorine, gold, mercury, and xenon itself.[31][37] Compounds of xenon bound to boron, hydrogen, bromine, iodine, beryllium, sulphur, titanium, copper, and silver have also been observed but only at low temperatures in noble gas matrices, or in supersonic noble gas jets.[31]

Structure of XeF
4, one of the first noble gas compounds to be discovered

In theory, radon is more reactive than xenon, and therefore should form chemical bonds more easily than xenon does. However, due to the high radioactivity and short half-life of radon isotopes, only a few fluorides and oxides of radon have been formed in practice.[38]

Krypton is less reactive than xenon, but several compounds have been reported with krypton in the oxidation state of +2.[31] Krypton difluoride is the most notable and easily characterized. Under extreme conditions, krypton reacts with fluorine to form KrF_2 according to the following equation:

$$Kr + F_2 \rightarrow KrF_2$$

Compounds in which krypton forms a single bond to nitrogen and oxygen have also been characterized,[39] but are only stable below −60 °C (−76 °F) and −90 °C (−130 °F) respectively.[31]

Krypton atoms chemically bound to other nonmetals (hydrogen, chlorine, carbon) as well as some late transition metals (copper, silver, gold) have also been observed, but only either at low temperatures in noble gas matrices, or in supersonic noble gas jets.[31] Similar conditions were used to obtain the first few compounds of argon in 2000, such as argon fluoro-hydride (HArF), and some bound to the late transition metals copper, silver, and gold.[31] As of 2007, no stable neutral molecules involving covalently bound helium or neon are known.[31]

The noble gases—including helium—can form stable molecular ions in the gas phase. The simplest is the helium hydride molecular ion, HeH+, discovered in 1925.[40] Because it is composed of the two most abundant elements in the universe, hydrogen and helium, it is believed to occur naturally in the interstellar medium, although it has not been detected yet.[41] In addition to these ions, there are many known neutral excimers of the noble gases. These are compounds such as ArF and KrF that are stable only when in an excited electronic state; some of them find application in excimer lasers.

In addition to the compounds where a noble gas atom is involved in a covalent bond, noble gases also form non-covalent compounds. The clathrates, first described in 1949,[42] consist of a noble gas atom trapped within cavities of crystal lattices of certain organic and inorganic substances. The essential condition for their formation is that the guest (noble gas) atoms must be of appropriate size to fit in the cavities of the host crystal lattice. For instance, argon, krypton, and xenon form clathrates with hydroquinone, but helium and neon do not because they are too small or insufficiently polarizable to be retained.[43] Neon, argon, krypton, and xenon also form clathrate hydrates, where the noble gas is trapped in ice.[44]

An endohedral fullerene compound containing a noble gas atom

Noble gases can form endohedral fullerene compounds, in which the noble gas atom is trapped inside a fullerene molecule. In 1993, it was discovered that when C
60, a spherical molecule consisting of 60 carbon atoms, is exposed to noble gases at high pressure, complexes such as

He@C

60 can be formed (the @ notation indicates He is contained inside C

60 but not covalently bound to it).[45] As of 2008, endohedral complexes with helium, neon, argon, krypton, and xenon have been obtained.[46] These compounds have found use in the study of the structure and reactivity of fullerenes by means of the nuclear magnetic resonance of the noble gas atom.[47]

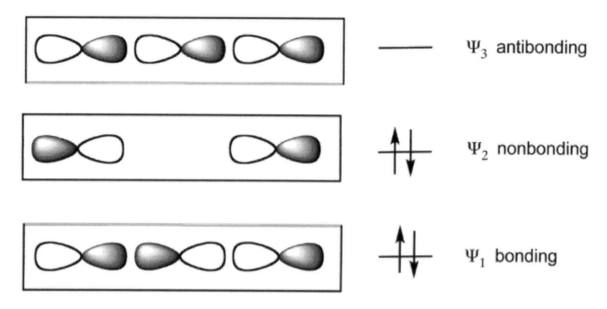

Bonding in XeF
2 according to the 3-center-4-electron bond model

Noble gas compounds such as xenon difluoride (XeF

2) are considered to be hypervalent because they violate the octet rule. Bonding in such compounds can be explained using a three-center four-electron bond model.[48][49] This model, first proposed in 1951, considers bonding of three collinear atoms. For example, bonding in XeF

2 is described by a set of three molecular orbitals (MOs) derived from p-orbitals on each atom. Bonding results from the combination of a filled p-orbital from Xe with one half-filled p-orbital from each F atom, resulting in a filled bonding orbital, a filled non-bonding orbital, and an empty antibonding orbital. The highest occupied molecular orbital is localized on the two terminal atoms. This represents a localization of charge which is facilitated by the high electronegativity of fluorine.[50]

The chemistry of heavier noble gases, krypton and xenon, are well established. The chemistry of the lighter ones, argon and helium, is still at an early stage, while a neon compound is yet to be identified.

9.4 Occurrence and production

The abundances of the noble gases in the universe decrease as their atomic numbers increase. Helium is the most common element in the universe after hydrogen, with a mass fraction of about 24%. Most of the helium in the universe was formed during Big Bang nucleosynthesis, but the amount of helium is steadily increasing due to the fusion of hydrogen in stellar nucleosynthesis (and, to a very slight degree, the alpha decay of heavy elements).[51][52] Abundances on Earth follow different trends; for example, helium is only the third most abundant noble gas in the atmosphere. The reason is that there is no primordial helium in the atmosphere; due to the small mass of the atom, helium cannot be retained by the Earth's gravitational field.[53] Helium on Earth comes from the alpha decay of heavy elements such as uranium and thorium found in the Earth's crust, and tends to accumulate in natural gas deposits.[53] The abundance of argon, on the other hand, is increased as a result of the beta decay of potassium-40, also found in the Earth's crust, to form argon-40, which is the most abundant isotope of argon on Earth despite being relatively rare in the Solar System. This process is the base for the potassium-argon dating method.[54] Xenon has an unexpectedly low abundance in the atmosphere, in what has been called the *missing xenon problem*; one theory is that the missing xenon may be trapped in minerals inside the Earth's crust.[55] After the discovery of xenon dioxide, a research showed that Xe can substitute for Si in the quartz.[56] Radon is formed in the lithosphere as from the alpha decay of radium. It can seep into buildings through cracks in their foundation and accumulate in areas that are not well ventilated. Due to its high radioactivity, radon presents a significant health hazard; it is implicated in an estimated 21,000 lung cancer deaths per year in the United States alone.[57]

For large-scale use, helium is extracted by fractional distillation from natural gas, which can contain up to 7% helium.[62]

Neon, argon, krypton, and xenon are obtained from air using the methods of liquefaction of gases, to convert elements to a liquid state, and fractional distillation, to separate mixtures into component parts. Helium is typically produced by separating it from natural gas, and radon is isolated from the radioactive decay of radium compounds.[111] The prices of the noble gases are influenced by their natural abundance, with argon being the cheapest and xenon the most expensive. As an example, the table to the right lists the 2004 prices in the United States for laboratory quantities of each gas.

9.5   Applications

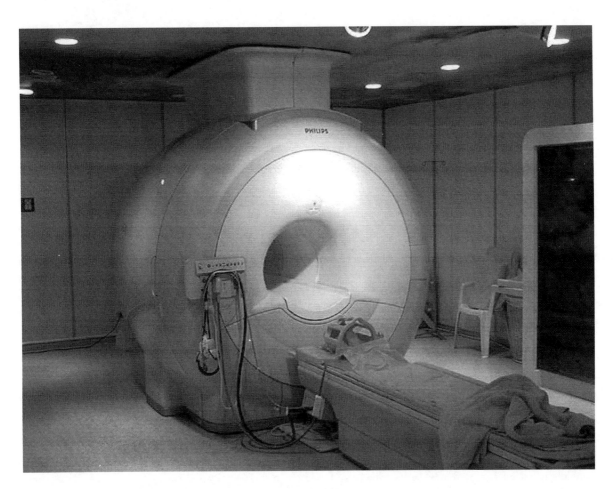

Liquid helium is used to cool the superconducting magnets in modern MRI scanners

Noble gases have very low boiling and melting points, which makes them useful as cryogenic refrigerants.[63] In particular, liquid helium, which boils at 4.2 K (−268.95 °C; −452.11 °F), is used for superconducting magnets, such as those needed in nuclear magnetic resonance imaging and nuclear magnetic resonance.[64] Liquid neon, although it does not reach temperatures as low as liquid helium, also finds use in cryogenics because it has over 40 times more refrigerating capacity than liquid helium and over three times more than liquid hydrogen.[60]

Helium is used as a component of breathing gases to replace nitrogen, due its low solubility in fluids, especially in lipids. Gases are absorbed by the blood and body tissues when under pressure like in scuba diving, which causes an anesthetic effect known as nitrogen narcosis.[65] Due to its reduced solubility, little helium is taken into cell membranes, and when helium is used to replace part of the breathing mixtures, such as in trimix or heliox, a decrease in the narcotic effect of the gas at depth is obtained.[66] Helium's reduced solubility offers further advantages for the condition known as decompression sickness, or *the bends*.[11][67] The reduced amount of dissolved gas in the body means that fewer gas bubbles form during the decrease in pressure of the ascent. Another noble gas, argon, is considered the best option for use as a drysuit inflation gas for scuba diving.[68] Helium is also used as filling gas in nuclear fuel rods for nuclear reactors.[69]

Goodyear Blimp

Since the *Hindenburg* disaster in 1937,[70] helium has replaced hydrogen as a lifting gas in blimps and balloons due to its lightness and incombustibility, despite an 8.6%[71] decrease in buoyancy.[11]

In many applications, the noble gases are used to provide an inert atmosphere. Argon is used in the synthesis of air-sensitive compounds that are sensitive to nitrogen. Solid argon is also used for the study of very unstable compounds, such as reactive intermediates, by trapping them in an inert matrix at very low temperatures.[72] Helium is used as the carrier medium in gas chromatography, as a filler gas for thermometers, and in devices for measuring radiation, such as the Geiger counter and the bubble chamber.[61] Helium and argon are both commonly used to shield welding arcs and the surrounding base metal from the atmosphere during welding and cutting, as well as in other metallurgical processes and in the production of silicon for the semiconductor industry.[60]

Noble gases are commonly used in lighting because of their lack of chemical reactivity. Argon, mixed with nitrogen, is used as a filler gas for incandescent light bulbs.[60] Krypton is used in high-performance light bulbs, which have higher color temperatures and greater efficiency, because it reduces the rate of evaporation of the filament more than argon; halogen lamps, in particular, use krypton mixed with small amounts of compounds of iodine or bromine.[60] The noble gases glow in distinctive colors when used inside gas-discharge lamps, such as "neon lights". These lights are called after neon but often contain other gases and phosphors, which add various hues to the orange-red color of neon. Xenon is commonly used in xenon arc lamps which, due to their nearly continuous spectrum that resembles daylight, find application in film projectors and as automobile headlamps.[60]

The noble gases are used in excimer lasers, which are based on short-lived electronically excited molecules known as excimers. The excimers used for lasers may be noble gas dimers such as Ar_2, Kr_2 or Xe_2, or more commonly, the noble gas is combined with a halogen in excimers such as ArF, KrF, XeF, or XeCl. These lasers produce ultraviolet light which, due to its short wavelength (193 nm for ArF and 248 nm for KrF), allows for high-precision imaging. Excimer lasers have many industrial, medical, and scientific applications. They are used for microlithography and microfabrication, which are essential for integrated circuit manufacture, and for laser surgery, including laser angioplasty and eye surgery.[73]

15,000-watt xenon short-arc lamp used in IMAX projectors

Some noble gases have direct application in medicine. Helium is sometimes used to improve the ease of breathing of asthma sufferers.[60] Xenon is used as an anesthetic because of its high solubility in lipids, which makes it more potent than the usual nitrous oxide, and because it is readily eliminated from the body, resulting in faster recovery.[74] Xenon finds application in medical imaging of the lungs through hyperpolarized MRI.[75] Radon, which is highly radioactive and is only available in minute amounts, is used in radiotherapy.[11]

9.6 Discharge color

The color of gas discharge emission depends on several factors, including the following:[76]

- discharge parameters (local value of current density and electric field, temperature, etc. – note the color variation along the discharge in the top row);

- gas purity (even small fraction of certain gases can affect color);

- material of the discharge tube envelope – note suppression of the UV and blue components in the bottom-row tubes made of thick household glass.

9.7 See also

- Noble gas (data page), for extended tables of physical properties.

- Noble metal, for metals that are resistant to corrosion or oxidation.

- Inert gas, for any gas that is not reactive under normal circumstances.

- Industrial gas

- Neutronium

- Noble gas configuration

9.8 Notes

[1] "Flerov laboratory of nuclear reactions" (PDF). JINR. Retrieved 2009-08-08.

[2] Nash, Clinton S. (2005). "Atomic and Molecular Properties of Elements 112, 114, and 118". *J. Phys. Chem. A* **109** (15): 3493–3500. doi:10.1021/jp050736o. PMID 16833687.

[3] Renouf, Edward (1901). "Noble gases". *Science* **13**(320):268–270.:1901Sci....13..268R.doi:10.1126/science.13.320.268.

[4] Ojima 2002, p. 30

[5] Ojima 2002, p. 4

[6] "argon". *Encyclopædia Britannica*. 2008.

[7] *Oxford English Dictionary* (1989), s.v. "helium". Retrieved December 16, 2006, from Oxford English Dictionary Online. Also, from quotation there: Thomson, W. (1872). *Rep. Brit. Assoc.* xcix: "Frankland and Lockyer find the yellow prominences to give a very decided bright line not far from D, but hitherto not identified with any terrestrial flame. It seems to indicate a new substance, which they propose to call Helium."

[8] Ojima 2002, p. 1

[9] Mendeleev 1903, p. 497

[10] Partington, J. R. (1957). "Discovery of Radon". *Nature* **179** (4566): 912. Bibcode:1957Natur.179..912P. doi:10.1038/179912a0.

[11] "Noble Gas". *Encyclopædia Britannica*. 2008.

[12] Cederblom, J. E. (1904). "The Nobel Prize in Physics 1904 Presentation Speech".

[13] Cederblom, J. E. (1904). "The Nobel Prize in Chemistry 1904 Presentation Speech".

[14] Gillespie, R. J.; Robinson, E. A. (2007). "Gilbert N. Lewis and the chemical bond: the electron pair and the octet rule from 1916 to the present day". *J Comput Chem* **28** (1): 87–97. doi:10.1002/jcc.20545. PMID 17109437.

[15] Bartlett, N. (1962). "Xenon hexafluoroplatinate $Xe^+[PtF_6]^-$".*Proceedings of the Chemical Society*(6):218.doi:10.1039/PS96.

[16] Fields, Paul R.; Stein, Lawrence; Zirin, Moshe H. (1962). "Radon Fluoride". *Journal of the American Chemical Society* **84** (21): 4164–4165. doi:10.1021/ja00880a048.

[17] Grosse, A. V.; Kirschenbaum, A. D.; Streng, A. G.; Streng, L. V. (1963). "Krypton Tetrafluoride: Preparation and Some Properties". *Science* **139** (3559): 1047–1048. Bibcode:1963Sci...139.1047G. doi:10.1126/science.139.3559.1047. PMID 17812982.

[18] Khriachtchev, Leonid; Pettersson, Mika; Runeberg, Nino; Lundell, Jan; Räsänen, Markku (2000). "A stable argon compound". *Nature* **406** (6798): 874–876. doi:10.1038/35022551. PMID 10972285.

[19] Oganessian, Yu. Ts.; Utyonkov, V.; Lobanov, Yu.; Abdullin, F.; Polyakov, A.; et al. (1999). "Synthesis of Superheavy Nuclei in the ^{48}Ca + ^{244}Pu Reaction".*Physical Review Letters*(American Physical Society)**83**(16):3154–3157.Bibcode:1999PhRvL. doi:10.1103/PhysRevLett.83.3154.

[20] Woods, Michael (2003-05-06). "Chemical element No. 110 finally gets a name—darmstadtium". *Pittsburgh Post-Gazette*. Retrieved 2008-06-26.

[21] "Gas Phase Chemistry of Superheavy Elements" (PDF). Texas A&M University. Retrieved 2008-05-31.

[22] Robert C. Barber, Paul J. Karol, Hiromichi Nakahara, Emanuele Vardaci, and Erich W. Vogt (2011). "Discovery of the elements with atomic numbers greater than or equal to 113 (IUPAC Technical Report)*" (PDF). *Pure Appl. Chem.* (IUPAC) **83** (7). doi:10.1515/ci.2011.33.5.25b. Retrieved 2014-05-30.

[23] Oganessian, Yu. Ts.; Utyonkov, V.; Lobanov, Yu.; Abdullin, F.; Polyakov, A.,; et al. (2006). "Synthesis of the isotopes of elements 118 and 116 in the 249Cf and 245Cm + 48Ca fusion reactions". *Physical Review C* **74** (4): 44602. Bibcode:2006PhRvC..74d4602O.doi:10.1103/PhysRevC.74.044602.

[24] Greenwood 1997, p. 891

[25] Allen, Leland C. (1989). "Electronegativity is the average one-electron energy of the valence-shell electrons in ground-state free atoms". *Journal of the American Chemical Society* **111** (25): 9003–9014. doi:10.1021/ja00207a003.

[26] "Solid Helium". University of Alberta. Retrieved 2008-06-22.

[27] Wheeler, John C. (1997). "Electron Affinities of the Alkaline Earth Metals and the Sign Convention for Electron Affinity". *Journal of Chemical Education* **74**: 123–127. Bibcode:1997JChEd..74..123W. doi:10.1021/ed074p123.; Kalcher, Josef; Sax, Alexander F. (1994). "Gas Phase Stabilities of Small Anions: Theory and Experiment in Cooperation". *Chemical Reviews* **94** (8): 2291–2318. doi:10.1021/cr00032a004.

[28] Mott, N. F. (1955). "John Edward Lennard-Jones. 1894–1954". *Biographical Memoirs of Fellows of the Royal Society* **1**: 175–184. doi:10.1098/rsbm.1955.0013.

[29] Ojima 2002, p. 35

[30] CliffsNotes 2007, p. 15

[31] Grochala, Wojciech (2007). "Atypical compounds of gases, which have been called noble". *Chemical Society Reviews* **36** (10): 1632–1655. doi:10.1039/b702109g. PMID 17721587.

[32] Pauling, Linus (1933). "The Formulas of Antimonic Acid and the Antimonates". *Journal of the American Chemical Society* **55** (5): 1895–1900. doi:10.1021/ja01332a016.

[33] Holloway 1968

[34] Seppelt, Konrad (1979). "Recent developments in the Chemistry of Some Electronegative Elements". *Accounts of Chemical Research* **12** (6): 211–216. doi:10.1021/ar50138a004.

[35] Moody, G.J. (1974). "A Decade of Xenon Chemistry". *Journal of Chemical Education* **51**(10):628–630.Bibcode:1974JChEd..51. doi:10.1021/ed051p628. Retrieved 2007-10-16.

[36] Zupan, Marko; Iskra, Jernej; Stavber, Stojan (1998). "Fluorination with XeF$_2$. 44. Effect of Geometry and Heteroatom on the Regioselectivity of Fluorine Introduction into an Aromatic Ring". *J. Org. Chem* **63** (3): 878–880. doi:10.1021/jo971496e. PMID 11672087.

[37] Harding 2002, pp. 90–99

[38] .Avrorin, V. V.; Krasikova, R. N.; Nefedov, V. D.; Toropova, M. A. (1982). "The Chemistry of Radon". *Russian Chemical Review* **51** (1): 12–20. Bibcode:1982RuCRv..51...12A. doi:10.1070/RC1982v051n01ABEH002787.

[39] Lehmann, J (2002). "The chemistry of krypton". *Coordination Chemistry Reviews*. 233–234: 1–39. doi:10.1016/S0010-8545(02)00202-3.

[40] Hogness, T. R.; Lunn, E. G. (1925). "The Ionization of Hydrogen by Electron Impact as Interpreted by Positive Ray Analysis". *Physical Review* **26**: 44–55. Bibcode:1925PhRv...26...44H. doi:10.1103/PhysRev.26.44.

[41] Fernandez, J.; Martin, F. (2007). "Photoionization of the HeH$_2^+$ molecular ion". *J. Phys. B: At. Mol. Opt. Phys* **40** (12): 2471–2480. Bibcode:2007JPhB...40.2471F. doi:10.1088/0953-4075/40/12/020.

[42] H. M. Powell and M. Guter (1949). "An Inert Gas Compound". *Nature* **164** (4162): 240–241. Bibcode:1949Natur.164..240P. doi:10.1038/164240b0.

[43] Greenwood 1997, p. 893

[44] Dyadin, Yuri A.; et al. (1999). "Clathrate hydrates of hydrogen and neon". *Mendeleev Communications* **9** (5): 209–210. doi:10.1070/MC1999v009n05ABEH001104.

[45] Saunders, M.; Jiménez-Vázquez, H. A.; Cross, R. J.; Poreda, R. J. (1993). "Stable compounds of helium and neon. He@C60 and Ne@C60". *Science* **259** (5100): 1428–1430. Bibcode:1993Sci...259.1428S. doi:10.1126/science.259.5100.1428. PMID 17801275.

[46] Saunders, Martin; Jimenez-Vazquez, Hugo A.; Cross, R. James; Mroczkowski, Stanley; Gross, Michael L.; Giblin, Daryl E.; Poreda, Robert J. (1994). "Incorporation of helium, neon, argon, krypton, and xenon into fullerenes using high pressure". *J. Am. Chem. Soc.* **116** (5): 2193–2194. doi:10.1021/ja00084a089.

[47] Frunzi, Michael; Cross, R. Jame; Saunders, Martin (2007). "Effect of Xenon on Fullerene Reactions". *Journal of the American Chemical Society* **129** (43): 13343–6. doi:10.1021/ja075568n. PMID 17924634.

[48] Greenwood 1997, p. 897

[49] Weinhold 2005, pp. 275–306

[50] Pimentel, G. C. (1951). "The Bonding of Trihalide and Bifluoride Ions by the Molecular Orbital Method". *The Journal of Chemical Physics* **19** (4): 446–448. Bibcode:1951JChPh..19..446P. doi:10.1063/1.1748245.

[51] Weiss, Achim. "Elements of the past: Big Bang Nucleosynthesis and observation". Max Planck Institute for Gravitational Physics. Retrieved 2008-06-23.

[52] Coc, A.; et al. (2004). "Updated Big Bang Nucleosynthesis confronted to WMAP observations and to the Abundance of Light Elements". *Astrophysical Journal* **600**(2):544–552.arXiv:astro-ph/0309480.Bibcode:2004ApJ...600..544C.doi:10.1086/3801.

[53] Morrison, P.; Pine, J. (1955). "Radiogenic Origin of the Helium Isotopes in Rock". *Annals of the New York Academy of Sciences* **62** (3): 71–92. Bibcode:1955NYASA..62...71M. doi:10.1111/j.1749-6632.1955.tb35366.x.

[54] Scherer, Alexandra (2007-01-16). "^{40}Ar/^{39}Ar dating and errors". Technische Universität Bergakademie Freiberg. Archived from the original on 2007-10-14. Retrieved 2008-06-26.

[55] Sanloup, Chrystèle; Schmidt, Burkhard C.; et al. (2005). "Retention of Xenon in Quartz and Earth's Missing Xenon". *Science* **310** (5751): 1174–1177. Bibcode:2005Sci...310.1174S. doi:10.1126/science.1119070. PMID 16293758.

[56] Tyler Irving (May 2011). "Xenon Dioxide May Solve One of Earth's Mysteries". L'Actualité chimique canadienne (Canadian Chemical News). Retrieved 2012-05-18.

[57] "A Citizen's Guide to Radon". U.S. Environmental Protection Agency. 2007-11-26. Retrieved 2008-06-26.

[58] Lodders, Katharina (July 10, 2003). "Solar System Abundances and Condensation Temperatures of the Elements" (PDF). *The Astrophysical Journal*(The American Astronomical Society)**591**(2):1220–1247.Bibcode:2003ApJ...591.1220L.doi:10.1086/.

[59] "The Atmosphere". National Weather Service. Retrieved 2008-06-01.

[60] Häussinger, Peter; Glatthaar, Reinhard; Rhode, Wilhelm; Kick, Helmut; Benkmann, Christian; Weber, Josef; Wunschel, Hans-Jörg; Stenke, Viktor; Leicht, Edith; Stenger, Hermann (2002). "Noble gases". *Ullmann's Encyclopedia of Industrial Chemistry*. Wiley. doi:10.1002/14356007.a17_485.

[61] Hwang, Shuen-Chen; Lein, Robert D.; Morgan, Daniel A. (2005). "Noble Gases". *Kirk Othmer Encyclopedia of Chemical Technology*. Wiley. pp. 343–383. doi:10.1002/0471238961.0701190508230114.a01.

[62] Winter, Mark (2008). "Helium: the essentials". University of Sheffield. Retrieved 2008-07-14.

[63] "Neon". *Encarta*. 2008.

[64] Zhang, C. J.; Zhou, X. T.; Yang, L. (1992). "Demountable coaxial gas-cooled current leads for MRI superconducting magnets". *Magnetics, IEEE Transactions on* (IEEE) **28** (1): 957–959. Bibcode:1992ITM....28..957Z. doi:10.1109/20.120038.

[65] Fowler, B; Ackles, K. N.; Porlier, G. (1985). "Effects of inert gas narcosis on behavior—a critical review". *Undersea Biomed. Res.* **12** (4): 369–402. ISSN 0093-5387. OCLC 2068005. PMID 4082343. Retrieved 2008-04-08.

[66] Bennett 1998, p. 176

[67] Vann, R. D. (ed) (1989). "The Physiological Basis of Decompression". *38th Undersea and Hyperbaric Medical Society Workshop*. 75(Phys)6-1-89: 437. Retrieved 2008-05-31.

[68] Maiken, Eric (2004-08-01). "Why Argon?". Decompression. Retrieved 2008-06-26.

[69] Horhoianu, G; Ionescu, D.V; Olteanu, G (1999). "Thermal behaviour of CANDU type fuel rods during steady state and transient operating conditions". *Annals of Nuclear Energy* **26** (16): 1437–1445. doi:10.1016/S0306-4549(99)00022-5.

[70] "Disaster Ascribed to Gas by Experts". *The New York Times*. 1937-05-07. p. 1.

[71] Freudenrich, Craig (2008). "How Blimps Work". HowStuffWorks. Retrieved 2008-07-03.

[72] Dunkin, I. R. (1980). "The matrix isolation technique and its application to organic chemistry". *Chem. Soc. Rev.* **9**: 1–23. doi:10.1039/CS9800900001.

[73] Basting, Dirk; Marowsky, Gerd (2005). *Excimer Laser Technology*. Springer. ISBN 3-540-20056-8.

[74] Sanders, Robert D.; Ma, Daqing; Maze, Mervyn (2005). "Xenon: elemental anaesthesia in clinical practice". *British Medical Bulletin* **71** (1): 115–135. doi:10.1093/bmb/ldh034. PMID 15728132.

[75] Albert, M. S.; Balamore, D. (1998). "Development of hyperpolarized noble gas MRI". *Nuclear Instruments and Methods in Physics Research A* **402** (2–3): 441–453. Bibcode:1998NIMPA.402..441A. doi:10.1016/S0168-9002(97)00888-7. PMID 11543065.

[76] Ray, Sidney F. (1999). *Scientific photography and applied imaging*. Focal Press. pp. 383–384. ISBN 0-240-51323-1.

9.9 References

- Bennett, Peter B.; Elliott, David H. (1998). *The Physiology and Medicine of Diving*. SPCK Publishing. ISBN 0-7020-2410-4.

- Bobrow Test Preparation Services (2007-12-05). *CliffsAP Chemistry*. CliffsNotes. ISBN 0-470-13500-X.

- Greenwood, N. N.; Earnshaw, A. (1997). *Chemistry of the Elements* (2nd ed.). Oxford:Butterworth-Heinemann. ISBN 0-7506-3365-4.

- Harding, Charlie J.; Janes, Rob (2002). *Elements of the ƒ Block*. Royal Society of Chemistry. ISBN 0-85404-690-9.

- Holloway, John H. (1968). *Noble-Gas Chemistry*. London: Methuen Publishing. ISBN 0-412-21100-9.

- Mendeleev, D. (1902–1903). *Osnovy Khimii (The Principles of Chemistry)* (in Russian) (7th ed.).

- Ojima, Minoru; Podosek, Frank A. (2002). *Noble Gas Geochemistry*. Cambridge University Press. ISBN 0-521-80366-7.

- Weinhold, F.; Landis, C. (2005). *Valency and bonding*. Cambridge University Press. ISBN 0-521-83128-8.

- Scerri, Eric R. (2007). *The Periodic Table, Its Story and Its Significance*. Oxford University Press. ISBN 0-19-530573-6.

Chapter 10

Dmitri Mendeleev

For the Russian Prime Minister with a similar name, see Dmitry Medvedev.

This name uses Eastern Slavic naming customs; the patronymic is *Ivanovich* and the family name is *Mendeleev*.

Dmitri Ivanovich Mendeleev[3] (/ˌmɛndəlˈeɪəf/;[4] Russian: Дми́трий Ива́нович Менделе́ев; IPA: [ˈdmʲitrʲɪj ɪˈvanəvʲɪtɕ mʲɪndʲɪˈlʲejɪf]; 8 February 1834 – 2 February 1907 O.S. 27 January 1834 – 20 January 1907) was a Russian chemist and inventor. He formulated the Periodic Law, created his own version of the periodic table of elements, and used it to correct the properties of some already discovered elements and also to predict the properties of eight elements yet to be discovered.

10.1   Early life

Mendeleev was born in the village of Verkhnie Aremzyani, near Tobolsk in Siberia, to Ivan Pavlovich Mendeleev and Maria Dmitrievna Mendeleeva (née Kornilieva). His grandfather was Pavel Maximovich Sokolov, a priest of the Russian Orthodox Church from the Tver region.[5] Ivan, along with his brothers and sisters, obtained new family names while attending the theological seminary.[6] Mendeleev was raised as an Orthodox Christian, his mother encouraging him to "patiently search divine and scientific truth."[7] His son would later inform that he departed from the Church and embraced a form of deism.[8]

Mendeleev is thought to be the youngest of either 11, 13, 14 or 17 siblings;[9] the exact number differs among sources.[10] His father was a teacher of fine arts, politics and philosophy. Unfortunately for the family's financial well being, his father became blind and lost his teaching position. His mother was forced to work and she restarted her family's abandoned glass factory. At the age of 13, after the passing of his father and the destruction of his mother's factory by fire, Mendeleev attended the Gymnasium in Tobolsk.

In 1849, his mother took Mendeleev across the entire state of Russia from Siberia to Moscow with the aim of getting Mendeleev a higher education. The university in Moscow did not accept him. The mother and son continued to St. Petersburg to the father's alma mater. The now poor Mendeleev family relocated to Saint Petersburg, where he entered the Main Pedagogical Institute in 1850. After graduation, he contracted tuberculosis, causing him to move to the Crimean Peninsula on the northern coast of the Black Sea in 1855. While there he became a science master of the Simferopol gymnasium №1. In 1857, he returned to Saint Petersburg with fully restored health.

10.2   Later life

Between 1859 and 1861, he worked on the capillarity of liquids and the workings of the spectroscope in Heidelberg. In late August 1861 he wrote his first book on the spectroscope. On 4 April 1862 he became engaged to Feozva Nikitichna Leshcheva, and they married on 27 April 1862 at Nikolaev Engineering Institute's church in Saint Petersburg (where

Dmitri Mendeleev

he taught).[11] Mendeleev became a professor at the Saint Petersburg Technological Institute and Saint Petersburg State University in 1864 and 1865, respectively. In 1865 he became Doctor of Science for his dissertation "On the Combinations of Water with Alcohol". He achieved tenure in 1867, and by 1871 had transformed Saint Petersburg into an internationally recognized center for chemistry research. In 1876, he became obsessed with Anna Ivanova Popova and began courting her; in 1881 he proposed to her and threatened suicide if she refused. His divorce from Leshcheva was finalized one month after he had married Popova (on 2 April[12]) in early 1882. Even after the divorce, Mendeleev was technically a bigamist; the Russian Orthodox Church required at least seven years before lawful remarriage. His divorce and the surrounding controversy contributed to his failure to be admitted to the Russian Academy of Sciences (despite his international fame by that time). His daughter from his second marriage, Lyubov, became the wife of the famous Russian poet Alexander Blok. His other children were son Vladimir (a sailor, he took part in the notable Eastern journey of Nicholas II) and daughter Olga, from his first marriage to Feozva, and son Ivan and twins from Anna.

Though Mendeleev was widely honored by scientific organizations all over Europe, including (in 1882) the Davy Medal from the Royal Society of London (which later also awarded him the Copley Medal in 1905), he resigned from Saint Petersburg University on 17 August 1890. He was later elected a Foreign Member of the Royal Society (ForMemRS) in 1892.[2]

Mendeleev also investigated the composition of petroleum, and helped to found the first oil refinery in Russia. He recognized the importance of petroleum as a feedstock for petrochemicals. He is credited with a remark that burning petroleum as a fuel "would be akin to firing up a kitchen stove with bank notes."[13]

In 1905, Mendeleev was elected a member of the Royal Swedish Academy of Sciences. The following year the Nobel Committee for Chemistry recommended to the Swedish Academy to award the Nobel Prize in Chemistry for 1906 to Mendeleev for his discovery of the periodic system. The Chemistry Section of the Swedish Academy supported this recommendation. The Academy was then supposed to approve the Committee's choice, as it has done in almost every case. Unexpectedly, at the full meeting of the Academy, a dissenting member of the Nobel Committee, Peter Klason, proposed the candidacy of Henri Moissan whom he favored. Svante Arrhenius, although not a member of the Nobel Committee for Chemistry, had a great deal of influence in the Academy and also pressed for the rejection of Mendeleev, arguing that the periodic system was too old to acknowledge its discovery in 1906. According to the contemporaries, Arrhenius was motivated by the grudge he held against Mendeleev for his critique of Arrhenius's dissociation theory. After heated arguments, the majority of the Academy voted for Moissan. The attempts to nominate Mendeleev in 1907 were again frustrated by the absolute opposition of Arrhenius.[14]

In 1907, Mendeleev died at the age of 72 in Saint Petersburg from influenza. The crater Mendeleev on the Moon, as well as element number 101, the radioactive mendelevium, are named after him.

10.3 Periodic table

In 1863 there were 56 known elements with a new element being discovered at a rate of approximately one per year. Other scientists had previously identified periodicity of elements. John Newlands described a Law of Octaves, noting their periodicity according to relative atomic weight in 1864, publishing it in 1865. His proposal identified the potential for new elements such as germanium. The concept was criticized and his innovation was not recognized by the Society of Chemists until 1887. Another person to propose a periodic table was Lothar Meyer, who published a paper in 1864 describing 28 elements classified by their valence, but with no prediction of new elements.

After becoming a teacher, Mendeleev wrote the definitive textbook of his time: *Principles of Chemistry* (two volumes, 1868–1870). As he attempted to classify the elements according to their chemical properties, he noticed patterns that led him to postulate his periodic table; he claimed to have envisioned the complete arrangement of the elements in a dream:[15][16][17][18][19]

> "I saw in a dream a table where all elements fell into place as required. Awakening, I immediately wrote it down on a piece of paper, only in one place did a correction later seem necessary."
> — Mendeleev, as quoted by Inostrantzev[20][21]

Unaware of the earlier work on periodic tables going on in the 1860s, he made the following table:

By adding additional elements following this pattern, Dmitri developed his extended version of the periodic table.[22][23] On 6 March 1869, Mendeleev made a formal presentation to the Russian Chemical Society, entitled *The Dependence*

Reihen	Gruppo I. — $R'O$	Gruppo II. — RO	Gruppo III. — $R'O^3$	Gruppo IV. RH^4 RO'	Gruppo V. RH^3 $R'O^5$	Gruppo VI. RH^2 RO^3	Gruppo VII. RH $R'O'$	Gruppo VIII. — RO^4
1	H=1							
2	Li=7	Be=9,4	B=11	C=12	N=14	O=16	F=19	
3	Na=23	Mg=24	Al=27,3	Si=28	P=31	S=32	Cl=35,5	
4	K=39	Ca=40	—=44	Ti=48	V=51	Cr=52	Mn=55	Fe=56, Co=59, Ni=59, Cu=63.
5	(Cu=63)	Zn=65	—=68	—=72	As=75	Se=78	Br=80	
6	Rb=85	Sr=87	?Yt=88	Zr=90	Nb=94	Mo=96	—=100	Ru=104, Rh=104, Pd=106, Ag=108.
7	(Ag=108)	Cd=112	In=113	Sn=118	Sb=122	Te=125	J=127	
8	Cs=133	Ba=137	?Di=138	?Ce=140	—	—	—	— — — —
9	(—)	—	—	—	—	—	—	
10	—	—	?Er=178	?La=180	Ta=182	W=184	—	Os=195, Ir=197, Pt=198, Au=199.
11	(Au=199)	Hg=200	Tl=204	Pb=207	Bi=208	—	—	— — — —
12	—	—	—	Th=231	—	U=240	—	

Mendeleev's 1871 periodic table

Sculpture in honor of Mendeleev and the periodic table, located in Bratislava, Slovakia

between the Properties of the Atomic Weights of the Elements, which described elements according to both atomic weight and valence. This presentation stated that

1. The elements, if arranged according to their atomic weight, exhibit an apparent periodicity of properties.

2. Elements which are similar regarding their chemical properties have atomic weights which are either of nearly the same value (e.g., Pt, Ir, Os) or which increase regularly (e.g., K, Rb, Cs).

3. The arrangement of the elements in groups of elements in the order of their atomic weights corresponds to their so-called valencies, as well as, to some extent, to their distinctive chemical properties; as is apparent among other series in that of Li, Be, B, C, N, O, and F.

4. The elements which are the most widely diffused have small atomic weights.

5. The magnitude of the atomic weight determines the character of the element, just as the magnitude of the molecule determines the character of a compound body.

6. We must expect the discovery of many yet unknown elements–for example, two elements, analogous to aluminium and silicon, whose atomic weights would be between 65 and 75.

7. The atomic weight of an element may sometimes be amended by a knowledge of those of its contiguous elements. Thus the atomic weight of tellurium must lie between 123 and 126, and cannot be 128. *(Tellurium's atomic mass is 127.6, and Mendeleev was incorrect in his assumption that atomic mass must increase with position within a period.)*

8. Certain characteristic properties of elements can be foretold from their atomic weights.

Mendeleev published his periodic table of all known elements and predicted several new elements to complete the table. Only a few months after, Meyer published a virtually identical table. Some consider Meyer and Mendeleev the co-creators of the periodic table. Mendeleev has the distinction of accurately predicting of the qualities of what he called ekasilicon, ekaaluminium and ekaboron (germanium, gallium and scandium, respectively).

For his predicted eight elements, he used the prefixes of eka, dvi, and tri (Sanskrit one, two, three) in their naming. Mendeleev questioned some of the currently accepted atomic weights (they could be measured only with a relatively low accuracy at that time), pointing out that they did not correspond to those suggested by his Periodic Law. He noted that tellurium has a higher atomic weight than iodine, but he placed them in the right order, incorrectly predicting that the accepted atomic weights at the time were at fault. He was puzzled about where to put the known lanthanides, and predicted the existence of another row to the table which were the actinides which were some of the heaviest in atomic mass. Some people dismissed Mendeleev for predicting that there would be more elements, but he was proven to be correct when Ga (gallium) and Ge (germanium) were found in 1875 and 1886 respectively, fitting perfectly into the two missing spaces.[24]

By giving Sanskrit names to his "missing" elements, Mendeleev showed his appreciation and debt to the Sanskrit grammarians of ancient India, who had created sophisticated theories of language based on their discovery of the two-dimensional patterns in basic sounds. Mendeleev was a friend and colleague of the Sanskritist Böhtlingk, who was preparing the second edition of his book on Pāṇini[25] at about this time, and Mendeleev wished to honor Pāṇini with his nomenclature.[26] Noting that there are striking similarities between the periodic table and the introductory Śiva Sūtras in Pāṇini's grammar, Prof. Kiparsky says:

> "The analogies between the two systems are striking. Just as Panini found that the phonological patterning of sounds in the language is a function of their articulatory properties, so Mendeleev found that the chemical properties of elements are a function of their atomic weights. Like Panini, Mendeleev arrived at his discovery through a search for the "grammar" of the elements...[27]"

The original draft made by Mendeleev would be found years later and published under the name *Tentative System of Elements*.[28]

10.4   Other achievements

Mendeleev made other important contributions to chemistry. The Russian chemist and science historian Lev Chugaev has characterized him as "a chemist of genius, first-class physicist, a fruitful researcher in the fields of hydrodynamics, meteorology, geology, certain branches of chemical technology (explosives, petroleum, and fuels, for example) and other disciplines adjacent to chemistry and physics, a thorough expert of chemical industry and industry in general, and an original thinker in the field of economy." Mendeleev was one of the founders, in 1869, of the Russian Chemical Society. He worked on the theory and practice of protectionist trade and on agriculture.

In an attempt at a chemical conception of the Aether, he put forward a hypothesis that there existed two inert chemical elements of lesser atomic weight than hydrogen. Of these two proposed elements, he thought the lighter to be an all-penetrating, all-pervasive gas, and the slightly heavier one to be a proposed element, *coronium*.

Mendeleev devoted much study and made important contributions to the determination of the nature of such indefinite compounds as solutions.

In another department of physical chemistry, he investigated the expansion of liquids with heat, and devised a formula similar to Gay-Lussac's law of the uniformity of the expansion of gases, while in 1861 he anticipated Thomas Andrews' conception of the critical temperature of gases by defining the absolute boiling-point of a substance as the temperature at which cohesion and heat of vaporization become equal to zero and the liquid changes to vapor, irrespective of the pressure and volume.

Mendeleev is given credit for the introduction of the metric system to the Russian Empire.

He invented *pyrocollodion*, a kind of smokeless powder based on nitrocellulose. This work had been commissioned by the Russian Navy, which however did not adopt its use. In 1892 Mendeleev organized its manufacture.

Mendeleev studied petroleum origin and concluded hydrocarbons are abiogenic and form deep within the earth – see Abiogenic petroleum origin. He wrote: "*The capital fact to note is that petroleum was born in the depths of the earth, and it is only there that we must seek its origin.*" (Dmitri Mendeleev, 1877)[29]

10.5   Vodka myth

A very popular Russian story is that it was Mendeleev who came up with the 40% standard strength of vodka in 1894, after having been appointed Director of the Bureau of Weights and Measures with the assignment to formulate new state standards for the production of vodka. This story has, for instance, been used in marketing claims by the Russian Standard vodka brand that, "*In 1894, Dmitri Mendeleev, the greatest scientist in all Russia, received the decree to set the Imperial quality standard for Russian vodka and the 'Russian Standard' was born*",[30] or that the vodka is "*compliant with the highest quality of Russian vodka approved by the royal government commission headed by Mendeleev in 1894.*"[31]

While it is true that Mendeleev in 1892 became head of the Archive of Weights and Measures in Saint Petersburg, and evolved it into a government bureau the following year, that institution was never involved in setting any production quality standards, but was issued with standardising Russian trade weights and measuring instruments. Furthermore, the 40% standard strength was already introduced by the Russian government in 1843, when Mendeleev was nine years old.[31]

The basis for the whole story is a popular myth that Mendeleev's 1865 doctoral dissertation "A Discourse on the combination of alcohol and water" contained a statement that 38% is the ideal strength of vodka, and that this number was later rounded to 40% to simplify the calculation of alcohol tax. However, Mendeleev's dissertation was about alcohol concentrations over 70% and he never wrote anything about vodka.[31][32]

10.6   Commemoration

A number of places and objects are associated with the name and achievements of the scientist.

In Saint Petersburg his name was given to the National Metrology Institute[33] dealing with establishing and supporting national and worldwide standards for precise measurements. Next to it there is a monument to him pictured above that

Mendeleev Medal

consists of his sitting statue and a depiction of his periodic table on the wall of the establishment.

In the Twelve Collegia building, now being the centre of Saint Petersburg State University and in Mendeleev's time – Head Pedagogical Institute – there is Dmitry Mendeleev's Memorial Museum Apartment[34] with his archives. The street in front of these is named after him as Mendeleevskaya liniya (Mendeleev Line).

In Moscow, there is the D. Mendeleev University of Chemical Technology of Russia.[35]

After him was also named mendelevium, which is a synthetic chemical element with the symbol Md (formerly Mv) and the atomic number 101. It is a metallic radioactive transuranic element in the actinide series, usually synthesized by bombarding einsteinium with alpha particles.

A large lunar impact crater Mendeleev that is located on the far side of the Moon, as seen from the Earth, also bears the name of the scientist.

Russian Academy of Sciences yearly awards since 1998 Mendeleev Golden Medal.

10.7   See also

- List of Russian chemists
- Mendeleev's predicted elements
- Periodic Systems of Small Molecules

10.8   References

[1] Physics Tree profile Dmitri Ivanovich Mendeleev

[2] "Fellows of the Royal Society". London: Royal Society. Archived from the original on 2015-03-16.

[3] Also romanized **Mendeleyev** or **Mendeleef**

[4] "Mendeleev". *Random House Webster's Unabridged Dictionary.*

[5] Dmitriy Mendeleev: A Short CV, and A Story of Life, mendcomm.org

[6] Удомельские корни Дмитрия Ивановича Менделеева (1834–1907), starina.library.tver.ru

[7] Hiebert, Ray Eldon; Hiebert, Roselyn (1975). *Atomic Pioneers: From ancient Greece to the 19th century.* U.S. Atomic Energy Commission. Division of Technical Information. p. 25.

[8] Gordin, Michael D. (2004). *A Well-ordered Thing: Dmitrii Mendeleev And The Shadow Of The Periodic Table.* Basic Books. pp. 229–230. ISBN 9780465027750. Mendeleev seemed to have very few theological commitments. This was not for lack of exposure. His upbringing was actually heavily religious, and his mother — by far the dominating force in his youth - was exceptionally devout. One of his sisters even joined a fanatical religious sect for a time. Despite, or perhaps because of, this background, Mendeleev withheld comment on religious affairs for most of his life, reserving his few words for anti-clerical witticisms... Mendeleev's son Ivan later vehemently denied claims that his father was devoutly Orthodox: "I have also heard the view of my father's 'church religiosity' — and I must reject this categorically. From his earliest years Father practically split from the church — and if he tolerated certain simple everyday rites, then only as an innocent national tradition, similar to Easter cakes, which he didn't consider worth fighting against." ...Mendeleev's opposition to traditional Orthodoxy was not due to either atheism or a scientific materialism. Rather, he held to a form of romanticized deism.

[9] Johnson, George (3 January 2006). "The Nitpicking of the Masses vs. the Authority of the Experts". *New York Times.* Retrieved 14 March 2011.

[10] The number of Mendeleev's siblings is a matter of some historical dispute. When the Princeton historian of science Michael Gordin reviewed this article as part of an analysis of the accuracy of Wikipedia for the 14 December 2005 issue of *Nature*, he cited as one of Wikipedia's errors that "They say Mendeleev is the 14th child. He is the 14th surviving child of 17 total. 14 is right out." However in a *New York Times* article from January 2006, it was noted that in Gordin's own 2004 biography of Mendeleev, he also had the Russian chemist listed as the 17th child, and quoted Gordin's response to this as being: "That's curious. I believe that is a typographical error in my book. Mendeleyev was the final child, that is certain, and the number the reliable sources have is 13." Gordin's book specifically says that Mendeleev's mother bore her husband "seventeen children, of whom eight survived to young adulthood," with Mendeleev being the youngest. See: Johnson, George (3 January 2006). "The Nitpicking of the Masses vs. the Authority of the Experts". *The New York Times.* and Gordin, Michael (22 December 2005). "Supplementary information to accompany *Nature* news article 'Internet encyclopaedias go head to head' (*Nature* 438, 900–901; 2005)" (PDF). *Blogs.Nature.com* – via 2004, p. 178.

[11] "Rustest.spb.ru". Rustest.spb.ru. Retrieved 13 March 2010.

[12] "Gazeta.ua". Gazeta.ua. 9 March 2010. Retrieved 13 March 2010.

[13] John W. Moore, Conrad L. Stanitski, Peter C. Jurs. *Chemistry: The Molecular Science, Volume 1.* Retrieved 6 September 2011.

[14] Friedman, Robert M. (2001). *The politics of excellence: behind the Nobel Prize in science*. New York: Times Books. pp. 32–34. ISBN 0-7167-3103-7.

[15] John B. Arden (1998). "Science, Theology and Consciousness", Praeger Frederick A. p. 59: *The initial expression of the commonly used chemical periodic table was reportedly envisioned in a dream. In 1869, Dmitri Mendeleev claimed to have had a dream in which he envisioned a table in which all the chemical elements were arranged according to their atomic weight.*

[16] John Kotz, Paul Treichel, Gabriela Weaver (2005). "Chemistry and Chemical Reactivity," Cengage Learning. p. 333

[17] Gerard I. Nierenberg (1986). "The art of creative thinking", Simon & Schuster, p. 201: *Dmitri Mendeleev's solution for the arrangement of the elements that came to him in a dream.*

[18] Helen Palmer (1998). "Inner Knowing: Consciousness, Creativity, Insight, and Intuition". J.P. Tarcher/Putnam. p. 113: *The sewing machine, for instance, invented by Elias Howe, was developed from material appearing in a dream, as was Dmitri Mendeleev's periodic table of elements*

[19] Simon S. Godfrey (2003). "Dreams & Reality". Trafford Publishing. Chapter 2.: *'The Russian chemist, Dmitri Mendeleev (1834-1907), described a dream in which he saw the periodic table of elements in its complete form.* ISBN 1412011434

[20] "The Soviet Review Translations" Summer 1967. Vol. VIII, No. 2, M.E. Sharpe, Incorporated, p. 38

[21] Myron E. Sharpe, (1967). "Soviet Psychology". Volume 5, p. 30.

[22] A brief history of the development of the period table, wou.edu

[23] Mendeleev and the Periodic Table, chemsheets.co.uk

[24] Emsley, John (2001). *Nature's Building Blocks* ((Hardcover, First Edition) ed.). Oxford University Press. pp. 521–522. ISBN 0-19-850340-7.

[25] Otto Böhtlingk, Panini's Grammatik: Herausgegeben, Ubersetzt, Erlautert und MIT Verschiedenen Indices Versehe. St. Petersburg, 1839–40.

[26] Kiparsky, Paul. "Economy and the construction of the Sivasutras." In M. M. Deshpande and S. Bhate (eds.), *Paninian Studies*. Ann Arbor, Michigan, 1991.

[27] Kak, Subhash (2004). "Mendeleev and the Periodic Table of Elements". *Sandhan* **4** (2): 115–123. arXiv:physics/0411080

[28] "The Soviet Review Translations" Summer 1967. Vol. VIII, No. 2, M.E. Sharpe, Incorporated, p. 39

[29] Mendeleev, D., 1877. L'Origine du pétrole. Revue Scientifique, 2e Ser., VIII, p. 409-416.

[30] Sainsburys: *Russian Standard Vodka 1L* Linked 2014-06-28

[31] Evseev, Anton (2011-11-21). "Dmitry Mendeleev and 40 degrees of Russian vodka". *Science*. Moscow: English Pravda.Ru. Retrieved 2014-07-06.

[32] "Prominent Russians: Dmitry Mendeleev". *Prominent Russians: Science and technology*. Moscow: RT. 2011. Retrieved 2014-07-06.

[33] ВНИИМ Дизайн Груп (2011-04-13). "D.I.Mendeleyev Institute for Metrology". Vniim.ru. Retrieved 2012-08-20.

[34] Saint-PetersburgState University. "Museum-Archives n. a. Dmitry Mendeleev - Museums - Culture and Sport - University - Saint-Petersburg state university". Eng.spbu.ru. Retrieved 2012-08-19.

[35] University homepage in English

10.9   Further reading

- Gordin, Michael (2004). *A Well-Ordered Thing: Dmitrii Mendeleev and the Shadow of the Periodic Table*. New York: Basic Books. ISBN 0-465-02775-X.

- Mendeleyev, Dmitry Ivanovich; Jensen, William B. (2005). *Mendeleev on the Periodic Law: Selected Writings, 1869–1905*. Mineola, New York: Dover Publications. ISBN 0-486-44571-2.

- Strathern, Paul (2001). *Mendeleyev's Dream: The Quest For the Elements*. New York: St Martins Press. ISBN 0-241-14065-X.

- Mendeleev, Dmitrii Ivanovich (1901). *Principles of Chemistry*. New York: Collier.

10.10   External links

- Babaev, Eugene V. (February 2009). Dmitriy Mendeleev: A Short CV, and A Story of Life - 2009 biography on the occasion of Mendeleev's 175th anniversary

- Babaev, Eugene V., Moscow State University. Dmitriy Mendeleev Online

- Original Periodic Table, annotated.

- Mendeleev's first draft version of the Periodic Table, 17 February 1869.

- "Everything in its Place", essay by Oliver Sacks

- Works by or about Dmitri Mendeleev in libraries (WorldCat catalog)

Chapter 11

Radionuclide

Not to be confused with radionucleotide.

A **radionuclide** (**radioactive nuclide**, **radioisotope** or **radioactive isotope**) is an atom that has excess nuclear energy, making it unstable. This excess energy can either create and emit from the nucleus new radiation (gamma radiation), or a new particle (alpha particle or beta particle); or transfer this excess energy to one of its electrons, causing it to be ejected (conversion electron). During this process, the radionuclide is said to undergo radioactive decay.[1] These emissions constitute ionizing radiation. The unstable nucleus is more stable following the emission, but sometimes will undergo further decay. Radioactive decay is a random process at the level of single atoms: it is impossible to predict when one particular atom will decay.[2][3][4][5] However, for a collection of atoms of a single element the decay rate, and thus the half-life (t1/2) for that collection can be calculated from their measured decay constants. The duration of the half-lives of radioactive atoms have no known limits; the time range is over 55 orders of magnitude.

Radionuclides both occur naturally and are artificially made using nuclear reactors, cyclotrons, particle accelerators or radionuclide generators. There are about 650 radionuclides with half-lives longer than 60 minutes (see list of nuclides). Of these, 34 are primordial radionuclides that existed before the creation of the solar system, and there are another 50 radionuclides detectable in nature as daughters of these, or produced naturally on Earth by cosmic radiation. More than 2400 radionuclides have half-lives less than 60 minutes. Most of these are only produced artificially, and have very short half-lives. For comparison, there are about 254 stable nuclides.

All chemical elements have radionuclides. Even the lightest element, hydrogen, has a well-known radionuclide, tritium. Elements heavier than lead, and the elements technetium and promethium, exist only as radionuclides.

Radionuclides can have both beneficial and harmful effects on living organisms. Radionuclides with suitable half-lives are used in nuclear medicine for both diagnosis and treatment. An imaging tracer made with radionuclides is called a radioactive tracer. A pharmaceutical drug made with radionuclides is called a radiopharmaceutical.

11.1   Origin

11.1.1   Natural

Naturally occurring radionuclides fall into three categories: primordial radionuclides, secondary radionuclides, and cosmogenic radionuclides. Primordial radionuclides, such as uranium and thorium, originate mainly from the interior of stars, and exist in present time since their half-lives are so long they have not yet completely decayed. Secondary radionuclides are radiogenic isotopes derived from the decay of primordial radionuclides.They have shorter half-lives than primordial radionuclides. Cosmogenic isotopes,such ascarbon-14, are present because they are continually being formed in theatmosphere due tocosmic rays.[6]

11.1.2 Synthetic

Synthetic radionuclides are artificially produced by human activity using nuclear reactors, particle accelerators or radionuclide generators:

- Radioisotopes produced with nuclear reactors exploit the high flux of neutrons present. These neutrons activate elements placed within the reactor. A typical product from a nuclear reactor is thallium-201 and iridium-192. The elements that have a large propensity to take up the neutrons in the reactor are said to have a high neutron cross-section.

- Particle accelerators such as cyclotrons accelerate particles to bombard a target to produce radionuclides. Cyclotrons accelerate protons at a target to produce positron-emitting radionuclides, e.g., fluorine-18.

- Radionuclide generators contain a parent radionuclide that decays to produce a radioactive daughter. The parent is usually produced in a nuclear reactor. A typical example is the technetium-99m generator used in nuclear medicine. The parent produced in the reactor is molybdenum-99.

- Radionuclides are produced as an unavoidable side-effect of nuclear and thermonuclear explosions.

Trace radionuclides are those that occur in tiny amounts in nature either due to inherent rarity or due to half-lives that are significantly shorter than the age of the Earth.

11.2 Uses

Radionuclides are used in two major ways: for their chemical properties and as sources of radiation.

- In biology, radionuclides of carbon can serve as radioactive tracers because they are chemically very similar to the nonradioactive nuclides, so most chemical, biological, and ecological processes treat them in a nearly identical way. One can then examine the result with a radiation detector, such as a Geiger counter, to determine where the provided atoms were incorporated. For example, one might culture plants in an environment in which the carbon dioxide contained radioactive carbon; then the parts of the plant that incorporate atmospheric carbon would be radioactive.

- In nuclear medicine, radioisotopes are used for diagnosis, treatment, and research. Radioactive chemical tracers emitting gamma rays or positrons can provide diagnostic information about internal anatomy and the functioning of specific organs. This is used in some forms of tomography: single-photon emission computed tomography and positron emission tomography (PET) scanning and Cherenkov luminescence imaging. Radioisotopes are also a method of treatment in hemopoietic forms of tumors; the success for treatment of solid tumors has been limited. More powerful gamma sources sterilise syringes and other medical equipment.

- In biochemistry and genetics, radionuclides label molecules and allow tracing chemical and physiological processes occurring in living organisms, such as DNA replication or amino acid transport.

- In food preservation, radiation is used to stop the sprouting of root crops after harvesting, to kill parasites and pests, and to control the ripening of stored fruit and vegetables.

- In industry, and in mining, radionuclides examine welds, to detect leaks, to study the rate of wear, erosion and corrosion of metals, and for on-stream analysis of a wide range of minerals and fuels.

- In particle physics, radionuclides help discover new physics (physics beyond the Standard Model) by measuring the energy and momentum of their beta decay products.[7]

- In ecology, radionuclides are used to trace and analyze pollutants, to study the movement of surface water, and to measure water runoffs from rain and snow, as well as the flow rates of streams and rivers.

- In geology, archaeology, and paleontology, natural radionuclides are used to measure ages of rocks, minerals, and fossil materials.

11.3 Common examples

11.3.1 Americium-241

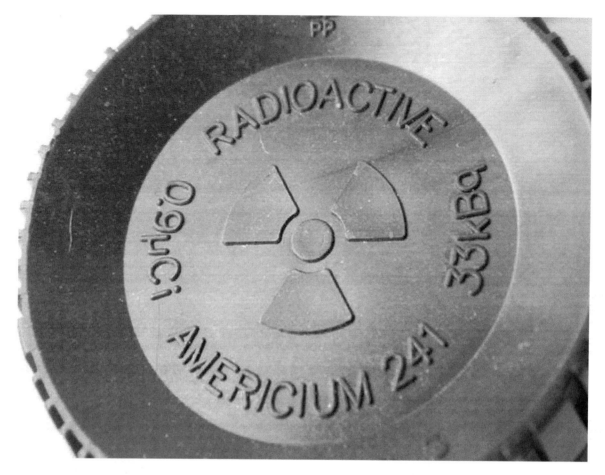

Americium-241 container in a smoke detector.

Most household smoke detectors contain americium produced in nuclear reactors. The radioisotope used is americium-241. The element americium is created by bombarding plutonium with neutrons in a nuclear reactor. Its isotope americium-241 decays by emitting alpha particles and gamma radiation to become neptunium-237. Most common household smoke detectors use a very small quantity of ^{241}Am (about 0.29 micrograms per smoke detector) in the form of americium dioxide. Smoke detectors use ^{241}Am since the alpha particles it emits collide with oxygen and nitrogen particles in the air. This occurs in the detector's ionization chamber where it produces charged particles or ions. Then, these charged particles are collected by a small electric voltage that will create an electric current that will pass between two electrodes. Then, the ions that are flowing between the electrodes will be neutralized when coming in contact with smoke, thereby decreasing the electric current between the electrodes, which will activate the detector's alarm.[8][9]

Steps for creating americium-241

The plutonium-241 is formed in any nuclear reactor by neutron capture from uranium-238.

1. 238U + n → 239U

2. 239U → 239Np + e− + ν
   e

Americium-241 capsule as found in smoke detector. The circle of darker metal in the center is americium-241; the surrounding casing is aluminium.

3. 239Np → 239Pu + e− + ν
   e

4. 239Pu + n → 240Pu

5. 240Pu + n → 241Pu

This will decay both in the reactor and subsequently to form ^{241}Am, which has a half-life of 432.2 years.[10][11]

11.3.2 Gadolinium-153

The ^{153}Gd isotope is used in X-ray fluorescence and osteoporosis screening. It is a gamma-emitter with an 8-month half-life, making it easier to use for medical purposes. In nuclear medicine, it serves to calibrate the equipment needed like single-photon emission computed tomography systems (SPECT) to make x-rays. It ensures that the machines work correctly to produce images of radioisotope distribution inside the patient. This isotope is produced in a nuclear reactor from europium or enriched gadolinium.[12] It can also detect the loss of calcium in the hip and back bones, allowing the ability to diagnose osteoporosis.[13]

11.4   Impacts on organisms

Radionuclides that find their way into the environment may cause harmful effects as radioactive contamination. They can also cause damage if they are excessively used during treatment or in other ways exposed to living beings, by radiation poisoning. Potential health damage from exposure to radionuclides depends on a number of factors, and "can damage the functions of healthy tissue/organs. Radiation exposure can produce effects ranging from skin redness and hair loss, to radiation burns and acute radiation syndrome. Prolonged exposure can lead to cells being damaged and in turn lead to cancer. Signs of cancerous cells might not show up until years, or even decades, after exposure."[14]

11.5   Summary table for classes of nuclides, "stable" and radioactive

Following is a summary table for the total list of nuclides with half-lives greater than one hour. Ninety of these 905 nuclides are theoretically stable, except to proton-decay (which has never been observed). About 254 nuclides have never been observed to decay, and are classically considered stable.

The remaining 650 radionuclides have half-lives longer than 1 hour, and are well-characterized (see list of nuclides for a complete tabulation). They include 28 nuclides with measured half-lives longer than the estimated age of the universe (13.8 billion years[15]), and another 6 nuclides with half-lives long enough (> 80 million years) that they are radioactive primordial nuclides, and may be detected on Earth, having survived from their presence in interstellar dust since before the formation of the solar system, about 4.6 billion years ago. Another ~51 short-lived nuclides can be detected naturally as daughters of longer-lived nuclides or cosmic-ray products. The remaining known nuclides are known solely from artificial nuclear transmutation.

Numbers are not exact, and may change slightly in the future, as "stable nuclides" are observed to be radioactive with very long half-lives.

This is a summary table [16] for the 905 nuclides with half-lives longer than one hour (including those that are stable), given in list of nuclides.

11.6   List of commercially available radionuclides

This list covers common isotopes, most of which are available in very small quantities to the general public in most countries. Others that are not publicly accessible are traded commercially in industrial, medical, and scientific fields and are subject to government regulation. For a complete list of all known isotopes for every element (minus activity data), see List of nuclides and Isotope lists. For a table, see Table of nuclides.

11.6.1 Gamma emission only

11.6.2 Beta emission only

11.6.3 Alpha emission only

11.6.4 Multiple radiation emitters

11.7 See also

- List of nuclides shows all radionuclides with half-life > 1 hour
- Hyperaccumulators table – 3
- Radioactivity in biology
- Radiometric dating
- Radionuclide cisternogram
- Uses of radioactivity in oil and gas wells

11.8 Notes

[1] R.H. Petrucci, W.S. Harwood and F.G. Herring, *General Chemistry* (8th ed., Prentice-Hall 2002), p.1025–26

[2] "Decay and Half Life". Retrieved 2009-12-14.

[3] Stabin, Michael G. (2007). "3". *Radiation Protection and Dosimetry: An Introduction to Health Physics*. Springer. doi:10.1007/978-0-387-49983-3. ISBN 978-0387499826.

[4] Best, Lara; Rodrigues, George; Velker, Vikram (2013). "1.3". *Radiation Oncology Primer and Review*. Demos Medical Publishing. ISBN 978-1620700044.

[5] Loveland, W.; Morrissey, D.; Seaborg, G.T. (2006). *Modern Nuclear Chemistry*. Wiley-Interscience. p. 57. ISBN 0-471-11532-0.

[6] Eisenbud, Merril; Gesell, Thomas F (1997-02-25). *Environmental Radioactivity: From Natural, Industrial, and Military Sources*. p. 134. ISBN 9780122351549.

[7] Severijns, Nathal; Beck, Marcus; Naviliat-Cuncic, Oscar (2006). "Tests of the standard electroweak model in nuclear beta decay".*Reviews of Modern Physics*78(3):991.arXiv:nucl-ex/0605029.Bibcode:2006RvMP...78..991S.doi:10.1103/RevModPh.

[8] "Smoke Detectors and Americium". *world-nuclear.org*.

[9] Office of Radiation Protection – Am 241 Fact Sheet – Washington State Department of Health

[10] "Smoke Detectors: Uses of Radioactive Isotopes". *Chemistry Tutorial : Radioisotopes in Smoke Detectors*. AUS-e-TUTE n.d. Retrieved March 30, 2011.

[11] Reaction in a Smoke Detector

[12] "PNNL: Isotope Sciences Program – Gadolinium-153". *pnl.gov*.

[13] "Gadolinium". *BCIT Chemistry Resource Center*. British Columbia Institute of Technology. Retrieved 30 March 2011.

[14] "Ionizing radiation, health effects and protective measures". World Health Organization. November 2012. Retrieved January 27, 2014.

[15] "Cosmic Detectives". The European Space Agency (ESA). 2013-04-02. Retrieved 2013-04-15.

[16] Table data is derived by counting members of the list; see WP:CALC. References for the list data itself are given below in the reference section in list of nuclides

11.9 References

- Carlsson, J.; Hietala, SO; Stigbrand, T; Tennvall, J; et al. (2003). "Tumour therapy with radionuclides: assessment of progress and problems". *Radiotherapy and Oncology* **66** (2): 107–117. doi:10.1016/S0167-8140(02)00374-2. PMID 12648782.

- "Radioisotopes in Industry". *World Nuclear Association.*

- Martin, James (2006). *Physics for Radiation Protection: A Handbook.* p. 130. ISBN 3527406115.

11.10 Further reading

- Luig, H.; Kellerer, A. M.; Griebel, J. R. (2011). "Radionuclides, 1. Introduction". *Ullmann's Encyclopedia of Industrial Chemistry.* doi:10.1002/14356007.a22_499.pub2. ISBN 3527306730.

11.11 External links

- EPA – Radionuclides – EPA's Radiation Protection Program: Information.

- FDA – Radionuclides – FDA's Radiation Protection Program: Information.

- Interactive Chart of Nuclides – A chart of all nuclides

- National Isotope Development Center – U.S. Government source of radionuclides – production, research, development, distribution, and information

- The Live Chart of Nuclides – IAEA

Chapter 12

Neutronium

Neutronium (sometimes shortened to **neutrium**[1]) is a proposed name for a substance composed purely of neutrons. The word was coined by scientist Andreas von Antropoff in 1926 (before the discovery of the neutron) for the conjectured "element of atomic number zero" that he placed at the head of the periodic table.[2][3] However, the meaning of the term has changed over time, and from the last half of the 20th century onward it has been also used legitimately to refer to extremely dense substances resembling the neutron-degenerate matter theorized to exist in the cores of neutron stars; henceforth "*degenerate* neutronium" will refer to this. Science fiction and popular literature frequently use the term "neutronium" to refer to a highly dense phase of matter composed primarily of neutrons.

12.1   Neutronium and neutron stars

Main article: Neutron star

Neutronium is used in popular literature to refer to the material present in the cores of neutron stars (stars which are too massive to be supported by electron degeneracy pressure and which collapse into a denser phase of matter). This term is very rarely used in scientific literature, for three reasons:

- There are multiple definitions for the term "neutronium".

- There is considerable uncertainty over the composition of the material in the cores of neutron stars (it could be neutron-degenerate matter, strange matter, quark matter, or a variant or combination of the above).

- The properties of neutron star material should depend on depth due to changing pressure (see below), and no sharp boundary between the crust (consisting primarily of atomic nuclei) and almost protonless inner layer is expected to exist.

When neutron star core material is presumed to consist mostly of free neutrons, it is typically referred to as neutron-degenerate matter in scientific literature.[4]

12.2   Neutronium and the periodic table

The term "neutronium" was coined in 1926 by Andreas von Antropoff for a conjectured form of matter made up of neutrons with no protons or electrons, which he placed as the chemical element of atomic number zero at the head of his new version of the periodic table. It was subsequently placed in the middle of several spiral representations of the periodic system for classifying the chemical elements, such as those of Charles Janet (1928), E. I. Emerson (1944), John D. Clark (1950) and in Philip Stewart's Chemical Galaxy (2005).

Although the term is not used in the scientific literature either for a condensed form of matter, or as an element, there have been reports that, besides the free neutron, there may exist two bound forms of neutrons without protons.[5] If neutronium were considered to be an element, then these neutron clusters could be considered to be the isotopes of that element. However, these reports have not been further substantiated.

- Mononeutron: An isolated neutron undergoes beta decay with a mean lifetime of approximately 15 minutes (half-life of approximately 10 minutes), becoming a proton (the nucleus of hydrogen), an electron and an antineutrino.

- Dineutron: The dineutron, containing two neutrons was unambiguously observed in the decay of beryllium-16, in 2012 by researchers at Michigan State University.[6][7] It is not a bound particle, but had been proposed as an extremely short-lived state produced by nuclear reactions involving tritium. It has been suggested to have a transitory existence in nuclear reactions produced by helions that result in the formation of a proton and a nucleus having the same atomic number as the target nucleus but a mass number two units greater. There had been evidence of dineutron emission from neutron-rich isotopes such as beryllium−16 where mononeutron decay would result in a less stable isotope. The dineutron hypothesis had been used in nuclear reactions with exotic nuclei for a long time.[8] Several applications of the dineutron in nuclear reactions can be found in review papers.[9] Its existence has been proven to be relevant for nuclear structure of exotic nuclei.[10] A system made up of only two neutrons is not bound, though the attraction between them is very nearly enough to make them so.[11] This has some consequences on nucleosynthesis and the abundance of the chemical elements.[9][12]

- Trineutron: A trineutron state consisting of three bound neutrons has not been detected, and is not expected to exist even for a short time.

- Tetraneutron: A tetraneutron is a hypothetical particle consisting of four bound neutrons. Reports of its existence have not been replicated. If confirmed, it would require revision of current nuclear models.[13][14]

- Pentaneutron: Calculations indicate that the hypothetical pentaneutron state, consisting of a cluster of five neutrons, would not be bound.[15]

Although not called "neutronium", the National Nuclear Data Center's *Nuclear Wallet Cards* lists as its first "isotope" an "element" with the symbol **n** and atomic number $Z = 0$ and mass number $A = 1$. This isotope is described as decaying to element **H** with a half life of 10.24 ± 0.02 min.

12.3 Properties

See also: Neutron star § Properties

Due to beta (β^-) decay of mononeutron and extreme instability of aforementioned heavier "isotopes", degenerate neutronium is not expected to be stable under ordinary pressures. Free neutrons decay with a half-life of 10 minutes, 11 seconds. A teaspoon of degenerate neutronium gas would have a mass of two billion tonnes, and if moved to standard temperature and pressure, would emit 57 billion joules of β^- decay energy in the first half-life (average of 95 MW of power).[16] This energy may be absorbed as the neutronium gas expands. Though, in the presence of atomic matter compressed to the state of electron degeneracy, the β^- decay may be inhibited due to Pauli exclusion principle, thus making free neutrons stable. Also, elevated pressures should make neutrons degenerate themselves. Compared to ordinary elements, neutronium should be more compressible due to the absence of electrically charged protons and electrons. This makes neutronium more energetically favorable than (positive-Z) atomic nuclei and leads to their conversion to (degenerate) neutronium through electron capture, a process which is believed to occur in stellar cores in the final seconds of the lifetime of massive stars, where it is facilitated by cooling via v
e emission. As a result, degenerate neutronium can have a density of 4×10^{17} kg/m^3,[17] roughly 13 magnitudes denser than the densest known ordinary substances. It was theorized that extreme pressures may deform the neutrons into a cubic symmetry, allowing tighter packing of neutrons,[18] or cause a strange matter formation.

12.4   In fiction

The term "neutronium" has been popular in science fiction since at least the middle of the 20th century. It typically refers to an extremely dense, incredibly strong form of matter. While presumably inspired by the concept of neutron-degenerate matter in the cores of neutron stars, the material used in fiction bears at most only a superficial resemblance, usually depicted as an extremely strong solid under Earth-like conditions, or possessing exotic properties such as the ability to manipulate time and space. In contrast, all proposed forms of neutron star core material are fluids and are extremely unstable at pressures lower than that found in stellar cores. According to one analysis, a neutron star with a mass below about 0.2 solar masses will explode.[19]

Noteworthy appearances of neutronium in fiction include the following:

- In Hal Clement's short story *Proof* (1942), neutronium is the only form of solid matter known to Solarians, the inhabitants of the Sun's interior.

- In Vladimir Savchenko's *Black Stars* (1960), neutronium is a mechanically and thermally indestructible substance. It is also used to make antimatter, which leads to an annihilation accident.

- In *Doctor Who* (1963), neutronium is a substance which can shield spaces from time-shear when used as shielding in time-vessels.

- In Larry Niven's *Known Space* fictional universe (1964), neutronium is actual neutron star core material, but it is stable in smaller quantities.

- In the *Star Trek* universe, neutronium is an extremely hard and durable substance, often used as armor, which conventional weapons cannot penetrate or even dent. The substance is referred to in the storyline dialogue of "The Doomsday Machine", "A Piece of the Action", "Evolution", "Relics", "To the Death", "What You Leave Behind", "Phage", "Prey", and "Think Tank".

- In Peter F. Hamilton's novel *The Neutronium Alchemist* (1997), neutronium is created by the "aggressive" setting off of a superweapon.

- In the *Stargate* universe, neutronium is a substance which is the basis of the technology of the advanced Asgard race, as well as a primary component of human-form Replicators.

- In Greg Bear's *The Forge of God* (1987), alien aggressors inject two high-mass weapons made of neutronium and antineutronium into the Earth which orbit the Earth's core until they meet and annihilate, destroying the planet.

- Action Comics #376 (May, 1969), "The Only Way to Kill Superman"[20] has Superman flying into a white dwarf star to grab a couple of handfuls of neutronium to fashion earplugs that will protect his ears from a hypersonic blast. In the panel on page seven, he states that each handful weighs a million tons.

- In the SF webcomic *Schlock Mercenary*, neutronium is used as a fuel.

12.5   See also

- Compact star

12.6   References

[1] "Neutrium: The Most Neutral Hypothetical State of Matter Ever". *io9.com*. 2012. Retrieved 2013-02-11.

[2] von Antropoff, A. (1926). "Eine neue Form des periodischen Systems der Elementen" (pdf). *Zeitschrift für Angewandte Chemie* **39** (23): 722–725. doi:10.1002/ange.19260392303.

[3] Stewart, P. J. (2007). "A century on from Dmitrii Mendeleev: Tables and spirals, noble gases and Nobel prizes". *Foundations of Chemistry* **9** (3): 235–245. doi:10.1007/s10698-007-9038-x.

[4] Angelo, J. A. (2006). *Encyclopedia of Space and Astronomy*. Infobase Publishing. p. 178. ISBN 978-0-8160-5330-8.

[5] Timofeyuk, N.K. (2003). "Do multineutrons exist?".*Journal of Physics G***29**(2):L9.arXiv:nucl-th/0301020.Bibcode:2003JPhG.. doi:10.1088/0954-3899/29/2/102.

[6] Schirber, M. (2012). "Nuclei Emit Paired-up Neutrons". *Physics* **5**: 30. Bibcode:2012PhyOJ...5...30S. doi:10.1103/Physics.5.30.

[7] Spyrou, A.; Kohley, Z.; Baumann, T.; Bazin, D.; et al. (2012). "First Observation of Ground State Dineutron Decay: ^{16}Be". *Physical Review Letters* **108** (10): 102501. Bibcode:2012PhRvL.108j2501S. doi:10.1103/PhysRevLett.108.102501. PMID 22463404.

[8] Bertulani, C. A.; Baur, G. (1986). "Coincidence Cross-sections for the Dissociation of Light Ions in High-energy Collisions" (pdf). *Nuclear Physics A* **480** (3–4): 615–628. Bibcode:1988NuPhA.480..615B. doi:10.1016/0375-9474(88)90467-8.

[9] Bertulani, C. A.; Canto, L. F.; Hussein, M. S. (1993). "The Structure And Reactions Of Neutron-Rich Nuclei" (pdf). *Physics Reports* **226** (6): 281–376. Bibcode:1993PhR...226..281B. doi:10.1016/0370-1573(93)90128-Z.

[10] Hagino, K.; Sagawa, H.; Nakamura, T.; Shimoura, S. (2009). "Two-particle correlations in continuum dipole transitions in Borromean nuclei".*Physical Review C***80**(3):1301.arXiv:0904.4775.Bibcode:2009PhRvC..80c1301H.doi:10.1103/PhysRevC.80..

[11] MacDonald, J.; Mullan, D. J. (2009). "Big Bang Nucleosynthesis: The Strong Nuclear Force meets the Weak Anthropic Principle". *Physical Review D* **80** (4): 3507. arXiv:0904.1807. Bibcode:2009PhRvD..80d3507M. doi:10.1103/PhysRevD.80.043507.

[12] Kneller, J. P.; McLaughlin, G. C. (2004). "The Effect of Bound Dineutrons upon BBN". *Physical Review D* **70** (4): 3512. arXiv:astro-ph/0312388. Bibcode:2004PhRvD..70d3512K. doi:10.1103/PhysRevD.70.043512.

[13] Bertulani, C. A.; Zelevinsky, V. (2002). "Is the tetraneutron a bound dineutron-dineutron molecule?". *Journal of Physics G* **29** (10): 2431. arXiv:nucl-th/0212060. Bibcode:2003JPhG...29.2431B. doi:10.1088/0954-3899/29/10/309.

[14] Timofeyuk, N. K. (2002). "On the existence of a bound tetraneutron". arXiv:nucl-th/0203003 [nucl-th].

[15] Bevelacqua, J.J. (1981). "Particle stability of the pentaneutron".*Physics Letters B***102**(2–3):79–80.Bibcode:1981PhLB..102. doi:10.1016/0370-2693(81)91033-9.

[16] "Neutrinos give neutron stars a chill". *Ars Technica OpenForum*. Retrieved 4 December 2013.

[17] Zarkonnen (2002). "Neutronium". *Everything2.com*. Retrieved 2013-02-11.

[18] Felipe J. Llanes-Estrada; Gaspar Moreno Navarro (2011). "Cubic neutrons". arXiv:1108.1859v1 [nucl-th].

[19] K. Sumiyoshi; S. Yamada; H. Suzuki; W. Hillebrandt (21 Jul 1997). "The fate of a neutron star just below the minimum mass: does it explode?". *Max-Planck-Institut für Astrophysik, Germany; RIKEN, U. Tokyo, and KEK, Japan*. arXiv:astro-ph/9707230. Given this assumption... the minimum possible mass of a neutron star is 0.189 (solar masses)

[20] Action Comics #376 (May, 1969), "The Only Way to Kill Superman"

12.7 Further reading

- Glendenning, N. K. (2000). *Compact Stars: Nuclear Physics, Particle Physics, and General Relativity* (2nd ed.). Springer. ISBN 978-0-387-98977-8.

Chapter 13

Chemical Galaxy

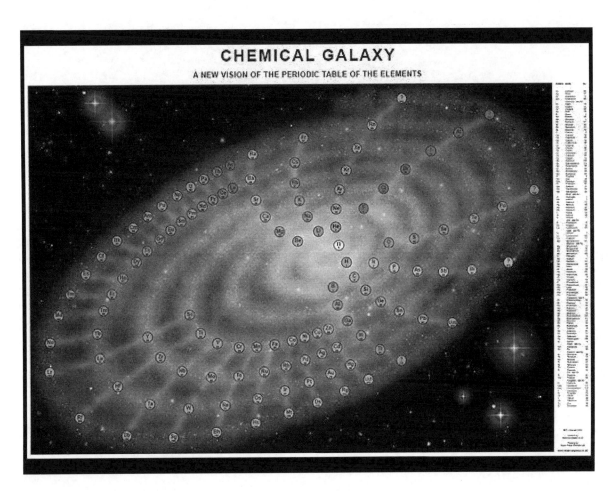

The Chemical Galaxy

Chemical Galaxy is a new representation by Philip Stewart of the periodic system of the elements, better known in tabular form as the periodic table, based on the cyclical nature of characteristics of the chemical elements (which depend principally on the valence electrons). Even before Dmitri Mendeleev produced the first satisfactory table, chemists were making spiral representations of the periodic system, and this has continued ever since, but these were usually circular in outline.

John Drury Clark was the first to present a spiral with an oval outline. His design was used as a vividly coloured two-page

104

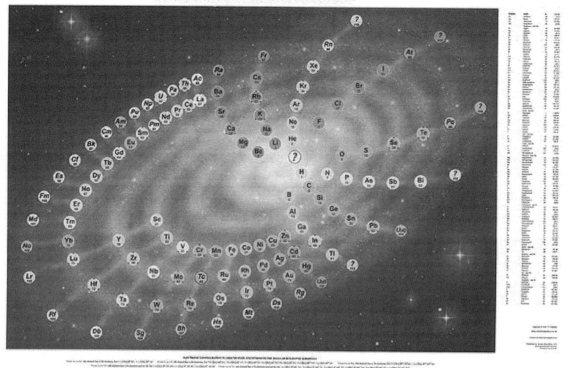

The Chemical Galaxy II

illustration in *Life* magazine for 16 May 1949. In 1951, Edgar Longman, an artist, not a chemist, painted a large mural, adapting the *Life* image by making the shape elliptical and tilting it to produce a dynamic effect. This inspired Stewart, then 12 years old, with a love of chemistry. Having just read Fred Hoyle's book *The Nature of the Universe*, he had the idea that Longman's design resembled a spiral galaxy. He returned to the idea many years later and published a first version of his "galaxy" in November 2004. His design seeks to express the link between the utterly minute world of atoms and the vastness of the stars, in the interior of which the elements were forged, as Hoyle was the first to demonstrate in detail.

Chemical Galaxy is intended primarily to excite an interest in chemistry among non-chemists, especially young people, but it is fully accurate scientifically in the information that it conveys about relationships between the elements, and it has the advantage over a table that it does not break up the continuous sequence of elements. A revised version, *Chemical Galaxy II*, introduces a new scheme, inspired by Michael Laing, for coloring the lanthanides and actinides, to bring out parallels with the transition metals. The design was translated into digital form by Carl Wenczek of Born Digital Ltd.

13.1 References

- Clark, John D. (November 1933). "A new periodic chart". *Journal of Chemical Education* **10** (11): 675–677. Bibcode:1933JChEd..10..675C. doi:10.1021/ed010p675.

- Stewart, P. J. (November 2004). "A new image of the periodic table". *Education in Chemistry* **41** (6): 156–158.

- Stewart, Philip J. (October 2007). "A century on from Dmitrii Mendeleev: tables and spirals, noble gases and Nobel prizes". *Foundations of Chemistry* **9** (3): 235–245. doi:10.1007/s10698-007-9038-x.

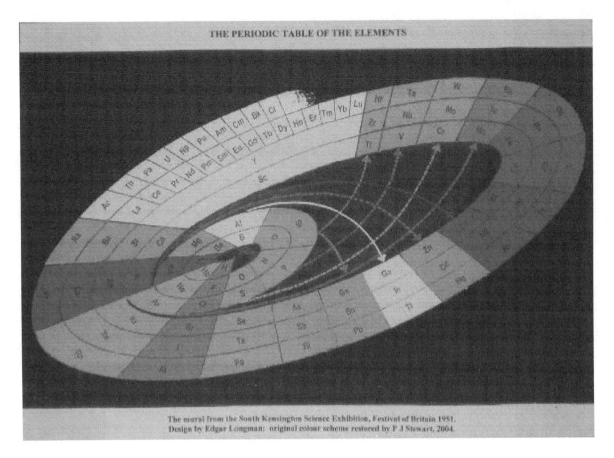

Longman Version

- Scerri, Eric (January–February 2008). "The past and future of the periodic table". *American Scientist* **96** (1): 52–58. doi:10.1511/2008.69.52.

13.2   External links

- Chemical Galaxy – Official Website

Chapter 14

Systematic element name

A **systematic element name** is the temporary name and symbol assigned to newly synthesized and not yet synthesized chemical elements. In chemistry, a transuranic element receives a permanent name and symbol only after its synthesis has been confirmed. In some cases, this has been a protracted and highly political process (see element naming controversy and Transfermium Wars). In order to discuss such elements without ambiguity, the International Union of Pure and Applied Chemistry (IUPAC) uses a set of rules to assign a temporary systematic name and symbol to each such element. This approach to naming originated in the successful development of regular rules for the naming of organic compounds.

14.1   The IUPAC rules

The temporary names are derived systematically from the element's atomic number. Each digit is translated to a 'numerical root', according to the table to the right. The roots are concatenated, and the name is completed with the ending suffix *-ium*. Some of the roots are Latin and others are Greek; the reason is to avoid two digits starting with the same letter (Ex: 0 = nil, 9 = enn, 4 = quad, 5 = pent, 6 = hex, 7 = sept) . There are two elision rules designed to prevent odd-looking names.

- If *bi* or *tri* is followed by the ending *ium* (i.e. the last digit is 2 or 3), the result is '-bium' or '-trium', not '-biium' or '-triium'.

- If *enn* is followed by *nil* (i.e. the sequence −90- occurs), the result is '-ennil-', not '-ennnil-'.

The systematic symbol is formed by taking the first letter of each root, converting the first to a capital.

The suffix *-ium* overrides traditional chemical suffix rules, thus 117 and 118 are *ununseptium* and *ununoctium*, not *ununseptine* and *ununocton*.

As of 2015, all elements up to atomic number 112, as well as elements 114 and 116, have received individual permanent names and symbols. So the systematic names and symbols are only used for unnamed elements 113, 115, 117, 118, and higher. The systematic names are exactly those with 3-letter symbols.

14.2   Notes

Examples in Period 8 of the periodic table:

14.3   External links

- The IUPAC recommendation. Untitled draft, March 2004. (PDF, 143 kB).

- http://media.iupac.org/publications/pac/2002/pdf/7405x0787.pdf

Chapter 15

Primordial nuclide

"Primordial element" redirects here. For a concept in algebra, see Primordial element (algebra).

In geochemistry and geonuclear physics, **primordial nuclides**, also known as **primordial isotopes**, are nuclides found

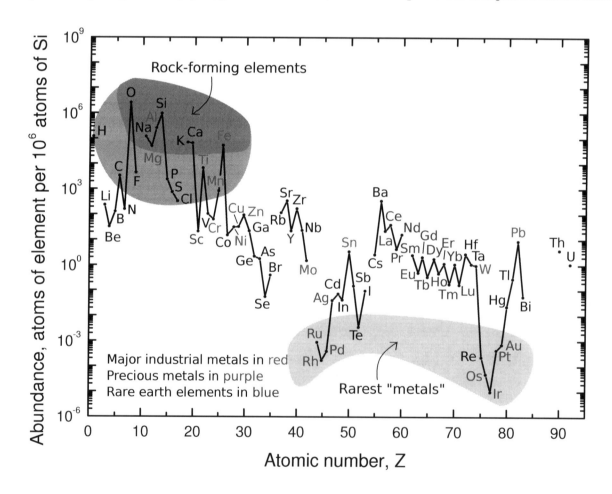

Relative abundance of the chemical elements in the Earth's upper continental crust, on a per-atom basis

on the Earth that have existed in their current form since before Earth was formed. Primordial nuclides are residues from the Big Bang, from cosmogenic sources, and from ancient supernova explosions which occurred before the formation of the Solar System. They are the stable nuclides plus the long-lived fraction of radionuclides surviving in the primordial solar nebula through planet accretion until the present. Only 288 such nuclides are known.

All of the known 254 stable nuclides occur as primordial nuclides, plus another 34 nuclides that have half-lives long enough to have survived from the formation of the Earth. These 34 primordial radionuclides represent isotopes of 28 separate elements. Cadmium, tellurium, neodymium and uranium each have two primordial radioisotopes (113Cd, 116Cd; 128Te, 130Te; 144Nd, 150Nd; and 235U, 238U), and samarium has three (146Sm, 147Sm, 148Sm).

Due to the age of the Earth of 4.58×10^9 years (4.6 billion years), this means that the half-life of the given nuclides must be greater than about 5×10^7 years (50 million years) for practical considerations. For example, for a nuclide with half-life 6×10^7 years (60 million years), this means 77 half-lives have elapsed, meaning that for each mole (6.02×10^{23} atoms) of that nuclide being present at the formation of Earth, only 4 atoms remain today.

The shortest-lived primordial nuclides (i.e. nuclides with shortest half-lives) are:

> ..., 232Th, 238U, 40K, 235U, 146Sm and 244Pu.

These are the 6 nuclides with half-lives comparable to, or less than, the estimated age of the universe. (In the case for ^{232}Th, it has a half life of more than 14 billion years, slightly longer than the age of the universe.) For a complete list of the 34 known primordial radionuclides, including the next 28 with half-lives much *longer* than the age of the universe, see the complete list in the section below.

The next longest-living nuclide after the end of the list given in the table is niobium-92 with a half-life of 3.47×10^7 years. (See list of nuclides for the list of all nuclides with half-lives longer than 60 minutes.) To be detected primordially, ^{92}Nb would have to survive at least 132 half-lives since the Earth's formation, meaning its original concentration will have decreased by a factor of 10^{40}. As of 2015, it has not been detected. It has been found that the next longer-lived nuclide, 244Pu, with a half-life of 8.08×10^7 years is primordial, although just barely, as its concentration in a few ores is nearly 10^{-18} weight parts.[1][2] Taking into account that all these nuclides must exist since at least 4.6×10^9 years, meaning survive 57 half-lives, their original number is now reduced by a factor of 2^{57} which equals more than 10^{17}.[3]

Although it is estimated that about 34 primordial nuclides are radioactive (list below), it becomes very difficult to determine the exact total number of radioactive primordials, because the total number of stable nuclides is uncertain. There exist many extremely long-lived nuclides whose half-lives are still unknown. For example, it is known theoretically that all isotopes of tungsten, including those indicated by even the most modern empirical methods to be stable, must be radioactive and can decay by alpha emission, but as of 2013 this could only be measured experimentally for 180W.[4] Nevertheless, the number of nuclides with half-lives so long that they cannot be measured with present instruments—and are considered from this viewpoint to be stable nuclides—is limited. Even when a "stable" nuclide is found to be radioactive, the fact merely moves it from the *stable* to the *unstable* list of primordial nuclides, and the total number of primordial nuclides remains unchanged.

Because **primordial chemical elements** often consist of more than one primordial isotope, there are only 84 distinct primordial chemical elements. Of these, 80 have at least one observationally stable isotope and four additional primordial elements have only radioactive isotopes.

15.1   Naturally occurring nuclides that are not primordial

Some unstable isotopes which occur naturally (such as 14C, 3H, and 239Pu) are not primordial, as they must be constantly regenerated. This occurs by cosmic radiation (in the case of cosmogenic nuclides such as 14C and 3H), or (rarely) by such processes as geonuclear transmutation (neutron capture of uranium in the case of 239Pu). Other examples of common naturally-occurring but non-primordial nuclides are radon, polonium, and radium, which are all radiogenic nuclide daughters of uranium decay and are found in uranium ores. A similar radiogenic series is derived from the long-lived radioactive primordial nuclide thorium-232. All of such nuclides have shorter half-lives than their parent radioactive primordial nuclides.

There are about 51 nuclides which are radioactive and exist naturally on Earth but are not primordial (making a total of fewer than 340 total nuclides to be found naturally on Earth).

15.2 Primordial elements

There are 254 stable primordial nuclides and 34 radioactive primordial nuclides, but only 80 primordial stable *elements* (1 through 82, i.e. hydrogen through lead, exclusive of 43 and 61, technetium and promethium respectively) and four radioactive primordial *elements* (bismuth, thorium, uranium, and plutonium). The numbers of elements are smaller, because many primordial elements are represented by more than one primordial nuclide. See chemical element for more information.

15.3 Naturally occurring stable nuclides

As noted, these number about 254. For a list, see the article list of stable isotopes. For a complete list noting which of the "stable" 254 nuclides may be in some respect unstable, see list of nuclides and stable isotope. These questions do not impact the question of whether a nuclide is primordial, since all "nearly stable" nuclides, with half-lives longer than the age of the universe, are primordial also.

15.4 List of 34 radioactive primordial nuclides and measured half-lives

These 34 primordial nuclides represent radioisotopes of 28 distinct chemical elements (cadmium, neodymium, tellurium, and uranium each have two primordial radioisotopes, and samarium has three). The radionuclides are listed in order of stability, with the longest half-life beginning the list. These radionuclides in many cases are so nearly stable that they compete for abundance with stable isotopes of their respective elements. For three chemical elements, a very long lived radioactive primordial nuclide is found to be the most abundant nuclide for an element that also has a stable nuclide. These unusual elements are tellurium, indium, and rhenium.

The longest has a half-life of 2.2×10^{24} years, which is 160 million million times the age of the Universe (the latter is about 4.32×10^{17} s). Only six of these 34 nuclides have half-lives shorter than, or equal to, the age of the universe. Most of the remaining 28 have half-lives much longer. The shortest-lived primordial isotope has a half-life of only 80 million years, less than 2% of the age of the Earth and Solar System.

15.4.1 List legends

no (number)

A running positive integer for reference. These numbers may change slightly in the future since there are 164 nuclides now classified as stable, but which are theoretically predicted to be unstable (see Stable nuclide#Still-unobserved decay), so that future experiments may show that some are in fact unstable. The number starts at 255, to follow the 254 nuclides (or stable isotopes) not yet found to be radioactive.

nuclide column

Nuclide identifiers are given by their mass number A and the symbol for the corresponding chemical element (implies a unique proton number).

energy column

The column labeled "energy" denotes the mass of the average nucleon of this nuclide relative to the mass of a neutron (so all nuclides get a positive value) in MeV, formally: $m_n - m_{nuclide} / A$.

half-life column

All times are given in years

decay mode column

decay energy column

Multiple values for (maximal) decay energy in MeV are mapped to decay modes in their order.

15.5   See also

- Table of nuclides sorted by half-life

- Table of nuclides

- Isotope geochemistry

- Radionuclide

- Mononuclidic element

- Monoisotopic element

- Stable isotope

- List of nuclides

- List of elements by stability of isotopes

- Big Bang nucleosynthesis

15.6   References

[1] D.C. Hoffman, F.O. Lawrence, J.L. Mewherter, F.M. Rourke (1971). "Detection of Plutonium-244 in Nature". *Nature* **234** (5325): 132–134. Bibcode:1971Natur.234..132H. doi:10.1038/234132a0.

[2] S. Maji, S. Lahiri, B. Wierczinski, G. Korschinek (2006). "Separation of samarium and neodymium: a prerequisite for getting signals from nuclear synthesis". *Analyst* **131** (12): 1332–1334. Bibcode:2006Ana...131.1332M. doi:10.1039/b608157f. PMID 17124541.

[3] P.K. Kuroda (1979). "Origin of the elements: pre-Fermi reactor and plutonium-244 in nature". *Accounts of Chemical Research* **12** (2): 73–78. doi:10.1021/ar50134a005.

[4] "Interactive Chart of Nuclides (Nudat2.5)". National Nuclear Data Center. Retrieved 2009-06-22.

Chapter 16

Electron shell

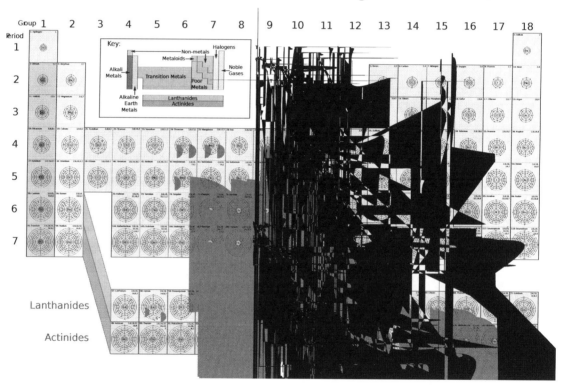

Periodic Table Of Elements Showing Electron Shells

Periodic table with electron shells

In chemistry and atomic physics, an **electron shell**, or a **principal energy level**, may be thought of as an orbit followed by electrons around an atom's nucleus. The closest shell to the nucleus is called the "1 shell" (also called "K shell"), followed by the "2 shell" (or "L shell"), then the "3 shell" (or "M shell"), and so on farther and farther from the nucleus. The shells correspond with the principal quantum numbers ($n = 1, 2, 3, 4 ...$) or are labeled alphabetically with letters used in the X-ray notation (K, L, M, …).

Each shell can contain only a fixed number of electrons: The first shell can hold up to two electrons, the second shell can hold up to eight ($2 + 6$) electrons, the third shell can hold up to 18 ($2 + 6 + 10$) and so on. The general formula is that the nth shell can in principle hold up to $2(n^2)$ electrons.[1] Since electrons are electrically attracted to the nucleus, an atom's electrons will generally occupy outer shells only if the more inner shells have already been completely filled by other electrons. However, this is not a strict requirement: atoms may have two or even three incomplete outer shells. (See Madelung rule for more details.) For an explanation of why electrons exist in these shells see electron configuration.[2]

The electrons in the outermost occupied shell (or shells) determine the chemical properties of the atom; it is called the *valence shell*.

Each shell consists of one or more *subshells*, and each subshell consists of one or more atomic orbitals.

16.1   History

The shell terminology comes from Arnold Sommerfeld's modification of the Bohr model. Sommerfeld retained Bohr's planetary model, but added mildly elliptical orbits (characterized by additional quantum numbers ℓ and m) to explain the fine spectroscopic structure of some elements.[3] The multiple electrons with the same principal quantum number (n) had close orbits that formed a "shell" of positive thickness instead of the infinitely thin circular orbit of Bohr's model.

The existence of electron shells was first observed experimentally in Charles Barkla's and Henry Moseley's X-ray absorption studies. Barkla labeled them with the letters K, L, M, N, O, P, and Q. The origin of this terminology was alphabetic. A "J" series was also suspected, though later experiments indicated that the K absorption lines are produced by the innermost electrons. These letters were later found to correspond to the n values 1, 2, 3, etc. They are used in the spectroscopic Siegbahn notation.

The physical chemist Gilbert Lewis was responsible for much of the early development of the theory of the participation of valence shell electrons in chemical bonding. Linus Pauling later generalized and extended the theory while applying insights from quantum mechanics.

16.2   Shells

The electron shells are labeled K, L, M, N, O, P, and Q; or 1, 2, 3, 4, 5, 6, and 7; going from innermost shell outwards. Electrons in outer shells have higher average energy and travel farther from the nucleus than those in inner shells. This makes them more important in determining how the atom reacts chemically and behaves as a conductor, because the pull of the atom's nucleus upon them is weaker and more easily broken. In this way, a given element's reactivity is highly dependent upon its electronic configuration.

16.3   Subshells

Each shell is composed of one or more subshells, which are themselves composed of atomic orbitals. For example, the first (K) shell has one subshell, called 1s; the second (L) shell has two subshells, called 2s and 2p; the third shell has 3s, 3p, and 3d; the fourth shell has 4s, 4p, 4d and 4f; the fifth shell has 5s, 5p, 5d, and 5f and can theoretically hold more but the 5f subshell, although partially occupied in actinides, is not filled in any element occurring naturally.[2] The various possible subshells are shown in the following table:

- The first column is the "subshell label", a lowercase-letter label for the type of subshell. For example, the "4s subshell" is a subshell of the fourth (N) shell, with the type (s) described in the first row.

- The second column is the azimuthal quantum number (ℓ) of the subshell. The precise definition involves quantum mechanics, but it is a number that characterizes the subshell.

- The third column is the maximum number of electrons that can be put into a subshell of that type. For example, the top row says that each s-type subshell (1s, 2s, etc.) can have at most two electrons in it. In each case the figure is 4 greater than the one above it.

- The fourth column says which shells have a subshell of that type. For example, looking at the top two rows, every shell has an s subshell, while only the second shell and higher have a p subshell (i.e., there is no "1p" subshell).

- The final column gives the historical origin of the labels s, p, d, and f. They come from early studies of atomic spectral lines. The other labels, namely g, h and i, are an alphabetical continuation following the last historically originated label of f.

Although it is commonly stated that all the electrons in a shell have the same energy, this is an approximation. However, the electrons in one *subshell* do have exactly the same level of energy,[5] with later subshells having more energy per electron than earlier ones. This effect is great enough that the energy ranges associated with shells can overlap (see *valence shells* and *Aufbau principle*).

16.4 Number of electrons in each shell

Each subshell is constrained to hold $4\ell + 2$ electrons at most, namely:

- Each s subshell holds at most 2 electrons

- Each p subshell holds at most 6 electrons

- Each d subshell holds at most 10 electrons

- Each f subshell holds at most 14 electrons

- Each g subshell holds at most 18 electrons

Therefore, the K shell, which contains only an s subshell, can hold up to 2 electrons; the L shell, which contains an s and a p, can hold up to $2 + 6 = 8$ electrons, and so forth; in general, the nth shell can hold up to $2n^2$ electrons.[1]

Although that formula gives the maximum in principle, in fact that maximum is only *achieved* (by known elements) for the first four shells (K, L, M, N). No known element has more than 32 electrons in any one shell.[6][7] This is because the subshells are filled according to the Aufbau principle. The first elements to have more than 32 electrons in one shell would belong to the g-block of period 8 of the periodic table. These elements would have some electrons in their 5g subshell and thus have more than 32 electrons in the O shell (fifth principal shell).

16.5 Valence shells

Main article: Valence electron

The **valence shell** is the outermost shell of an atom. It is usually (and misleadingly) said that the electrons in this shell make up its valence electrons, that is, the electrons that determine how the atom behaves in chemical reactions. Just as atoms with complete valence shells (noble gases) are the most chemically non-reactive, those with only one electron in their valence shells (alkali metals) or just missing one electron from having a complete shell (halogens) are the most reactive.[8]

However, this is a simplification of the truth. The electrons that determine how an atom reacts chemically are those that travel farthest from the nucleus, that is, those with the highest energy. For the transition elements, the partially filled $(n - 1)$d energy level is very close in energy to the ns level[9] and hence the d electrons in transition metals behave as valence electrons although they are not in the so-called valence shell.

16.6   List of elements with electrons per shell

The list below gives the elements arranged by increasing atomic number and shows the number of electrons per shell. At a glance, one can see that subsets of the list show obvious patterns. In particular, the seven elements (in electric blue) before a noble gas (group 18, in yellow) higher than helium have the number of electrons in the valence shell in arithmetic progression. (However, this pattern may break down in the seventh period due to relativistic effects.)

Sorting the table by chemical group shows additional patterns, especially with respect to the last two outermost shells. (Elements 57 to 71 belong to the lanthanides, while 89 to 103 are the actinides.)

The list below is primarily consistent with the Aufbau principle. However, there are a number of exceptions to the rule; for example palladium (atomic number 46) has no electrons in the fifth shell, unlike other atoms with *lower* atomic number. Some entries in the table are uncertain, when experimental data is unavailable. (For example, some atoms have such short half-life that it is impossible to measure their electron configurations).

16.7   See also

- Periodic table (electron configurations)

- Electron counting

- 18-Electron rule

- Core charge

16.8   References

[1] Re: Why do electron shells have set limits ? madsci.org, 17 March 1999, Dan Berger, Faculty Chemistry/Science, Bluffton College

[2] Electron Subshells. Corrosion Source. Retrieved on 1 December 2011.

[3] Donald Sadoway, *Introduction to Solid State Chemistry*, Lecture 5

[4] Jue, T. (2009). "Quantum Mechanic Basic to Biophysical Methods". *Fundamental Concepts in Biophysics*. Berlin: Springer. p. 33. ISBN 1-58829-973-2.

[5] The statement that the electrons in one subshell have exactly the same level of energy is true in an isolated atom, where it follows quantum-mechanically from the spherical symmetry of the system. When the atom is part of a molecule, this no longer holds; see, for example, crystal field theory.

[6] Orbitals. Chem4Kids. Retrieved on 1 December 2011.

[7] Electron & Shell Configuration. Chemistry.patent-invent.com. Retrieved on 1 December 2011.

[8] Chemical Reactions. Vision Learning (26 July 2011). Retrieved on 1 December 2011.

[9] THE ORDER OF FILLING 3d AND 4s ORBITALS. chemguide.co.uk

16.9   External links

Chapter 17

Group 4 element

Group 4 is a group of **elements** in the periodic table. It contains the elements titanium (Ti), zirconium (Zr), hafnium (Hf) and rutherfordium (Rf). This group lies in the d-block of the periodic table. The group itself has not acquired a trivial name; it belongs to the broader grouping of the transition metals.

The three Group 4 elements that occur naturally are titanium (Ti), zirconium (Zr) and hafnium (Hf). The first three members of the group share similar properties; all three are hard refractory metals under standard conditions. However, the fourth element rutherfordium (Rf), has been synthesized in the laboratory; none of its isotopes have been found occurring in nature. All isotopes of rutherfordium are radioactive. So far, no experiments in a supercollider have been conducted to synthesize the next member of the group, unpenthexium (Uph), and it is unlikely that they will be synthesized in the near future.

17.1   Characteristics

17.1.1   Chemistry

Like other groups, the members of this family show patterns in its electron configuration, especially the outermost shells resulting in trends in chemical behavior:

Most of the chemistry has been observed only for the first three members of the group. The chemistry of rutherfordium is not very established and therefore the rest of the section deals only with titanium, zirconium, and hafnium. All the elements of the group are reactive metals with a high melting point (1668 °C, 1855 °C, 2233 °C, 2100 °C?). The reactivity is not always obvious due to the rapid formation of a stable oxide layer, which prevents further reactions. The oxides TiO_2, ZrO_2 and HfO_2 are white solids with high melting points and unreactive against most acids.[1]

As tetravalent transition metals, all three elements form various inorganic compounds, generally in the oxidation state of +4. For the first three metals, it has been shown that they are resistant to concentrated alkalis, but halogens react with them to form tetrahalides. At higher temperatures, all three metals react with oxygen, nitrogen, carbon, boron, sulfur, and silicon. Because of the lanthanide contraction of the elements in the fifth period, zirconium and hafnium have nearly identical ionic radii. The ionic radius of Zr^{4+} is 79 picometers and that of Hf^{4+} is 78 pm.[1][2]

This similarity results in nearly identical chemical behavior and in the formation of similar chemical compounds.[2] The chemistry of hafnium is so similar to that of zirconium that a separation on chemical reactions was not possible; only the physical properties of the compounds differ. The melting points and boiling points of the compounds and the solubility in solvents are the major differences in the chemistry of these twin elements.[1] Titanium is considerably different from the other two owing to the effects of the lanthanide contraction.

17.1.2   Physical

The table below is a summary of the key physical properties of the group 4 elements. The four question-marked values are extrapolated.[3]

17.2   History

Crystal of the abundant mineral Ilmenite

17.2.1   Titanium

William Gregor, Franz Joseph Muller and Martin Heinrich Klaproth independently discovered titanium between 1791 and 1795. Klaproth named it for the Titans of Greek mythology.[4]

17.2.2   Zirconium

Klaproth also discovered zirconium in the mineral zircon in 1789 and named it after the already known Zirkonerde (zirconia).

17.2.3 Hafnium

Hafnium had been predicted by Dmitri Mendeleev in 1869 and Henry Moseley measured in 1914 the effective nuclear charge by X-ray spectroscopy to be 72, placing it between the already known elements lutetium and tantalum. Dirk Coster and Georg von Hevesy were the first to search for the new element in zirconium ores.[5] Hafnium was discovered by the two in 1923 in Copenhagen, Denmark, validating the original 1869 prediction of Mendeleev.[6] There has been some controversy surrounding the discovery of hafnium and the extent to which Coster and Hevesy were guided by Bohr's prediction that hafnium would be a transition metal rather than a rare earth element.[7] While titanium and zirconium, as relatively abundant elements, were discovered in the late 18th century, it took until 1923 for hafnium to be identified. This was only partly due to hafnium's relative scarcity. The chemical similarity between zirconium and hafnium made a separation difficult and, without knowing what to look for, hafnium was left undiscovered, although all samples of zirconium, and all of its compounds, used by chemists for over two centuries contained significant amounts of hafnium.[8]

17.2.4 Rutherfordium

Rutherfordium was reportedly first detected in 1966 at the Joint Institute of Nuclear Research at Dubna (then in the Soviet Union). Researchers there bombarded ^{242}Pu with accelerated ^{22}Ne ions and separated the reaction products by gradient thermochromatography after conversion to chlorides by interaction with $ZrCl_4$.[9]

242
94Pu + 22
10Ne → 264−x
104Rf → 264−x
104RfCl$_4$

17.3 Production

The production of the metals itself is difficult due to their reactivity. The formation of oxides, nitrides and carbides must be avoided to yield workable metals, this is normally achieved by the Kroll process. The oxides (MO_2) are reacted with coal and chlorine to form the chlorides (MCl_4).The chlorides of the metals are then reacted with magnesium, yielding magnesium chloride and the metals.

Further purification is done by a chemical transport reaction developed by Anton Eduard van Arkel and Jan Hendrik de Boer. In a closed vessel, the metal reacts with iodine at temperatures of above 500 °C forming metal(IV) iodide; at a tungsten filament of nearly 2000 °C the reverse reaction happens and the iodine and metal are set free. The metal forms a solid coating at the tungsten filament and the iodine can react with additional metal resulting in a steady turn over.[1][10]

$$M + 2\,I_2 \text{ (low temp.)} \rightarrow MI_4$$
$$MI_4 \text{ (high temp.)} \rightarrow M + 2\,I_2$$

17.4 Occurrence

If the abundance of elements in Earth's crust is compared for titanium, zirconium and hafnium, the abundance decreases with increase of atomic mass. Titanium is the seventh most abundant metal in Earth's crust and has an abundance of 6320 ppm, while zirconium has an abundance of 162 ppm and hafnium has only an abundance of 3 ppm.[11]

All three stable elements occur in heavy mineral sands ore deposits, which are placer deposits formed, most usually in beach environments, by concentration due to the specific gravity of the mineral grains of erosion material from mafic and ultramafic rock. The titanium minerals are mostly anatase and rutile, and zirconium occurs in the mineral zircon. Because of the chemical similarity, up to 5% of the zirconium in zircon is replaced by hafnium. The largest producers of the group 4 elements are Australia, South Africa and Canada.[12][13][14][15][16]

Heavy minerals (dark) in a quartz beach sand (Chennai, India).

17.5   Applications

Titanium metal and its alloys have a wide range of applications, where the corrosion resistance, the heat stability and the low density (light weight) are of benefit. The foremost use of corrosion-resistant hafnium and zirconium has been in nuclear reactors. Zirconium has a very low and hafnium has a high thermal neutron-capture cross-section. Therefore, zirconium (mostly as zircaloy) is used as cladding of fuel rods in nuclear reactors,[17] while hafnium is used as control rod for nuclear reactors, because each hafnium atom can absorb multiple neutrons.[18][19]

Smaller amounts of hafnium[20] and zirconium are used in super alloys to improve the properties of those alloys.[21]

17.6   Biological occurrences

The group 4 elements are not known to be involved in the biological chemistry of any living systems.[22] They are hard refractory metals with low aqueous solubility and low availability to the biosphere. Titanium is one of the few first row d-block transition metals with no known biological role. Rutherfordium's radioactivity would make it toxic to living cells.

17.7   Precautions

Titanium is non-toxic even in large doses and does not play any natural role inside the human body.[22] Zirconium powder can cause irritation, but only contact with the eyes requires medical attention.[23] OSHA recommends for zirconium are 5 mg/m^3 time weighted average limit and a 10 mg/m^3 short-term exposure limit.[24] Only limited data exists on the toxicology of hafnium.[25]

17.8 References

[1] Holleman, Arnold F.; Wiberg, Egon; Wiberg, Nils (1985). *Lehrbuch der Anorganischen Chemie* (in German) (91–100 ed.). Walter de Gruyter. pp. 1056–1057. ISBN 3-11-007511-3.

[2] "Los Alamos National Laboratory – Hafnium". Archived from the original on June 2, 2008. Retrieved 2008-09-10.

[3] Hoffman, Darleane C.; Lee, Diana M.; Pershina, Valeria (2006). "Transactinides and the future elements". In Morss; Edelstein, Norman M.; Fuger, Jean. *The Chemistry of the Actinide and Transactinide Elements* (3rd ed.). Dordrecht, The Netherlands: Springer Science+Business Media. ISBN 1-4020-3555-1.

[4] Weeks, Mary Elvira (1932). "III. Some Eighteenth-Century Metals". *Journal of Chemical Education* **9** (7): 1231–1243. Bibcode:1932JChEd...9.1231W. doi:10.1021/ed009p1231.

[5] Urbain, M. G. (1922). "Sur les séries L du lutécium et de l'ytterbium et sur l'identification d'un celtium avec l'élément de nombre atomique 72" [The L series from lutetium to ytterbium and the identification of element 72 celtium]. *Comptes rendus* (in French) **174**: 1347–1349. Retrieved 2008-10-30.

[6] Coster,D.;Hevesy,G. (1923-01-20). "On the Missing Element of Atomic Number72".*Nature***111**(2777):79–79.Bibcode:1923C. doi:10.1038/111079a0.

[7] *Scerri, Eric (2007). The Periodic System, Its Story and Its Significance. New York: Oxford University Press. ISBN 0-19-530573-6.* Missing or empty |title= (help)

[8] Barksdale, Jelks (1968). The Encyclopedia of the Chemical Elements. Skokie, Illinois: Reinhold Book Corporation. pp. 732–38 "Titanium". LCCCN 68-29938.

[9] Barber, R. C.; Greenwood, N. N.; Hrynkiewicz, A. Z.; Jeannin, Y. P.; Lefort, M.; Sakai, M.; Ulehla, I.; Wapstra, A. P.; et al. (1993). "Discovery of the transfermium elements. Part II: Introduction to discovery profiles. Part III: Discovery profiles of the transfermium elements". *Pure and Applied Chemistry* **65** (8): 1757–1814. doi:10.1351/pac199365081757.

[10] van Arkel, A. E.; de Boer, J. H. (1925). "Darstellung von reinem Titanium-, Zirkonium-, Hafnium- und Thoriummetall (Production of pure titanium, zirconium, hafnium and Thorium metal)". *Zeitschrift für anorganische und allgemeine Chemie* (in German) **148** (1): 345–350. doi:10.1002/zaac.19251480133.

[11] "Abundance in Earth's Crust". WebElements.com. Retrieved 2007-04-14.

[12] "Dubbo Zirconia Project Fact Sheet" (PDF). Alkane Resources Limited. June 2007. Retrieved 2008-09-10.

[13] "Zirconium and Hafnium" (PDF). *Mineral Commodity Summaries* (US Geological Survey): 192–193. January 2008. Retrieved 2008-02-24.

[14] Callaghan, R. (2008-02-21). "Zirconium and Hafnium Statistics and Information". US Geological Survey. Retrieved 2008-02-24.

[15] "Minerals Yearbook Commodity Summaries 2009: Titanium" (PDF). US Geological Survey. May 2009. Retrieved 2008-02-24.

[16] Gambogi, Joseph (January 2009). "Titanium and Titanium dioxide Statistics and Information" (PDF). US Geological Survey. Retrieved 2008-02-24.

[17] Schemel, J. H. (1977). *ASTM Manual on Zirconium and Hafnium.* ASTM International. pp. 1–5. ISBN 978-0-8031-0505-8.

[18] Hedrick, James B. "Hafnium" (PDF). United States Geological Survey. Retrieved 2008-09-10.

[19] Spink, Donald (1961). "Reactive Metals. Zirconium, Hafnium, and Titanium". *Industrial and Engineering Chemistry* **53** (2): 97–104. doi:10.1021/ie50614a019.

[20] Hebda, John (2001). "Niobium alloys and high Temperature Applications" (PDF). CBMM. Retrieved 2008-09-04.

[21] Donachie, Matthew J. (2002). *Superalloys.* ASTM International. pp. 235–236. ISBN 978-0-87170-749-9.

[22] Emsley, John (2001). "Titanium". *Nature's Building Blocks: An A-Z Guide to the Elements.* Oxford, England, UK: Oxford University Press. pp. 457–456. ISBN 0-19-850340-7.

[23] "International Chemical Safety Cards". International Labour Organization. October 2004. Retrieved 2008-03-30. lcontribu-
       tion= ignored (help)

[24] "Zirconium Compounds". National Institute for Occupational Health and Safety. 2007-12-17. Retrieved 2008-02-17.

[25] "Occupational Safety & Health Administration: Hafnium". U.S. Department of Labor. Archived from the original on 2008-
       03-13. Retrieved 2008-09-10.

Chapter 18

Carbon group

The **carbon group** is a periodic table group consisting of carbon (C), silicon (Si), germanium (Ge), tin (Sn), lead (Pb), and flerovium (Fl).

In modern IUPAC notation, it is called **Group 14**. In the field of semiconductor physics, it is still universally called **Group IV**. The group was once also known as the **tetrels** (from Greek *tetra*, four), stemming from the Roman numeral IV in the group names, or (not coincidentally) from the fact that these elements have four valence electrons (see below). The group is sometimes also referred to as **tetragens** or **crystallogens**.

18.1 Characteristics

18.1.1 Chemical

Like other groups, the members of this family show patterns in electron configuration, especially in the outermost shells, resulting in trends in chemical behavior:

Each of the elements in this group has 4 electrons in its outer orbital (the atom's top energy level). The last orbital of all these elements is the p^2 orbital. In most cases, the elements share their electrons. The tendency to lose electrons increases as the size of the atom increases, as it does with increasing atomic number. Carbon alone forms negative ions, in the form of carbide (C^{4-}) ions. Silicon and germanium, both metalloids, each can form +4 ions. Tin and lead both are metals while flerovium is a synthetic, radioactive (its half life is very short), element that may have a few noble gas-like properties, though it is still most likely a post-transition metal. Tin and lead are both capable of forming +2 ions.

Carbon forms tetrahalides with all the halogens. Carbon also forms three oxides: carbon monoxide, carbon suboxide (C_3O_2), and carbon dioxide. Carbon forms disulfides and diselenides.[1]

Silicon forms two hydrides: SiH_4 and Si_2H_6. Silicon forms tetrahalides with fluorine, chlorine, and iodine. Silicon also forms a dioxide and a disulfide.[2] Silicon nitride has the formula Si_3N_4.[3]

Germanium forms two hydrides: GeH_4 and Ge_2H_6. Germanium forms tetrahalides with all halogens except astatine and forms dihalides with all halogens except bromine and astatine. Germanium bonds to all natural single chalcogens except polonium, and forms dioxides, disulfides, and diselenides. Germanium nitride has the formula Ge_3N_4.[4]

Tin forms two hydrides: SnH_4 and Sn_2H_6. Tin forms dihalides and tetrahalides with all halogens except astatine. Tin forms chalcogenides with one of each naturally occurring chalcogen except polonium, and forms chalcogenides with two of each naturally occurring chalcogen except polonium and tellurium.[5]

Lead forms one hydride, which has the formula PbH_4. Lead forms dihalides and tetrahalides with fluorine and chlorine, and forms a tetrabromide and a lead diiodide. Lead forms four oxides, a sulfide, a selenide, and a telluride.[6]

There are no known compounds of flerovium.[7]

18.1.2 Physical

The boiling points of the carbon group tend to get lower with the heavier elements. Carbon, the lightest carbon group element, sublimates at 3825 °C. Silicon's boiling point is 3265 °C, germanium's is 2833 °C, tin's is 2602 °C, and lead's is 1749 °C. The melting points of the carbon group elements have roughly the same trend as their boiling points. Silicon melts at 1414 °C, germanium melts at 939 °C, tin melts at 232 °C, and lead melts at 328 °C.[8]

Carbon's crystal structure is hexagonal; at high pressures and temperatures it forms diamond (see below). Silicon and germanium have diamond cubic crystal structures, as does tin at low temperatures (below 13.2 °C). Tin at room temperature has a tetragonal crystal structure. Lead has a face-centered cubic crystal structure.[8]

The densities of the carbon group elements tend to increase with increasing atomic number. Carbon has a density of 2.26 grams per cubic centimeter, silicon has a density of 2.33 grams per cubic centimeter, germanium has a density of 5.32 grams per cubic centimeter. Tin has a density of 7.26 grams per cubic centimeter, and lead has a density of 11.3 grams per cubic centimeter.[8]

The atomic radii of the carbon group elements tend to increase with increasing atomic number. Carbon's atomic radius is 77 picometers, silicon's is 118 picometers, germanium's is 123 picometers, tin's is 141 picometers, and lead's is 175 picometers.[8]

Allotropes

Main article: Allotropes of carbon

Carbon has multiple allotropes. The most common is graphite, which is carbon in the form of stacked sheets. Another form of carbon is diamond, but this is relatively rare. Amorphous carbon is a third allotrope of carbon; it is a component of soot. Another allotrope of carbon is a fullerene, which has the form of sheets of carbon atoms folded into a sphere. A fifth allotrope of carbon, discovered in 2003, is called graphene, and is in the form of a layer of carbon atoms arranged in a honeycomb-shaped formation.[3][9][10]

Silicon has two known allotropes that exist at room temperature. These allotropes are known as the amorphous and the crystalline allotropes. The amorphous allotrope is a brown powder. The crystalline allotrope is gray and has a metallic luster.[11]

Tin has two allotropes: α-tin, also known as gray tin, and β-tin. Tin is typically found in the β-tin form, a silvery metal. However, at standard pressure, β-tin converts to α-tin, a gray powder, at temperatures below 56° Fahrenheit. This can cause tin objects in cold temperatures to crumble to gray powder in a process known as tin rot.[3][12]

18.1.3 Nuclear

At least two of the carbon group elements (tin and lead) have magic nuclei, meaning that these elements are more common and more stable than elements that do not have a magic nucleus.[12]

Isotopes

There are 15 known isotopes of carbon. Of these, three are naturally occurring. The most common is stable carbon-12, followed by stable carbon-13.[8] Carbon-14 is a natural radioactive isotope with a half-life of 5,730 years.[13]

23 isotopes of silicon have been discovered. Five of these are naturally occurring. The most common is stable silicon-28, followed by stable silicon-29 and stable silicon-30. Silicon-32 is a radioactive isotope that occurs naturally as a result of radioactive decay of actinides, and via spallation in the upper atmosphere. Silicon-34 also occurs naturally as the result of radioactive decay of actinides.[13]

32 isotopes of germanium have been discovered. Five of these are naturally occurring. The most common is the stable isotope germanium-74, followed by the stable isotope germanium-72, the stable isotope germanium-70, and the stable isotope germanium-73. The isotope germanium-76 is a primordial radioisotope.[13]

40 isotopes of tin have been discovered. 14 of these occur in nature. The most common is the stable isotope tin-120, followed by the stable isotope tin-118, the stable isotope tin-116, the stable isotope tin-119, the stable isotope tin-117, the primordial radioisotope tin-124, the stable isotope tin-122, the stable isotope tin-112, and the stable isotope tin-114. Tin also has four radioisotopes that occur as the result of the radioactive decay of uranium. These isotopes are tin-121, tin-123, tin-125, and tin-126.[13]

38 isotopes of lead have been discovered. 9 of these are naturally occurring. The most common isotope is the primordial radioisotope lead-208, followed by the primordial radioisotope lead-206, the primordial radioisotope lead-207, and the primordial radioisotope lead-204. 4 isotopes of lead occur from the radioactive decay of uranium and thorium. These isotopes are lead-209, lead-210, lead-211, and lead-212.[13]

6 isotopes of flerovium (flerovium-284, flerovium-285, flerovium-286, flerovium-287, flerovium-288, and flerovium-289) have been discovered. None of these are naturally occurring. Flerovium's most stable isotope is flerovium-289, which has a half-life of 2.6 seconds.[13]

18.2 Occurrence

Carbon accumulates as the result of stellar fusion in most stars, even small ones.[12] Carbon is present in the earth's crust in concentrations of 480 parts per million, and is present in seawater at concentrations of 28 parts per million. Carbon is present in the atmosphere in the form of carbon monoxide, carbon dioxide, and methane. Carbon is a key constituent of carbonate minerals, and is in hydrogen carbonate, which is common in seawater. Carbon forms 22.8% of a typical human.[13]

Silicon is present in the earth's crust at concentrations of 28%, making it the second most abundant element there. Silicon's concentration in seawater can vary from 30 parts per billion on the surface of the ocean to 2000 parts per billion deeper down. Silicon dust occurs in trace amounts in earth's atmosphere. Silicate minerals are the most common type of mineral on earth. Silicon makes up 14.3 parts per million of the human body on average.[13] Only the largest stars produce silicon via stellar fusion.[12]

Germanium makes up 2 parts per million of the earth's crust, making it the 52nd most abundant element there. On average, germanium makes up 1 part per million of soil. Germanium makes up 0.5 parts per trillion of seawater. Organogermanium compounds are also found in seawater. Germanium occurs in the human body at concentrations of 71.4 parts per billion. Germanium has been found to exist in some very faraway stars.[13]

Tin makes up 2 parts per million of the earth's crust, making it the 49th most abundant element there. On average, tin makes up 1 part per million of soil. Tin exists in seawater at concentrations of 4 parts per trillion. Tin makes up 428 parts per million of the human body. Tin (IV) oxide occurs at concentrations of 0.1 to 300 parts per million in soils.[13] Tin also occurs in concentrations of one part per thousand in igneous rocks.[14]

Lead makes up 14 parts per million of the earth's crust, making it the 36th most abundant element there. On average, lead makes up 23 parts per million of soil, but the concentration can reach 20000 parts per million (2 percent) near old lead mines. Lead exists in seawater at concentrations of 2 parts per trillion. Lead makes up 0.17% of the human body by weight. Human activity releases more lead into the environment than any other metal.[13]

Flerovium only occurs in particle accelerators.[13]

18.3 History

18.3.1 Discoveries and uses in antiquity

Carbon, tin, and lead are a few of the elements well known in the ancient world—together with sulfur, iron, copper, mercury, silver, and gold.[15]

Carbon as an element was used by the first human to handle charcoal from a fire.

Silicon as silica in the form of rock crystal was familiar to the predynastic Egyptians, who used it for beads and small vases; to the early Chinese; and probably to many others of the ancients. The manufacture of glass containing silica was carried out both by the Egyptians — at least as early as 1500 BCE — and by the Phoenicians. Many of the naturally occurring compounds or silicate minerals were used in various kinds of mortar for construction of dwellings by the earliest people.

The origins of tin seem to be lost in history. It appears that bronzes, which are alloys of copper and tin, were used by prehistoric man some time before the pure metal was isolated. Bronzes were common in early Mesopotamia, the Indus Valley, Egypt, Crete, Israel, and Peru. Much of the tin used by the early Mediterranean peoples apparently came from the Scilly Isles and Cornwall in the British Isles,[16] where mining of the metal dates from about 300–200 BCE. Tin mines were operating in both the Inca and Aztec areas of South and Central America before the Spanish conquest.

Lead is mentioned often in early Biblical accounts. The Babylonians used the metal as plates on which to record inscriptions. The Romans used it for tablets, water pipes, coins, and even cooking utensils; indeed, as a result of the last use, lead poisoning was recognized in the time of Augustus Caesar. The compound known as white lead was apparently prepared as a decorative pigment at least as early as 200 BCE.

18.3.2   Modern discoveries

Amorphous elemental silicon was first obtained pure in 1824 by the Swedish chemist Jöns Jacob Berzelius; impure silicon had already been obtained in 1811. Crystalline elemental silicon was not prepared until 1854, when it was obtained as a product of electrolysis.

Germanium is one of three elements the existence of which was predicted in 1869 by the Russian chemist Dmitri Mendeleev when he first devised his periodic table. However, the element was not actually discovered for some time. In September 1885, a miner discovered a mineral sample in a silver mine and gave it to the mine manager, who determined that it was a new mineral and sent the mineral to Clemens A. Winkler. Winkler realized that the sample was 75% silver, 18% sulfur, and 7% of an undiscovered element. After several months, Winkler isolated the element and determined that it was element 32.[13]

The first attempt to discover flerovium (then referred to as "element 114") was in 1969, at the Joint Institute for Nuclear Research, but it was unsuccessful. In 1977, researchers at the Joint Institute for Nuclear Research bombarded plutonium-244 atoms with calcium-48, but were again unsuccessful. This nuclear was repeated in 1998, this time successfully.[13]

18.3.3   Etymologies

The word "carbon" comes from the Latin word *carbo*, meaning "charcoal".The word "silicon" comes from the Latin word *silex* or *silicis*, which mean "flint". The word "germanium" comes from the word *germania*, which is Latin for Germany, the county where germanium was discovered. The word "tin" derives from the Old English word *tin*. The word "lead" comes from the Old English word *lead*.[13]

18.4   Applications

Carbon is most commonly used in its amorphous form. In this form, carbon is used for steelmaking, as carbon black, as a filling in tires, in respirators, and as activated charcoal. Carbon is also used in the form of graphite is commonly used as the lead in pencils. Diamond, another form of carbon, is commonly used in jewelery.[13] Carbon fibers are used in numerous applications, such as satellite struts, because the fibers are highly strong yet elastic.[17]

Silicon dioxide has a wide variety of applications, including toothpaste, construction fillers, and silica is a major component of glass. 50% of pure silicon is devoted to the manufacture of metal alloys. 45% of silicon is devoted to the manufacture of silicones. Silicon is also commonly used in semiconductors since the 1950s.[12][17]

Germanium was used in semiconductors until the 1950s, when it was replaced by silicon.[12] Radiation detectors contain germanium. Germanium oxide is used in fiber optics and wide-angle camera lenses. A small amount of germanium mixed with silver can make silver tarnish-proof. The resulting alloy is known as argentium.[13]

Solder is the most important use of tin; 50% of all tin produced goes into this application. 20% of all tin produced is used in tin plate. 20% of tin is also used by the chemical industry. Tin is also a constituent of numerous alloys, including pewter. Tin (IV) oxide has been commonly used in ceramics for thousands of years. Cobalt stannate is a tin compound which is used as a cerulean blue pigment.[13]

80% of all lead produced goes into lead-acid batteries. Other applications for lead include weights, pigments, and shielding against radioactive materials. Lead was historically used in gasoline in the form of tetraethyl lead, but this application has been discontinued due to concerns of toxicity.[18]

18.5 Production

Carbon's allotrope diamond is produced mostly by Russia, Botswana, Congo, Canada, and South Africa. 80% of all synthetic diamonds are produced by Russia. China produces 70% of the world's graphite. Other graphite-mining countries are Brazil, Canada, and Mexico.[13]

Silicon can be produced by heating silica with carbon.[17]

There are some germanium ores, such as germanite, but these are not mined on account of being rare. Instead, germanium is extracted from the ores of metals such as zinc. In Russia and China, germanium is also separated from coal deposits. Germanium-containing ores are first treated with chlorine to form germanium tetrachloride, which is mixed with hydrogen gas. Then the germanium is further refined by zone refining Roughly 140 metric tons of germanium are produced each year.[13]

Mines output 300,000 metric tons of tin each year. China, Indonesia, Peru, Bolivia, and Brazil are the main producers of tin. The method by which tin is produced is to head the tin mineral cassiterite (SnO_2) with coke.[13]

The most commonly mined lead ore is galena (lead sulfide). 4 million metric tons of lead are newly mined each year, mostly in China, Australia, the United States, and Peru. The ores are mixed with coke and limestone and roasted to produce pure lead. Most lead is recycled from lead batteries. The total amount of lead ever mined by humans amounts to 350 million metric tons.[13]

18.6 Biological role

Carbon is a key element to all known life. It is in all organic compounds, for example, DNA, steroids, and proteins.[3] Carbon's importance to life is primarily due to its ability to form numerous bonds with other elements.[12] There are 16 kilograms of carbon in a typical 70-kilogram human.[13]

Silicon-based life's feasibility is commonly discussed. However, it is less able than carbon to form elaborate rings and chains.[3] Silicon in the form of silicon dioxide is used by diatoms and sea sponges to form their cell walls and skeletons. Silicon is essential for bone growth in chickens and rats and may also be essential in humans. Humans consume on average between 20 and 1200 milligrams of silicon per day, mostly from cereals. There is 1 gram of silicon in a typical 70-kilogram human.[13]

A biological role for germanium is not known, although it does stimulate metabolism. In 1980, germanium was reported by Kazuhiko Asai to benefit health, but the claim has not been proven. Some plants take up germanium from the soil in the form of germanium oxide. These plants, which include grains and vegetables contain roughly 0.05 parts per million of germanium. The estimated human intake of germanium is 1 milligram per day. There are 5 milligrams of germanium in a typical 70-kilogram human.[13]

Tin has been shown to be essential for proper growth in rats, but there is, as of 2013, no evidence to indicate that humans need tin in their diet. Plants do not require tin. However, plants do collect tin in their roots. Wheat and corn contain seven and three parts per million respectively. However, the level of tin in plants can reach 2000 parts per million if the plants are near a tin smelter. On average, humans consume 0.3 milligrams of tin per day. There are 30 milligrams of tin in a typical 70-kilogram human.[13]

Lead has no known biological role, and is in fact highly toxic, but some microbes are able to survive in lead-contaminated environments. Some plants, such as cucumbers contain up to tens of parts per million of lead. There are 120 milligrams of lead in a typical 70-kilogram human.[13]

18.6.1 Toxicity

Elemental carbon is not generally toxic, but many of its compounds are, such as carbon monoxide and hydrogen cyanide. However, carbon dust can be dangerous because it lodges in the lungs in a manner similar to asbestos.[13]

Silicon minerals are not typically poisonous. However, silicon dioxide dust, such as that emitted by volcanoes can cause adverse health effects if it enters the lungs.[12]

Germanium can interfere with such enzymes as lactate and alcohol dehydrogenase. Organic germanium compounds are more toxic than inorganic germanium compounds. Germanium has a low degree of oral toxicity in animals. Severe germanium poisoning can cause death by respiratory paralysis.[19]

Some tin compounds are toxic to ingest, but most inorganic compounds of tin are considered nontoxic. Organic tin compounds, such as trimethyl tin and triethyl tin are highly toxic, and can disrupt metabolic processes inside cells.[13]

Lead and its compounds, such as lead acetate are highly toxic. Lead poisoning can cause headaches, stomach pain, constipation, and gout.[13]

18.7 References

[1] *Carbon compounds*, retrieved January 24, 2013

[2] *Silicon compounds*, retrieved January 24, 2013

[3] Gray, Theodore (2011), *The Elements*

[4] *Germanium compounds*, retrieved January 24, 2013

[5] *Tin compounds*, retrieved January 24, 2013

[6] *Lead compounds*, retrieved January 24, 2013

[7] *Flerovium compounds*, retrieved January 24, 2013

[8] Jackson, Mark (2001), *Periodic Table Advanced*

[9] *Graphene*, retrieved January 2013

[10] *Carbon:Allotropes*, retrieved January 2013

[11] Gagnon, Steve, *The Element Silicon*, retrieved January 20, 2013

[12] Kean, Sam (2011), *The Disappearing Spoon*

[13] Emsley, John (2011), *Nature's Building Blocks*

[14] *tin (Sn)*, Encyclopedia Britannica, 2013, retrieved February 24, 2013

[15] *Chemical Elements*, retrieved January 2013

[16] *Online Encyclopaedia Britannica, Tin*

[17] Galan, Mark (1992), *Structure of Matter*, ISBN 0-809-49663-1

[18] Blum, Deborah (2010), *The Poisoner's Handbook*

[19] *Risk Assessment* (PDF), 2003, retrieved January 19, 2013

Chapter 19

Group 10 element

"Group 10" redirects here. For the rugby league competition, see Group 10 Rugby League.

Group 10, numbered by current IUPAC style, is the group of chemical elements in the periodic table that consists of nickel (Ni), palladium (Pd), platinum (Pt), and darmstadtium (Ds). All are d-block transition metals. All known isotopes of Ds are radioactive with short half-lives, and are not known to occur in nature; only minute quantities have been synthesized in laboratories.

Like other groups, the members of this group show patterns in electron configuration, especially in the outermost shells, although for this group they are particularly weak, with palladium being an exceptional case. The relativistic stabilization of the 7s orbital is the explanation to the predicted electron configuration of darmstadtium, which, unusually for this group, conforms to that predicted by the Aufbau principle.

19.1   Chemistry

Darmstadtium has not been isolated in pure form, and its properties have not been conclusively observed; only nickel, palladium, and platinum have had their properties experimentally confirmed. All three elements are typical silvery-white transition metals, hard, and refractory, with high melting and boiling points.

19.2   Properties

Group 10 metals are white to light grey in color, and possess a high luster, a resistance to tarnish (oxidation) at STP, are highly ductile, and enter into oxidation states of +2 and +4, with +1 being seen in special conditions. The existence of a +3 state is debated, as the state could be an illusory state created by +2 and +4 states. Theory suggests that group 10 metals may produce a +6 oxidation state under precise conditions, but this remains to be proven conclusively in the laboratory other than for platinum.

19.3   Occurrence

19.4   Production

19.5   Applications

The group 10 metals share several uses. These include:

- Decorative purposes, in the form of jewelry and electroplating.

- Catalysts in a variety of chemical reactions.

- Metal alloys.

- Electrical components, due to their predictable changes in electrical resistivity with regard to temperature.

- Superconductors, as components in alloys with other metals.

19.6   Biological role and toxicity

Nickel has an important role in the biochemistry of organisms, as part of the active center of enzymes. None of the other group 10 elements have a known biological role, but platinum compounds have widely been used as anticancer drugs. Aside from nickel, the elements are toxic for organisms.

19.7   See also

- Platinum group

19.8   Notes and references

[1] http://www.chemistry-reference.com/pdictable/q_elements.asp?language=en&Symbol=Ds

Chapter 20

Group 3 element

Group 3 is a group of **elements** in the periodic table. This group, like other d-block groups, should contain four elements, but it is not agreed what elements belong in the group. Scandium (Sc) and yttrium (Y) are always included, but the other two spaces are usually occupied by lanthanum (La) and actinium (Ac), or by lutetium (Lu) and lawrencium (Lr); less frequently, it is considered the group should be expanded to 32 elements (with all the lanthanides and actinides included) or contracted to contain only scandium and yttrium. The group itself has not acquired a trivial name; however, scandium, yttrium and the lanthanides are sometimes called rare earth metals.

Three group 3 elements occur naturally, scandium, yttrium, and either lanthanum or lutetium. Lanthanum continues the trend started by two lighter members in general chemical behavior, while lutetium behaves more similarly to yttrium. This is in accordance with the trend for period 6 transition metals to behave more similarly to their upper periodic table neighbors. This trend is seen from hafnium, which is almost identical chemically to zirconium, to mercury, which is quite distant chemically from cadmium, but still shares with it almost equal atomic size and other similar properties. They all are silvery-white metals under standard conditions. The fourth element, either actinium or lawrencium, has only radioactive isotopes. Actinium, which occurs only in trace amounts, continues the trend in chemical behavior for metals that form tripositive ions with a noble gas configuration; synthetic lawrencium is calculated and partially shown to be more similar to lutetium and yttrium. So far, no experiments have been conducted to synthesize any element that could be the next group 3 element. Unbiunium (Ubu), which could be considered a group 3 element if preceded by lanthanum and actinium, might be synthesized in the near future, it being only three spaces away from the current heaviest element known, ununoctium.

20.1 History

In 1787, Swedish part-time chemist Carl Axel Arrhenius found a heavy black rock near the Swedish village of Ytterby, Sweden (part of the Stockholm Archipelago).[1] Thinking that it was an unknown mineral containing the newly discovered element tungsten,[2] he named it ytterbite.[note 1] Finnish scientist Johan Gadolin identified a new oxide or "earth" in Arrhenius' sample in 1789, and published his completed analysis in 1794;[3] in 1797, the new oxide was named *yttria*.[4] In the decades after French scientist Antoine Lavoisier developed the first modern definition of chemical elements, it was believed that earths could be reduced to their elements, meaning that the discovery of a new earth was equivalent to the discovery of the element within, which in this case would have been *yttrium*.[note 2] Until the early 1920s, the chemical symbol "Yt" was used for the element, after which "Y" came into common use.[5] Yttrium metal was first isolated in 1828 when Friedrich Wöhler heated anhydrous yttrium(III) chloride with potassium to form metallic yttrium and potassium chloride.[6][7]

In 1869, Russian chemist Dmitri Mendeleev published his periodic table, which had empty spaces for elements directly above and under yttrium.[8] Mendeleev made several predictions on the upper neighbor of ytttrium, which he called *eka-boron*. Swedish chemist Lars Fredrik Nilson and his team discovered the missing element in the minerals euxenite and gadolinite and prepared 2 grams of scandium(III) oxide of high purity.[9][10] He named it scandium, from the Latin *Scandia* meaning "Scandinavia". Chemical experiments on the element proved that Mendeleev's suggestions were correct;

along with discovery and characterization of gallium and germanium this proved the correctness of the whole periodic table and periodic law. Nilson was apparently unaware of Mendeleev's prediction, but Per Teodor Cleve recognized the correspondence and notified Mendeleev.[11] Metallic scandium was produced for the first time in 1937 by electrolysis of a eutectic mixture, at 700–800 °C, of potassium, lithium, and scandium chlorides.[12]

Lutetium was independently discovered in 1907 by French scientist Georges Urbain,[13] Austrian mineralogist Baron Carl Auer von Welsbach, and American chemist Charles James[14] as an impurity in the mineral ytterbia, which was thought by most chemists to consist entirely of ytterbium. Welsbach proposed the names *cassiopeium* for element 71 (after the constellation Cassiopeia) and *aldebaranium* (after the star Aldebaran) for the new name of ytterbium but these naming proposals were rejected, although many German scientists in the 1950s called the element 71 cassiopeium. Urbain chose the names *neoytterbium* (Latin for "new ytterbium") for ytterbium and *lutecium* (from Latin Lutetia, for Paris) for the new element. The dispute on the priority of the discovery is documented in two articles in which Urbain and von Welsbach accuse each other of publishing results influenced by the published research of the other.[15][16] The Commission on Atomic Mass, which was responsible for the attribution of the names for the new elements, settled the dispute in 1909 by granting priority to Urbain and adopting his names as official ones. An obvious problem with this decision was that Urbain was one of the four members of the commission.[17] The separation of lutetium from ytterbium was first described by Urbain and the naming honor therefore went to him, but *neoytterbium* was eventually reverted to ytterbium and in 1949, the spelling of element 71 was changed to lutetium.[18][19] Ironically, Charles James, who had modestly stayed out of the argument as to priority, worked on a much larger scale than the others, and undoubtedly possessed the largest supply of lutetium at the time.[20]

Lawrencium was first synthesized by the Albert Ghiorso and his team on February 14, 1961, at the Lawrence Radiation Laboratory (now called the Lawrence Berkeley National Laboratory) at the University of California in Berkeley, California, United States. The first atoms of lawrencium were produced by bombarding a three-milligram target consisting of three isotopes of the element californium with boron−10 and boron-11 nuclei from the Heavy Ion Linear Accelerator (HILAC).[21] The nuclide $^{257}103$ was originally reported, but then this was reassigned to $^{258}103$. The team at the University of California suggested the name *lawrencium* (after Ernest O. Lawrence, the inventor of cyclotron particle accelerator) and the symbol "Lw",[21] for the new element, but "Lw" was not adopted, and "Lr" was officially accepted instead. Nuclear-physics researchers in Dubna, Soviet Union (now Russia), reported in 1967 that they were not able to confirm American scientists' data on $^{257}103$.[22] Two years earlier, the Dubna team reported $^{256}103$.[23] In 1992, the IUPAC Trans-fermium Working Group officially recognized element 103, confirmed its naming as lawrencium, with symbol "Lr", and named the nuclear physics teams at Dubna and Berkeley as the co-discoverers of lawrencium.[24]

So far, no experiments were conducted to synthesize any element that could be the next group 3 element; if lutetium and lawrencium are considered to be group 3 elements, then the next element in the group should be element 153, unpenttrium (Upt). However, after element 120, filling electronic configurations stops obeying Aufbau principle. According to the principle, unpenttrium should have an electronic configuration of $[Uuo]8s^25g^{18}6f^{14}7d^{1}$ [note 3] and filling the 5g-subshell should be stopped at element 138. However, the 7d-orbitals are calculated to start being filled on element 137, while the 5g-subshell closes only at element 144, *after* filling of 7d-subshell begins. Therefore, it is hard to calculate which element should be the next group 3 element.[25] Calculations suggest that unpentpentium (Upp, element 155) could also be the next group 3 element.[26] If lanthanum and actinium are considered group 3 elements, then element 121, unbiunium (Ubu), should be the fifth group 3 element. The element is calculated have electronic configuration of $[Uuo]8s^28p_{1/2}^{1}$, which is not associated with transition metals, without having a partially filled d-subshell.[25] No experiments have been performed to create unpenttrium, unbiunium or any element that could be considered the next group 3 element; however, unbiunium is the element with the lowest atomic number that has not been tried to be created and thus has chances to be,[27] while unpenttrium, unpentpentium or any other element considered if preceded by lawrencium is very unlikely to be created due to drip instabilities that imply that the periodic table ends soon after the island of stability at unbihexium.[28]

20.2 Characteristics

20.2.1 Chemical

Like other groups, the members of this family show patterns in their electron configurations, especially the outermost shells, resulting in trends in chemical behavior. However, lawrencium is an exception, since its last electron is transferred to the $7p_1/_2$ subshell due to relativistic effects.[29][30]

Most of the chemistry has been observed only for the first three members of the group; chemical properties of both actinium and especially lawrencium are not well-characterized. The remaining elements of the group (scandium, yttrium, lutetium) are reactive metals with high melting points (1541 °C, 1526 °C, 1652 °C respectively). They are usually oxidized to the +3 oxidation state, even through scandium,[31] yttrium[32][33] and lanthanum[34] can form lower oxidation states. The reactivity of the elements, especially yttrium, is not always obvious due to the formation of a stable oxide layer, which prevents further reactions. Scandium(III) oxide, yttrium(III) oxide, lanthanum(III) oxide and lutetium(III) oxide are white high-temperature-melting solids. Yttrium(III) oxide and lutetium(III) oxide exhibit weak basic character, but scandium(III) oxide is amphoteric.[35] Lanthanum(III) oxide is strongly basic.

20.2.2 Physical

Elements that show tripositive ions with electronic configuration of a noble gas (scandium, yttrium, lanthanum, actinium) show a clear trend in their physical properties, such as hardness. At the same time, if group 3 is continued with lutetium and lawrencium, several trends are broken. For example, scandium and yttrium are both soft metals. Lanthanum is soft as well; all these elements have their outermost electrons quite far from the nucleus compared to the nuclei charges. Due to the lanthanide contraction, lutetium, the last in the lanthanide series, has a significantly smaller atomic radius and a higher nucleus charge,[36] thus making the extraction of the electrons from the atom to form metallic bonding more difficult, and thus making the metal harder. However, lutetium suits the previous elements better in several other properties, such as melting[37] and boiling points.[38] Very little is known about lawrencium, and none of its physical properties have been confirmed.

20.3 Group borders

It is disputed whether lutetium and lawrencium should be included in group 3, rather than lanthanum and actinium. Other d-block groups are composed of four transition metals,[note 6] and group 3 is sometimes considered to follow suit. Scandium and yttrium are always classified as group 3 elements, but it is controversial which elements should follow them in group 3, lanthanum and actinium or lutetium and lawrencium. Scerri has proposed a resolution to this debate on the basis of moving to a 32-column table and consideration of which option results in a continuous sequence of atomic number increase. He thereby finds that group 3 should consist of Sc, Y, Lu, Lr.[43] The current IUPAC definition of the term "lanthanoid" includes fifteen elements including both lanthanum and lutetium, and that of "transition element"[44] applies to lanthanum and actinium, as well as lutetium but *not* lawrencium, since it does not correctly follow the Aufbau principle. Normally, the 103rd electron would enter the d-subshell, but quantum mechanical research has found that the configuration is actually $[Rn]7s^25f^{14}7p^1$[note 7] due to relativistic effects.[29][30] IUPAC thus has not recommended a specific format for the in-line-f-block periodic table, leaving the dispute open.

- Lanthanum and actinium are sometimes considered the remaining members of group 3.[45] In their most commonly encountered tripositive ion forms, these elements do not possess any partially filled f-orbitals, thus continuing the scandium—yttrium—lanthanum—actinium trend, in which all the elements have relationship similar to that of elements of the calcium—strontium—barium—radium series, the elements' left neighbors in s-block. However, different behavior is observed in other d-block groups, especially in group 4, in which zirconium, hafnium and rutherfordium share similar chemical properties lacking a clear trend.

- In other tables, lutetium and lawrencium are classified as the remaining members of group 3.[46] In these tables, lutetium and lawrencium end (or sometimes succeed) the lanthanide and actinide series, respectively. Since the f-shell is nominally full in the ground state electron configuration for both of these metals, they behave most similarly to other period 6 and period 7 transition metals compared to the other lanthanides and actinides, and thus logically exhibit properties similar to those of scandium and yttrium.

- Some tables, including the official IUPAC table[47] refer to *all* lanthanides and actinides by a marker in group 3. This sometimes is believed to be the inclusion of all 30 lanthanide and actinide elements as included in group 3. Lanthanides, as electropositive trivalent metals, all have a closely related chemistry, and all show many similarities to scandium and yttrium, but they also show additional properties characteristic of their partially filled f-orbitals which are not common to scandium and yttrium.

- Exclusion of all elements is based on properties of earlier actinides, which show a much wider variety of chemistry (for instance, in range of oxidation states) within their series than the lanthanides, and comparisons to scandium and yttrium are even less useful.[48] However, these elements are destabilized,[49] and if they were stabilized to more closely match chemistry laws, they would be similar to lanthanides as well. Also, the later actinides from californium onwards behave more like the corresponding lanthanides, with only the valence +3 (and sometimes +2) shown.[48]

20.4   Occurrence

Scandium, yttrium, and lutetium tend to occur together with other lanthanides (except promethium) tend to occur together in the Earth's crust, and are often harder to extract from their ores. The abundance of elements in Earth's crust for group 3 is quite low — all the elements in the group are uncommon, the most abundant being yttrium with abundance of approximately 30 parts per million (ppm); the abundance of scandium is 16 ppm, while that of lutetium is about 0.5 ppm. For comparison, the abundance of copper is 50 ppm, that of chromium is 160 ppm, and that of molybdenum is 1.5 ppm.[45]

Scandium is distributed sparsely and occurs in trace amounts in many minerals.[50] Rare minerals from Scandinavia[51] and Madagascar[52] such as gadolinite, euxenite, and thortveitite are the only known concentrated sources of this element, the latter containing up to 45% of scandium in the form of scandium(III) oxide.[51] Yttrium has the same trend in occurrence places; it is found in lunar rock samples collected during the American Apollo Project in a relatively high content as well.[53]

The principal commercially viable ore of lutetium is the rare earth phosphate mineral monazite, $(Ce,La,etc.)PO_4$, which contains 0.003% of the element. The main mining areas are China, United States, Brazil, India, Sri Lanka and Australia. Pure lutetium metal is one of the rarest and most expensive of the rare earth metals with the price about US$10,000/kg, or about one-fourth that of gold.[54][55]

20.5   Production

The most available element in group 3 is yttrium, with annual production of 8,900 tonnes in 2010. Yttrium is mostly produced as oxide, by a single country, China (99%).[56] Lutetium and scandium are also mostly obtained as oxides, and their annual production by 2001 was about 10 and 2 tonnes, respectively.[57]

Group 3 elements are mined only as a byproduct from the extraction of other elements.[58] The metallic elements are extremely rare; the production of metallic yttrium is about a few tonnes, and that of scandium is in the order of 10 kg per year;[58][59] production of lutetium is not calculated, but it is certainly small. The elements, after purification from other rare earth metals, are isolated as oxides; the oxides are converted to fluorides during reactions with hydrofluoric acid.[60] The resulting fluorides are reduced with alkaline earth metals or alloys of the metals; metallic calcium is used most frequently.[60] For example:

$$Sc_2O_3 + 3\,HF \rightarrow 2\,ScF_3 + 3\,H_2O$$

$$2\,ScF_3 + 3\,Ca \rightarrow 3\,CaF_2 + 2\,Sc$$

20.6 Applications

20.7 Biological chemistry

Group 3 elements are generally hard metals with low aqueous solubility, and have low availability to the biosphere. No group 3 element has any documented biological role in living organisms. The radioactivity of the actinides generally makes them highly toxic to living cells, causing radiation poisoning.

Scandium has no biological role, but it is found in living organisms. Once reached a human, scandium concentrates in the liver and is a threat to it; some its compounds are possibly carcinogenic, even through in general scandium is not toxic.[61] Scandium is known to have reached the food chain, but in trace amounts only; a typical human takes in less than 0.1 micrograms per day.[61] Once released into the environment, scandium gradually accumulates in soils, which leads to increased concentrations in soil particles, animals and humans. Scandium is mostly dangerous in the working environment, due to the fact that damps and gases can be inhaled with air. This can cause lung embolisms, especially during long-term exposure. The element is known to damage cell membranes of water animals, causing several negative influences on reproduction and on the functions of the nervous system.[61]

Yttrium has no known biological role, though it is found in most, if not all, organisms and tends to concentrate in the liver, kidney, spleen, lungs, and bones of humans.[62] There is normally as little as 0.5 milligrams found within the entire human body; human breast milk contains 4 ppm.[63] Yttrium can be found in edible plants in concentrations between 20 ppm and 100 ppm (fresh weight), with cabbage having the largest amount.[63] With up to 700 ppm, the seeds of woody plants have the highest known concentrations.[63]

Lutetium has no biological role as well, but it is found even in the highest known organism, the humans, concentrating in bones, and to a lesser extent in the liver and kidneys.[64] Lutetium salts are known to cause metabolism and they occur together with other lanthanide salts in nature; the element is the least abundant in the human body of all lanthanides.[64] Human diets have not been monitored for lutetium content, so it is not known how much the average human takes in, but estimations show the amount is only about several micrograms per year, all coming from tiny amounts taken by plants. Soluble lutetium salts are mildly toxic, but insoluble ones are not.[64]

The high radioactivity of lawrencium would make it highly toxic to living cells, causing radiation poisoning.

20.8 Notes

[1] *Ytterbite* was named after the village it was discovered near, plus the -ite ending to indicate it was a mineral.

[2] Earths were given an -a ending and new elements are normally given an -ium ending.

[3] Unpenttrium, according to calculations, should have an electronic configuration of $[Uuo]8s^2 5g^{18} 6f^{11} 7d^2 8p_{1/2}^2$.[25]

[4] If lanthanum and actinium are included instead, the table ends with the following lines:

[5] If lanthanum and actinium are included instead, the table ends with the following lines (some of the data for actinium is approximate):

[6] However, the group 12 elements are not always considered to be transition metals.

[7] The expected configuration of lawrencium if it did obey the Aufbau principle would be $[Rn]7s^2 5f^{14} 6d^1$, with the normal incomplete 6d-subshell in the neutral state.

20.9 References

[1] van der Krogt, Peter. "39 Yttrium – Elementymology & Elements Multidict". Elements.vanderkrogt.net. Retrieved 2008-08-06.

[2] Emsley 2001, p. 496

[3] Gadolin, Johan (1794). "Undersökning af en svart tung Stenart ifrån Ytterby Stenbrott i Roslagen". *Kongl. Vetenskaps Academiens Nya Handlingar* (in Swedish) **15**: 137–155.

[4] Greenwood, N. N.; Earnshaw, A. (1997). *Chemistry of the Elements* (2nd ed.). Oxford: Butterworth-Heinemann. p. 944. ISBN 0-7506-3365-4.

[5] Coplen, Tyler B.; Peiser, H. S. (1998). "History of the Recommended Atomic-Weight Values from 1882 to 1997: A Comparison of Differences from Current Values to the Estimated Uncertainties of Earlier Values (Technical Report)". *Pure Appl. Chem.* (IUPAC's Inorganic Chemistry Division Commission on Atomic Weights and Isotopic Abundances) **70** (1): 237–257. doi:10.1351/pac199870010237.

[6] Heiserman, David L. (1992). "Element 39: Yttrium". *Exploring Chemical Elements and their Compounds*. New York: TAB Books. pp. 150–152. ISBN 0-8306-3018-X.

[7] Wöhler,Friedrich(1828). "Über das Beryllium und Yttrium".*Annalen der Physik*(in German)**89**(8):577–582.Bibcode:1828W. doi:10.1002/andp.18280890805.

[8] Ball, Philip (2002). *The Ingredients: A Guided Tour of the Elements*. Oxford University Press. pp. 100–102. ISBN 0-19-284100-9.

[9] Nilson, Lars Fredrik (1879). "Sur l'ytterbine, terre nouvelle de M. Marignac". *Comptes Rendus* (in French) **88**: 642–647.

[10] Nilson, Lars Fredrik (1879). "Ueber Scandium, ein neues Erdmetall". *Berichte der deutschen chemischen Gesellschaft* (in German) **12** (1): 554–557. doi:10.1002/cber.187901201157.

[11] Cleve, Per Teodor (1879). "Sur le scandium". *Comptes Rendus* (in French) **89**: 419–422.

[12] Fischer, Werner; Brünger, Karl; Grieneisen, Hans (1937). "Über das metallische Scandium". *Zeitschrift für anorganische und allgemeine Chemie* (in German) **231** (1–2): 54–62. doi:10.1002/zaac.19372310107.

[13] Urbain, M. G. (1908). "Un nouvel élément, le lutécium, résultant du dédoublement de l'ytterbium de Marignac". *Comptes rendus* (in French) **145**: 759–762.

[14] "Separation of Rare Earth Elements by Charles James". *National Historic Chemical Landmarks*. American Chemical Society. Retrieved 2014-02-21.

[15] von Welsbach, Carl Auer (1908). "Die Zerlegung des Ytterbiums in seine Elemente". *Monatshefte für Chemie* (in German) **29** (2): 181–225. doi:10.1007/BF01558944.

[16] Urbain, G. (1909). "Lutetium und Neoytterbium oder Cassiopeium und Aldebaranium – Erwiderung auf den Artikel des Herrn Auer v. Welsbach". *Monatshefte für Chemie* (in German) **31** (10): I. doi:10.1007/BF01530262.

[17] Clarke, F. W.; Ostwald, W.; Thorpe, T. E. and Urbain, G. (1909). "Bericht des Internationalen Atomgewichts-Ausschusses für 1909". *Berichte der deutschen chemischen Gesellschaft* (in German) **42** (1): 11–17. doi:10.1002/cber.19090420104.

[18] van der Krogt, Peter. "70. Ytterbium – Elementymology & Elements Multidict". Elements.vanderkrogt.net. Retrieved 4 July 2011.

[19] van der Krogt, Peter. "71. Lutetium – Elementymology & Elements Multidict". Elements.vanderkrogt.net. Retrieved 4 July 2011.

[20] Emsley, John (2001). *Nature's building blocks: an A-Z guide to the elements*. US: Oxford University Press. pp. 240–242. ISBN 0-19-850341-5.

[21] Ghiorso, Albert; Sikkeland, T.; Larsh, A. E.; Latimer, R. M. (1961). "New Element, Lawrencium, Atomic Number 103". *Phys. Rev. Lett.* **6** (9): 473. Bibcode:1961PhRvL...6..473G. doi:10.1103/PhysRevLett.6.473.

[22] Flerov, G. N. (1967). *At. En.* **106**: 476. Missing or empty |title= (help)

[23] Donets, E. D.; Shchegolev, V.A.; Ermakov, V. A. (1965). *Nucl. Phys.* (in Russian) **19**: 109. Missing or empty |title= (help)

[24] Greenwood, Norman N. (1997). "Recent developments concerning the discovery of elements 101–111". *Pure & Appl. Chem* **69** (1): 179–184. doi:10.1351/pac199769010179.

[25] Hoffman, Darleane C.; Lee, Diana M.; Pershina, Valeria (2006). "Transactinides and the future elements". In Morss; Edelstein, Norman M.; Fuger, Jean. *The Chemistry of the Actinide and Transactinide Elements* (3rd ed.). Dordrecht, The Netherlands: Springer Science+Business Media. ISBN 1-4020-3555-1.

[26] Pyykkö, Pekka (2011). "A suggested periodic table up to Z ≤ 172, based on Dirac–Fock calculations on atoms and ions". *Physical Chemistry Chemical Physics* **13** (1): 161–8. Bibcode:2011PCCP...13..161P. doi:10.1039/c0cp01575j. PMID 20967377.

[27] van der Krogt, Peter. "Elementymology & Elements Multidict". Elements.vanderkrogt.net. Retrieved 4 July 2011.

[28] Seaborg, G. T. (c. 2006). "transuranium element (chemical element)". Encyclopædia Britannica. Retrieved 2010-03-16.

[29] Eliav, E.; Kaldor, U.; Ishikawa, Y. (1995). "Transition energies of ytterbium, lutetium, and lawrencium by the relativistic coupled-cluster method". *Phys. Rev. A* **52**: 291–296. Bibcode:1995PhRvA..52..291E. doi:10.1103/PhysRevA.52.291.

[30] Zou, Yu; Froese, Fischer C. (2002). "Resonance Transition Energies and Oscillator Strengths in Lutetium and Lawrencium". *Phys. Rev. Lett.* **88** (18): 183001. Bibcode:2002PhRvL..88b3001M. doi:10.1103/PhysRevLett.88.023001. PMID 12005680.

[31] Corbett, J.D. (1981). "Extended metal-metal bonding in halides of the early transition metals". *Acc. Chem. Res.* **14** (8): 239–246. doi:10.1021/ar00068a003.

[32] Nikolai B., Mikheev; Auerman, L N; Rumer, Igor A; Kamenskaya, Alla N; Kazakevich, M Z (1992). "The anomalous stabilisation of the oxidation state 2+ of lanthanides and actinides". *Russian Chemical Reviews* **61** (10): 990–998. Bibcode:199 M.doi:10.1070/RC1992v061n10ABEH001011.

[33] Kang, Weekyung; Bernstein, E. R. (2005). "Formation of Yttrium Oxide Clusters Using Pulsed Laser Vaporization". *Bull. Korean Chem. Soc.* **26** (2): 345–348. doi:10.5012/bkcs.2005.26.2.345.

[34] Patnaik, Pradyot (2003). *Handbook of Inorganic Chemical Compounds*. McGraw-Hill. pp. 444–446. ISBN 0-07-049439-8. Retrieved 2009-06-06.

[35] Cotton, S. A. (1994). "Scandium, Yttrium and the Lanthanides: Inorganic and Coordination Chemistry". *Encyclopedia of Inorganic Chemistry*. John Wiley & Sons. ISBN 0-471-93620-0.

[36] Dean, John A. (1999). *Lange's handbook of chemistry (Fifteenth edition)*. McGraw-Hill, Inc. pp. 589–592. ISBN 0-07-016190-9.

[37] Barbalace, Kenneth. "Periodic Table of Elements Sorted by Melting Point". Environmental Chemistry.com. Retrieved 2011-05-18.

[38] Barbalace, Kenneth. "Periodic Table of Elements Sorted by Boiling Point". Environmental Chemistry.com. Retrieved 2011-05-18.

[39] Lide, D. R., ed. (2003). *CRC Handbook of Chemistry and Physics* (84th ed.). Boca Raton, FL: CRC Press.

[40] Barbalace, Kenneth. "Scandium". Chemical Book. Retrieved 2011-05-18.

[41] Barbalace, Kenneth. "Yttrium". Chemical Book. Retrieved 2011-05-18.

[42] Barbalace, Kenneth. "Lutetium". Chemical Book. Retrieved 2011-05-18.

[43] Scerri, Eric (2012). "Mendeleev's Periodic Table Is Finally Completed and What To Do about Group ?". *Chem. Int.* **34** (4): 28–31. doi:10.1515/ci.2012.34.4.28.

[44] IUPAC, *Compendium of Chemical Terminology*, 2nd ed. (the "Gold Book") (1997). Online corrected version: (2006–) "transition element".

[45] Barbalace, Kenneth. "Periodic Table of Elements". Environmental Chemistry.com. Retrieved 2007-04-14.

[46] "WebElements Periodic Table of the Elements". Webelements.com. Retrieved 2010-04-03.

[47] "Periodic Table of the Elements". International Union of Pure and Applied Chemistry. Retrieved 2010-04-03.

[48] "Visual Elements". Royal Society of Chemistry. Retrieved 4 July 2011.

[49] Dolg, Michael. "Lanthanides and Actinides" (PDF). *Max-Planck-Institut für Physik komplexer Systeme, Dresden, Germany*. CLA01. Retrieved 4 July 2011.

[50] Bernhard, F. (2001). "Scandium mineralization associated with hydrothermal lazurite-quartz veins in the Lower Austroalpie Grobgneis complex, East Alps, Austria". *Mineral Deposits in the Beginning of the 21st Century*. Lisse: Balkema. ISBN 90-265-1846-3.

[51] Kristiansen, Roy (2003). "Scandium – Mineraler I Norge" (PDF). *Stein* (in Norwegian): 14–23.

[52] von Knorring,O.;Condliffe,E. (1987). "Mineralized pegmatites in Africa".*Geological Journal*22:253.doi:10.1002/gj.33502206.

[53] Stwertka, Albert (1998). "Yttrium". *Guide to the Elements* (Revised ed.). Oxford University Press. pp. 115–116. ISBN 0-19-508083-1.

[54] Hedrick, James B. "Rare-Earth Metals" (PDF). USGS. Retrieved 2009-06-06.

[55] Castor, Stephen B. and Hedrick, James B. "Rare Earth Elements" (PDF). Retrieved 2009-06-06.

[56] "Mineral Commodity Summaries 2010: Yttrium" (PDF). United States Geological Survey. Retrieved 2011-07-07.

[57] Emsley 2001, p. 241

[58] Deschamps, Y. "Scandium" (PDF). mineralinfo.com. Retrieved 2008-10-21.

[59] "Mineral Commodity Summaries 2010: Scandium" (PDF). United States Geological Survey. Retrieved 2011-07-07.

[60] Holleman, Arnold F.; Wiberg, Egon; Wiberg, Nils (1985). *Lehrbuch der Anorganischen Chemie* (in German) (91–100 ed.). Walter de Gruyter. pp. 1056–1057. ISBN 3-11-007511-3.

[61] Lenntech (1998). "Scandium (Sc) — chemical properties of scandium, health effects of scandium, environmental effects of scandium". Lenntech. Retrieved 2011-05-21.

[62] MacDonald, N. S.; Nusbaum, R. E.; Alexander, G. V. (1952). "The Skeletal Deposition of Yttrium" (PDF). *Journal of Biological Chemistry* **195** (2): 837–841. PMID 14946195.

[63] Emsley 2001, pp. 495–498

[64] Emsley 2001, p. 240

20.10   Bibliography

- Emsley, John (2001). *Nature's building blocks: an A-Z guide to the elements*. US: Oxford University Press. ISBN 0-19-850341-5.

20.11   Links to related articles

Lawrencium, the only synthetic element in the group, was named after American physicist Ernest Lawrence, the inventor of the cyclotron atom-smasher and founder of discovery place, then-called Lawrence Radiation Laboratory (now Lawrence Berkeley National Laboratory)

Monazite, the most important lutetium ore

Chapter 21

Group 11 element

Group 11, by modern IUPAC numbering,[1] is a group of chemical elements in the periodic table, consisting of copper (Cu), silver (Ag), and gold (Au). Roentgenium (Rg) is also placed in this group in the periodic table, although no chemical experiments have yet been carried out to confirm that it behaves like the heavier homologue to gold. Group 11 is also known as the *coinage metals*, due to their former usage. They were most likely the first three elements discovered.[2] Copper, silver, and gold all occur naturally in elemental form.

21.1 History

All the elements of the group except roentgenium have been known since prehistoric times, as all of them occur in metallic form in nature and no extraction metallurgy is necessary to produce them.

21.2 Characteristics

Like other groups, the members of this family show patterns in electron configuration, especially in the outermost shells, resulting in trends in chemical behavior, although roentgenium is probably an exception:

All Group 11 elements are relatively inert, corrosion-resistant metals. Copper and gold are colored.

These elements have low electrical resistivity so they are used for wiring. Copper is the cheapest and most widely used. Bond wires for integrated circuits are usually gold. Silver and silver plated copper wiring are found in some special applications.

21.3 Occurrence

Copper occurs in its native form in Chile, China, Mexico, Russia and the USA. Various natural ores of copper are: copper pyrites ($CuFeS_2$), cuprite or ruby copper (Cu_2O), copper glance (Cu_2S), malachite, ($Cu(OH)_2CuCO_3$), and azurite ($Cu(OH)_22CuCO_3$).

Copper pyrite is the principal ore, and yields nearly 76% of the world production of copper.

21.4 Production

Silver is found in native form, as an alloy with gold (electrum), and in ores containing sulfur, arsenic, antimony or chlorine. Ores include argentite (Ag_2S), chlorargyrite (AgCl) which includes horn silver, and pyrargyrite (Ag_3SbS_3). Silver is extracted using the Parkes process.

21.5   Applications

These metals, especially silver, have unusual properties that make them essential for industrial applications outside of their monetary or decorative value. They are all excellent conductors of electricity. The most conductive of all metals are silver, copper and gold in that order. Silver is also the most thermally conductive element, and the most light reflecting element. Silver also has the unusual property that the tarnish that forms on silver is still highly electrically conductive.

Copper is used extensively in electrical wiring and circuitry. Gold contacts are sometimes found in precision equipment for their ability to remain corrosion-free. Silver is used widely in mission-critical applications as electrical contacts, and is also used in photography (because silver nitrate reverts to metal on exposure to light), agriculture, medicine, audiophile and scientific applications.

Gold, silver, and copper are quite soft metals and so are easily damaged in daily use as coins. Precious metal may also be easily abraded and worn away through use. In their numismatic functions these metals must be alloyed with other metals to afford coins greater durability. The alloying with other metals makes the resulting coins harder, less likely to become deformed and more resistant to wear.

Gold coins: Gold coins are typically produced as either 90% gold (e.g. with pre-1933 US coins), or 22 carat (91.66%) gold (e.g. current collectible coins and Krugerrands), with copper and silver making up the remaining weight in each case. Bullion gold coins are being produced with up to 99.999% gold (in the Canadian Gold Maple Leaf series).

Silver coins: Silver coins are typically produced as either 90% silver – in the case of pre 1965 US minted coins (which were circulated in many countries), or sterling silver (92.5%) coins for pre-1920 British Commonwealth and other silver coinage, with copper making up the remaining weight in each case. Old European coins were commonly produced with 83.5% silver. Modern silver bullion coins are often produced with purity varying between 99.9% to 99.999%.

Copper coins: Copper coins are often of quite high purity, around 97%, and are usually alloyed with small amounts of zinc and tin.

Inflation has caused the face value of coins to fall below the hard currency value of the historically used metals. This had led to most modern coins being made of base metals – copper nickel (around 80:20, silver in color) is popular as are nickel-brass (copper (75), nickel (5) and zinc (20), gold in color), manganese-brass (copper, zinc, manganese, and nickel), bronze, or simple plated steel.

21.6   Biological role and toxicity

Copper, although potentially toxic in excessive amounts, is essential for life. Copper is shown to have antimicrobial properties which make it useful for hospital doorknobs to keep diseases from being spread. Eating food in copper containers is known to increase the risk of copper toxicity.

Elemental gold and silver have no known toxic effects or biological use, although gold salts can be toxic to liver and kidney tissue.[3][4] Like copper, silver also has antimicrobial properties. The prolonged use of preparations containing gold or silver can also lead to the accumulation of these metals in body tissue; the results are the irreversible but apparently harmless pigmentation conditions known as chrysiasis and argyria respectively.

Due to being short lived and radioactive, roentgenium has no biological use but it is likely extremely harmful due to its radioactivity.

21.7   References

[1]Fluck,E. (1988).“New Notations in the Periodic Table”(PDF).*Pure Appl.Chem.*(IUPAC)**60**(3):431–436.doi:10.1351/pac.
      Retrieved 24 March 2012.

[2] Greenwood and Earnshaw, p. 1173

[3] Wright, I. H.; Vesey, C. J. (1986). "Acute poisoning with gold cyanide". *Anaesthesia* **41** (79): 936–939. doi:10.1111/j.1365-2044.1986.tb12920.x. PMID 3022615.

[4] Wu, Ming-Ling; Tsai, Wei-Jen; Ger, Jiin; Deng, Jou-Fang; Tsay, Shyh-Haw; Yang, Mo-Hsiung. (2001). "Cholestatic Hepatitis Caused by Acute Gold Potassium Cyanide Poisoning". *Clinical toxicology* **39** (7): 739–743. doi:10.1081/CLT-100108516. PMID 11778673.

21.8 Bibliography

• Greenwood, Norman N.; Earnshaw, Alan (1997). *Chemistry of the Elements* (2nd ed.). Butterworth-Heinemann. ISBN 0080379419.

21.9 See also

Chapter 22

Metalloid

A **metalloid** is a chemical element with properties in between, or that are a mixture of, those of metals and nonmetals. There is no standard definition of a metalloid, nor is there complete agreement as to which elements are appropriately classified as such. Despite this lack of specificity, the term remains in use in the literature of chemistry.

The six commonly recognised metalloids are boron, silicon, germanium, arsenic, antimony, and tellurium. Elements less commonly recognised as metalloids include carbon, aluminium, selenium, polonium, and astatine. On a standard periodic table all of these elements may be found in a diagonal region of the p-block, extending from boron at one end, to astatine at the other. Some periodic tables include a dividing line between metals and nonmetals and the metalloids may be found close to this line.

Typical metalloids have a metallic appearance, but they are brittle and only fair conductors of electricity. Chemically, they mostly behave as nonmetals. They can form alloys with metals. Most of their other physical and chemical properties are intermediate in nature. Metalloids are usually too brittle to have any structural uses. They and their compounds are used in alloys, biological agents, catalysts, flame retardants, glasses, optical storage and optoelectronics, pyrotechnics, semiconductors, and electronics.

The electrical properties of silicon and germanium enabled the establishment of the semiconductor industry in the 1950s and the development of solid-state electronics from the early 1960s.[1]

The term *metalloid* originally referred to nonmetals. Its more recent meaning, as a category of elements with intermediate or hybrid properties, became widespread in 1940–1960. Metalloids sometimes are called semimetals, a practice that has been discouraged,[2] as the term *semimetal* has a different meaning in physics than in chemistry. In physics it more specifically refers to the electronic band structure of a substance.

22.1   Definitions

See also: List of metalloid lists

22.1.1   Judgement-based

A metalloid is an element with properties in between, or that are a mixture of, those of metals and nonmetals, and which is therefore hard to classify as either a metal or a nonmetal. This is a generic definition that draws on metalloid attributes consistently cited in the literature.[n 2] Difficulty of categorisation is a key attribute. Most elements have a mixture of metallic and nonmetallic properties,[9] and can be classified according to which set of properties is more pronounced.[10][n 3] Only the elements at or near the margins, lacking a sufficiently clear preponderance of either metallic or nonmetallic properties, are classified as metalloids.[14]

Boron, silicon, germanium, arsenic, antimony, and tellurium are recognised commonly as metalloids.[15][n 4] Depending on the author, one or more from selenium, polonium, or astatine sometimes are added to the list.[17] Boron sometimes is excluded, by itself, or with silicon.[18] Sometimes tellurium is not regarded as a metalloid.[19] The inclusion of antimony, polonium, and astatine as metalloids also has been questioned.[20]

Other elements occasionally are classified as metalloids. These elements include,[21] hydrogen,[22] beryllium,[23] nitrogen, phosphorus,[25] sulfur,[26] zinc,[27] gallium,[28] tin, iodine,[29] lead,[30] bismuth,[19] and radon.[31] The term metalloid also has been used for elements that exhibit metallic lustre and electrical conductivity, and that are amphoteric, such as arsenic, antimony, vanadium, chromium, molybdenum, tungsten, tin, lead, and aluminium.[32] The p-block metals,[33] and non-metals (such as carbon or nitrogen) that can form alloys with metals[34] or modify their properties[35] also have occasionally been considered as metalloids.

22.1.2 Criteria-based

No widely accepted definition of a metalloid exists, nor any division of the periodic table into metals, metalloids and nonmetals;[38] Hawkes[39] questioned the feasibility of establishing a specific definition, noting that anomalies can be found in several attempted constructs. Classifying an element as a metalloid has been described by Sharp[40] as "arbitrary".

The number and identities of metalloids depend on what classification criteria are used. Emsley[41] recognised four metalloids (germanium, arsenic, antimony and tellurium); James et al.[42] listed twelve (Emsley's plus boron, carbon, silicon, selenium, bismuth, polonium, ununpentium and livermorium). On average, seven elements are included in such lists; individual classification arrangements tend to share common ground and vary in the ill-defined[43] margins.[n 5][n 6]

A single quantitative criterion such as electronegativity is commonly used,[46] metalloids having electronegativity values from 1.8 or 1.9 to 2.2.[47] Further examples include packing efficiency (the fraction of volume in a crystal structure occupied by atoms) and the Goldhammer-Herzfeld criterion ratio.[48] The commonly recognised metalloids have packing efficiencies of between 34% and 41%.[n 7] The Goldhammer-Herzfeld ratio, roughly equal to the cube of the atomic radius divided by the molar volume,[56][n 8] is a simple measure of how metallic an element is, the recognised metalloids having ratios from around 0.85 to 1.1 and averaging 1.0.[58][n 9] Other authors have relied on, for example, atomic conductance[n 10][62] or bulk coordination number.[63]

Jones, writing on the role of classification in science, observed that "[classes] are usually defined by more than two attributes".[64] Masterton and Slowinski[65] used three criteria to describe the six elements commonly recognised as metalloids: metalloids have ionization energies around 200 kcal/mol (837 kJ/mol) and electronegativity values close to 2.0. They also said that metalloids are typically semiconductors, though antimony and arsenic (semimetals from a physics perspective) have electrical conductivities approaching those of metals. Selenium and polonium are suspected as not in this scheme, while astatine's status is uncertain.[n 11]

22.2 Periodic table territory

22.2.1 Location

Metalloids lie on either side of the dividing line between metals and nonmetals. This can be found, in varying configurations, on some periodic tables. Elements to the lower left of the line generally display increasing metallic behaviour; elements to the upper right display increasing nonmetallic behaviour.[68] When presented as a regular stairstep, elements with the highest critical temperature for their groups (Li, Be, Al, Ge, Sb, Po) lie just below the line.[69]

The diagonal positioning of the metalloids represents an exception to the observation that elements with similar properties tend to occur in vertical groups.[70] A related effect can be seen in other diagonal similarities between some elements and their lower right neighbours, specifically lithium-magnesium, beryllium-aluminium, and boron-silicon. Rayner-Canham[71] has argued that these similarities extend to carbon-phosphorus, nitrogen-sulfur, and into three d-block series.

This exception arises due to competing horizontal and vertical trends in the nuclear charge. Going along a period, the nuclear charge increases with atomic number as do the number of electrons. The additional pull on outer electrons as nuclear charge increases generally outweighs the screening effect of having more electrons. With some irregularities, atoms therefore become smaller, ionization energy increases, and there is a gradual change in character, across a period, from strongly metallic, to weakly metallic, to weakly nonmetallic, to strongly nonmetallic elements.[72] Going down a main group, the effect of increasing nuclear charge is generally outweighed by the effect of additional electrons being further away from the nucleus. Atoms generally become larger, ionization energy falls, and metallic character increases.[73] The net effect is that the location of the metal–nonmetal transition zone shifts to the right in going down a group,[70] and analogous diagonal similarities are seen elsewhere in the periodic table, as noted.[74]

22.2.2   Alternative treatments

Depictions of metalloids vary according to the author. Some do not classify elements bordering the metal–nonmetal dividing line as metalloids, noting that a binary classification can facilitate the establishment of rules for determining bond types between metals and nonmetals.[75] Metalloids are variously grouped with metals,[76] regarded as nonmetals[77] or treated as a sub-category of nonmetals.[78][n 12] Other authors have suggested that classifying some elements as metalloids "emphasizes that properties change gradually rather than abruptly as one moves across or down the periodic table".[80] Some periodic tables distinguish elements that are metalloids and display no formal dividing line between metals and nonmetals. Metalloids are shown as occurring in a diagonal band[81] or diffuse region.[82]

22.3   Properties

Metalloids usually look like metals but behave largely like nonmetals. Physically, they are shiny, brittle solids with intermediate to relatively good electrical conductivity and the electronic band structure of a semimetal or semiconductor. Chemically, they mostly behave as (weak) nonmetals, have intermediate ionization energies and electronegativity values, and amphoteric or weakly acidic oxides. They can form alloys with metals. Most of their other physical and chemical properties are intermediate in nature.

22.3.1   Compared to metals and nonmetals

Main article: Properties of metals, metalloids and nonmetals

Characteristic properties of metals, metalloids and nonmetals are summarized in the table.[83] Physical properties are listed in order of ease of determination; chemical properties run from general to specific, and then to descriptive.

The above table reflects the hybrid nature of metalloids. The properties of *form, appearance,* and *behaviour when mixed with metals* are more like metals. *Elasticity* and *general chemical behaviour* are more like nonmetals. *Electrical conductivity, band structure, ionization energy, electronegativity,* and *oxides* are intermediate between the two.

22.4   Common applications

> *The focus of this section is on the recognised metalloids. Elements less often recognised as metalloids are ordinarily classified as either metals or nonmetals; some of these are included here for comparative purposes.*

Metalloids are too brittle to have any structural uses in their pure forms.[104] They and their compounds are used as (or in) alloying components, biological agents (toxicological, nutritional and medicinal), catalysts, flame retardants, glasses (oxide and metallic), optical storage media and optoelectronics, pyrotechnics, semiconductors and electronics.[n 18]

22.4.1   Alloys

Writing early in the history of intermetallic compounds, the British metallurgist Cecil Desch observed that "certain non-metallic elements are capable of forming compounds of distinctly metallic character with metals, and these elements may

Copper-germanium alloy pellets, likely ~84% Cu; 16% Ge.[106] *When combined with silver the result is a tarnish resistant sterling silver.*
Also shown are two silver pellets.

therefore enter into the composition of alloys". He associated silicon, arsenic and tellurium, in particular, with the alloy-forming elements.[107] Phillips and Williams[108] suggested that compounds of silicon, germanium, arsenic and antimony with B metals, "are probably best classed as alloys".

Among the lighter metalloids, alloys with transition metals are well-represented. Boron can form intermetallic compounds and alloys with such metals of the composition MnB, if $n > 2$.[109] Ferroboron (15% boron) is used to introduce boron into steel; nickel-boron alloys are ingredients in welding alloys and case hardening compositions for the engineering industry. Alloys of silicon with iron and with aluminium are widely used by the steel and automotive industries, respectively. Germanium forms many alloys, most importantly with the coinage metals.[110]

The heavier metalloids continue the theme. Arsenic can form alloys with metals, including platinum and copper;[111] it is also added to copper and its alloys to improve corrosion resistance[112] and appears to confer the same benefit when added to magnesium.[113] Antimony is well known as an alloy-former, including with the coinage metals. Its alloys include pewter (a tin alloy with up to 20% antimony) and type metal (a lead alloy with up to 25% antimony).[114] Tellurium readily alloys with iron, as ferrotellurium (50–58% tellurium), and with copper, in the form of copper tellurium (40–50% tellurium).[115] Ferrotellurium is used as a stabilizer for carbon in steel casting.[116] Of the non-metallic elements less often recognised as metalloids, selenium—in the form of ferroselenium (50–58% selenium)—is used to improve the machinability of stainless steels.[117]

22.4.2 Biological agents

All six of the elements commonly recognised as metalloids have toxic, dietary or medicinal properties.[119] Arsenic and antimony compounds are especially toxic; boron, silicon, and possibly arsenic, are essential trace elements. Boron, silicon, arsenic and antimony have medical applications, and germanium and tellurium are thought to have potential.

Arsenic trioxide or white arsenic, *one of the most toxic and prevalent forms of arsenic. The antileukaemic properties of white arsenic were first reported in 1878.*[118]

Boron is used in insecticides[120] and herbicides.[121] It is an essential trace element.[122] As boric acid, it has antiseptic, antifungal, and antiviral properties.[123]

Silicon is present in silatrane, a highly toxic rodenticide.[124] Long-term inhalation of silica dust causes silicosis, a fatal disease of the lungs. Silicon is an essential trace element.[122] Silicone gel can be applied to badly burned patients to reduce scarring.[125]

Salts of germanium are potentially harmful to humans and animals if ingested on a prolonged basis.[126] There is interest in the pharmacological actions of germanium compounds but no licensed medicine as yet.[127]

Arsenic is notoriously poisonous and may also be an essential element in ultratrace amounts.[128] It has been used as a pharmaceutical agent since antiquity, including for the treatment of syphilis before the development of antibiotics.[129] Arsenic is also a component of melarsoprol, a medicinal drug used in the treatment of human African trypanosomiasis or sleeping sickness. In 2003, arsenic trioxide (under the trade name Trisenox) was re-introduced for the treatment of acute promyelocytic leukaemia, a cancer of the blood and bone marrow.[129] Arsenic in drinking water, which causes lung and bladder cancer, has been associated with a reduction in breast cancer mortality rates.[130]

Metallic antimony is relatively non-toxic, but most antimony compounds are poisonous.[131] Two antimony compounds, sodium stibogluconate and stibophen, are used as antiparasitical drugs.[132]

Elemental tellurium is not considered particularly toxic; two grams of sodium tellurate, if administered, can be lethal.[133] People exposed to small amounts of airborne tellurium exude a foul and persistent garlic-like odour.[134] Tellurium dioxide has been used to treat seborrhoeic dermatitis; other tellurium compounds were used as antimicrobial agents before the development of antibiotics.[135] In the future, such compounds may need to be substituted for antibiotics that have become ineffective due to bacterial resistance.[136]

Of the elements less often recognised as metalloids, beryllium and lead are noted for their toxicity; lead arsenate has been extensively used as an insecticide.[137] Sulfur is one of the oldest of the fungicides and pesticides. Phosphorus, sulfur, zinc, selenium and iodine are essential nutrients, and aluminium, tin and lead may be.[128] Sulfur, gallium, selenium, iodine and bismuth have medicinal applications. Sulfur is a constituent of sulfonamide drugs, still widely used for conditions such as acne and urinary tract infections.[138] Gallium nitrate is used to treat the side effects of cancer;[139] gallium citrate, a radiopharmaceutical, facilitates imaging of inflamed body areas.[140] Selenium sulfide is used in medicinal shampoos and to treat skin infections such as tinea versicolor.[141] Iodine is used as a disinfectant in various forms. Bismuth is an ingredient in some antibacterials.[142]

22.4.3 Catalysts

Boron trifluoride and trichloride are used as catalysts in organic synthesis and electronics; the tribromide is used in the manufacture of diborane.[143] Non-toxic boron ligands can replace toxic phosphorus ligands in transition metal catalysts.[144] Silica sulfuric acid (SiO_2OSO_3H) is used in organic reactions.[145] Germanium dioxide is sometimes used as a catalyst in the production of PET plastic for containers;[146] cheaper antimony compounds, such as the trioxide or triacetate, are more commonly employed for the same purpose[147] despite concerns about antimony contamination of food and drinks.[148] Arsenic trioxide has been used in the production of natural gas, to boost the removal of carbon dioxide, as have selenous acid and tellurous acid.[149] Selenium acts as a catalyst in some microorganisms.[150] Tellurium, and its dioxide and tetrachloride, are strong catalysts for air oxidation of carbon above 500 °C.[151] Graphite oxide can be used as a catalyst in the synthesis of imines and their derivatives.[152] Activated carbon and alumina have been used as catalysts for the removal of sulfur contaminants from natural gas.[153] Titanium doped aluminium has been identified as a substitute for expensive noble metal catalysts used in the production of industrial chemicals.[154]

22.4.4 Flame retardants

Compounds of boron, silicon, arsenic and antimony have been used as flame retardants. Boron, in the form of borax, has been used as a textile flame retardant since at least the 18th century.[155] Silicon compounds such as silicones, silanes, silsesquioxane, silica and silicates, some of which were developed as alternatives to more toxic halogenated products, can considerably improve the flame retardancy of plastic materials.[156] Arsenic compounds such as sodium arsenite or sodium arsenate are effective flame retardants for wood but have been less frequently used due to their toxicity.[157] Antimony trioxide is a flame retardant.[158] Aluminium hydroxide has been used as a wood-fibre, rubber, plastic and textile flame retardant since the 1890s.[159] Apart from aluminium hydroxide, use of phosphorus based flame-retardants—in the form of, for example, organophosphates—now exceeds that of any of the other main retardant types. These employ boron, antimony or halogenated hydrocarbon compounds.[160]

22.4.5 Glass formation

The oxides B_2O_3, SiO_2, GeO_2, As_2O_3 and Sb_2O_3 readily form glasses. TeO_2 forms a glass but this requires a "heroic quench rate"[161] or the addition of an impurity, otherwise the crystalline form results.[161] These compounds are used in chemical, domestic and industrial glassware[162] and optics.[163] Boron trioxide is used as a glass fibre additive,[164] and is also a component of borosilicate glass, widely used for laboratory glassware and domestic ovenware for its low thermal expansion.[165] Most ordinary glassware is made from silicon dioxide.[166] Germanium dioxide is used as a glass fibre additive, as well as in infrared optical systems.[167] Arsenic trioxide is used in the glass industry as a decolourizing and fining agent (for the removal of bubbles),[168] as is antimony trioxide.[169] Tellurium dioxide finds application in laser and nonlinear optics.[170]

Amorphous metallic glasses are generally most easily prepared if one of the components is a metalloid or "near metalloid" such as boron, carbon, silicon, phosphorus or germanium.[171][n 19] Aside from thin films deposited at very low temperatures, the first known metallic glass was an alloy of composition $Au_{75}Si_{25}$ reported in 1960.[173] A metallic glass having a strength and toughness not previously seen, of composition $Pd_{82.5}P_6Si_{9.5}Ge_2$, was reported in 2011.[174]

Optical fibres, usually made of pure silicon dioxide glass, with additives such as boron trioxide or germanium dioxide for increased sensitivity

Phosphorus, selenium and lead, which are less often recognised as metalloids, are also used in glasses. Phosphate glass has a substrate of phosphorus pentoxide (P_2O_5), rather than the silica (SiO_2) of conventional silicate glasses. It is used, for example, to make sodium lamps.[175] Selenium compounds can be used both as decolourising agents and to add a red colour to glass.[176] Decorative glassware made of traditional lead glass contains at least 30% lead(II) oxide (PbO); lead glass used for radiation shielding may have up to 65% PbO.[177] Lead-based glasses have also been extensively used in electronics components; enamelling; sealing and glazing materials; and solar cells. Bismuth based oxide glasses have emerged as a less toxic replacement for lead in many of these applications.[178]

22.4.6   Optical storage and optoelectronics

Varying compositions of GeSbTe ("GST alloys") and Ag- and In- doped Sb_2Te ("AIST alloys"), being examples of phase-change materials, are widely used in rewritable optical discs and phase-change memory devices. By applying heat, they can be switched between amorphous (glassy) and crystalline states. The change in optical and electrical properties can be used for information storage purposes.[179] Future applications for GeSbTe may include, "ultrafast, entirely solid-state displays with nanometre-scale pixels, semi-transparent "smart" glasses, "smart" contact lenses and artificial retina devices."[180]

22.4.7   Pyrotechnics

The recognised metalloids have either pyrotechnic applications or associated properties. Boron and silicon are commonly encountered;[182] they act somewhat like metal fuels.[183] Boron is used in pyrotechnic initiator compositions (for igniting other hard-to-start compositions), and in delay compositions that burn at a constant rate.[184] Boron carbide has been identified as a possible replacement for more toxic barium or hexachloroethane mixtures in smoke munitions, signal flares and fireworks.[185] Silicon, like boron, is a component of initiator and delay mixtures.[184] Doped germanium can act

Archaic blue light signal, fuelled by a mixture of sodium nitrate, sulfur and (red) arsenic trisulfide[181]

as a variable speed thermite fuel.[n 20] Arsenic trisulfide As$_2$S$_3$ was used in old naval signal lights; in fireworks to make white stars;[187] in yellow smoke screen mixtures; and in initiator compositions.[188] Antimony trisulfide Sb$_2$S$_3$ is found in white-light fireworks and in flash and sound mixtures.[189] Tellurium has been used in delay mixtures and in blasting cap initiator compositions.[190]

Carbon, aluminium, phosphorus and selenium continue the theme. Carbon, in black powder, is a constituent of fireworks rocket propellants, bursting charges, and effects mixtures, and military delay fuses and igniters.[191][n 21] Aluminium is a common pyrotechnic ingredient,[182] and is widely employed for its capacity to generate light and heat,[193] including in thermite mixtures.[194] Phosphorus can be found in smoke and incendiary munitions, paper caps used in toy guns, and party poppers.[195] Selenium has been used in the same way as tellurium.[190]

22.4.8   Semiconductors and electronics

Semiconductor-based electronic components. From left to right: a transistor, an integrated circuit and an LED. The elements commonly recognised as metalloids find widespread use in such devices, as elemental or compound semiconductor constituents (Si, Ge or GaAs, for example) or as doping agents (B, Sb, Te, for example).

All the elements commonly recognised as metalloids (or their compounds) have been used in the semiconductor or solid-state electronic industries.[196]

Some properties of boron have limited its use as a semiconductor. It has a high melting point, single crystals are relatively hard to obtain, and introducing and retaining controlled impurities is difficult.[197]

Silicon is the leading commercial semiconductor; it forms the basis of modern electronics (including standard solar cells)[198] and information and communication technologies.[199] This was despite the study of semiconductors, early in the 20th century, having been regarded as the "physics of dirt" and not deserving of close attention.[200]

Germanium has largely been replaced by silicon in semiconducting devices, being cheaper, more resilient at higher operating temperatures, and easier to work during the microelectronic fabrication process.[106] Germanium is still a constituent of semiconducting silicon-germanium "alloys" and these have been growing in use, particularly for wireless communication devices; such alloys exploit the higher carrier mobility of germanium.[106] The synthesis of gram-scale quantities of semiconducting germanane was reported in 2013. This comprises one-atom thick sheets of hydrogen-terminated germanium atoms, analogous to graphane. It conducts electrons more than ten times faster than silicon and five times faster than germanium, and is thought to have potential for optoelectronic and sensing applications.[201] The development of a germanium-wire based anode that more than doubles the capacity of lithium-ion batteries was reported in 2014.[202] In the same year, Lee at al. reported that defect-free crystals of graphene large enough to have electronic uses could be grown on, and removed from, a germanium substrate.[203]

Arsenic and antimony are not semiconductors in their standard states. Both form type III-V semiconductors (such as GaAs, AlSb or GaInAsSb) in which the average number of valence electrons per atom is the same as that of Group 14 elements. These compounds are preferred for some special applications.[204] Antimony nanocrystals may enable lithium-ion batteries to be replaced by more powerful sodium ion batteries.[205]

Tellurium, which is a semiconductor in its standard state, is used mainly as a component in type II/VI semiconducting-chalcogenides; these have applications in electro-optics and electronics.[206] Cadmium telluride (CdTe) is used in solar modules for its high conversion efficiency, low manufacturing costs, and large band gap of 1.44 eV, letting it absorb a wide range of wavelengths.[198] Bismuth telluride (Bi_2Te_3), alloyed with selenium and antimony, is a component of thermoelectric devices used for refrigeration or portable power generation.[207]

Five metalloids—boron, silicon, germanium, arsenic and antimony—can be found in cell phones (along with at least 39 other metals and nonmetals).[208] Tellurium is expected to find such use.[209] Of the less often recognised metalloids, phosphorus, gallium (in particular) and selenium have semiconductor applications. Phosphorus is used in trace amounts as a dopant for n-type semiconductors.[210] The commercial use of gallium compounds is dominated by semiconductor applications—in integrated circuits; cell phones; laser diodes; light-emitting diodes; photodetectors; and solar cells.[211] Selenium is used in the production of solar cells[212] and in high-energy surge protectors.[213]

Boron, silicon, germanium, antimony and tellurium,[214] as well as heavier metals and metalloids such as Sm, Hg, Tl, Pb, Bi and Se,[215] can be found in topological insulators. These are alloys[216] or compounds which, at ultracold temperatures or room temperature (depending on their composition), are metallic conductors on their surfaces but insulators through their interiors.[217] Cadmium arsenide Cd_3As_2, at about 1 K, is a Dirac-semimetal—a bulk electronic analogue of graphene—in which electrons travel effectively as massless particles.[218] These two classes of material are thought to have potential quantum computing applications.[219]

22.5 Nomenclature and history

22.5.1 Derivation and other names

The word metalloid comes from the Latin *metallum* ("metal") and the Greek *oeides* ("resembling in form or appearance").] Several names are sometimes used synonymously although some of these have other meanings that are not necessarily interchangeable: *amphoteric element*,[221] *boundary element*,[222] *half-metal*,[223] *half-way element*,[224] *near metal*,[225] *meta-metal*,[226] *semiconductor*,[227] *semimetal*[228] and *submetal*.[229] "Amphoteric element" is sometimes used more broadly to include transition metals capable of forming oxyanions, such as chromium and manganese.[230] "Half-metal" is used in physics to refer to a compound (such as chromium dioxide) or alloy that can act as a conductor and an insulator. "Meta-metal" is sometimes used instead to refer to certain metals (Be, Zn, Cd, Hg, In, Tl, β-Sn, Pb) located just to the left of the metalloids on standard periodic tables.[223] These metals are mostly diamagnetic[231] and tend to have distorted crystalline structures, electrical conductivity values at the lower end of those of metals, and amphoteric (weakly basic) oxides.[232] "Semimetal" sometimes refers, loosely or explicitly, to metals with incomplete metallic character in crystalline structure, electrical conductivity or electronic structure. Examples include gallium,[233] ytterbium,[234] bismuth[235] and neptunium.[236] The names *amphoteric element* and *semiconductor* are problematic as some elements referred to as metalloids do not show marked amphoteric behaviour (bismuth, for example)[237] or semiconductivity (polonium)[238] in their most stable forms.

22.5.2 Origin and usage

Main article: Metalloid (nomenclature origin and usage)

The origin and usage of the term *metalloid* is convoluted. Its origin lies in attempts, dating from antiquity, to describe metals and to distinguish between typical and less typical forms. It was first applied in the early 19th century to metals that floated on water (sodium and potassium), and then more popularly to nonmetals. Earlier usage in mineralogy, to describe a mineral having a metallic appearance, can be sourced to as early as 1800.[239] Since the mid-20th century it has been used to refer to intermediate or borderline chemical elements.[240][n 22] The International Union of Pure and Applied Chemistry (IUPAC) previously recommended abandoning the term metalloid, and suggested using the term *semimetal* instead.[242] Use of this latter term has more recently been discouraged by Atkins et al.[2] as it has a different meaning in physics—one that more specifically refers to the electronic band structure of a substance rather than the overall classification of an element. The most recent IUPAC publications on nomenclature and terminology do not include any recommendations on the usage of the terms metalloid or semimetal.[243]

22.6 Elements commonly recognised as metalloids

Properties noted in this section refer to the elements in their most thermodynamically stable forms under ambient conditions.

22.6.1 Boron

Main article: Boron

Pure boron is a shiny, silver-grey crystalline solid.[245] It is less dense than aluminium (2.34 vs. 2.70 g/cm^3), and is hard

Boron, shown here in the form of its β-rhombohedral phase (its most thermodynamically stable allotrope)[244]

and brittle. It is barely reactive under normal conditions, except for attack by fluorine,[246] and has a melting point of

2076 °C (cf. steel ~1370 °C).[247] Boron is a semiconductor;[248] its room temperature electrical conductivity is 1.5×10^{-6} S•cm^{-1}[249] (about 200 times less than that of tap water)[250] and it has a band gap of about 1.56 eV.[251][n 23]

The structural chemistry of boron is dominated by its small atomic size, and relatively high ionization energy. With only three valence electrons per boron atom, simple covalent bonding cannot fulfil the octet rule.[253] Metallic bonding is the usual result among the heavier congenors of boron but this generally requires low ionization energies.[254] Instead, because of its small size and high ionization energies, the basic structural unit of boron (and nearly all of its allotropes)[n 24] is the icosahedral B_{12} cluster. Of the 36 electrons associated with 12 boron atoms, 26 reside in 13 delocalized molecular orbitals; the other 10 electrons are used to form two- and three-centre covalent bonds between icosahedra.[256] The same motif can be seen, as are deltahedral variants or fragments, in metal borides and hydride derivatives, and in some halides.[257]

The bonding in boron has been described as being characteristic of behaviour intermediate between metals and nonmetallic covalent network solids (such as diamond).[258] The energy required to transform B, C, N, Si and P from nonmetallic to metallic states has been estimated as 30, 100, 240, 33 and 50 kJ/mol, respectively. This indicates the proximity of boron to the metal-nonmetal borderline.[259]

Most of the chemistry of boron is nonmetallic in nature.[259] Unlike its heavier congeners, it is not known to form a simple B^{3+} or hydrated $[B(H_2O)_4]^{3+}$ cation.[260] The small size of the boron atom enables the preparation of many interstitial alloy-type borides.[261] Analogies between boron and transition metals have been noted in the formation of complexes,[262] and adducts (for example, BH_3 + CO →BH_3CO and, similarly, $Fe(CO)_4$ + CO →$Fe(CO)_5$),[n 25] as well as in the geometric and electronic structures of cluster species such as $[B_6H_6]^{2-}$ and $[Ru_6(CO)_{18}]^{2-}$.[264][n 26] The aqueous chemistry of boron is characterised by the formation of many different polyborate anions.[266] Given its high charge-to-size ratio, boron bonds covalently in nearly all of its compounds;[267] the exceptions are the borides as these include, depending on their composition, covalent, ionic and metallic bonding components.[268] Simple binary compounds, such as boron trichloride are Lewis acids as the formation of three covalent bonds leaves a hole in the octet which can be filled by an electron-pair donated by a Lewis base.[253] Boron has a strong affinity for oxygen and a duly extensive borate chemistry.[261] The oxide B_2O_3 is polymeric in structure,[269] weakly acidic,[270][n 27] and a glass former.[276] Organometallic compounds of boron[n 28] have been known since the 19th century (see organoboron chemistry).[278]

22.6.2 Silicon

Main article: Silicon

Silicon is a crystalline solid with a blue-grey metallic lustre.[279] Like boron, it is less dense (at 2.33 g/cm^3) than aluminium, and is hard and brittle.[280] It is a relatively unreactive element.[279] According to Rochow,[281] the massive crystalline form (especially if pure) is "remarkably inert to all acids, including hydrofluoric".[n 29] Less pure silicon, and the powdered form, are variously susceptible to attack by strong or heated acids, as well as by steam and fluorine.[285] Silicon dissolves in hot aqueous alkalis with the evolution of hydrogen, as do metals[286] such as beryllium, aluminium, zinc, gallium or indium.[287] It melts at 1414 °C. Silicon is a semiconductor with an electrical conductivity of 10^{-4} S•cm^{-1}[288] and a band gap of about 1.11 eV.[282] When it melts, silicon becomes a reasonable metal[289] with an electrical conductivity of 1.0–1.3×10^4 S•cm^{-1}, similar to that of liquid mercury.[290]

The chemistry of silicon is generally nonmetallic (covalent) in nature.[291] It is not known to form a cation.[292] Silicon can form alloys with metals such as iron and copper.[293] It shows fewer tendencies to anionic behaviour than ordinary nonmetals.[294] Its solution chemistry is characterised by the formation of oxyanions.[295] The high strength of the silicon-oxygen bond dominates the chemical behaviour of silicon.[296] Polymeric silicates, built up by tetrahedral SiO_4 units sharing their oxygen atoms, are the most abundant and important compounds of silicon.[297] The polymeric borates, comprising linked trigonal and tetrahedral BO_3 or BO_4 units, are built on similar structural principles.[298] The oxide SiO_2 is polymeric in structure,[269] weakly acidic,[299][n 30] and a glass former.[276] Traditional organometallic chemistry includes the carbon compounds of silicon (see organosilicon).[303]

22.6.3 Germanium

Main article: Germanium

Germanium is a shiny grey-white solid.[304] It has a density of 5.323 g/cm^3 and is hard and brittle.[305] It is mostly unreactive at room temperature[n 31] but is slowly attacked by hot concentrated sulfuric or nitric acid.[307] Germanium

Silicon has a blue-grey metallic lustre.

also reacts with molten caustic soda to yield sodium germanate Na_2GeO_3 and hydrogen gas.[308] It melts at 938 °C. Germanium is a semiconductor with an electrical conductivity of around 2×10^{-2} S•cm^{-1}[307] and a band gap of 0.67 eV.[309] Liquid germanium is a metallic conductor, with an electrical conductivity similar to that of liquid mercury.[310]

Most of the chemistry of germanium is characteristic of a nonmetal.[311] Whether or not germanium forms a cation is unclear, aside from the reported existence of the Ge^{2+} ion in a few esoteric compounds.[n 32] It can form alloys with metals such as aluminium and gold.[324] It shows fewer tendencies to anionic behaviour than ordinary nonmetals.[294] Its solution chemistry is characterised by the formation of oxyanions.[295] Germanium generally forms tetravalent (IV) compounds, and it can also form less stable divalent (II) compounds, in which it behaves more like a metal.[325] Germanium analogues of all of the major types of silicates have been prepared.[326] The metallic character of germanium is also suggested by the formation of various oxoacid salts. A phosphate [$(HPO_4)_2Ge·H_2O$] and highly stable trifluoroacetate $Ge(OCOCF_3)_4$ have been described, as have $Ge_2(SO_4)_2$, $Ge(ClO_4)_4$ and $GeH_2(C_2O_4)_3$.[327] The oxide GeO_2 is polymeric,[269] amphoteric,[328] and a glass former.[276] The dioxide is soluble in acidic solutions (the monoxide GeO, is even more so), and this is sometimes used to classify germanium as a metal.[329] Up to the 1930s germanium was considered to be a poorly conducting metal;[330] it has occasionally been classified as a metal by later writers.[331] As with all the elements commonly recognised as metalloids, germanium has an established organometallic chemistry (see organogermanium chemistry).[332]

22.6.4 Arsenic

Main article: Arsenic

Arsenic is a grey, metallic looking solid. It has a density of 5.727 g/cm^3 and is brittle, and moderately hard (more than aluminium; less than iron).[333] It is stable in dry air but develops a golden bronze patina in moist air, which blackens on further exposure. Arsenic is attacked by nitric acid and concentrated sulfuric acid. It reacts with fused caustic soda to give

Germanium is sometimes described as a metal

the arsenate Na_3AsO_3 and hydrogen gas.[334] Arsenic sublimes at 615 °C. The vapour is lemon-yellow and smells like garlic.[335] Arsenic only melts under a pressure of 38.6 atm, at 817 °C.[336] It is a semimetal with an electrical conductivity of around 3.9×10^4 S•cm^{-1}[337] and a band overlap of 0.5 eV.[338][n 33] Liquid arsenic is a semiconductor with a band gap of 0.15 eV.[340]

The chemistry of arsenic is predominately nonmetallic.[341] Whether or not arsenic forms a cation is unclear.[n 34] Its many metal alloys are mostly brittle.[349] It shows fewer tendencies to anionic behaviour than ordinary nonmetals.[294] Its solution chemistry is characterised by the formation of oxyanions.[295] Arsenic generally forms compounds in which it has an oxidation state of +3 or +5.[350] The halides, and the oxides and their derivatives are illustrative examples.[297] In the trivalent state, arsenic shows some incipient metallic properties.[351] The halides are hydrolysed by water but these reactions, particularly those of the chloride, are reversible with the addition of a hydrohalic acid.[352] The oxide is acidic but, as noted below, (weakly) amphoteric. The higher, less stable, pentavalent state has strongly acidic (nonmetallic) properties.[353] Compared to phosphorus, the stronger metallic character of arsenic is indicated by the formation of oxoacid salts such as $AsPO_4$, $As_2(SO_4)_3$[n 35] and arsenic acetate $As(CH_3COO)_3$.[356] The oxide As_2O_3 is polymeric,[269] amphoteric,[357][n 36]

Arsenic, sealed in a container to prevent tarnishing

and a glass former.[276] Arsenic has an extensive organometallic chemistry (see organoarsenic chemistry).[360]

22.6.5 Antimony

Main article: Antimony

Antimony is a silver-white solid with a blue tint and a brilliant lustre.[334] It has a density of 6.697 g/cm^3 and is brittle, and moderately hard (more so than arsenic; less so than iron; about the same as copper).[333] It is stable in air and moisture at room temperature. It is attacked by concentrated nitric acid, yielding the hydrated pentoxide Sb_2O_5. Aqua regia gives the pentachloride $SbCl_5$ and hot concentrated sulfuric acid results in the sulfate $Sb_2(SO_4)_3$.[361] It is not affected by molten alkali.[362] Antimony is capable of displacing hydrogen from water, when heated: $2\,Sb + 3\,H_2O \rightarrow Sb_2O_3 + 3\,H_2$.[363] It melts at 631 °C. Antimony is a semimetal with an electrical conductivity of around 3.1×10^4 S•cm^{-1}[364] and a band overlap of 0.16 eV.[338][n 37] Liquid antimony is a metallic conductor with an electrical conductivity of around 5.3×10^4 S•cm^{-1}.[366]

Most of the chemistry of antimony is characteristic of a nonmetal.[367] Antimony has some definite cationic chemistry,[368] SbO^+ and $Sb(OH)_2^+$ being present in acidic aqueous solution;[369][n 38] the compound $Sb_8(GaCl_4)_2$, which contains the homopolycation, Sb_8^{2+}, was prepared in 2004.[371] It can form alloys with one or more metals such as aluminium,[372] iron, nickel, copper, zinc, tin, lead and bismuth.[373] Antimony has fewer tendencies to anionic behaviour than ordinary nonmetals.[294] Its solution chemistry is characterised by the formation of oxyanions.[295] Like arsenic, antimony generally forms compounds in which it has an oxidation state of +3 or +5.[350] The halides, and the oxides and their derivatives are illustrative examples.[297] The +5 state is less stable than the +3, but relatively easier to attain than with arsenic. This is explained by the poor shielding afforded the arsenic nucleus by its 3d^{10} electrons. In comparison, the tendency of antimony (being a heavier atom) to oxidize more easily partially offsets the effect of its 4d^{10} shell.[374] Tripositive antimony is amphoteric; pentapositive antimony is (predominately) acidic.[375] Consistent with an increase in metallic character down group 15, antimony forms salts or salt-like compounds including a nitrate $Sb(NO_3)_3$, phosphate $SbPO_4$, sulfate $Sb_2(SO_4)_3$ and perchlorate $Sb(ClO_4)_3$.[376] The otherwise acidic pentoxide Sb_2O_5 shows some basic (metallic) behaviour in that it can be dissolved in very acidic solutions, with the formation of the oxycation SbO+ 2.[377] The oxide Sb_2O_3 is polymeric,[269] amphoteric,[378] and a glass former.[276] Antimony has an extensive organometallic chemistry (see organoantimony chemistry).[379]

Antimony, showing its brilliant lustre

22.6.6 Tellurium

Main article: Tellurium

Tellurium is a silvery-white shiny solid.[381] It has a density of 6.24 g/cm^3, is brittle, and is the softest of the commonly recognised metalloids, being marginally harder than sulfur.[333] Large pieces of tellurium are stable in air. The finely powdered form is oxidized by air in the presence of moisture. Tellurium reacts with boiling water, or when freshly precipitated even at 50 °C, to give the dioxide and hydrogen: $Te + 2\ H_2O \rightarrow TeO_2 + 2\ H_2$.[382] It reacts (to varying degrees) with nitric, sulfuric and hydrochloric acids to give compounds such as the sulfoxide $TeSO_3$ or tellurous acid H_2TeO_3,[383] the basic nitrate $(Te_2O_4H)^+(NO_3)^-$,[384] or the oxide sulfate $Te_2O_3(SO_4)$.[385] It dissolves in boiling alkalis, to give the tellurite and telluride: $3\ Te + 6\ KOH = K_2TeO_3 + 2\ K_2Te + 3\ H_2O$, a reaction that proceeds or is reversible with increasing or decreasing temperature.[386]

At higher temperatures tellurium is sufficiently plastic to extrude.[387] It melts at 449.51 °C. Crystalline tellurium has a structure consisting of parallel infinite spiral chains. The bonding between adjacent atoms in a chain is covalent, but

Tellurium, described by Dmitri Mendeleev as forming a transition between metals and nonmetals[380]

there is evidence of a weak metallic interaction between the neighbouring atoms of different chains.[388] Tellurium is a semiconductor with an electrical conductivity of around 1.0 S•cm^{-1}[389] and a band gap of 0.32 to 0.38 eV.[390] Liquid tellurium is a semiconductor, with an electrical conductivity, on melting, of around 1.9×10^3 S•cm^{-1}[390] Superheated liquid tellurium is a metallic conductor.[391]

Most of the chemistry of tellurium is characteristic of a nonmetal.[392] It shows some cationic behaviour. The dioxide dissolves in acid to yield the trihydroxotellurium (IV) $Te(OH)_3^+$ ion;[393][n 39] the red Te_4^{2+} and yellow-orange Te_6^{2+} ions form when tellurium is oxidized in fluorosulfuric acid (HSO_3F), or liquid sulfur dioxide (SO_2), respectively.[396] It can form alloys with aluminium, silver and tin.[397] Tellurium shows fewer tendencies to anionic behaviour than ordinary nonmetals.[294] Its solution chemistry is characterised by the formation of oxyanions.[295] Tellurium generally forms compounds in which it has an oxidation state of −2, +4 or +6. The +4 state is the most stable.[382] Tellurides of composition X_xTe_y are easily formed with most other elements and represent the most common tellurium minerals. Nonstoichiometry is pervasive, especially with transition metals. Many tellurides can be regarded as metallic alloys.[398] The increase in metallic character evident in tellurium, as compared to the lighter chalcogens, is further reflected in the reported formation of various other oxyacid salts, such as a basic selenate $2TeO_2 \cdot SeO_3$ and an analogous perchlorate and periodate $2TeO_2 \cdot HXO_4$.[399] Tellurium forms a polymeric,[269] amphoteric,[378] glass-forming oxide[276] TeO_2. It is a "conditional" glass-forming oxide—it forms a glass with a very small amount of additive.[276] Tellurium has an extensive organometallic chemistry (see organotellurium chemistry).[400]

22.7 Elements less commonly recognised as metalloids

22.7.1 Carbon

Main article: Carbon

Carbon is ordinarily classified as a nonmetal[402] but has some metallic properties and is occasionally classified as a

Carbon (as graphite). Delocalized valence electrons within the layers of graphite give it a metallic appearance.[401]

metalloid.[403] Hexagonal graphitic carbon (graphite) is the most thermodynamically stable allotrope of carbon under ambient conditions.[404] It has a lustrous appearance[405] and is a fairly good electrical conductor.[406] Graphite has a layered structure. Each layer comprises carbon atoms bonded to three other carbon atoms in a honeycomb lattice arrangement. The layers are stacked together and held loosely by van der Waals forces and delocalized valence electrons.[407]

Like a metal, the conductivity of graphite in the direction of its planes decreases as the temperature is raised;[408][n 40] it has the electronic band structure of a semimetal.[408] The allotropes of carbon, including graphite, can accept foreign atoms or compounds into their structures via substitution, intercalation or doping. The resulting materials are referred to as "carbon alloys".[412] Carbon can form ionic salts, including a hydrogen sulfate, perchlorate, and nitrate (C+
24X−.2HX, where X = HSO_4, ClO_4; and C+
24NO−
3.3HNO_3).[413][n 41] In organic chemistry, carbon can form complex cations—termed *carbocations*—in which the positive charge is on the carbon atom; examples are CH+
3 and CH+
5, and their derivatives.[414]

Carbon is brittle,[415] and behaves as a semiconductor in a direction perpendicular to its planes.[408] Most of its chemistry is nonmetallic;[416] it has a relatively high ionization energy[417] and, compared to most metals, a relatively high electronegativity.[418] Carbon can form anions such as C^{4-} (methanide), C2−
2 (acetylide) and C3−
4 (sesquicarbide or allylenide), in compounds with metals of main groups 1–3, and with the lanthanides and actinides.[419] Its oxide CO_2 forms carbonic acid H_2CO_3.[420][n 42]

22.7.2   Aluminium

Main article: Aluminium
 Aluminium is ordinarily classified as a metal.[423] It is lustrous, malleable and ductile, and has high electrical and thermal conductivity. Like most metals it has a close-packed crystalline structure,[424] and forms a cation in aqueous solution.[425]

It has some properties that are unusual for a metal; taken together,[426] these are sometimes used as a basis to classify aluminium as a metalloid.[427] Its crystalline structure shows some evidence of directional bonding.[428] Aluminium bonds covalently in most compounds.[429] The oxide Al_2O_3 is amphoteric,[430] and a conditional glass-former.[276] Aluminium can form anionic aluminates,[426] such behaviour being considered nonmetallic in character.[68]

Classifying aluminium as a metalloid has been disputed[431] given its many metallic properties. It is therefore, arguably, an exception to the mnemonic that elements adjacent to the metal–nonmetal dividing line are metalloids.[432][n 43]

Stott[434] labels aluminium as a weak metal. It has the physical properties of a metal but some of the chemical properties of a nonmetal. Steele[435] notes the paradoxical chemical behaviour of aluminium: "It resembles a weak metal in its amphoteric oxide and in the covalent character of many of its compounds ... Yet it is a highly electropositive metal ... [with] a high negative electrode potential".

22.7.3   Selenium

Main article: Selenium
 Selenium shows borderline metalloid or nonmetal behaviour.[437][n 44]

Its most stable form, the grey trigonal allotrope, is sometimes called "metallic" selenium because its electrical conductivity is several orders of magnitude greater than that of the red monoclinic form.[440] The metallic character of selenium is further shown by its lustre,[441] and its crystalline structure, which is thought to include weakly "metallic" interchain bonding.[442] Selenium can be drawn into thin threads when molten and viscous.[443] It shows reluctance to acquire "the high positive oxidation numbers characteristic of nonmetals".[444] It can form cyclic polycations (such as Se2+
8) when dissolved in oleums[445] (an attribute it shares with sulfur and tellurium), and a hydrolysed cationic salt in the form of trihydroxoselenium (IV) perchlorate [Se(OH)$_3$]+·ClO−
4.[446]

The nonmetallic character of selenium is shown by its brittleness[441] and the low electrical conductivity ($\sim 10^{-9}$ to 10^{-12} S•cm^{-1}) of its highly purified form.[92] This is comparable to or less than that of bromine (7.95×10^{-12} S•cm^{-1}),[447] a

High purity aluminium is very much softer than its familiar alloys. People who handle it for the first time often ask if it is the real thing.[422]

nonmetal. Selenium has the electronic band structure of a semiconductor[448] and retains its semiconducting properties in liquid form.[448] It has a relatively high[449] electronegativity (2.55 revised Pauling scale). Its reaction chemistry is mainly that of its nonmetallic anionic forms Se^{2-}, $SeO2-$
3 and $SeO2-$
4.[450]

Selenium is commonly described as a metalloid in the environmental chemistry literature.[451] It moves through the aquatic environment similarly to arsenic and antimony;[452] its water-soluble salts, in higher concentrations, have a similar toxicological profile to that of arsenic.[453]

Grey selenium, being a photoconductor, conducts electricity around 1,000 times better when light falls on it, a property used since the mid-1870s in various light-sensing applications[436]

22.7.4 Polonium

Main article: Polonium

Polonium is "distinctly metallic" in some ways.[238] Both of its allotropic forms are metallic conductors.[238] It is soluble in acids, forming the rose-coloured Po^{2+} cation and displacing hydrogen: $Po + 2 H^+ \rightarrow Po^{2+} + H_2$.[454] Many polonium salts are known.[455] The oxide PoO_2 is predominantly basic in nature.[456] Polonium is a reluctant oxidizing agent, unlike its lighter congener oxygen: highly reducing conditions are required for the formation of the Po^{2-} anion in aqueous solution.[457]

Whether polonium is ductile or brittle is unclear. It is predicted to be ductile based on its calculated elastic constants.[458] It has a simple cubic crystalline structure. Such a structure has few slip systems and "leads to very low ductility and hence low fracture resistance".[459]

Polonium shows nonmetallic character in its halides, and by the existence of polonides. The halides have properties generally characteristic of nonmetal halides (being volatile, easily hydrolyzed, and soluble in organic solvents).[460] Many metal polonides, obtained by heating the elements together at 500–1,000 °C, and containing the Po^{2-} anion, are also known.[461]

22.7.5 Astatine

Main article: Astatine

As a halogen, astatine tends to be classified as a nonmetal.[462] It has some marked metallic properties[463] and is sometimes instead classified as either a metalloid[464] or (less often) as a metal.[n 45] Immediately following its production in 1940, early investigators considered it a metal.[466] In 1949 it was called the most noble (difficult to reduce) nonmetal as well as being a relatively noble (difficult to oxidize) metal.[467] In 1950 astatine was described as a halogen and (therefore) a reactive nonmetal.[468] In 2013, on the basis of relativistic modelling, astatine was predicted to be a monatomic metal, with a face-centred cubic crystalline structure.[469]

Several authors have commented on the metallic nature of some of the properties of astatine. Since iodine is a semiconductor in the direction of its planes, and since the halogens become more metallic with increasing atomic number, it has been presumed that astatine would be a metal if it could form a condensed phase.[470][n 46] Astatine may be metallic in the liquid state on the basis that elements with an enthalpy of vaporization (ΔH_{vap}) greater than ~42 kJ/mol are metallic when liquid.[472] Such elements include boron,[n 47] silicon, germanium, antimony, selenium and tellurium. Estimated values for ΔH_{vap} of diatomic astatine are 50 kJ/mol or higher;[476] diatomic iodine, with a ΔH_{vap} of 41.71,[477] falls just short of the threshold figure.

"Like typical metals, it [astatine] is precipitated by hydrogen sulfide even from strongly acid solutions and is displaced in a free form from sulfate solutions; it is deposited on the cathode on electrolysis."[478][n 48] Further indications of a tendency for astatine to behave like a (heavy) metal are: "... the formation of pseudohalide compounds ... complexes of astatine cations ... complex anions of trivalent astatine ... as well as complexes with a variety of organic solvents".[480] It has also been argued that astatine demonstrates cationic behaviour, by way of stable At^+ and AtO^+ forms, in strongly acidic aqueous solutions.[481]

Some of astatine's reported properties are nonmetallic. It has the narrow liquid range ordinarily associated with nonmetals (mp 302 °C; bp 337 °C).[482] Batsanov gives a calculated band gap energy for astatine of 0.7 eV;[483] this is consistent with nonmetals (in physics) having separated valence and conduction bands and thereby being either semiconductors or insulators.[484] The chemistry of astatine in aqueous solution is mainly characterised by the formation of various anionic species.[485] Most of its known compounds resemble those of iodine,[486] which is a halogen and a nonmetal.[487] Such compounds include astatides (XAt), astatates ($XAtO_3$), and monovalent interhalogen compounds.[488]

Restrepo et al.[489] reported that astatine appeared to be more polonium-like than halogen-like. They did so on the basis of detailed comparative studies of the known and interpolated properties of 72 elements.

22.8   Related concepts

22.8.1   Near metalloids

Iodine crystals, showing a metallic lustre. Iodine is a semiconductor in the direction of its planes, with a band gap of ~1.3 eV. It has an electrical conductivity of 1.7 × 10⁻⁸ S•cm⁻¹ at room temperature.[490] This is higher than selenium but lower than boron, the least electrically conducting of the recognised metalloids.[n 49]

In the periodic table, some of the elements adjacent to the commonly recognised metalloids, although usually classified as either metals or nonmetals, are occasionally referred to as *near-metalloids*[493] or noted for their metalloidal character. To the left of the metal–nonmetal dividing line, such elements include gallium,[494] tin[495] and bismuth.[496] They show unusual packing structures,[497] marked covalent chemistry (molecular or polymeric),[498] and amphoterism.[499] To the right of the dividing line are carbon,[500] phosphorus,[501] selenium[502] and iodine.[503] They exhibit metallic lustre, semiconducting properties[n 50] and bonding or valence bands with delocalized character. This applies to their most thermodynamically stable forms under ambient conditions: carbon as graphite; phosphorus as black phosphorus;[n 51] and selenium as grey selenium.

22.8.2   Allotropes

Different crystalline forms of an element are called allotropes. Some allotropes, particularly those of elements located (in periodic table terms) alongside or near the notional dividing line between metals and nonmetals, exhibit more pronounced metallic, metalloidal or nonmetallic behaviour than others.[509] The existence of such allotropes can complicate the classification of the elements involved.[510]

Tin, for example, has two allotropes: tetragonal "white" β-tin and cubic "grey" α-tin. White tin is a very shiny, ductile and malleable metal. It is the stable form at or above room temperature and has an electrical conductivity of 9.17 × 10⁴ S·cm⁻¹ (~1/6th that of copper).[511] Grey tin usually has the appearance of a grey micro-crystalline powder, and can also

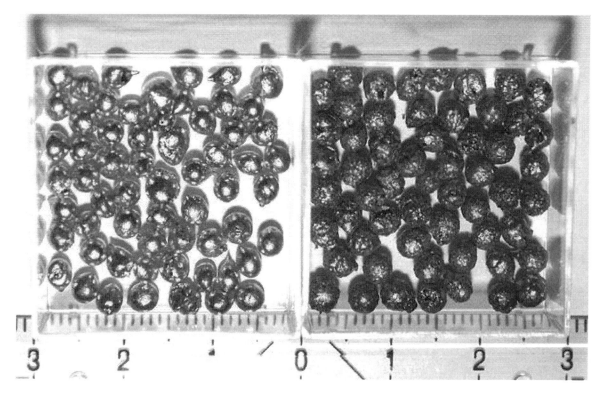

White tin (left) and grey tin (right). Both forms have a metallic appearance.

be prepared in brittle semi-lustrous crystalline or polycrystalline forms. It is the stable form below 13.2 °C and has an electrical conductivity of between $(2\text{--}5) \times 10^2$ S·cm^{-1} (~1/250th that of white tin).[512] Grey tin has the same crystalline structure as that of diamond. It behaves as a semiconductor (with a band gap of 0.08 eV), but has the electronic band structure of a semimetal.[513] It has been referred to as either a very poor metal,[514] a metalloid,[515] a nonmetal[516] or a near metalloid.[496]

The diamond allotrope of carbon is clearly nonmetallic, being translucent and having a low electrical conductivity of 10^{-14} to 10^{-16} S·cm^{-1}.[517] Graphite has an electrical conductivity of 3×10^4 S·cm^{-1},[518] a figure more characteristic of a metal. Phosphorus, sulfur, arsenic, selenium, antimony and bismuth also have less stable allotropes that display different behaviours.[519]

22.9   Abundance, extraction and cost

22.9.1   Abundance

The table gives crustal abundances of the elements commonly to rarely recognised as metalloids.[520] Some other elements are included for comparison: oxygen and xenon (the most and least abundant elements with stable isotopes); iron and the coinage metals copper, silver and gold; and rhenium, the least abundant stable metal (aluminium is normally the most abundant metal). Various abundance estimates have been published; these often disagree to some extent.[521]

22.9.2   Extraction

The recognised metalloids can be obtained by chemical reduction of either their oxides or their sulfides. Simpler or more complex extraction methods may be employed depending on the starting form and economic factors.[522] Boron is routinely obtained by reducing the trioxide with magnesium: $B_2O_3 + 3\,Mg \rightarrow 2\,B + 3MgO$; after secondary processing the resulting brown powder has a purity of up to 97%.[523] Boron of higher purity (> 99%) is prepared by heating volatile boron compounds, such as BCl_3 or BBr_3, either in a hydrogen atmosphere ($2\,BX_3 + 3\,H_2 \rightarrow 2\,B + 6\,HX$) or to the point of thermal decomposition. Silicon and germanium are obtained from their oxides by heating the oxide with carbon or hydrogen: $SiO_2 + C \rightarrow Si + CO_2$; $GeO_2 + 2\,H_2 \rightarrow Ge + 2\,H_2O$. Arsenic is isolated from its pyrite (FeAsS) or arsenical pyrite (FeAs$_2$) by heating; alternatively, it can be obtained from its oxide by reduction with carbon: $2\,As_2O_3 + 3\,C \rightarrow 2\,As + 3\,CO_2$.[524] Antimony is derived from its sulfide by reduction with iron: $Sb_2S_3 \rightarrow 2\,Sb + 3\,FeS$. Tellurium is prepared from its oxide by dissolving it in aqueous NaOH, yielding tellurite, then by electrolytic reduction: $TeO_2 + 2\,NaOH \rightarrow Na_2TeO_3 + H_2O$;[525] $Na_2TeO_3 + H_2O \rightarrow Te + 2\,NaOH + O_2$.[526] Another option is reduction of the oxide by roasting with carbon: $TeO_2 + C \rightarrow Te + CO_2$.[527]

Production methods for the elements less frequently recognised as metalloids involve natural processing, electrolytic or chemical reduction, or irradiation. Carbon (as graphite) occurs naturally and is extracted by crushing the parent rock and floating the lighter graphite to the surface. Aluminium is extracted by dissolving its oxide Al_2O_3 in molten cryolite Na_3AlF_6 and then by high temperature electrolytic reduction. Selenium is produced by roasting its coinage metal selenides X_2Se (X = Cu, Ag, Au) with soda ash to give the selenite: $X_2Se + O_2 + Na_2CO_3 \rightarrow Na_2SeO_3 + 2\,X + CO_2$; the selenide is neutralized by sulfuric acid H_2SO_4 to give selenous acid H_2SeO_3; this is reduced by bubbling with SO_2 to yield elemental selenium. Polonium and astatine are produced in minute quantities by irradiating bismuth.[528]

22.9.3   Cost

The recognised metalloids and their closer neighbours mostly cost less than silver; only polonium and astatine are more expensive than gold. As of 5 April 2014, prices for small samples (up to 100 g) of silicon, antimony and tellurium, and graphite, aluminium and selenium, average around one third the cost of silver (US$1.5 per gram or about $45 an ounce). Boron, germanium and arsenic samples average about three-and-a-half times the cost of silver.[n 52] Polonium is available for about $100 per microgram, which is $100,000,000 a gram.[529] Zalutsky and Pruszynski[530] estimate a similar cost for producing astatine. Prices for the applicable elements traded as commodities tend to range from two to three times cheaper than the sample price (Ge), to nearly three thousand times cheaper (As).[n 53]

22.10   See also

- Properties of metals, metalloids and nonmetals

22.11   Notes

[1] For a related commentary see also: Vernon RE 2013, 'Which Elements Are Metalloids?', *Journal of Chemical Education,* vol. 90, no. 12, pp. 1703–1707, doi:10.1021/ed3008457

[2] Definitions and extracts by different authors, illustrating aspects of the generic definition, follow:

- "In chemistry a metalloid is an element with properties intermediate between those of metals and nonmetals."[3]

- "Between the metals and nonmetals in the periodic table we find elements ... [that] share some of the characteristic properties of both the metals and nonmetals, making it difficult to place them in either of these two main categories"[4]

- "Chemists sometimes use the name metalloid ... for these elements which are difficult to classify one way or the other."[5]

- "Because the traits distinguishing metals and nonmetals are qualitative in nature, some elements do not fall unambiguously in either category. These elements ... are called metalloids ..."[6]

More broadly, metalloids have been referred to as:

- "elements that ... are somewhat of a cross between metals and nonmetals";[7] or

- "weird in-between elements".[8]

[3] Gold, for example, has mixed properties but is still recognised as "king of metals". Besides metallic behaviour (such as high electrical conductivity, and cation formation), gold shows nonmetallic behaviour:

- It has the highest electrode potential
- It has the third-highest ionization energy among the metals (after zinc and mercury)
- It has the lowest electron affinity
- Its electronegativity of 2.54 is highest among the metals and exceeds that of some nonmetals (hydrogen 2.2; phosphorus 2.19; and radon 2.2)
- It forms the Au⁻ auride anion, acting in this way like a halogen
- It sometimes has a tendency, known as "aurophilicity", to bond to itself;[11]

on halogen character, see also Belpassi et al,[12] who conclude that in the aurides MAu (M = Li–Cs) gold "behaves as a halogen, intermediate between Br and I"; on aurophilicity, see also Schmidbaur and Schier.[13]

[4] Mann et al.[16] refer to these elements as "the recognized metalloids".

[5] Jones[44] writes: "Though classification is an essential feature in all branches of science, there are always hard cases at the boundaries. Indeed, the boundary of a class is rarely sharp."

[6] The lack of a standard division of the elements into metals, metalloids and nonmetals is not necessarily an issue. There is more or less, a continuous progression from the metallic to the nonmetallic. Potentially, a specified subset of this continuum can serve its particular purpose as well as any other.[45]

[7] The packing efficiency of boron is 38%; silicon and germanium 34; arsenic 38.5; antimony 41; and tellurium 36.4.[49] These values are lower than in most metals (80% of which have a packing efficiency of at least 68%),[50] but higher than those of elements usually classified as nonmetals. (Gallium is unusual, for a metal, in having a packing efficiency of just 39%.[51] Other notable values for metals are 42.9 for bismuth[52] and 58.5 for liquid mercury.[53]) Packing efficiencies for nonmetals are: graphite 17%,[54] sulfur 19.2,[55] iodine 23.9,[55] selenium 24.2,[55] and black phosphorus 28.5.[52]

[8] More specifically, the *Goldhammer-Herzfeld criterion* is the ratio of the force holding an individual atom's valence electrons in place with the forces on the same electrons from interactions *between* the atoms in the solid or liquid element. When the interatomic forces are greater than, or equal to, the atomic force, valence electron itinerancy is indicated and metallic behaviour is predicted.[57] Otherwise nonmetallic behaviour is anticipated.

[9] As the ratio is based on classical arguments[59] it does not accommodate the finding that polonium, which has a value of ~0.95, adopts a metallic (rather than covalent) crystalline structure, on relativistic grounds.[60] Even so it offers a first order rationalization for the occurrence of metallic character amongst the elements.[61]

[10] Atomic conductance is the electrical conductivity of one mole of a substance. It is equal to electrical conductivity divided by molar volume.[5]

[11] Selenium has an ionization energy (IE) of 225 kcal/mol (941 kJ/mol) and is sometimes described as a semiconductor. It has a relatively high 2.55 electronegativity (EN). Polonium has an IE of 194 kcal/mol (812 kJ/mol) and a 2.0 EN, but has a metallic band structure.[66] Astatine has an IE of 215 kJ/mol (899 kJ/mol) and an EN of 2.2.[67] Its electronic band structure is not known with any certainty.

[12] Oderberg[79] argues on ontological grounds that anything not a metal is therefore a nonmetal, and that this includes semi-metals (i.e. metalloids).

[13] Copernicium is reportedly the only metal thought to be a gas at room temperature.[85]

[14] Metals have electrical conductivity values of from 6.9×10^3 S•cm⁻¹ for manganese to 6.3×10^5 for silver.[89]

[15] Metalloids have electrical conductivity values of from 1.5×10^{-6} S•cm⁻¹ for boron to 3.9×10^4 for arsenic.[91] If selenium is included as a metalloid the applicable conductivity range would start from ~10^{-9} to 10^{-12} S•cm⁻¹.[92]

[16] Nonmetals have electrical conductivity values of from ~10^{-18} S•cm⁻¹ for the elemental gases to 3×10^4 in graphite.[93]

[17] Chedd[100] defines metalloids as having electronegativity values of 1.8 to 2.2 (Allred-Rochow scale). He included boron, silicon, germanium, arsenic, antimony, tellurium, polonium and astatine in this category. In reviewing Chedd's work, Adler[101] described this choice as arbitrary, as other elements whose electronegativities lie in this range include copper, silver, phosphorus, mercury and bismuth. He went on to suggest defining a metalloid as "a semiconductor or semimetal" and to include bismuth and selenium in this category.

[18] Olmsted and Williams[105] commented that, "Until quite recently, chemical interest in the metalloids consisted mainly of isolated curiosities, such as the poisonous nature of arsenic and the mildly therapeutic value of borax. With the development of metalloid semiconductors, however, these elements have become among the most intensely studied".

[19] Research published in 2012 suggests that metal-metalloid glasses can be characterised by an interconnected atomic packing scheme in which metallic and covalent bonding structures coexist.[172]

[20] The reaction involved is $Ge + 2 MoO_3 \rightarrow GeO_2 + 2 MoO_2$. Adding arsenic or antimony (n-type electron donors) increases the rate of reaction; adding gallium or indium (p-type electron acceptors) decreases it.[186]

[21] Ellern, writing in *Military and Civilian Pyrotechnics* (1968), comments that carbon black "has been specified for and used in a nuclear air-burst simulator."[192]

[22] For a post-1960 example of the former use of the term metalloid to refer to nonmetals see Zhdanov,[241] who divides the elements into metals; intermediate elements (H, B, C, Si, Ge, Se, Te); and metalloids (of which the most typical are given as O, F and Cl).

[23] Boron, at 1.56 eV, has the largest band gap amongst the commonly recognised (semiconducting) metalloids. Of nearby elements in periodic table terms, selenium has the next highest band gap (close to 1.8 eV) followed by white phosphorus (around 2.1 eV).[252]

[24] The synthesis of B_{40} borospherene, a "distorted fullerene with a hexagonal hole on the top and bottom and four heptagonal holes around the waist" was announced in 2014.[255]

[25] The BH_3 and $Fe(CO_4)$ species in these reactions are short-lived reaction intermediates.[263]

[26] On the analogy between boron and metals, Greenwood[265] commented that: "The extent to which metallic elements mimic boron (in having fewer electrons than orbitals available for bonding) has been a fruitful cohering concept in the development of metalloborane chemistry ... Indeed, metals have been referred to as "honorary boron atoms" or even as "flexiboron atoms". The converse of this relationship is clearly also valid ..."

[27] Boron trioxide B_2O_3 is sometimes described as being (weakly) amphoteric.[271] It reacts with alkalies to give various borates.[272] In its hydrated form (as H_3BO_3, boric acid) it reacts with sulfur trioxide, the anhydride of sulphuric acid, to form a bisulfate $B(HSO_3)_4$.[273] In its pure (anhydrous) form it reacts with phosphoric acid to form a "phosphate" BPO_4.[274] The latter compound may be regarded as a mixed oxide of B_2O_3 and P_2O_5.[275]

[28] Organic derivatives of metalloids are traditionally counted as organometallic compounds.[277]

[29] In air, silicon forms a thin coating of amorphous silicon dioxide, 2 to 3 nm thick.[282] This coating is dissolved by hydrogen fluoride at a very low pace—on the order of two to three hours per nanometre.[283] Silicon dioxide, and silicate glasses (of which silicon dioxide is a major component), are otherwise readily attacked by hydrofluoric acid.[284]

[30] Although SiO_2 is classified as an acidic oxide, and hence reacts with alkalis to give silicates, it reacts with phosphoric acid to yield a silicon oxide orthophosphate $Si_5O(PO_4)_6$,[300] and with hydrofluoric acid to give hexafluorosilicic acid H_2SiF_6.[301] The latter reaction "is sometimes quoted as evidence of basic [that is, metallic] properties".[302]

[31] Temperatures above 400 °C are required to form a noticeable surface oxide layer.[306]

[32] Sources mentioning germanium cations include: Powell & Brewer[312] who state that the cadmium iodide CdI_2 structure of germanous iodide GeI_2 establishes the existence of the Ge^{++} ion (the CdI_2 structure being found, according to Ladd,[313] in "many metallic halides, hydroxides and chalcides"); Everest[314] who comments that, "it seems probable that the Ge^{++} ion can also occur in other crystalline germanous salts such as the phosphite, which is similar to the salt-like stannous phosphite and germanous phosphate, which resembles not only the stannous phosphates, but the manganous phosphates also"; Pan, Fu & Huang[315] who presume the formation of the simple Ge^{++} ion when $Ge(OH)_2$ is dissolved in a perchloric acid solution, on the basis that, "$ClO4^-$ has little tendency to enter complex formation with a cation"; Monconduit et al.[316] who prepared the layer compound or phase $Nb_3Ge_xTe_6$ ($x \simeq 0.9$), and reported that this contained a Ge^{II} cation; Richens[317] who records that, "Ge^{2+} (aq) or possibly $Ge(OH)^+$(aq) is said to exist in dilute air-free aqueous suspensions of the yellow hydrous monoxide...however both are unstable with respect to the ready formation of $GeO_2 . nH_2O$"; Rupar et al.[318] who synthesized a cryptand compound containing a Ge^{2+} cation; and Schwietzer and Pesterfield[319] who write that, "the monoxide GeO dissolves in dilute acids to give Ge^{+2} and in dilute bases to produce GeO_2^{-2}, all three entities being unstable in water". Sources dismissing germanium cations or further qualifying their presumed existence include: Jolly and Latimer[320] who assert that, "the germanous ion cannot be studied directly because no germanium (II) species exists in any appreciable concentration in noncomplexing aqueous solutions"; Lidin[321] who says that, "[germanium] forms no aquacations"; Ladd[322] who notes that the CdI_2 structure is "intermediate in type between ionic and molecular compounds"; and Wiberg[323] who states that, "no germanium cations are known".

[33] Arsenic also exists as a naturally occurring (but rare) allotrope *(arsenolamprite)*, a crystalline semiconductor with a band gap of around 0.3 eV or 0.4 eV. It can also be prepared in a semiconducting amorphous form, with a band gap of around 1.2–1.4 eV.[339]

[34] Sources mentioning cationic arsenic include: Gillespie & Robinson[342] who find that, "in very dilute solutions in 100% sulphuric acid, arsenic (III) oxide forms arsonyl (III) hydrogen sulphate, $AsO.HO_4$, which is partly ionized to give the AsO^+ cation. Both these species probably exist mainly in solvated forms, e.g., $As(OH)(SO_4H)_2$, and $As(OH)(SO_4H)^+$ respectively"; Paul et al.[343] who reported spectroscopic evidence for the presence of As_4^{2+} and As_2^{2+} cations when arsenic was oxidized with peroxydisulfuryl difluoride $S_2O_6F_2$ in highly acidic media (Gillespie and Passmore[344] noted the spectra of these species were very similar to S_4^{2+} and S_8^{2+} and concluded that, "at present" there was no reliable evidence for any homopolycations of arsenic); Van Muylder and Pourbaix,[345] who write that, "As_2O_3 is an amphoteric oxide which dissolves in water and in solutions of pH between 1 and 8 with the formation of undissociated arsenious acid $HAsO_2$; the solubility…increases at pH's below 1 with the formation of 'arsenyl' ions AsO^+…"; Kolthoff and Elving[346] who write that, "the As^{3+} cation exists to some extent only in strongly acid solutions; under less acid conditions the tendency is toward hydrolysis, so that the anionic form predominates"; Moody[347] who observes that, "arsenic trioxide, As_4O_6, and arsenious acid, H_3AsO_3, are apparently amphoteric but no cations, As^{3+}, $As(OH)^{2+}$ or $As(OH)_2^+$ are known"; and Cotton et al.[348] who write that (in aqueous solution) the simple arsenic cation As^{3+} "may occur to some slight extent [along with the AsO^+ cation]" and that, "Raman spectra show that in acid solutions of As_4O_6 the only detectable species is the pyramidal $As(OH)_3$".

[35] The formulae of $AsPO_4$ and $As_2(SO_4)_3$ suggest straightforward ionic formulations, with As^{3+}, but this is not the case. $AsPO_4$, "which is virtually a covalent oxide", has been referred to as a double oxide, of the form $As_2O_3 \cdot P_2O_5$. It comprises AsO_3 pyramids and PO_4 tetrahedra, joined together by all their corner atoms to form a continuous polymeric network.[354] $As_2(SO_4)_3$ has a structure in which each SO_4 tetrahedron is bridged by two AsO_3 trigonal pyramida.[355]

[36] As_2O_3 is usually regarded as being amphoteric but a few sources say it is (weakly)[358] acidic. They describe its "basic" properties (its reaction with concentrated hydrochloric acid to form arsenic trichloride) as being alcoholic, in analogy with the formation of covalent alkyl chlorides by covalent alcohols (e.g., R-OH + HCl → RCl + H_2O)[359]

[37] Antimony can also be prepared in an amorphous semiconducting black form, with an estimated (temperature-dependent) band gap of 0.06–0.18 eV.[365]

[38] Lidin[370] asserts that SbO^+ does not exist and that the stable form of Sb(III) in aqueous solution is an incomplete hydrocomplex $[Sb(H_2O)_4(OH)_2]^+$.

[39] Cotton et al.[394] note that TeO_2 appears to have an ionic lattice; Wells[395] suggests that the Te–O bonds have "considerable covalent character".

[40] Liquid carbon may[409] or may not[410] be a metallic conductor, depending on pressure and temperature; see also.[411]

[41] For the sulfate, the method of preparation is (careful) direct oxidation of graphite in concentrated sulfuric acid by an oxidising agent, such as nitric acid, chromium trioxide or ammonium persulfate; in this instance the concentrated sulfuric acid is acting as an inorganic nonaqueous solvent.

[42] Only a small fraction of dissolved CO_2 is present in water as carbonic acid so, even though H_2CO_3 is a medium-strong acid, solutions of carbonic acid are only weakly acidic.[421]

[43] A mnemonic that captures the elements commonly recognised as metalloids goes: *Up, up-down, up-down, up … are the metalloids!*[433]

[44] Rochow,[438] who later wrote his 1966 monograph *The metalloids*,[439] commented that, "In some respects selenium acts like a metalloid and tellurium certainly does".

[45] A further option is to include astatine both as a nonmetal and as a metalloid.[465]

[46] A visible piece of astatine would be immediately and completely vaporized because of the heat generated by its intense radioactivity.[471]

[47] The literature is contradictory as to whether boron exhibits metallic conductivity in liquid form. Krishnan et al.[473] found that liquid boron behaved like a metal. Glorieux et al.[474] characterised liquid boron as a semiconductor, on the basis of its low electrical conductivity. Millot et al.[475] reported that the emissivity of liquid boron was not consistent with that of a liquid metal.

[48] Korenman[479] similarly noted that "the ability to precipitate with hydrogen sulfide distinguishes astatine from other halogens and brings it closer to bismuth and other heavy metals".

[49] The separation between molecules in the layers of iodine (350 pm) is much less than the separation between iodine layers (427 pm; cf. twice the van der Waals radius of 430 pm).[491] This is thought to be caused by electronic interactions between the molecules in each layer of iodine, which in turn give rise to its semiconducting properties and shiny appearance.[492]

[50] For example: intermediate electrical conductivity;[504] a relatively narrow band gap;[505] light sensitivity.[504]

[51] White phosphorus is the least stable and most reactive form.[506] It is also the most common, industrially important,[507] and easily reproducible allotrope, and for these three reasons is regarded as the standard state of the element.[508]

[52] Sample prices of gold, in comparison, start at roughly thirty-five times that of silver. Based on sample prices for B, C, Al, Si, Ge, As, Se, Ag, Sb, Te and Au available on-line from Alfa Aesa; Goodfellow; Metallium; and United Nuclear Scientific.

[53] Based on spot prices for Al, Si, Ge, As, Sb, Se, and Te available on-line from FastMarkets: Minor Metals; Fast Markets: Base Metals; EnergyTrend: PV Market Status, Polysilicon; and Metal-Pages: Arsenic metal prices, news and information.

22.12   Sources

22.12.1   Citations

[1] Chedd 1969, pp. 58, 78; National Research Council 1984, p. 43

[2] Atkins et al. 2010, p. 20

[3] Cusack 1987, p. 360

[4] Kelter, Mosher & Scott 2009, p. 268

[5] Hill & Holman 2000, p. 41

[6] King 1979, p. 13

[7] Moore 2011, p. 81

[8] Gray 2010

[9] Hopkins & Bailar 1956, p. 458

[10] Glinka 1965, p. 77

[11] Wiberg 2001, p. 1279

[12] Belpassi et al. 2006, pp. 4543–4

[13] Schmidbaur & Schier 2008, pp. 1931–51

[14] Tyler Miller 1987, p. 59

[15] Goldsmith 1982, p. 526; Kotz, Treichel & Weaver 2009, p. 62; Bettelheim et al. 2010, p. 46

[16] Mann et al. 2000, p. 2783

[17] Hawkes 2001, p. 1686; Segal 1989, p. 965; McMurray & Fay 2009, p. 767

[18] Bucat 1983, p. 26; Brown c. 2007

[19] Swift & Schaefer 1962, p. 100

[20] Hawkes 2001, p. 1686; Hawkes 2010; Holt, Rinehart & Wilson c. 2007

[21] Dunstan 1968, pp. 310, 409. Dunstan lists Be, Al, Ge (maybe), As, Se (maybe), Sn, Sb, Te, Pb, Bi, and Po as metalloids (pp. 310, 323, 409, 419).

[22] Tilden 1876, pp. 172, 198–201; Smith 1994, p. 252; Bodner & Pardue 1993, p. 354

[23] Bassett et al. 1966, p. 127

[24] Rausch 1960

[25] Thayer 1977, p. 604; Warren & Geballe 1981; Masters & Ela 2008, p. 190

[26] Warren & Geballe 1981; Chalmers 1959, p. 72; US Bureau of Naval Personnel 1965, p. 26

[27] Siebring 1967, p. 513

[28] Wiberg 2001, p. 282

[29] Rausch 1960; Friend 1953, p. 68

[30] Murray 1928, p. 1295

[31] Hampel & Hawley 1966, p. 950; Stein 1985; Stein 1987, pp. 240, 247–8

[32] Hatcher 1949, p. 223; Secrist & Powers 1966, p. 459

[33] Taylor 1960, p. 614

[34] Considine & Considine 1984, p. 568; Cegielski 1998, p. 147; *The American heritage science dictionary 2005* p. 397

[35] Woodward 1948, p. 1

[36] NIST 2010. Values shown in the above table have been converted from the NIST values, which are given in eV.

[37] Berger 1997; Lovett 1977, p. 3

[38] Goldsmith 1982, p. 526; Hawkes 2001, p. 1686

[39] Hawkes 2001, p. 1687

[40] Sharp 1981, p. 299

[41] Emsley 1971, p. 1

[42] James et al. 2000, p. 480

[43] Chatt 1951, p. 417 "The boundary between metals and metalloids is indefinite ..."; Burrows et al. 2009, p. 1192: "Although the elements are conveniently described as metals, metalloids, and nonmetals, the transitions are not exact ..."

[44] Jones 2010, p. 170

[45] Kneen, Rogers & Simpson 1972, pp. 218–220

[46] Rochow 1966, pp. 1, 4–7

[47] Rochow 1977, p. 76; Mann et al. 2000, p. 2783

[48] Askeland, Phulé & Wright 2011, p. 69

[49] Van Setten et al. 2007, pp. 2460–1; Russell & Lee 2005, p. 7 (Si, Ge); Pearson 1972, p. 264 (As, Sb, Te; also black P)

[50] Russell & Lee 2005, p. 1

[51] Russell & Lee 2005, pp. 6–7, 387

[52] Pearson 1972, p. 264

[53] Okajima & Shomoji 1972, p. 258

[54] Kitaĭgorodskiĭ 1961, p. 108

[55] Neuburger 1936

[56] Edwards & Sienko 1983, p. 693

[57] Herzfeld 1927; Edwards 2000, pp. 100–3

[58] Edwards & Sienko 1983, p. 695; Edwards et al. 2010

[59] Edwards 1999, p. 416

[60] Steurer 2007, p. 142; Pyykkö 2012, p. 56

[61] Edwards & Sienko 1983, p. 695

[62] Hill & Holman 2000, p. 41. They characterise metalloids (in part) on the basis that they are "poor conductors of electricity with atomic conductance usually less than 10^{-3} but greater than 10^{-5} ohm^{-1} cm^{-4}".

[63] Bond 2005, p. 3: "One criterion for distinguishing semi-metals from true metals under normal conditions is that the bulk coordination number of the former is never greater than eight, while for metals it is usually twelve (or more, if for the body-centred cubic structure one counts next-nearest neighbours as well)."

[64] Jones 2010, p. 169

[65] Masterton & Slowinski 1977, p. 160 list B, Si, Ge, As, Sb and Te as metalloids, and comment that Po and At are ordinarily classified as metalloids but add that this is arbitrary as so little is known about them.

[66] Kraig, Roundy & Cohen 2004, p. 412; Alloul 2010, p. 83

[67] Vernon 2013, pp. 1704

[68] Hamm 1969, p. 653

[69] Horvath 1973, p. 336

[70] Gray 2009, p. 9

[71] Rayner-Canham 2011

[72] Booth & Bloom 1972, p. 426; Cox 2004, pp. 17, 18, 27–8; Silberberg 2006, pp. 305–13

[73] Cox 2004, pp. 17–18, 27–8; Silberberg 2006, p. 305–13

[74] Rodgers 2011, pp. 232–3; 240–1

[75] Roher 2001, pp. 4–6

[76] Tyler 1948, p. 105; Reilly 2002, pp. 5–6

[77] Hampel & Hawley 1976, p. 174

[78] Goodrich 1844, p. 264; *The Chemical News* 1897, p. 189; Hampel & Hawley 1976, p. 191; Lewis 1993, p. 835; Hérold 2006, pp. 149–50

[79] Oderberg 2007, p. 97

[80] Brown & Holme 2006, p. 57

[81] Wiberg 2001, p. 282; Simple Memory Art c. 2005

[82] Chedd 1969, pp. 12–13

[83] Kneen, Rogers & Simpson, 1972, p. 263. Columns 2 and 4 are sourced from this reference unless otherwise indicated.

[84] Stoker 2010, p. 62; Chang 2002, p. 304. Chang speculates that the melting point of francium would be about 23 °C.

[85] New Scientist 1975; Soverna 2004; Eichler et al. 2007; Austen 2012

[86] Rochow 1966, p. 4

[87] Hunt 2000, p. 256

[88] McQuarrie & Rock 1987, p. 85

[89] Desai, James & Ho 1984, p. 1160; Matula 1979, p. 1260

[90] Choppin & Johnsen 1972, p. 351

[91] Schaefer 1968, p. 76; Carapella 1968, p. 30

[92] Kozyrev 1959, p. 104; Chizhikov & Shchastlivyi 1968, p. 25; Glazov, Chizhevskaya & Glagoleva 1969, p. 86

[93] Bogoroditskii & Pasynkov 1967, p. 77; Jenkins & Kawamura 1976, p. 88

[94] Hampel & Hawley 1976, p. 191; Wulfsberg 2000, p. 620

[95] Swalin 1962, p. 216

[96] Bailar et al. 1989, p. 742

[97] Metcalfe, Williams & Castka 1974, p. 86

[98] Chang 2002, p. 306

[99] Pauling 1988, p. 183

[100] Chedd 1969, pp. 24–5

[101] Adler 1969, pp. 18–19

[102] Hultgren 1966, p. 648; Young & Sessine 2000, p. 849; Bassett et al. 1966, p. 602

[103] Rochow 1966, p. 4; Atkins et al. 2006, pp. 8, 122–3

[104] Russell & Lee 2005, pp. 421, 423; Gray 2009, p. 23

[105] Olmsted & Williams 1997, p. 975

[106] Russell & Lee 2005, p. 401; Büchel, Moretto & Woditsch 2003, p. 278

[107] Desch 1914, p. 86

[108] Phillips & Williams 1965, p. 620

[109] Van der Put 1998, p. 123

[110] Klug & Brasted 1958, p. 199

[111] Good et al. 1813

[112] Sequeira 2011, p. 776

[113] Gary 2013

[114] Russell & Lee 2005, pp. 423–4; 405–6

[115] Davidson & Lakin 1973, p. 627

[116] Wiberg 2001, p. 589

[117] Greenwood & Earnshaw 2002, p. 749; Schwartz 2002, p. 679

[118] Antman 2001

[119] Řezanka & Sigler 2008; Sekhon 2012

[120] Emsley 2001, p. 67

[121] Zhang et al. 2008, p. 360

[122] Science Learning Hub 2009

[123] Skinner et al. 1979; Tom, Elden & Marsh 2004, p. 135

[124] Büchel 1983, p. 226

[125] Emsley 2001, p. 391

[126] Schauss 1991; Tao & Bolger 1997

[127] Eagleson 1994, p. 450; EVM 2003, pp. 197–202

[128] Nielsen 1998

[129] Jaouen & Gibaud 2010

[130] Smith et al. 2014

[131] Stevens & Klarner, p. 205

[132] Sneader 2005, pp. 57–59

[133] Keall, Martin and Tunbridge 1946

[134] Emsley 2001, p. 426

[135] Oldfield et al. 1974, p. 65; Turner 2011

[136] Ba et al. 2010; Daniel-Hoffmann, Sredni & Nitzan 2012; Molina-Quiroz et al. 2012

[137] Peryea 1998

[138] Hager 2006, p. 299

[139] Apseloff 1999

[140] Trivedi, Yung & Katz 2013, p. 209

[141] Emsley 2001, p. 382; Burkhart, Burkhart & Morrell 2011

[142] Thomas, Bialek & Hensel 2013, p. 1

[143] Perry 2011, p. 74

[144] UCR Today 2011; Wang & Robinson 2011; Kinjo et al. 2011

[145] Kauthale et al. 2015

[146] Gunn 2014, pp. 188, 191

[147] Gupta, Mukherjee & Cameotra 1997, p. 280; Thomas & Visakh 2012, p. 99

[148] Muncke 2013

[149] Mokhatab & Poe 2012, p. 271

[150] Craig, Eng & Jenkins 2003, p. 25

[151] McKee 1984

[152] Hai et al 2012

[153] Kohl & Nielsen 1997, pp. 699–700

[154] Chopra et al. 2011

[155] Le Bras, Wilkie & Bourbigot 2005, p. v

[156] Wilkie & Morgan 2009, p. 187

[157] Locke et al. 1956, p. 88

[158] Carlin 2011, p. 6.2

[159] Evans 1993, pp. 257–8

[160] Corbridge 2013, p. 1149

[161] Kaminow & Li 2002, p. 118

[162] Deming 1925, pp. 330 (As_2O_3), 418 (B_2O_3; SiO_2; Sb_2O_3); Witt & Gatos 1968, p. 242 (GeO_2)

[163] Eagleson 1994, p. 421 (GeO_2); Rothenberg 1976, 56, 118–19 (TeO_2)

[164] Geckeler 1987, p. 20

[165] Kreith & Goswami 2005, p. 12–109

[166] Russell & Lee 2005, p. 397

[167] Butterman & Jorgenson 2005, pp. 9–10

[168] Shelby 2005, p. 43

[169] Butterman & Carlin 2004, p. 22; Russell & Lee 2005, p. 422

[170] Träger 2007, pp. 438, 958; Eranna 2011, p. 98

[171] Rao 2002, p. 552; Löffler, Kündig & Dalla Torre 2007, p. 17–11

[172] Guan et al. 2012; WPI-AIM 2012

[173] Klement, Willens & Duwez 1960; Wanga, Dongb & Shek 2004, p. 45

[174] Demetriou et al 2011; Oliwenstein 2011

[175] Karabulut et al. 2001, p. 15; Haynes 2012, p. 4–26

[176] Schwartz 2002, pp. 679–680

[177] Carter & Norton 2013, p. 403

[178] Maeder 2013, pp. 3, 9–11

[179] Tominaga 2006, p. 327–8; Chung 2010, p. 285–6; Kolobov & Tominaga 2012, p. 149

[180] New Scientist 2014; Hosseini, Wright & Bhaskaran 2014; Farandos et al. 2014

[181] Ordnance Office 1863, p. 293

[182] Kosanke 2002, p. 110

[183] Ellern 1968, pp. 246, 326–7

[184] Conkling & Mocella 2010, p. 82

[185] Crow 2011; DailyRecord 2014

[186] Schwab & Gerlach 1967; Yetter 2012, pp. 81; Lipscomb 1972, pp. 2–3, 5–6, 15

[187] Ellern 1968, p. 135; Weingart 1947, p. 9

[188] Conkling & Mocella 2010, p. 83

[189] Conkling & Mocella 2010, pp. 181, 213

[190] Ellern 1968, pp. 209–10; 322

[191] Russell 2009, pp. 15, 17, 41, 79–80

[192] Ellern 1968, p. 324

[193] Ellern 1968, p. 328

[194] Conkling & Mocella 2010, p. 171

[195] Conkling & Mocella 2011, pp. 83–4

[196] Berger 1997, p. 91; Hampel 1968, passim

[197] Rochow 1966, p. 41; Berger 1997, pp. 42–3

[198] Bomgardner 2013, p. 20

[199] Russell & Lee 2005, p. 395; Brown et al. 2009, p. 489

[200] Haller 2006, p. 4: "The study and understanding of the physics of semiconductors progressed slowly in the 19th and early 20th centuries ... Impurities and defects ... could not be controlled to the degree necessary to obtain reproducible results. This led influential physicists, including W. Pauli and I. Rabi, to comment derogatorily on the 'Physics of Dirt'."; Hoddeson 2007, pp. 25–34 (29)

[201] Bianco et. al. 2013

[202] University of Limerick 2014; Kennedy et al. 2014

[203] Lee et al. 2014

[204] Russell & Lee 2005, pp. 421–2, 424

[205] He et al. 2014

[206] Berger 1997, p. 91

[207] ScienceDaily 2012

[208] Reardon 2005; Meskers, Hagelüken & Van Damme 2009, p. 1131

[209] The Economist 2012

[210] Whitten 2007, p. 488

[211] Jaskula 2013

[212] German Energy Society 2008, p. 43–44

[213] Patel 2012, p. 248

[214] Moore 2104; University of Utah 2014; Xu et al. 2014

[215] Yang et al. 2012, p. 614

[216] Moore 2010, p. 195

[217] Moore 2011

[218] Liu 2014

[219] Bradley 2014; University of Utah 2014

[220] *Oxford English Dictionary* 1989, 'metalloid'; Gordh, Gordh & Headrick 2003, p. 753

[221] Foster 1936, pp. 212–13; Brownlee et al. 1943, p. 293

[222] Calderazzo, Ercoli & Natta 1968, p. 257

[223] Klemm 1950, pp. 133–42; Reilly 2004, p. 4

[224] Walters 1982, pp. 32–3

[225] Tyler 1948, p. 105

[226] Foster & Wrigley 1958, p. 218: "The elements may be grouped into two classes: those that are *metals* and those that are *nonmetals*. There is also an intermediate group known variously as *metalloids, meta-metals, semiconductors,* or *semimetals*."

[227] Slade 2006, p. 16

[228] Corwin 2005, p. 80

[229] Barsanov & Ginzburg 1974, p. 330

[230] Bradbury et al. 1957, pp. 157, 659

[231] Miller, Lee & Choe 2002, p. 21

[232] King 2004, pp. 196–8; Ferro & Saccone 2008, p. 233

[233] Pashaey & Seleznev 1973, p. 565; Gladyshev & Kovaleva 1998, p. 1445; Eason 2007, p. 294

[234] Johansen & Mackintosh 1970, pp. 121–4; Divakar, Mohan & Singh 1984, p. 2337; Dávila et al. 2002, p. 035411-3

[235] Jezequel & Thomas 1997, pp. 6620–6

[236] Hindman 1968, p. 434: "The high values obtained for the [electrical] resistivity indicate that the metallic properties of neptunium are closer to the semimetals than the true metals. This is also true for other metals in the actinide series."; Dunlap et al. 1970, pp. 44, 46: "... α-Np is a semimetal, in which covalency effects are believed to also be of importance ... For a semimetal having strong covalent bonding, like α-Np ..."

[237] Lister 1965, p. 54

[238] Cotton et al. 1999, p. 502

[239] Pinkerton 1800, p. 81

[240] Goldsmith 1982, p. 526

[241] Zhdanov 1965, pp. 74–5

[242] Friend 1953, p. 68; IUPAC 1959, p. 10; IUPAC 1971, p. 11

[243] IUPAC 2005; IUPAC 2006–

[244] Van Setten et al. 2007, pp. 2460–1; Oganov et al. 2009, pp. 863–4

[245] Housecroft & Sharpe 2008, p. 331; Oganov 2010, p. 212

[246] Housecroft & Sharpe 2008, p. 333

[247] Kross 2011

[248] Berger 1997, p. 37

[249] Greenwood & Earnshaw 2002, p. 144

[250] Kopp, Lipták & Eren 2003, p. 221

[251] Prudenziati 1977, p. 242

[252] Berger 1997, pp. 87, 84

[253] Rayner-Canham & Overton 2006, p. 291

[254] Siekierski & Burgess 2002, p. 63

[255] Wogan 2014

[256] Siekierski & Burgess 2002, p. 86

[257] Greenwood & Earnshaw 2002, p. 141; Henderson 2000, p. 58; Housecroft & Sharpe 2008, pp. 360–72

[258] Parry et al. 1970, pp. 438, 448–51

[259] Fehlner 1990, p. 202

[260] Owen & Brooker 1991, p. 59; Wiberg 2001, p. 936

[261] Greenwood & Earnshaw 2002, p. 145

[262] Houghton 1979, p. 59

[263] Fehlner 1990, pp. 205

[264] Fehlner 1990, pp. 204–205, 207

[265] Greenwood 2001, p. 2057

[266] Salentine 1987, pp. 128–32; MacKay, MacKay & Henderson 2002, pp. 439–40; Kneen, Rogers & Simpson 1972, p. 394;
       Hiller & Herber 1960, inside front cover; p. 225

[267] Sharp 1983, p. 56

[268] Fokwa 2014, p. 10

[269] Puddephatt & Monaghan 1989, p. 59

[270] Mahan 1965, p. 485

[271] Danaith 2008, p. 81.

[272] Lidin 1996, p. 28

[273] Kondrat'ev & Mel'nikova 1978

[274] Holderness & Berry 1979, p. 111; Wiberg 2001, p. 980

[275] Toy 1975, p. 506

[276] Rao 2002, p. 22

[277] Fehlner 1992, p. 1

[278] Haiduc & Zuckerman 1985, p. 82

[279] Greenwood & Earnshaw 2002, p. 331

[280] Wiberg 2001, p. 824

[281] Rochow 1973, p. 1337–38

[282] Russell & Lee 2005, p. 393

[283] Zhang 2002, p. 70

[284] Sacks 1998, p. 287

[285] Rochow 1973, p. 1337, 1340

[286] Allen & Ordway 1968, p. 152

[287] Eagleson 1994, pp. 48, 127, 438, 1194; Massey 2000, p. 191

[288] Orton 2004, p. 7. This is a typical value for high-purity silicon.

[289] Coles & Caplin 1976, p. 106

[290] Glazov, Chizhevskaya & Glagoleva 1969, pp. 59–63; Allen & Broughton 1987, p. 4967

[291] Cotton, Wilkinson & Gaus 1995, p. 393

[292] Wiberg 2001, p. 834

[293] Partington 1944, p. 723

[294] Cox 2004, p. 27

[295] Hiller & Herber 1960, inside front cover; p. 225

[296] Kneen, Rogers and Simpson 1972, p. 384

[297] Bailar, Moeller & Kleinberg 1965, p. 513

[298] Cotton, Wilkinson & Gaus 1995, pp. 319, 321

[299] Smith 1990, p. 175

[300] Poojary, Borade & Clearfield 1993

[301] Wiberg 2001, pp. 851, 858

[302] Barmett & Wilson 1959, p. 332

[303] Powell 1988, p. 1

[304] Greenwood & Earnshaw 2002, p. 371

[305] Cusack 1967, p. 193

[306] Russell & Lee 2005, pp. 399–400

[307] Greenwood & Earnshaw 2002, p. 373

[308] Moody 1991, p. 273

[309] Russell & Lee 2005, p. 399

[310] Berger 1997, pp. 71–2

[311] Jolly 1966, pp. 125–6

[312] Powell & Brewer 1938

[313] Ladd 1999, p. 55

[314] Everest 1953, p. 4120

[315] Pan, Fu and Huang 1964, p. 182

[316] Monconduit et al. 1992

[317] Richens 1997, p. 152

[318] Rupar et al. 2008

[319] Schwietzer & Pesterfield 2010, pp. 190

[320] Jolly & Latimer, p. 2

[321] Lidin 1996, p. 140

[322] Ladd 1999, p. 56

[323] Wiberg 2001, p. 896

[324] Schwartz 2002, p. 269

[325] Eggins 1972, p. 66; Wiberg 2001, p. 895

[326] Greenwood & Earnshaw 2002, p. 383

[327] Glockling 1969, p. 38; Wells 1984, p. 1175

[328] Cooper 1968, pp. 28–9

[329] Steele 1966, pp. 178, 188–9

[330] Haller 2006, p. 3

[331] See, for example, Walker & Tarn 1990, p. 590

[332] Wiberg 2001, p. 742

[333] Gray, Whitby & Mann 2011

[334] Greenwood & Earnshaw 2002, p. 552

[335] Parkes & Mellor 1943, p. 740

[336] Russell & Lee 2005, p. 420

[337] Carapella 1968, p. 30

[338] Barfuß et al. 1981, p. 967

[339] Greaves, Knights & Davis 1974, p. 369; Madelung 2004, pp. 405, 410

[340] Bailar & Trotman-Dickenson 1973, p. 558; Li 1990

[341] Bailar, Moeller & Kleinberg 1965, p. 477

[342] Gillespie & Robinson 1963, p. 450

[343] Paul et al. 1971; see also Ahmeda & Rucka 2011, pp. 2893, 2894

[344] Gillespie & Passmore 1972, p. 478

[345] Van Muylder & Pourbaix 1974, p. 521

[346] Kolthoff & Elving 1978, p. 210

[347] Moody 1991, p. 248–249

[348] Cotton & Wilkinson 1999, pp. 396, 419

[349] Eagleson 1994, p. 91

[350] Massey 2000, p. 267

[351] Timm 1944, p. 454

[352] Partington 1944, p. 641; Kleinberg, Argersinger & Griswold 1960, p. 419

[353] Morgan 1906, p. 163; Moeller 1954, p. 559

[354] Corbridge 2013, pp. 122, 215

[355] Douglade 1982

[356] Zingaro 1994, p. 197; Emeléus & Sharpe 1959, p. 418; Addison & Sowerby 1972, p. 209; Mellor 1964, p. 337

[357] Pourbaix 1974, p. 521; Eagleson 1994, p. 92; Greenwood & Earnshaw 2002, p. 572

[358] Wiberg 2001, pp. 750, 975; Silberberg 2006, p. 314

[359] Sidgwick 1950, p. 784; Moody 1991, pp. 248–9, 319

[360] Krannich & Watkins 2006

[361] Greenwood & Earnshaw 2002, p. 553

[362] Dunstan 1968, p. 433

[363] Parise 1996, p. 112

[364] Carapella 1968a, p. 23

[365] Moss 1952, pp. 174, 179

[366] Dupree, Kirby & Freyland 1982, p. 604; Mhiaoui, Sar, & Gasser 2003

[367] Kotz, Treichel & Weaver 2009, p. 62

[368] Cotton et al. 1999, p. 396

[369] King 1994, p. 174

[370] Lidin 1996, p. 372

[371] Lindsjö, Fischer & Kloo 2004

[372] Friend 1953, p. 87

[373] Fesquet 1872, pp. 109–14

[374] Greenwood & Earnshaw 2002, p. 553; Massey 2000, p. 269

[375] King 1994, p.171

[376] Turova 2011, p. 46

[377] Pourbaix 1974, p. 530

[378] Wiberg 2001, p. 764

[379] House 2008, p. 497

[380] Mendeléeff 1897, p. 274

[381] Emsley 2001, p. 428

[382] Kudryavtsev 1974, p. 78

[383] Bagnall 1966, pp. 32–3, 59, 137

[384] Swink et al. 1966; Anderson et al. 1980

[385] Ahmed, Fjellvåg & Kjekshus 2000

[386] Chizhikov & Shchastlivyi 1970, p. 28

[387] Kudryavtsev 1974, p. 77

[388] Stuke 1974, p. 178; Donohue 1982, pp. 386–7; Cotton et al. 1999, p. 501

[389] Becker, Johnson & Nussbaum 1971, p. 56

[390] Berger 1997, p. 90

[391] Chizhikov & Shchastlivyi 1970, p. 16

[392] Jolly 1966, pp. 66–7

[393] Schwietzer & Pesterfield 2010, p. 239

[394] Cotton et al. 1999, p. 498

[395] Wells 1984, p. 715

[396] Wiberg 2001, p. 588

[397]   Mellor 1964a, p. 30; Wiberg 2001, p. 589

[398]   Greenwood & Earnshaw 2002, p. 765–6

[399]   Bagnall 1966, p. 134–51; Greenwood & Earnshaw 2002, p. 786

[400]   Detty & O'Regan 1994, pp. 1–2

[401]   Hill & Holman 2000, p. 124

[402]   Chang 2002, p. 314

[403]   Kent 1950, pp. 1–2; Clark 1960, p. 588; Warren & Geballe 1981

[404]   Housecroft & Sharpe 2008, p. 384; IUPAC 2006–, rhombohedral graphite entry

[405]   Mingos 1998, p. 171

[406]   Wiberg 2001, p. 781

[407]   Charlier, Gonze & Michenaud 1994

[408]   Atkins et al. 2006, pp. 320–1

[409]   Savvatimskiy 2005, p. 1138

[410]   Togaya 2000

[411]   Savvatimskiy 2009

[412]   Inagaki 2000, p. 216; Yasuda et al. 2003, pp. 3–11

[413]   O'Hare 1997, p. 230

[414]   Traynham 1989, pp. 930–1; Prakash & Schleyer 1997

[415]   Olmsted & Williams 1997, p. 436

[416]   Bailar et al. 1989, p. 743

[417]   Moore et al. 1985

[418]   House & House 2010, p. 526

[419]   Wiberg 2001, p. 798

[420]   Eagleson 1994, p. 175

[421]   Atkins et al. 2006, p. 121

[422]   Russell & Lee 2005, pp. 358–9

[423]   Keevil 1989, p. 103

[424]   Russell & Lee 2005, pp. 358–60 et seq

[425]   Harding, Janes & Johnson 2002, pp. 118

[426]   Metcalfe, Williams & Castka 1974, p. 539

[427]   Cobb & Fetterolf 2005, p. 64; Metcalfe, Williams & Castka 1974, p. 539

[428]   Ogata, Li & Yip 2002; Boyer et al. 2004, p. 1023; Russell & Lee 2005, p. 359

[429]   Cooper 1968, p. 25; Henderson 2000, p. 5; Silberberg 2006, p. 314

[430]   Wiberg 2001, p. 1014

[431] Daub & Seese 1996, pp. 70, 109: "Aluminum is not a metalloid but a metal because it has mostly metallic properties.";
Denniston, Topping & Caret 2004, p. 57: "Note that aluminum (Al) is classified as a metal, not a metalloid."; Hasan 2009, p.
16: "Aluminum does not have the characteristics of a metalloid but rather those of a metal."

[432] Holt, Rinehart & Wilson c. 2007

[433] Tuthill 2011

[434] Stott 1956, p. 100

[435] Steele 1966, p. 60

[436] Emsley 2001, p. 382

[437] Young et al. 2010, p. 9; Craig & Maher 2003, p. 391. Selenium is "near metalloidal".

[438] Rochow 1957

[439] Rochow 1966, p. 224

[440] Moss 1952, p. 192

[441] Glinka 1965, p. 356

[442] Evans 1966, pp. 124–5

[443] Regnault 1853, p. 208

[444] Scott & Kanda 1962, p. 311

[445] Cotton et al. 1999, pp. 496, 503–4

[446] Arlman 1939; Bagnall 1966, pp. 135, 142–3

[447] Chao & Stenger 1964

[448] Berger 1997, pp. 86–7

[449] Snyder 1966, p. 242

[450] Fritz & Gjerde 2008, p. 235

[451] Meyer et al. 2005, p. 284; Manahan 2001, p. 911; Szpunar et al. 2004, p. 17

[452] US Environmental Protection Agency 1988, p. 1; Uden 2005, pp. 347–8

[453] De Zuane 1997, p. 93; Dev 2008, pp. 2–3

[454] Wiberg 2001, p. 594

[455] Greenwood & Earnshaw 2002, p. 786; Schwietzer & Pesterfield 2010, pp. 242–3

[456] Bagnall 1966, p. 41; Nickless 1968, p. 79

[457] Bagnall 1990, pp. 313–14; Lehto & Hou 2011, p. 220; Siekierski & Burgess 2002, p. 117: "The tendency to form X^{2-} anions
decreases down the Group [16 elements] ..."

[458] Legit, Friák & Šob 2010, p. 214118-18

[459] Manson & Halford 2006, pp. 378, 410

[460] Bagnall 1957, p. 62; Fernelius 1982, p. 741

[461] Bagnall 1966, p. 41; Barrett 2003, p. 119

[462] Hawkes 2010; Holt, Rinehart & Wilson c. 2007; Hawkes 1999, p. 14; Roza 2009, p. 12

[463] Keller 1985

[464] Harding, Johnson & Janes 2002, p. 61

[465] Long & Hentz 1986, p. 58

[466] Vasáros & Berei 1985, p. 109

[467] Haissinsky & Coche 1949, p. 400

[468] Brownlee et al. 1950, p. 173

[469] Hermann, Hoffmann & Ashcroft 2013

[470] Siekierski & Burgess 2002, pp. 65, 122

[471] Emsley 2001, p. 48

[472] Rao & Ganguly 1986

[473] Krishnan et al. 1998

[474] Glorieux, Saboungi & Enderby 2001

[475] Millot et al. 2002

[476] Vasáros & Berei 1985, p. 117

[477] Kaye & Laby 1973, p. 228

[478] Samsonov 1968, p. 590

[479] Korenman 1959, p. 1368

[480] Rossler 1985, pp. 143–4

[481] Champion et al. 2010

[482] Borst 1982, pp. 465, 473

[483] Batsanov 1971, p. 811

[484] Swalin 1962, p. 216; Feng & Lin 2005, p. 157

[485] Schwietzer & Pesterfield 2010, pp. 258–60

[486] Hawkes 1999, p. 14

[487] Olmsted & Williams 1997, p. 328; Daintith 2004, p. 277

[488] Eberle1985, pp. 213–16, 222–7

[489] Restrepo et al. 2004, p. 69; Restrepo et al. 2006, p. 411

[490] Greenwood & Earnshaw 2002, p. 804

[491] Greenwood & Earnshaw 2002, p. 803

[492] Wiberg 2001, p. 416

[493] Craig & Maher 2003, p. 391; Schroers 2013, p. 32; Vernon 2013, pp. 1704–1705

[494] Cotton et al. 1999, p. 42

[495] Marezio & Licci 2000, p. 11

[496] Vernon 2013, p. 1705

[497] Russell & Lee 2005, p. 5

[498] Parish 1977, pp. 178, 192–3

[499] Eggins 1972, p. 66; Rayner-Canham & Overton 2006, pp. 29–30

[500] Atkins et al. 2006, pp. 320–1; Bailar et al. 1989, p. 742–3

[501] Rochow 1966, p. 7; Taniguchi et al. 1984, p. 867: "... black phosphorus ... [is] characterized by the wide valence bands with rather delocalized nature."; Morita 1986, p. 230; Carmalt & Norman 1998, p. 7: "Phosphorus ... should therefore be expected to have some metalloid properties."; Du et al. 2010. Interlayer interactions in black phosphorus, which are attributed to van der Waals-Keesom forces, are thought to contribute to the smaller band gap of the bulk material (calculated 0.19 eV; observed 0.3 eV) as opposed to the larger band gap of a single layer (calculated ~0.75 eV).

[502] Stuke 1974, p. 178; Cotton et al. 1999, p. 501; Craig & Maher 2003, p. 391

[503] Steudel 1977, p. 240: "... considerable orbital overlap must exist, to form intermolecular, many-center ... [sigma] bonds, spread through the layer and populated with delocalized electrons, reflected in the properties of iodine (lustre, color, moderate electrical conductivity)."; Segal 1989, p. 481: "Iodine exhibits some metallic properties ..."

[504] Lutz et al. 2011, p. 17

[505] Yacobi & Holt 1990, p. 10; Wiberg 2001, p. 160

[506] Greenwood & Earnshaw 2002, pp. 479, 482

[507] Eagleson 1994, p. 820

[508] Oxtoby, Gillis & Campion 2008, p. 508

[509] Brescia et al. 1980, pp. 166–71

[510] Fine & Beall 1990, p. 578

[511] Wiberg 2001, p. 901

[512] Berger 1997, p. 80

[513] Lovett 1977, p. 101

[514] Cohen & Chelikowsky 1988, p. 99

[515] Taguena-Martinez, Barrio & Chambouleyron 1991, p. 141

[516] Ebbing & Gammon 2010, p. 891

[517] Asmussen & Reinhard 2002, p. 7

[518] Deprez & McLachan 1988

[519] Addison 1964 (P, Se, Sn); Marković, Christiansen & Goldman 1998 (Bi); Nagao et al. 2004

[520] Lide 2005; Wiberg 2001, p. 423: At

[521] Cox 1997, pp. 182–86

[522] MacKay, MacKay & Henderson 2002, p. 204

[523] Baudis 2012, pp. 207–8

[524] Wiberg 2001, p. 741

[525] Chizhikov & Shchastlivyi 1968, p. 96

[526] Greenwood & Earnshaw 2002, pp. 140–1, 330, 369, 548–9, 749: B, Si, Ge, As, Sb, Te

[527] Kudryavtsev 1974, p. 158

[528] Greenwood & Earnshaw 2002, pp. 271, 219, 748–9, 886: C, Al, Se, Po, At; Wiberg 2001, p. 573: Se

[529] United Nuclear 2013

[530] Zalutsky & Pruszynski 2011, p. 181

22.12.2   References

- Addison WE 1964, *The Allotropy of the Elements,* Oldbourne Press, London

- Addison CC & Sowerby DB 1972, *Main Group Elements: Groups V and VI,* Butterworths, London, ISBN 0-8391-1005-7

- Adler D 1969, 'Half-way Elements: The Technology of Metalloids', book review, *Technology Review,* vol. 72, no. 1, Oct/Nov, pp. 18–19, ISSN 00401692

- Ahmed MAK, Fjellvåg H & Kjekshus A 2000, 'Synthesis, Structure and Thermal Stability of Tellurium Oxides and Oxide Sulfate Formed from Reactions in Refluxing Sulfuric Acid', *Journal of the Chemical Society, Dalton Transactions,* no. 24, pp. 4542–9, doi:10.1039/B005688J

- Ahmeda E & Rucka M 2011, 'Homo- and heteroatomic polycations of groups 15 and 16. Recent advances in synthesis and isolation using room temperature ionic liquids', *Coordination Chemistry Reviews,* vol. 255, nos 23–24, pp. 2892–2903, doi:10.1016/j.ccr.2011.06.011

- Allen DS & Ordway RJ 1968, *Physical Science,* 2nd ed., Van Nostrand, Princeton, New Jersey, ISBN 978-0-442-00290-9

- Allen PB & Broughton JQ 1987, 'Electrical Conductivity and Electronic Properties of Liquid Silicon', *Journal of Physical Chemistry,* vol. 91, no. 19, pp. 4964–70, doi:10.1021/j100303a015

- Alloul H 2010, *Introduction to the Physics of Electrons in Solids,* Springer-Verlag, Berlin, ISBN 3-642-13564-1

- Anderson JB, Rapposch MH, Anderson CP & Kostiner E 1980, 'Crystal Structure Refinement of Basic Tellurium Nitrate: A Reformulation as $(Te_2O_4H)^+(NO_3)^-$', *Monatshefte für Chemie/ Chemical Monthly,* vol. 111, no. 4, pp. 789–96, doi:10.1007/BF00899243

- Antman KH 2001, 'Introduction: The History of Arsenic Trioxide in Cancer Therapy', *The Oncologist,* vol. 6, suppl. 2, pp. 1–2, doi:10.1634/theoncologist.6-suppl_2-1

- Apseloff G 1999, 'Therapeutic Uses of Gallium Nitrate: Past, Present, and Future', *American Journal of Therapeutics,* vol. 6, no. 6, pp. 327–39, ISSN 15363686

- Arlman EJ 1939, 'The Complex Compounds $P(OH)_4.ClO_4$ and $Se(OH)_3.ClO_4$', *Recueil des Travaux Chimiques des Pays-Bas,* vol. 58, no. 10, pp. 871–4, ISSN 01650513

- Askeland DR, Phulé PP & Wright JW 2011, *The Science and Engineering of Materials,* 6th ed., Cengage Learning, Stamford, CT, ISBN 0-495-66802-8

- Asmussen J & Reinhard DK 2002, *Diamond Films Handbook,* Marcel Dekker, New York, ISBN 0-8247-9577-6

- Atkins P, Overton T, Rourke J, Weller M & Armstrong F 2006, *Shriver & Atkins' Inorganic Chemistry,* 4th ed., Oxford University Press, Oxford, ISBN 0-7167-4878-9

- Atkins P, Overton T, Rourke J, Weller M & Armstrong F 2010, *Shriver & Atkins' Inorganic Chemistry,* 5th ed., Oxford University Press, Oxford, ISBN 1-4292-1820-7

- Austen K 2012, 'A Factory for Elements that Barely Exist', *New Scientist,* 21 Apr, p. 12

- Ba LA, Döring M, Jamier V & Jacob C 2010, 'Tellurium: an Element with Great Biological Potency and Potential', *Organic & Biomolecular Chemistry,* vol. 8, pp. 4203–16, doi:10.1039/C0OB00086H

- Bagnall KW 1957, *Chemistry of the Rare Radioelements: Polonium-actinium,* Butterworths Scientific Publications, London

- Bagnall KW 1966, *The Chemistry of Selenium, Tellurium and Polonium,* Elsevier, Amsterdam

- Bagnall KW 1990, 'Compounds of Polonium', in KC Buschbeck & C Keller (eds), *Gmelin Handbook of Inorganic and Organometallic Chemistry,* 8th ed., Po Polonium, Supplement vol. 1, Springer-Verlag, Berlin, pp. 285–340, ISBN 3-540-93616-5

- Bailar JC, Moeller T & Kleinberg J 1965, *University Chemistry,* DC Heath, Boston

- Bailar JC & Trotman-Dickenson AF 1973, *Comprehensive Inorganic Chemistry,* vol. 4, Pergamon, Oxford

- Bailar JC, Moeller T, Kleinberg J, Guss CO, Castellion ME & Metz C 1989, *Chemistry,* 3rd ed., Harcourt Brace Jovanovich, San Diego, ISBN 0-15-506456-8

- Barfuß H, Böhnlein G, Freunek P, Hofmann R, Hohenstein H, Kreische W, Niedrig H and Reimer A 1981, 'The Electric Quadrupole Interaction of ^{111}Cd in Arsenic Metal and in the System $Sb_{1-x}In_x$ and $Sb_{1-x}Cd_x$', *Hyperfine Interactions,* vol. 10, nos 1–4, pp. 967–72, doi:10.1007/BF01022038

- Barnett EdB & Wilson CL 1959, *Inorganic Chemistry: A Text-book for Advanced Students,* 2nd ed., Longmans, London

- Barrett J 2003, *Inorganic Chemistry in Aqueous Solution,* The Royal Society of Chemistry, Cambridge, ISBN 0-85404-471-X

- Barsanov GP & Ginzburg AI 1974, 'Mineral', in AM Prokhorov (ed.), *Great Soviet Encyclopedia,* 3rd ed., vol. 16, Macmillan, New York, pp. 329–32

- Bassett LG, Bunce SC, Carter AE, Clark HM & Hollinger HB 1966, *Principles of Chemistry,* Prentice-Hall, Englewood Cliffs, New Jersey

- Batsanov SS 1971, 'Quantitative Characteristics of Bond Metallicity in Crystals', *Journal of Structural Chemistry,* vol. 12, no. 5, pp. 809–13, doi:10.1007/BF00743349

- Baudis U & Fichte R 2012, 'Boron and Boron Alloys', in F Ullmann (ed.), *Ullmann's Encyclopedia of Industrial Chemistry,* vol. 6, Wiley-VCH, Weinheim, pp. 205–17, doi:10.1002/14356007.a04_281

- Becker WM, Johnson VA & Nussbaum 1971, 'The Physical Properties of Tellurium', in WC Cooper (ed.), *Tellurium,* Van Nostrand Reinhold, New York

- Belpassi L, Tarantelli F, Sgamellotti A & Quiney HM 2006, 'The Electronic Structure of Alkali Aurides. A Four-Component Dirac–Kohn–Sham study', *The Journal of Physical Chemistry A,* vol. 110, no. 13, April 6, pp. 4543–54, doi:10.1021/jp054938w

- Berger LI 1997, *Semiconductor Materials,* CRC Press, Boca Raton, Florida, ISBN 0-8493-8912-7

- Bettelheim F, Brown WH, Campbell MK & Farrell SO 2010, *Introduction to General, Organic, and Biochemistry,* 9th ed., Brooks/Cole, Belmont CA, ISBN 0-495-39112-3

- Bianco E, Butler S, Jiang S, Restrepo OD, Windl W & Goldberger JE 2013, 'Stability and Exfoliation of Germanane: A Germanium Graphane Analogue,' *ACS Nano,* March 19 (web), doi:10.1021/nn4009406

- Bodner GM & Pardue HL 1993, *Chemistry, An Experimental Science,* John Wiley & Sons, New York, ISBN 0-471-59386-9

- Bogoroditskii NP & Pasynkov VV 1967, *Radio and Electronic Materials,* Iliffe Books, London

- Bomgardner MM 2013, 'Thin-Film Solar Firms Revamp To Stay In The Game', *Chemical & Engineering News,* vol. 91, no. 20, pp. 20–1, ISSN 00092347

- Bond GC 2005, *Metal-Catalysed Reactions of Hydrocarbons, Springer, New York, ISBN 0-387-24141-8*

- Booth VH & Bloom ML 1972, *Physical Science: A Study of Matter and Energy,* Macmillan, New York

- Borst KE 1982, 'Characteristic Properties of Metallic Crystals', *Journal of Educational Modules for Materials Science and Engineering,* vol. 4, no. 3, pp. 457–92, ISSN 01973940

- Boyer RD, Li J, Ogata S & Yip S 2004, 'Analysis of Shear Deformations in Al and Cu: Empirical Potentials Versus Density Functional Theory', *Modelling and Simulation in Materials Science and Engineering,* vol. 12, no. 5, pp. 1017–29, doi:10.1088/0965-0393/12/5/017

- Bradbury GM, McGill MV, Smith HR & Baker PS 1957, *Chemistry and You,* Lyons and Carnahan, Chicago

- Bradley D 2014, *Resistance is Low: New Quantum Effect,* spectroscopyNOW, viewed 15 December 2014-12-15

- Brescia F, Arents J, Meislich H & Turk A 1980, *Fundamentals of Chemistry,* 4th ed., Academic Press, New York, ISBN 0-12-132392-7

- Brown L & Holme T 2006, *Chemistry for Engineering Students, Thomson Brooks/Cole, Belmont California, ISBN 0-495-01718-3*

- Brown WP c. 2007 'The Properties of Semi-Metals or Metalloids,' *Doc Brown's Chemistry: Introduction to the Periodic Table,* viewed 8 February 2013

- Brown TL, LeMay HE, Bursten BE, Murphy CJ, Woodward P 2009, *Chemistry: The Central Science,* 11th ed., Pearson Education, Upper Saddle River, New Jersey, ISBN 978-0-13-235848-4

- Brownlee RB, Fuller RW, Hancock WJ, Sohon MD & Whitsit JE 1943, *Elements of Chemistry,* Allyn and Bacon, Boston

- Brownlee RB, Fuller RT, Whitsit JE Hancock WJ & Sohon MD 1950, *Elements of Chemistry,* Allyn and Bacon, Boston

- Bucat RB (ed.) 1983, *Elements of Chemistry: Earth, Air, Fire & Water, vol. 1,* Australian Academy of Science, Canberra, ISBN 0-85847-113-2

- Büchel KH (ed.) 1983, *Chemistry of Pesticides,* John Wiley & Sons, New York, ISBN 0-471-05682-0

- Büchel KH, Moretto H-H, Woditsch P 2003, *Industrial Inorganic Chemistry,* 2nd ed., Wiley-VCH, ISBN 3-527-29849-5

- Burkhart CN, Burkhart CG & Morrell DS 2011, 'Treatment of Tinea Versicolor', in HI Maibach & F Gorouhi (eds), *Evidence Based Dermatology,* 2nd ed., People's Medical Publishing House-USA, Shelton, CT, pp. 365–72, ISBN 978-1-60795-039-4

- Burrows A, Holman J, Parsons A, Pilling G & Price G 2009, *Chemistry[3]: Introducing Inorganic, Organic and Physical Chemistry,* Oxford University, Oxford, ISBN 0-19-927789-3

- Butterman WC & Carlin JF 2004, *Mineral Commodity Profiles: Antimony,* US Geological Survey

- Butterman WC & Jorgenson JD 2005, *Mineral Commodity Profiles: Germanium,* US Geological Survey

- Calderazzo F, Ercoli R & Natta G 1968, 'Metal Carbonyls: Preparation, Structure, and Properties', in I Wender & P Pino (eds), *Organic Syntheses via Metal Carbonyls: Volume 1,* Interscience Publishers, New York, pp. 1–272

- Carapella SC 1968a, 'Arsenic' in CA Hampel (ed.), *The Encyclopedia of the Chemical Elements,* Reinhold, New York, pp. 29–32

- Carapella SC 1968, 'Antimony' in CA Hampel (ed.), *The Encyclopedia of the Chemical Elements,* Reinhold, New York, pp. 22–5

- Carlin JF 2011, *Minerals Year Book: Antimony,* United States Geological Survey

- Carmalt CJ & Norman NC 1998, 'Arsenic, Antimony and Bismuth: Some General Properties and Aspects of Periodicity', in NC Norman (ed.), *Chemistry of Arsenic, Antimony and Bismuth,* Blackie Academic & Professional, London, pp. 1–38, ISBN 0-7514-0389-X

- Carter CB & Norton MG 2013, *Ceramic Materials: Science and Engineering,* 2nd ed., Springer Science+Business Media, New York, ISBN 978-1-4614-3523-5

- Cegielski C 1998, *Yearbook of Science and the Future,* Encyclopædia Britannica, Chicago, ISBN 0-85229-657-6

- Chalmers B 1959, *Physical Metallurgy,* John Wiley & Sons, New York

- Champion J, Alliot C, Renault E, Mokili BM, Chérel M, Galland N & Montavon G 2010, 'Astatine Standard Redox Potentials and Speciation in Acidic Medium', *The Journal of Physical Chemistry A,* vol. 114, no. 1, pp. 576–82, doi:10.1021/jp9077008

- Chang R 2002, *Chemistry,* 7th ed., McGraw Hill, Boston, ISBN 0-07-246533-6

- Chao MS & Stenger VA 1964, 'Some Physical Properties of Highly Purified Bromine', *Talanta,* vol. 11, no. 2, pp. 271–81, doi:10.1016/0039-9140(64)80036-9

- Charlier J-C, Gonze X, Michenaud J-P 1994, First-principles Study of the Stacking Effect on the Electronic Properties of Graphite(s), *Carbon,* vol. 32, no. 2, pp. 289–99, doi:10.1016/0008-6223(94)90192-9

- Chatt J 1951, 'Metal and Metalloid Compounds of the Alkyl Radicals', in EH Rodd (ed.), *Chemistry of Carbon Compounds: A Modern Comprehensive Treatise,* vol. 1, part A, Elsevier, Amsterdam, pp. 417–58

- Chedd G 1969, *Half-Way Elements: The Technology of Metalloids,* Doubleday, New York

- Chizhikov DM & Shchastlivyi VP 1968, *Selenium and Selenides,* translated from the Russian by EM Elkin, Collet's, London

- Chizhikov DM & Shchastlivyi 1970, *Tellurium and the Tellurides,* Collet's, London

- Choppin GR & Johnsen RH 1972, *Introductory Chemistry,* Addison-Wesley, Reading, Massachusetts

- Chopra IS, Chaudhuri S, Veyan JF & Chabal YJ 2011, 'Turning Aluminium into a Noble-metal-like Catalyst for Low-temperature Activation of Molecular Hydrogen', *Nature Materials,* vol. 10, pp. 884–889, doi:10.1038/nmat3123

- Chung DDL 2010, *Composite Materials: Science and Applications,* 2nd ed., Springer-Verlag, London, ISBN 978-1-84882-830-8

- Clark GL 1960, *The Encyclopedia of Chemistry,* Reinhold, New York

- Cobb C & Fetterolf ML 2005, *The Joy of Chemistry,* Prometheus Books, New York, ISBN 1-59102-231-2

- Cohen ML & Chelikowsky JR 1988, *Electronic Structure and Optical Properties of Semiconductors,* Springer Verlag, Berlin, ISBN 3-540-18818-5

- Coles BR & Caplin AD 1976, *The Electronic Structures of Solids,* Edward Arnold, London, ISBN 0-8448-0874-1

- Conkling JA & Mocella C 2011, *Chemistry of Pyrotechnics: Basic Principles and Theory,* 2nd ed., CRC Press, Boca Raton, FL, ISBN 978-1-57444-740-8

- Considine DM & Considine GD (eds) 1984, 'Metalloid', in *Van Nostrand Reinhold Encyclopedia of Chemistry,* 4th ed., Van Nostrand Reinhold, New York, ISBN 0-442-22572-5

- Cooper DG 1968, *The Periodic Table,* 4th ed., Butterworths, London

- Corbridge DEC 2013, *Phosphorus: Chemistry, Biochemistry and Technology,* 6th ed., CRC Press, Boca Raton, Florida, ISBN 978-1-4398-4088-7

- Corwin CH 2005, *Introductory Chemistry: Concepts & Connections,* 4th ed., Prentice Hall, Upper Saddle River, New Jersey, ISBN 0-13-144850-1

- Cotton FA, Wilkinson G & Gaus P 1995, *Basic Inorganic Chemistry,* 3rd ed., John Wiley & Sons, New York, ISBN 0-471-50532-3

- Cotton FA, Wilkinson G, Murillo CA & Bochmann 1999, *Advanced Inorganic Chemistry,* 6th ed., John Wiley & Sons, New York, ISBN 0-471-19957-5

- Cox PA 1997, *The Elements: Their Origin, Abundance and Distribution,* Oxford University, Oxford, ISBN 0-19-855298-X

- Cox PA 2004, *Inorganic Chemistry,* 2nd ed., Instant Notes series, Bios Scientific, London, ISBN 1-85996-289-0

- Craig PJ, Eng G & Jenkins RO 2003, 'Occurrence and Pathways of Organometallic Compounds in the Environment—General Considerations' in PJ Craig (ed.), *Organometallic Compounds in the Environment,* 2nd ed., John Wiley & Sons, Chichester, West Sussex, pp. 1–56, ISBN 0471899933

- Craig PJ & Maher WA 2003, 'Organoselenium compounds in the environment', in *Organometallic Compounds in the Environment,* PJ Craig (ed.), John Wiley & Sons, New York, pp. 391–398, ISBN 0-471-89993-3

- Crow JM 2011, 'Boron Carbide Could Light Way to Less-toxic Green Pyrotechnics', *Nature News,* 8 April, doi:10.1038/news.2011.22?

- Cusack N 1967, *The Electrical and Magnetic Properties of Solids: An Introductory Textbook*, 5th ed., John Wiley & Sons, New York

- Cusack N E 1987, *The Physics of Structurally Disordered Matter: An Introduction,* A Hilger in association with the University of Sussex Press, Bristol, ISBN 0-85274-591-5

- Daily Record 2014, 'Picatinny Chemist Recognized for Work on Smoke Grenades', *Daily Record,* 2 April, viewed 7 April 2014

- Daintith J (ed.) 2004, *Oxford Dictionary of Chemistry,* 5th ed., Oxford University, Oxford, ISBN 0-19-920463-2

- Danaith J (ed.) 2008, *Oxford Dictionary of Chemistry,* Oxford University Press, Oxford, ISBN 978-0-19-920463-2

- Daniel-Hoffmann M, Sredni B & Nitzan Y 2012, 'Bactericidal Activity of the Organo-Tellurium Compound AS101 Against *Enterobacter Cloacae,'* *Journal of Antimicrobial Chemotherapy,* vol. 67, no. 9, pp. 2165–72, doi:10.1093/jac/dks185

- Daub GW & Seese WS 1996, *Basic Chemistry,* 7th ed., Prentice Hall, New York, ISBN 0-13-373630-X

- Davidson DF & Lakin HW 1973, 'Tellurium', in DA Brobst & WP Pratt (eds), *United States Mineral Resources,* Geological survey professional paper 820, United States Government Printing Office, Washington, pp. 627–30

- Dávila ME, Molotov SL, Laubschat C & Asensio MC 2002, 'Structural Determination of Yb Single-Crystal Films Grown on W(110) Using Photoelectron Diffraction', *Physical Review B,* vol. 66, no. 3, p. 035411–18, doi:10.1103/PhysRevB.66.035411

- Demetriou MD, Launey ME, Garrett G, Schramm JP, Hofmann DC, Johnson WL & Ritchie RO 2011, 'A Damage-Tolerant Glass', *Nature Materials,* vol. 10, February, pp. 123–8, doi:10.1038/nmat2930

- Deming HG 1925, *General Chemistry: An Elementary Survey,* 2nd ed., John Wiley & Sons, New York

- Denniston KJ, Topping JJ & Caret RL 2004, *General, Organic, and Biochemistry,* 5th ed., McGraw-Hill, New York, ISBN 0-07-282847-1

- Deprez N & McLachan DS 1988, 'The Analysis of the Electrical Conductivity of Graphite Conductivity of Graphite Powders During Compaction', *Journal of Physics D: Applied Physics,* vol. 21, no. 1, doi:10.1088/0022-3727/21/1/015

- Desai PD, James HM & Ho CY 1984, 'Electrical Resistivity of Aluminum and Manganese', *Journal of Physical and Chemical Reference Data,* vol. 13, no. 4, pp. 1131–72, doi:10.1063/1.555725

- Desch CH 1914, *Intermetallic Compounds,* Longmans, Green and Co., New York

- Detty MR & O'Regan MB 1994, *Tellurium-Containing Heterocycles,* (The Chemistry of Heterocyclic Compounds, vol. 53), John Wiley & Sons, New York

- Dev N 2008, 'Modelling Selenium Fate and Transport in Great Salt Lake Wetlands', PhD dissertation, University of Utah, ProQuest, Ann Arbor, Michigan, ISBN 0-549-86542-X

- De Zuane J 1997, *Handbook of Drinking Water Quality,* 2nd ed., John Wiley & Sons, New York, ISBN 0-471-28789-X

- Di Pietro P 2014, *Optical Properties of Bismuth-Based Topological Insulators,* Springer International Publishing, Cham, Switzerland, ISBN 978-3-319-01990-1

- Divakar C, Mohan M & Singh AK 1984, 'The Kinetics of Pressure-Induced Fcc-Bcc Transformation in Ytterbium', *Journal of Applied Physics,* vol. 56, no. 8, pp. 2337–40, doi:10.1063/1.334270

- Donohue J 1982, *The Structures of the Elements,* Robert E. Krieger, Malabar, Florida, ISBN 0-89874-230-7

- Douglade J & Mercier R 1982, 'Structure Cristalline et Covalence des Liaisons dans le Sulfate d'Arsenic(III), $As_2(SO_4)_3$', *Acta Crystallographica Section B,* vol. 38, no. 3, pp. 720–3, doi:10.1107/S056774088200394X

- Du Y, Ouyang C, Shi S & Lei M 2010, 'Ab Initio Studies on Atomic and Electronic Structures of Black Phosphorus', *Journal of Applied Physics,* vol. 107, no. 9, pp. 093718–1–4, doi:10.1063/1.3386509

- Dunlap BD, Brodsky MB, Shenoy GK & Kalvius GM 1970, 'Hyperfine Interactions and Anisotropic Lattice Vibrations of ^{237}Np in α-Np Metal', *Physical Review B,* vol. 1, no. 1, pp. 44–9, doi:10.1103/PhysRevB.1.44

- Dunstan S 1968, *Principles of Chemistry,* D. Van Nostrand Company, London

- Dupree R, Kirby DJ & Freyland W 1982, 'N.M.R. Study of Changes in Bonding and the Metal-Non-metal Transition in Liquid Caesium-Antimony Alloys', *Philosophical Magazine Part B,* vol.46 no.6, pp.595–606, doi:10.1083546

- Eagleson M 1994, *Concise Encyclopedia Chemistry,* Walter de Gruyter, Berlin, ISBN 3-11-011451-8

- Eason R 2007, *Pulsed Laser Deposition of Thin Films: Applications-Led Growth of Functional Materials,* Wiley-Interscience, New York

- Ebbing DD & Gammon SD 2010, *General Chemistry,* 9th ed. enhanced, Brooks/Cole, Belmont, California, ISBN 978-0-618-93469-0

- Eberle SH 1985, 'Chemical Behavior and Compounds of Astatine', pp. 183–209, in Kugler & Keller

- Edwards PP & Sienko MJ 1983, 'On the Occurrence of Metallic Character in the Periodic Table of the Elements', *Journal of Chemical Education,* vol. 60, no. 9, pp. 691–6, doi:10.1021ed060p691

- Edwards PP 1999, 'Chemically Engineering the Metallic, Insulating and Superconducting State of Matter' in KR Seddon & M Zaworotko (eds), Crystal Engineering: The Design and Application of Functional Solids, Kluwer Academic, Dordrecht, pp. 409–431, ISBN 0-7923-5905-4

- Edwards PP 2000, 'What, Why and When is a metal?', in N Hall (ed.), The New Chemistry, Cambridge University, Cambridge, pp. 85–114, ISBN 0-521-45224-4

- Edwards PP, Lodge MTJ, Hensel F & Redmer R 2010, '... A Metal Conducts and a Non-metal Doesn't', *Philosophical Transactions of the Royal Society A: Mathematical, Physical and Engineering Sciences,* vol. 368, pp. 941–65, doi:10.1098/rsta.2009.0282

- Eggins BR 1972, *Chemical Structure and Reactivity,* MacMillan, London, ISBN 0-333-08145-5

- Eichler R, Aksenov NV, Belozerov AV, Bozhikov GA, Chepigin VI, Dmitriev SN, Dressler R, Gäggeler HW, Gorshkov VA, Haenssler F, Itkis MG, Laube A, Lebedev VY, Malyshev ON, Oganessian YT, Petrushkin OV, Piguet D, Rasmussen P, Shishkin SV, Shutov, AV, Svirikhin AI, Tereshatov EE, Vostokin GK, Wegrzecki M & Yeremin AV 2007, 'Chemical Characterization of Element 112,' *Nature,* vol. 447, pp. 72–5, doi:10.1038/nature05761

- Ellern H 1968, *Military and Civilian Pyrotechnics,* Chemical Publishing Company, New York

- Emeléus HJ & Sharpe AG 1959, *Advances in Inorganic Chemistry and Radiochemistry,* vol. 1, Academic Press, New York

- Emsley J 1971, *The Inorganic Chemistry of the Non-metals,* Methuen Educational, London, ISBN 0-423-86120-4

- Emsley J 2001, *Nature's Building Blocks: An A–Z guide to the Elements, Oxford University Press, Oxford, ISBN 0-19-850341-5*

- Eranna G 2011, *Metal Oxide Nanostructures as Gas Sensing Devices,* Taylor & Francis, Boca Raton, Florida, ISBN 1-4398-6340-7

- Evans KA 1993, 'Properties and Uses of Oxides and Hydroxides,' in AJ Downs (ed.), *Chemistry of Aluminium, Gallium, Indium, and Thallium,* Blackie Academic & Professional, Bishopbriggs, Glasgow, pp. 248–91, ISBN 0-7514-0103-X

- Evans RC 1966, *An Introduction to Crystal Chemistry,* Cambridge University, Cambridge

- Everest DA 1953, 'The Chemistry of Bivalent Germanium Compounds. Part IV. Formation of Germanous Salts by Reduction with Hydrophosphorous Acid.'*Journal of the Chemical Society,*pp.4117–4120,:10.1039/JR9530004117

- EVM (Expert Group on Vitamins and Minerals) 2003, *Safe Upper Levels for Vitamins and Minerals*, UK Food Standards Agency, London, ISBN 1-904026-11-7

- Farandos NM, Yetisen AK, Monteiro MJ, Lowe CR & Yun SH 2014, 'Contact Lens Sensors in Ocular Diagnostics', *Advanced Healthcare Materials,* doi:10.1002/adhm.201400504, viewed 23 November 2014

- Fehlner TP 1992, 'Introduction', in TP Fehlner (ed.), *Inorganometallic chemistry*, Plenum, New York, pp. 1–6, ISBN 0-306-43986-7

- Fehlner TP 1990, 'The Metallic Face of Boron,' in AG Sykes (ed.), *Advances in Inorganic Chemistry,* vol. 35, Academic Press, Orlando, pp. 199–233

- Feng & Jin 2005, *Introduction to Condensed Matter Physics: Volume 1,* World Scientific, Singapore, ISBN 1-84265-347-4

- Fernelius WC 1982, 'Polonium', *Journal of Chemical Education,* vol. 59, no. 9, pp. 741–2, doi:10.1021/ed059p741

- Ferro R & Saccone A 2008, *Intermetallic Chemistry,* Elsevier, Oxford, p. 233, ISBN 0-08-044099-1

- Fesquet AA 1872, *A Practical Guide for the Manufacture of Metallic Alloys,* trans. A. Guettier, Henry Carey Baird, Philadelphia

- Fine LW & Beall H 1990, *Chemistry for Engineers and Scientists,* Saunders College Publishing, Philadelphia, ISBN 0-03-021537-4

- Fokwa BPT 2014, 'Borides: Solid-state Chemistry', in *Encyclopedia of Inorganic and Bioinorganic Chemistry,* John Wiley and Sons, doi:10.1002/9781119951438.eibc0022.pub2

- Foster W 1936, *The Romance of Chemistry,* D Appleton-Century, New York

- Foster LS & Wrigley AN 1958, 'Periodic Table', in GL Clark, GG Hawley & WA Hamor (eds), *The Encyclopedia of Chemistry (Supplement),* Reinhold, New York, pp. 215–20

- Friend JN 1953, *Man and the Chemical Elements,* 1st ed., Charles Scribner's Sons, New York

- Fritz JS & Gjerde DT 2008, *Ion Chromatography, John Wiley & Sons, New York, ISBN 3-527-61325-0*

- Gary S 2013, 'Poisoned Alloy' the Metal of the Future', *News in science,* viewed 28 August 2013

- Geckeler S 1987, *Optical Fiber Transmission Systems*, Artech Hous, Norwood, Massachusetts, ISBN 0-89006-226-9

- Geman Energy Society 2008, *Planning and Installing Photovoltaic Systems: A Guide for Installers, Architects and Engineers*, 2nd ed., Earthscan, London, ISBN 978-1-84407-442-6

- Gordh G, Gordh G & Headrick D 2003, *A Dictionary of Entomology,* CABI Publishing, Wallingford, ISBN 0-85199-655-8

- Gillespie RJ & Robinson EA 1963, 'The Sulphuric Acid Solvent System. Part IV. Sulphato Compounds of Arsenic (III)', *Canadian Journal of Chemistry,* vol. 41, no. 2, pp. 450–458

- Gillespie RJ & Passmore J 1972, 'Polyatomic Cations', *Chemistry in Britain,* vol. 8, pp. 475–479

- Gladyshev VP & Kovaleva SV 1998, 'Liquidus Shape of the Mercury–Gallium System', *Russian Journal of Inorganic Chemistry,* vol. 43, no. 9, pp. 1445–6

- Glazov VM, Chizhevskaya SN & Glagoleva NN 1969, *Liquid Semiconductors,* Plenum, New York

- Glinka N 1965, *General Chemistry,* trans. D Sobolev, Gordon & Breach, New York

- Glockling F 1969, *The Chemistry of Germanium,* Academic, London

- Glorieux B, Saboungi ML & Enderby JE 2001, 'Electronic Conduction in Liquid Boron', *Europhysics Letters (EPL),* vol. 56, no. 1, pp. 81–5, doi:10.1209/epl/i2001-00490-0

- Goldsmith RH 1982, 'Metalloids',*Journal of Chemical Education*,vol.59,no.6,pp.526–7,:10.1021/ed059p526

- Good JM, Gregory O & Bosworth N 1813, 'Arsenicum', in *Pantologia: A New Cyclopedia ... of Essays, Treatises, and Systems ... with a General Dictionary of Arts, Sciences, and Words ...* , Kearsely, London

- Goodrich BG 1844, *A Glance at the Physical Sciences,* Bradbury, Soden & Co., Boston

- Gray T 2009, *The Elements: A Visual Exploration of Every Known Atom in the Universe,* Black Dog & Leventhal, New York, ISBN 978-1-57912-814-2

- Gray T 2010, 'Metalloids (7)', viewed 8 February 2013

- Gray T, Whitby M & Mann N 2011, *Mohs Hardness of the Elements,* viewed 12 Feb 2012

- Greaves GN, Knights JC & Davis EA 1974, 'Electronic Properties of Amorphous Arsenic', in J Stuke & W Brenig (eds), *Amorphous and Liquid Semiconductors: Proceedings,* vol. 1, Taylor & Francis, London, pp. 369–74, ISBN 978-0-470-83485-5

- Greenwood NN 2001, 'Main Group Element Chemistry at the Millennium', *Journal of the Chemical Society, Dalton Transactions,* issue 14, pp. 2055–66, doi:10.1039/b103917m

- Greenwood NN & Earnshaw A 2002, *Chemistry of the Elements,* 2nd ed., Butterworth-Heinemann, ISBN 0-7506-3365-4

- Guan PF, Fujita T, Hirata A, Liu YH & Chen MW 2012, 'Structural Origins of the Excellent Glass-forming Ability of $Pd_{40}Ni_{40}P_{20}$', *Physical Review Letters,* vol. 108, no. 17, pp. 175501–1–5, doi:10.1103/PhysRevLett.108.175501

- Gunn G (ed.) 2014, *Critical Metals Handbook,*John Wiley & Sons, Chichester, West Sussex, ISBN 9780470671719

- Gupta VB, Mukherjee AK & Cameotra SS 1997, 'Poly(ethylene Terephthalate) Fibres', in MN Gupta & VK Kothari (eds), *Manufactured Fibre Technology,* Springer Science+Business Media, Dordrecht, pp. 271–317, ISBN 9789401064736

- Hager T 2006, *The Demon under the Microscope,* Three Rivers Press, New York, ISBN 978-1-4000-8214-8

- Hai H, Jun H, Yong-Mei L, He-Yong H, Yong C & Kang-Nian F 2012, 'Graphite Oxide as an Efficient and Durable Metal-free Catalyst for Aerobic Oxidative Coupling of Amines to Imines', *Green Chemistry,* vol. 14, pp. 930–934, doi:10.1039/C2GC16681J

- Haiduc I & Zuckerman JJ 1985, *Basic Organometallic Chemistry,* Walter de Gruyter, Berlin, ISBN 0-89925-006-8

- Haissinsky M & Coche A 1949, 'New Experiments on the Cathodic Deposition of Radio-elements', *Journal of the Chemical Society,* pp. S397–400

- Manson SS & Halford GR 2006, *Fatigue and Durability of Structural Materials,* ASM International, Materials Park, OH, ISBN 0-87170-825-6

- Haller EE 2006, 'Germanium: From its Discovery to SiGe Devices', *Materials Science in Semiconductor Processing,* vol. 9, nos 4–5, doi:10.1016/j.mssp.2006.08.063, viewed 8 February 2013

- Hamm DI 1969, *Fundamental Concepts of Chemistry,* Meredith Corporation, New York, ISBN 0-390-40651-1

- Hampel CA & Hawley GG 1966, *The Encyclopedia of Chemistry,* 3rd ed., Van Nostrand Reinhold, New York

- Hampel CA (ed.) 1968, *The Encyclopedia of the Chemical Elements,* Reinhold, New York

- Hampel CA & Hawley GG 1976, *Glossary of Chemical Terms,* Van Nostrand Reinhold, New York, ISBN 0-442-23238-1

- Harding C, Johnson DA & Janes R 2002, *Elements of the p Block, Royal Society of Chemistry, Cambridge, ISBN 0-85404-690-9*

- Hasan H 2009, *The Boron Elements: Boron, Aluminum, Gallium, Indium, Thallium, The Rosen Publishing Group, New York, ISBN 1-4358-5333-4*

- Hatcher WH 1949, *An Introduction to Chemical Science,* John Wiley & Sons, New York

- Hawkes SJ 1999, 'Polonium and Astatine are not Semimetals', *Chem 13 News,* February, p. 14, ISSN 07031157

- Hawkes SJ 2001, 'Semimetallicity', *Journal of Chemical Education,* vol.78, no.12, pp.1686–7, doi:10.1021/ed078p86

- Hawkes SJ 2010, 'Polonium and Astatine are not Semimetals', *Journal of Chemical Education,* vol. 87, no. 8, p. 783, doi:10.1021ed100308w

- Haynes WM (ed.) 2012, *CRC Handbook of Chemistry and Physics,* 93rd ed., CRC Press, Boca Raton, Florida, ISBN 1-4398-8049-2

- He M, Kravchyk K, Walter M & Kovalenko MV 2014, 'Monodisperse Antimony Nanocrystals for High-Rate Li-ion and Na-ion Battery Anodes:Nano versus Bulk', *Nano Letters,* vol.14, no.3, pp.1255–1262, doi:10.1021/nl4041c

- Henderson M 2000, *Main Group Chemistry, The Royal Society of Chemistry, Cambridge, ISBN 0-85404-617-8*

- Hermann A, Hoffmann R & Ashcroft NW 2013, 'Condensed Astatine: Monatomic and Metallic', *Physical Review Letters,* vol. 111, pp. 11604–1–11604-5, doi:10.1103/PhysRevLett.111.116404

- Hérold A 2006, 'An Arrangement of the Chemical Elements in Several Classes Inside the Periodic Table According to their Common Properties', *Comptes Rendus Chimie,* vol. 9, no. 1, pp. 148–53, doi:10.1016/j.crci.2005.10.002

- Herzfeld K 1927, 'On Atomic Properties Which Make an Element a Metal', *Physical Review,* vol. 29, no. 5, pp. 701–705, doi:10.1103PhysRev.29.701

- Hill G & Holman J 2000, *Chemistry in Context, 5th ed., Nelson Thornes, Cheltenham, ISBN 0-17-448307-4*

- Hiller LA & Herber RH 1960, *Principles of Chemistry,* McGraw-Hill, New York

- Hindman JC 1968, 'Neptunium', in CA Hampel (ed.), *The Encyclopedia of the Chemical Elements,* Reinhold, New York, pp. 432–7

- Hoddeson L 2007, 'In the Wake of Thomas Kuhn's Theory of Scientific Revolutions: The Perspective of an Historian of Science,' in S Vosniadou, A Baltas & X Vamvakoussi (eds), *Reframing the Conceptual Change Approach in Learning and Instruction,* Elsevier, Amsterdam, pp. 25–34, ISBN 978-0-08-045355-2

- Holderness A & Berry M 1979, *Advanced Level Inorganic Chemistry,* 3rd ed., Heinemann Educational Books, London, ISBN 0-435-65435-7

- Holt, Rinehart & Wilson c. 2007 'Why Polonium and Astatine are not Metalloids in HRW texts', viewed 8 February 2013

- Hopkins BS & Bailar JC 1956, *General Chemistry for Colleges,* 5th ed., D. C. Heath, Boston

- Horvath 1973, 'Critical Temperature of Elements and the Periodic System', *Journal of Chemical Education,* vol. 50, no. 5, pp. 335–6, doi:10.1021/ed050p335

- Hosseini P, Wright CD & Bhaskaran H 2014, 'An optoelectronic framework enabled by low-dimensional phase-change films,' *Nature,* vol. 511, pp. 206–211, doi:10.1038/nature13487

- Houghton RP 1979, *Metal Complexes in Organic Chemistry,* Cambridge University Press, Cambridge, ISBN 0-521-21992-2

- House JE 2008, *Inorganic Chemistry,* Academic Press (Elsevier), Burlington, Massachusetts, ISBN 0-12-356786-6

- House JE & House KA 2010, *Descriptive Inorganic Chemistry,* 2nd ed., Academic Press, Burlington, Massachusetts, ISBN 0-12-088755-X

- Housecroft CE & Sharpe AG 2008, *Inorganic Chemistry*, 3rd ed., Pearson Education, Harlow, ISBN 978-0-13-175553-6

- Hultgren HH 1966, 'Metalloids', in GL Clark & GG Hawley (eds), *The Encyclopedia of Inorganic Chemistry,* 2nd ed., Reinhold Publishing, New York

- Hunt A 2000, *The Complete A-Z Chemistry Handbook,* 2nd ed., Hodder & Stoughton, London, ISBN 0-340-77218-2

- Inagaki M 2000, *New Carbons: Control of Structure and Functions,* Elsevier, Oxford, ISBN 0-08-043713-3

- IUPAC 1959, *Nomenclature of Inorganic Chemistry,* 1st ed., Butterworths, London

- IUPAC 1971, *Nomenclature of Inorganic Chemistry, 2nd ed., Butterworths, London, ISBN 0-408-70168-4*

- IUPAC 2005, *Nomenclature of Inorganic Chemistry* (the "Red Book"), NG Connelly & T Damhus eds, RSC Publishing, Cambridge, ISBN 0-85404-438-8

- IUPAC 2006–, *Compendium of Chemical Terminology* (the "Gold Book"), 2nd ed., by M Nic, J Jirat & B Kosata, with updates compiled by A Jenkins, ISBN 0-9678550-9-8, doi:10.1351/goldbook

- James M, Stokes R, Ng W & Moloney J 2000, *Chemical Connections 2: VCE Chemistry Units 3 & 4,* John Wiley & Sons, Milton, Queensland, ISBN 0-7016-3438-3

- Jaouen G & Gibaud S 2010, 'Arsenic-based Drugs: From Fowler's solution to Modern Anticancer Chemotherapy', *Medicinal Organometallic Chemistry,* vol. 32, pp. 1–20, doi:10.1007/978-3-642-13185-1_1

- Jaskula BW 2013, *Mineral Commodity Profiles: Gallium,* US Geological Survey

- Jenkins GM & Kawamura K 1976, *Polymeric Carbons—Carbon Fibre, Glass and Char,* Cambridge University Press, Cambridge, ISBN 0-521-20693-6

- Jezequel G & Thomas J 1997, 'Experimental Band Structure of Semimetal Bismuth', *Physical Review B,* vol. 56, no. 11, pp. 6620–6, doi:10.1103/PhysRevB.56.6620

- Johansen G & Mackintosh AR 1970, 'Electronic Structure and Phase Transitions in Ytterbium', *Solid State Communications,* vol. 8, no. 2, pp. 121–4

- Jolly WL & Latimer WM 1951, 'The Heat of Oxidation of Germanous Iodide and the Germanium Oxidation Potentials', University of California Radiation Laboratory, Berkeley

- Jolly WL 1966, *The Chemistry of the Non-metals,* Prentice-Hall, Englewood Cliffs, New Jersey

- Jones BW 2010, *Pluto: Sentinel of the Outer Solar System,* Cambridge University, Cambridge, ISBN 978-0-521-19436-5

- Kaminow IP & Li T 2002 (eds), *Optical Fiber Telecommunications,* Volume IVA, Academic Press, San Diego, ISBN 0-12-395172-0

- Karabulut M, Melnik E, Stefan R, Marasinghe GK, Ray CS, Kurkjian CR & Day DE 2001, 'Mechanical and Structural Properties of Phosphate Glasses', *Journal of Non-Crystalline Solids,* vol. 288, nos. 1–3, pp. 8–17, doi:10.1016/S0022-3093(01)00615-9

- Kauthale SS, Tekali SU, Rode AB, Shinde SV, Ameta KL & Pawar RP 2015, 'Silica Sulfuric Acid: A Simple and Powerful Heterogenous Catalyst in Organic Synthesis', in KL Ameta & A Penoni, *Heterogeneous Catalysis: A Versatile Tool for the Synthesis of Bioactive Heterocycles,* CRC Press, Boca Raton, Florida, pp. 133–162, ISBN 9781466594821

- Kaye GWC & Laby TH 1973, *Tables of Physical and Chemical Constants,* 14th ed., Longman, London, ISBN 0-582-46326-2

- Keall JHH, Martin NH & Tunbridge RE 1946, 'A Report of Three Cases of Accidental Poisoning by Sodium Tellurite', *British Journal of Industrial Medicine,* vol. 3, no. 3, pp. 175–6

- Keevil D 1989, 'Aluminium', in MN Patten (ed.), *Information Sources in Metallic Materials,* Bowker–Saur, London, pp. 103–119, ISBN 0-408-01491-1

- Keller C 1985, 'Preface', in Kugler & Keller

- Kelter P, Mosher M & Scott A 2009, *Chemistry: the Practical Science,* Houghton Mifflin, Boston, ISBN 0-547-05393-2

- Kennedy T, Mullane E, Geaney H, Osiak M, O'Dwyer C & Ryan KM 2014, 'High-Performance Germanium Nanowire-Based Lithium-Ion Battery Anodes Extending over 1000 Cycles Through in Situ Formation of a Continuous Porous Network', *Nano-letters,* vol. 14, no. 2, pp. 716–723, doi:10.1021/nl403979s

- Kent W 1950, *Kent's Mechanical Engineers' Handbook,* 12th ed., vol. 1, John Wiley & Sons, New York

- King EL 1979, *Chemistry,* Painter Hopkins, Sausalito, California, ISBN 0-05-250726-2

- King RB 1994, 'Antimony: Inorganic Chemistry', in RB King (ed), *Encyclopedia of Inorganic Chemistry,* John Wiley, Chichester, pp. 170–5, ISBN 0-471-93620-0

- King RB 2004, 'The Metallurgist's Periodic Table and the Zintl-Klemm Concept', in DH Rouvray & RB King (eds), *The Periodic Table: Into the 21st Century,* Research Studies Press, Baldock, Hertfordshire, pp. 191–206, ISBN 0-86380-292-3

- Kinjo R, Donnadieu B, Celik MA, Frenking G & Bertrand G 2011, 'Synthesis and Characterization of a Neutral Tricoordinate Organoboron Isoelectronic with Amines', *Science,* pp. 610–613, doi:10.1126/science.1207573

- Kitaĭgorodskiĭ AI 1961, *Organic Chemical Crystallography,* Consultants Bureau, New York

- Kleinberg J, Argersinger WJ & Griswold E 1960, *Inorganic Chemistry,* DC Health, Boston

- Klement W, Willens RH & Duwez P 1960, 'Non-Crystalline Structure in Solidified Gold–Silicon Alloys', *Nature,* vol. 187, pp. 869–70, doi:10.1038/187869b0

- Klemm W 1950, 'Einige Probleme aus der Physik und der Chemie der Halbmetalle und der Metametalle', *Angewandte Chemie,* vol. 62, no. 6, pp. 133–42

- Klug HP & Brasted RC 1958, *Comprehensive Inorganic Chemistry: The Elements and Compounds of Group IV A,* Van Nostrand, New York

- Kneen WR, Rogers MJW & Simpson P 1972, *Chemistry: Facts, Patterns, and Principles,* Addison-Wesley, London, ISBN 0-201-03779-3

- Kohl AL & Nielsen R 1997, *Gas Purification,* 5th ed., Gulf Valley Publishing, Houston, Texas, ISBN 0884152200

- Kolobov AV & Tominaga J 2012, *Chalcogenides: Metastability and Phase Change Phenomena,* Springer-Verlag, Heidelberg, ISBN 978-3-642-28705-3

- Kolthoff IM & Elving PJ 1978, *Treatise on Analytical Chemistry. Analytical Chemistry of Inorganic and Organic Compounds: Antimony, Arsenic, Boron, Carbon, Molybenum, Tungsten,* Wiley Interscience, New York, ISBN 0-471-49998-6

- Kondrat'ev SN & Mel'nikova SI 1978, 'Preparation and Various Characteristics of Boron Hydrogen Sulfates', *Russian Journal of Inorganic Chemistry,* vol. 23, no. 6, pp. 805–807

- Kopp JG, Lipták BG & Eren H 000, 'Magnetic Flowmeters', in BG Lipták (ed.), *Instrument Engineers' Handbook,* 4th ed., vol. 1, Process Measurement and Analysis, CRC Press, Boca Raton, Florida, pp. 208–224, ISBN 0-8493-1083-0

- Korenman IM 1959, 'Regularities in Properties of Thallium', *Journal of General Chemistry of the USSR,* English translation, Consultants Bureau, New York, vol. 29, no. 2, pp. 1366–90, ISSN 00221279

- Kosanke KL, Kosanke BJ & Dujay RC 2002, 'Pyrotechnic Particle Morphologies—Metal Fuels', in *Selected Pyrotechnic Publications of K.L. and B.J. Kosanke Part 5 (1998 through 2000),* Journal of Pyrotechnics, Whitewater, CO, ISBN 1-889526-13-4

- Kotz JC, Treichel P & Weaver GC 2009, *Chemistry and Chemical Reactivity,* 7th ed., Brooks/Cole, Belmont, California, ISBN 1-4390-4131-8

- Kozyrev PT 1959, 'Deoxidized Selenium and the Dependence of its Electrical Conductivity on Pressure. II', *Physics of the Solid State,* translation of the journal Solid State Physics (Fizika tverdogo tela) of the Academy of Sciences of the USSR, vol. 1, pp. 102–10

- Kraig RE, Roundy D & Cohen ML 2004, 'A Study of the Mechanical and Structural Properties of Polonium', *Solid State Communications,* vol. 129, issue 6, Feb, pp. 411–13, doi:10.1016/j.ssc.2003.08.001

- Krannich LK & Watkins CL 2006, 'Arsenic: Organoarsenic chemistry,' *Encyclopedia of inorganic chemistry,* viewed 12 Feb 2012

- Kreith F & Goswami DY (eds) 2005, *The CRC Handbook of Mechanical Engineering,* 2nd ed., Boca Raton, Florida, ISBN 0-8493-0866-6

- Krishnan S, Ansell S, Felten J, Volin K & Price D 1998, 'Structure of Liquid Boron', *Physical Review Letters,* vol. 81, no. 3, pp. 586–9, doi:10.1103/PhysRevLett.81.586

- Kross B 2011, 'What's the melting point of steel?', *Questions and Answers,* Thomas Jefferson National Accelerator Facility, Newport News, VA

- Kudryavtsev AA 1974, *The Chemistry & Technology of Selenium and Tellurium,* translated from the 2nd Russian edition and revised by EM Elkin, Collet's, London, ISBN 0-569-08009-6

- Kugler HK & Keller C (eds) 1985, *Gmelin Handbook of Inorganic and Organometallic chemistry,* 8th ed., 'At, Astatine', system no. 8a, Springer-Verlag, Berlin, ISBN 3-540-93516-9

- Ladd M 1999, *Crystal Structures: Lattices and Solids in Stereoview,* Horwood Publishing, Chichester, ISBN 1-898563-63-2

- Le Bras M, Wilkie CA & Bourbigot S (eds) 2005, *Fire Retardancy of Polymers: New Applications of Mineral Fillers,* Royal Society of Chemistry, Cambridge, ISBN 0-85404-582-1

- Lee J, Lee EK, Joo W, Jang Y, Kim B, Lim JY, Choi S, Ahn SJ, Ahn JR, Park M, Yang C, Choi BL, Hwang S & Whang D 2014, 'Wafer-Scale Growth of Single-Crystal Monolayer Graphene on Reusable Hydrogen-Terminated Germanium', *Science,* vol. 344, no. 6181, pp. 286–289, doi:10.1126/science.1252268

- Legit D, Friák M & Šob M 2010, 'Phase Stability, Elasticity, and Theoretical Strength of Polonium from First Principles,' *Physical Review B,* vol. 81, pp. 214118-1–19, doi:10.1103/PhysRevB.81.214118

- Lehto Y & Hou X 2011, *Chemistry and Analysis of Radionuclides: Laboratory Techniques and Methodology,* Wiley-VCH, Weinheim, ISBN 978-3-527-32658-7

- Lewis RJ 1993, *Hawley's Condensed Chemical Dictionary,* 12th ed., Van Nostrand Reinhold, New York, ISBN 0-442-01131-8

- Li XP 1990, 'Properties of Liquid Arsenic: A Theoretical Study', *Physical Review B,* vol. 41, no. 12, pp. 8392–406, doi:10.1103/PhysRevB.41.8392

- Lide DR (ed.) 2005, 'Section 14, Geophysics, Astronomy, and Acoustics; Abundance of Elements in the Earth's Crust and in the Sea', in *CRC Handbook of Chemistry and Physics,* 85th ed., CRC Press, Boca Raton, FL, pp. 14–17, ISBN 0-8493-0485-7

- Lidin RA 1996, *Inorganic Substances Handbook,* Begell House, New York, ISBN 1-56700-065-7

- University of Limerick 2014, 'Researchers make breakthrough in battery technology,' 7 February, viewed 2 March 2014

- Lindsjö M, Fischer A & Kloo L 2004, 'Sb8(GaCl4)2: Isolation of a Homopolyatomic Antimony Cation', *Angewandte Chemie,* vol. 116, no. 19, pp. 2594–2597, doi:10.1002/ange.200353578

- Lipscomb CA 1972 *Pyrotechnics in the '70's A Materials Approach*, Naval Ammunition Depot, Research and Development Department, Crane, IN

- Lister MW 1965, *Oxyacids,* Oldbourne Press, London

- Liu ZK, Jiang J, Zhou B, Wang ZJ, Zhang Y, Weng HM, Prabhakaran D, Mo S-K, Peng H, Dudin P, Kim T, Hoesch M, Fang Z, Dai X, Shen ZX, Feng DL, Hussain Z & Chen YL 2014, 'A Stable Three-dimensional Topological Dirac Semimetal Cd_3As_2', *Nature Materials,* vol. 13, pp. 677–681, doi:10.1038/nmat3990

- Locke EG, Baechler RH, Beglinger E, Bruce HD, Drow JT, Johnson KG, Laughnan DG, Paul BH, Rietz RC, Saeman JF & Tarkow H 1956, 'Wood', in RE Kirk & DF Othmer (eds), *Encyclopedia of Chemical Technology,* vol. 15, The Interscience Encyclopedia, New York, pp. 72–102

- Löffler JF, Kündig AA & Dalla Torre FH 2007, 'Rapid Solidification and Bulk Metallic Glasses—Processing and Properties,' in JR Groza, JF Shackelford, EJ Lavernia EJ & MT Powers (eds), *Materials Processing Handbook,* CRC Press, Boca Raton, Florida, pp. 17-1–44, ISBN 0-8493-3216-8

- Long GG & Hentz FC 1986, *Problem Exercises for General Chemistry,* 3rd ed., John Wiley & Sons, New York, ISBN 0-471-82840-8

- Lovett DR 1977, *Semimetals & Narrow-Bandgap Semi-conductors,* Pion, London, ISBN 0-85086-060-1

- Lutz J, Schlangenotto H, Scheuermann U, De Doncker R 2011, *Semiconductor Power Devices: Physics, Characteristics, Reliability, Springer-Verlag, Berlin, ISBN 3-642-11124-6*

- Masters GM & Ela W 2008, *Introduction to Environmental Engineering and Science,* 3rd ed., Prentice Hall, Upper Saddle River, New Jersey, ISBN 978-0-13-148193-0

- MacKay KM, MacKay RA & Henderson W 2002, *Introduction to Modern Inorganic Chemistry,* 6th ed., Nelson Thornes, Cheltenham, ISBN 0-7487-6420-8

- Madelung O 2004, *Semiconductors: Data Handbook,* 3rd ed., Springer-Verlag, Berlin, ISBN 978-3-540-40488-0

- Maeder T 2013, 'Review of Bi_2O_3 Based Glasses for Electronics and Related Applications, *International Materials Reviews,* vol. 58, no. 1, pp. 3–40, doi:10.1179/1743280412Y.0000000010

- Mahan BH 1965, *University Chemistry,* Addison-Wesley, Reading, Massachusetts

- Manahan SE 2001, *Fundamentals of Environmental Chemistry,* 2nd ed., CRC Press, Boca Raton, Florida, ISBN 1-56670-491-X

- Mann JB, Meek TL & Allen LC 2000, 'Configuration Energies of the Main Group Elements', *Journal of the American Chemical Society,* vol. 122, no. 12, pp. 2780–3, doi:10.1021ja992866e

- Marezio M & Licci F 2000, 'Strategies for Tailoring New Superconducting Systems', in X Obradors, F Sandiumenge & J Fontcuberta (eds), *Applied Superconductivity 1999: Large scale applications,* volume 1 of Applied Superconductivity 1999: Proceedings of EUCAS 1999, the Fourth European Conference on Applied Superconductivity, held in Sitges, Spain, 14–17 September 1999, Institute of Physics, Bristol, pp. 11–16, ISBN 0-7503-0745-5

- Marković N, Christiansen C & Goldman AM 1998, 'Thickness-Magnetic Field Phase Diagram at the Superconductor-Insulator Transition in 2D', *Physical Review Letters,* vol. 81, no. 23, pp. 5217–20, doi:10.1103/PhysRevLett.81.5217

- Massey AG 2000, *Main Group Chemistry,* 2nd ed., John Wiley & Sons, Chichester, ISBN 0-471-49039-3

- Masterton WL & Slowinski EJ 1977, *Chemical Principles,* 4th ed., W. B. Saunders, Philadelphia, ISBN 0-7216-6173-4

- Matula RA 1979, 'Electrical Resistivity of Copper, Gold, Palladium, and Silver,' *Journal of Physical and Chemical Reference Data,* vol. 8, no. 4, pp. 1147–298, doi:10.1063/1.555614

- McKee DW 1984, 'Tellurium—An Unusual Carbon Oxidation Catalyst', *Carbon,* vol. 22, no. 6, doi:10.1016/0008-6223(84)90084-8, pp. 513–516

- McMurray J & Fay RC 2009, *General Chemistry: Atoms First,* Prentice Hall, Upper Saddle River, New Jersey, ISBN 0-321-57163-0

- McQuarrie DA & Rock PA 1987, *General Chemistry,* 3rd ed., WH Freeman, New York, ISBN 0-7167-2169-4

- Mellor JW 1964, *A Comprehensive Treatise on Inorganic and Theoretical Chemistry,* vol. 9, John Wiley, New York

- Mellor JW 1964a, *A Comprehensive Treatise on Inorganic and Theoretical Chemistry,* vol. 11, John Wiley, New York

- Mendeléeff DI 1897, *The Principles of Chemistry,* vol. 2, 5th ed., trans. G Kamensky, AJ Greenaway (ed.), Longmans, Green & Co., London

- Meskers CEM, Hagelüken C & Van Damme G 2009, 'Green Recycling of EEE: Special and Precious Metal EEE', in SM Howard, P Anyalebechi & L Zhang (eds), *Proceedings of Sessions and Symposia Sponsored by the Extraction and Processing Division (EPD) of The Minerals, Metals and Materials Society (TMS),* held during the TMS 2009 Annual Meeting & Exhibition San Francisco, California, February 15–19, 2009, The Minerals, Metals and Materials Society, Warrendale, Pennsylvania, ISBN 978-0-87339-732-2, pp. 1131–6

- Metcalfe HC, Williams JE & Castka JF 1974, *Modern Chemistry,* Holt, Rinehart and Winston, New York, ISBN 0-03-089450-6

- Meyer JS, Adams WJ, Brix KV, Luoma SM, Mount DR, Stubblefield WA & Wood CM (eds) 2005, *Toxicity of Dietborne Metals to Aquatic Organisms,* Proceedings from the Pellston Workshop on Toxicity of Dietborne Metals to Aquatic Organisms, 27 July–1 August 2002, Fairmont Hot Springs, British Columbia, Canada, Society of Environmental Toxicology and Chemistry, Pensacola, Florida, ISBN 1-880611-70-8

- Mhiaoui S, Sar F, Gasser J 2003, 'Influence of the History of a Melt on the Electrical Resistivity of Cadmium–Antimony Liquid Alloys', *Intermetallics,* vol. 11, nos 11–12, pp. 1377–82, doi:10.1016/j.intermet.2003.09.008

- Miller GJ, Lee C & Choe W 2002, 'Structure and Bonding Around the Zintl border', in G Meyer, D Naumann & L Wesermann (eds), *Inorganic chemistry highlights,* Wiley-VCH, Weinheim, pp. 21–53, ISBN 3-527-30265-4

- Millot F, Rifflet JC, Sarou-Kanian V & Wille G 2002, 'High-Temperature Properties of Liquid Boron from Contactless Techniques',*International Journal of Thermophysics,*vol.23,no.5,pp.1185–95,doi:10.1023/A:10198361026

- Mingos DMP 1998, *Essential Trends in Inorganic Chemistry,* Oxford University, Oxford, ISBN 0-19-850108-0

- Moeller T 1954, *Inorganic Chemistry: An Advanced Textbook,* John Wiley & Sons, New York

- Mokhatab S & Poe WA 2012, *Handbook of Natural Gas Transmission and Processing,* 2nd ed., Elsevier, Kidlington, Oxford, ISBN 9780123869142

- Molina-Quiroz RC, Muñoz-Villagrán CM, de la Torre E, Tantaleán JC, Vásquez CC & Pérez-Donoso JM 2012, 'Enhancing the Antibiotic Antibacterial Effect by Sub Lethal Tellurite Concentrations: Tellurite and Cefotaxime Act Synergistically in*Escherichia Coli',PloS*(Public Library of Science)*ONE,*vol.7,no.4,doi:10.1371/journal..0035452

- Monconduit L, Evain M, Boucher F, Brec R & Rouxel J 1992, 'Short Te … Te Bonding Contacts in a New Layered Ternary Telluride: Synthesis and crystal structure of 2D $Nb_3Ge_xTe_6$ ($x \simeq 0.9$)', *Zeitschrift für Anorganische und Allgemeine Chemie,* vol. 616, no. 10, pp. 177–182, doi:10.1002/zaac.19926161028

- Moody B 1991, *Comparative Inorganic Chemistry,* 3rd ed., Edward Arnold, London, ISBN 0-7131-3679-0

- Moore LJ, Fassett JD, Travis JC, Lucatorto TB & Clark CW 1985, 'Resonance-Ionization Mass Spectrometry of Carbon', *Journal of the Optical Society of America B,* vol. 2, no. 9, pp. 1561–5, doi:10.1364/JOSAB.2.001561

- Moore JE 2010, 'The Birth of Topological Insulators,' *Nature,* vol. 464, pp. 194–198, doi:10.1038/nature08916

- Moore JE 2011, *Topological insulators,* IEEE Spectrum, viewed 15 December 2014

- Moore JT 2011, *Chemistry for Dummies,* 2nd ed., John Wiley & Sons, New York, ISBN 1-118-09292-9

- Moore NC 2014, '45-year Physics Mystery Shows a Path to Quantum Transistors', *Michigan News,* viewed 17 December 2014

- Morgan WC 1906, *Qualitative Analysis as a Laboratory Basis for the Study of General Inorganic Chemistry,* The Macmillan Company, New York

- Morita A 1986, 'Semiconducting Black Phosphorus', *Journal of Applied Physics A,* vol. 39, no. 4, pp. 227–42, doi:10.1007/BF00617267

- Moss TS 1952, *Photoconductivity in the Elements,* London, Butterworths

- Muncke J 2013, 'Antimony Migration from PET: New Study Investigates Extent of Antimony Migration from Polyethylene Terephthalate (PET) Using EU Migration Testing Rules', Food Packaging Forum, April 2

- Murray JF 1928, 'Cable-Sheath Corrosion', *Electrical World,* vol. 92, Dec 29, pp. 1295–7, ISSN 00134457

- Nagao T, Sadowski1 JT, Saito M, Yaginuma S, Fujikawa Y, Kogure T, Ohno T, Hasegawa Y, Hasegawa S & Sakurai T 2004, 'Nanofilm Allotrope and Phase Transformation of Ultrathin Bi Film on Si(111)−7×7', *Physical Review Letters,* vol. 93, no. 10, pp. 105501-1–4, doi:10.1103/PhysRevLett.93.105501

- Neuburger MC 1936, 'Gitterkonstanten für das Jahr 1936' (in German), *Zeitschrift für Kristallographie,* vol. 93, pp. 1–36, ISSN 00442968

- Nickless G 1968, *Inorganic Sulphur Chemistry,* Elsevier, Amsterdam

- Nielsen FH 1998, 'Ultratrace Elements in Nutrition: Current Knowledge and Speculation', *The Journal of Trace Elements in Experimental Medicine,* vol. 11, pp. 251–74, doi:10.1002/(SICI)1520-670X(1998)11:2/3<251::AID-JTRA15>3.0.CO;2-Q

- NIST (National Institute of Standards and Technology) 2010, *Ground Levels and Ionization Energies for Neutral Atoms*, by WC Martin, A Musgrove, S Kotochigova & JE Sansonetti, viewed 8 February 2013

- National Research Council 1984, *The Competitive Status of the U.S. Electronics Industry: A Study of the Influences of Technology in Determining International Industrial Competitive Advantage*, National Academy Press, Washington, DC, ISBN 0-309-03397-7

- *New Scientist* 1975, 'Chemistry on the Islands of Stability', 11 Sep, p. 574, ISSN 10321233

- New Scientist 2014, 'Colour-changing metal to yield thin, flexible displays', vol. 223, no. 2977

- Oderberg DS 2007, *Real Essentialism*, Routledge, New York, ISBN 1-134-34885-1

- *Oxford English Dictionary* 1989, 2nd ed., Oxford University, Oxford, ISBN 0-19-861213-3

- Oganov AR, Chen J, Gatti C, Ma Y, Ma Y, Glass CW, Liu Z, Yu T, Kurakevych OO & Solozhenko VL 2009, 'Ionic High-Pressure Form of Elemental Boron', *Nature,* vol. 457, 12 Feb, pp. 863–8, doi:10.1038/nature07736

- Oganov AR 2010, 'Boron Under Pressure: Phase Diagram and Novel High Pressure Phase,' in N Ortovoskaya N & L Mykola L (eds), *Boron Rich Solids: Sensors, Ultra High Temperature Ceramics, Thermoelectrics, Armor,* Springer, Dordrecht, pp. 207–25, ISBN 90-481-9823-2

- Ogata S, Li J & Yip S 2002, 'Ideal Pure Shear Strength of Aluminium and Copper', *Science,* vol. 298, no. 5594, 25 October, pp. 807–10, doi:10.1126/science.1076652

- O'Hare D 1997, 'Inorganic intercalation compounds' in DW Bruce & D O'Hare (eds), *Inorganic materials,* 2nd ed., John Wiley & Sons, Chichester, pp. 171–254, ISBN 0-471-96036-5

- Okajima Y & Shomoji M 1972, Viscosity of Dilute Amalgams', *Transactions of the Japan Institute of Metals,* vol. 13, no. 4, pp. 255–8, ISSN 00214434

- Oldfield JE, Allaway WH, HA Laitinen, HW Lakin & OH Muth 1974, 'Tellurium', in *Geochemistry and the Environment*, Volume 1: The Relation of Selected Trace Elements to Health and Disease, US National Committee for Geochemistry, Subcommittee on the Geochemical Environment in Relation to Health and Disease, National Academy of Sciences, Washington, ISBN 0-309-02223-1

- Oliwenstein L 2011, 'Caltech-Led Team Creates Damage-Tolerant Metallic Glass', California Institute of Technology, 12 January, viewed 8 February 2013

- Olmsted J & Williams GM 1997, *Chemistry, the Molecular Science, 2nd ed., Wm C Brown, Dubuque, Iowa, ISBN 0-8151-8450-6*

- Ordnance Office 1863, *The Ordnance Manual for the use of the Officers of the Confederate States Army,* 1st ed., Evans & Cogswell, Charleston, SC

- Orton JW 2004, *The Story of Semiconductors,* Oxford University, Oxford, ISBN 0-19-853083-8

- Owen SM & Brooker AT 1991, *A Guide to Modern Inorganic Chemistry,* Longman Scientific & Technical, Harlow, Essex, ISBN 0-582-06439-2

- Oxtoby DW, Gillis HP & Campion A 2008, *Principles of Modern Chemistry, 6th ed., Thomson Brooks/Cole, Belmont, California, ISBN 0-534-49366-1*

- Pan K, Fu Y & Huang T 1964, 'Polarographic Behavior of Germanium(II)-Perchlorate in Perchloric Acid Solutions', *Journal of the Chinese Chemical Society,* pp. 176–184, doi:10.1002/jccs.196400020

- Parise JB, Tan K, Norby P, Ko Y & Cahill C 1996, 'Examples of Hydrothermal Titration and Real Time X-ray Diffraction in the Synthesis of Open Frameworks', *MRS Proceedings*, vol. 453, pp. 103–14, doi:10.1557/PROC-453-103

- Parish RV 1977, *The Metallic Elements,* Longman, London, ISBN 0-582-44278-8

- Parkes GD & Mellor JW 1943, *Mellor's Nodern Inorganic Chemistry,* Longmans, Green and Co., London

- Parry RW, Steiner LE, Tellefsen RL & Dietz PM 1970, *Chemistry: Experimental Foundations,* Prentice-Hall/Martin Educational, Sydney, ISBN 0-7253-0100-7

- Partington 1944, *A Text-book of Inorganic Chemistry,* 5th ed., Macmillan, London

- Pashaey BP & Seleznev VV 1973, 'Magnetic Susceptibility of Gallium-Indium Alloys in Liquid State', *Russian Physics Journal,* vol. 16, no. 4, pp. 565–6, doi:10.1007/BF00890855

- Patel MR 2012, *Introduction to Electrical Power and Power Electronics* CRC Press, Boca Raton, ISBN 978-1-4665-5660-7

- Paul RC, Puri JK, Sharma RD & Malhotra KC 1971, 'Unusual Cations of Arsenic', Inorganic and Nuclear Chemistry Letters, vol. 7, no. 8, pp. 725–728, doi:10.1016/0020-1650(71)80079-X

- Pauling L 1988, *General Chemistry, Dover Publications, New York, ISBN 0-486-65622-5*

- Pearson WB 1972, *The Crystal Chemistry and Physics of Metals and Alloys,* Wiley-Interscience, New York, ISBN 0-471-67540-7

- Perry DL 2011,*Handbook of Inorganic Compounds,*2nd ed., CRC Press, Boca Raton, Florida,ISBN 978143981461

- Peryea FJ 1998, 'Historical Use of Lead Arsenate Insecticides, Resulting Soil Contamination and Implications for Soil Remediation, Proceedings', *16th World Congress of Soil Science,* Montpellier, France, 20–26 August

- Phillips CSG & Williams RJP 1965, *Inorganic Chemistry, I: Principles and Non-metals,* Clarendon Press, Oxford

- Pinkerton J 1800, *Petralogy. A Treatise on Rocks,* vol. 2, White, Cochrane, and Co., London

- Poojary DM, Borade RB & Clearfield A 1993, 'Structural Characterization of Silicon Orthophosphate', *Inorganica Chimica Acta,* vol. 208, no. 1, pp. 23–9, doi:10.1016/S0020-1693(00)82879-0

- Pourbaix M 1974, *Atlas of Electrochemical Equilibria in Aqueous Solutions,* 2nd English edition, National Association of Corrosion Engineers, Houston, ISBN 0-915567-98-9

- Powell HM & Brewer FM 1938, 'The Structure of Germanous Iodide', *Journal of the Chemical Society,*, pp. 197–198, doi:10.1039/JR9380000197

- Powell P 1988, *Principles of Organometallic Chemistry,* Chapman and Hall, London, ISBN 0-412-42830-X

- Prakash GKS & Schleyer PvR (eds) 1997, *Stable Carbocation Chemistry,* John Wiley & Sons, New York, ISBN 0-471-59462-8

- Prudenziati M 1977, IV. 'Characterization of Localized States in β-Rhombohedral Boron', in VI Matkovich (ed.), *Boron and Refractory Borides,* Springer-Verlag, Berlin, pp. 241–61, ISBN 0-387-08181-X

- Puddephatt RJ & Monaghan PK 1989, *The Periodic Table of the Elements,* 2nd ed., Oxford University, Oxford, ISBN 0-19-855516-4

- Pyykkö P 2012, 'Relativistic Effects in Chemistry: More Common Than You Thought', *Annual Review of Physical Chemistry,* vol. 63, pp. 45–64 (56), doi:10.1146/annurev-physchem-032511-143755

- Rao CNR & Ganguly P 1986, 'A New Criterion for the Metallicity of Elements', *Solid State Communications,* vol. 57, no. 1, pp. 5–6, doi:10.1016/0038-1098(86)90659-9

- Rao KY 2002, *Structural Chemistry of Glasses,* Elsevier, Oxford, ISBN 0-08-043958-6

- Rausch MD 1960, 'Cyclopentadienyl Compounds of Metals and Metalloids', *Journal of Chemical Education,* vol. 37, no. 11, pp. 568–78, doi:10.1021/ed037p568

- Rayner-Canham G & Overton T 2006, *Descriptive Inorganic Chemistry,* 4th ed., WH Freeman, New York, ISBN 0-7167-8963-9

- Rayner-Canham G 2011, 'Isodiagonality in the Periodic Table', *Foundations of chemistry,* vol. 13, no. 2, pp. 121–9, doi:10.1007/s10698-011-9108-y

- Reardon M 2005, 'IBM Doubles Speed of Germanium chips', *CNET News,* August 4, viewed 27 December 2013

- Regnault MV 1853, *Elements of Chemistry,* vol. 1, 2nd ed., Clark & Hesser, Philadelphia

- Reilly C 2002, *Metal Contamination of Food,* Blackwell Science, Oxford, ISBN 0-632-05927-3

- Reilly 2004, *The Nutritional Trace Metals,* Blackwell, Oxford, ISBN 1-4051-1040-6

- Restrepo G, Mesa H, Llanos EJ & Villaveces JL 2004, 'Topological Study of the Periodic System', *Journal of Chemical Information and Modelling,* vol. 44, no. 1, pp. 68–75, doi:10.1021/ci034217z

- Restrepo G, Llanos EJ & Mesa H 2006, 'Topological Space of the Chemical Elements and its Properties', *Journal of Mathematical Chemistry,* vol. 39, no. 2, pp. 401–16, doi:10.1007/s10910-005-9041-1

- Řezanka T & Sigler K 2008, 'Biologically Active Compounds of Semi-Metals', *Studies in Natural Products Chemistry,* vol. 35, pp. 585–606, doi:10.1016/S1572-5995(08)80018-X

- Richens DT 1997, *The Chemistry of Aqua Ions,* John Wiley & Sons, Chichester, ISBN 0-471-97058-1

- Rochow EG 1957, *The Chemistry of Organometallic Compounds,* John Wiley & Sons, New York

- Rochow EG 1966, *The Metalloids,* DC Heath and Company, Boston

- Rochow EG 1973, 'Silicon', in JC Bailar, HJ Eméleus, R Nyholm & AF Trotman-Dickenson (eds), *Comprehensive Inorganic Chemistry,* vol. 1, Pergamon, Oxford, pp. 1323–1467, ISBN 0-08-015655-X

- Rochow EG 1977, *Modern Descriptive Chemistry,* Saunders, Philadelphia, ISBN 0-7216-7628-6

- Rodgers G 2011, *Descriptive Inorganic, Coordination, & Solid-state Chemistry,* Brooks/Cole, Belmont, CA, ISBN 0-8400-6846-8

- Roher GS 2001, *Structure and Bonding in Crystalline Materials*, Cambridge University Press, Cambridge, ISBN 0-521-66379-2

- Rossler K 1985, 'Handling of Astatine', pp. 140–56, in Kugler & Keller

- Rothenberg GB 1976, *Glass Technology, Recent Developments,* Noyes Data Corporation, Park Ridge, New Jersey, ISBN 0-8155-0609-0

- Roza G 2009, *Bromine, Rosen Publishing, New York, ISBN 1-4358-5068-8*

- Rupar PA, Staroverov VN & Baines KM 2008, 'A Cryptand-Encapsulated Germanium(II) Dication', *Science,* vol. 322, no. 5906, pp. 1360–1363, doi:10.1126/science.1163033

- Russell AM & Lee KL 2005, *Structure-Property Relations in Nonferrous Metals, Wiley-Interscience, New York, ISBN 0-471-64952-X*

- Russell MS 2009, *The Chemistry of Fireworks,* 2nd ed., Royal Society of Chemistry, ISBN 978-0-85404-127-5

- Sacks MD 1998, 'Mullitization Behavior of Alpha Alumina Silica Microcomposite Powders', in AP Tomsia & AM Glaeser (eds), *Ceramic Microstructures: Control at the Atomic Level,* proceedings of the International Materials Symposium on Ceramic Microstructures '96: Control at the Atomic Level, June 24–27, 1996, Berkeley, CA, Plenum Press, New York, pp. 285–302, ISBN 0-306-45817-9

- Salentine CG 1987, 'Synthesis, Characterization, and Crystal Structure of a New Potassium Borate, $KB_3O_5 \cdot 3H_2O$', *Inorganic Chemistry,* vol. 26, no. 1, pp. 128–32, doi:10.1021/ic00248a025

- Samsonov GV 1968, *Handbook of the Physiochemical Properties of the Elements,* I F I/Plenum, New York

- Savvatimskiy AI 2005, 'Measurements of the Melting Point of Graphite and the Properties of Liquid Carbon (a review for 1963–2003)', *Carbon,* vol. 43, no. 6, pp. 1115–42, doi:10.1016/j.carbon.2004.12.027

- Savvatimskiy AI 2009, 'Experimental Electrical Resistivity of Liquid Carbon in the Temperature Range from 4800 to ~20,000 K', *Carbon,* vol. 47, no. 10, pp. 2322–8, doi:10.1016/j.carbon.2009.04.009

- Schaefer JC 1968, 'Boron' in CA Hampel (ed.), *The Encyclopedia of the Chemical Elements,* Reinhold, New York, pp. 73–81

- Schauss AG 1991, 'Nephrotoxicity and Neurotoxicity in Humans from Organogermanium Compounds and Germanium Dioxide', *Biological Trace Element Research,* vol. 29, no. 3, pp. 267–80, doi:10.1007/BF03032683

- Schmidbaur H & Schier A 2008, 'A Briefing on Aurophilicity,' *Chemical Society Reviews,* vol. 37, pp. 1931–51, doi:10.1039/B708845K

- Schroers J 2013, 'Bulk Metallic Glasses', *Physics Today,* vol. 66, no. 2, pp. 32–7, doi:10.1063/PT.3.1885

- Schwab GM & Gerlach J 1967, 'The Reaction of Germanium with Molybdenum(VI) Oxide in the Solid State' (in German), *Zeitschrift für Physikalische Chemie,* vol. 56, pp. 121–132, doi:10.1524/zpch.1967.56.3_4.121

- Schwartz MM 2002, *Encyclopedia of Materials, Parts, and Finishes,* 2nd ed., CRC Press, Boca Raton, Florida, ISBN 1-56676-661-3

- Schwietzer GK and Pesterfield LL 2010, *The Aqueous Chemistry of the Elements,* Oxford University, Oxford, ISBN 0-19-539335-X

- *ScienceDaily* 2012, 'Recharge Your Cell Phone With a Touch? New nanotechnology converts body heat into power', February 22, viewed 13 January 2013

- Scott EC & Kanda FA 1962, *The Nature of Atoms and Molecules: A General Chemistry,* Harper & Row, New York

- Secrist JH & Powers WH 1966, *General Chemistry,* D. Van Nostrand, Princeton, New Jersey

- Segal BG 1989, *Chemistry: Experiment and Theory,* 2nd ed., John Wiley & Sons, New York, ISBN 0-471-84929-4

- Sekhon BS 2012, 'Metalloid Compounds as Drugs', *Research in Pharmaceutical Sciences,* vol. 8, no. 3, pp. 145–58, ISSN 17359414

- Sequeira CAC 2011, 'Copper and Copper Alloys', in R Winston Revie (ed.), *Uhlig's Corrosion Handbook,* 3rd ed., John Wiley & Sons, Hoboken, New Jersey, pp. 757–86, ISBN 1-118-11003-X

- Sharp DWA 1981, 'Metalloids', in *Miall's Dictionary of Chemistry,* 5th ed, Longman, Harlow, ISBN 0-582-35152-9

- Sharp DWA 1983, *The Penguin Dictionary of Chemistry,* 2nd ed., Harmondsworth, Middlesex, ISBN 0-14-051113-X

- Shelby JE 2005, *Introduction to Glass Science and Technology,* 2nd ed., Royal Society of Chemistry, Cambridge, ISBN 0-85404-639-9

- Sidgwick NV 1950, *The Chemical Elements and Their Compounds,* vol. 1, Clarendon, Oxford

- Siebring BR 1967, *Chemistry,* MacMillan, New York

- Siekierski S & Burgess J 2002, *Concise Chemistry of the Elements,* Horwood, Chichester, ISBN 1-898563-71-3

- Silberberg MS 2006, *Chemistry: The Molecular Nature of Matter and Change,* 4th ed., McGraw-Hill, New York, ISBN 0-07-111658-3

- Simple Memory Art c. 2005, *Periodic Table,* EVA vinyl shower curtain, San Francisco

- Skinner GRB, Hartley CE, Millar D & Bishop E 1979, 'Possible Treatment for Cold Sores,' *British Medical Journal,* vol 2, no. 6192, p. 704, doi:10.1136/bmj.2.6192.704

- Slade S 2006, *Elements and the Periodic Table, The Rosen Publishing Group, New York, ISBN 1-4042-2165-4*

- Science Learning Hub 2009, 'The Essential Elements', The University of Waikato, viewed 16 January 2013

- Smith DW 1990, *Inorganic Substances: A Prelude to the Study of Descriptive Inorganic Chemistry,* Cambridge University, Cambridge, ISBN 0-521-33738-0

- Smith R 1994, *Conquering Chemistry,* 2nd ed., McGraw-Hill, Sydney, ISBN 0-07-470146-0

- Smith AH, Marshall G, Yuan Y, Steinmaus C, Liaw J, Smith MT, Wood L, Heirich M, Fritzemeier RM, Pegram MD & Ferreccio C 2014, 'Rapid Reduction in Breast Cancer Mortality with Inorganic Arsenic in Drinking Water', "EBioMedicine," doi:10.1016/j.ebiom.2014.10.005

- Sneader W 2005, *Drug Discovery: A History,* John Wiley & Sons, New York, ISBN 0-470-01552-7

- Snyder MK 1966, *Chemistry: Structure and Reactions,* Holt, Rinehart and Winston, New York

- Soverna S 2004, 'Indication for a Gaseous Element 112', in U Grundinger (ed.), *GSI Scientific Report 2003,* GSI Report 2004-1, p. 187, ISSN 01740814

- Steele D 1966, *The Chemistry of the Metallic Elements,* Pergamon Press, Oxford

- Stein L 1985, 'New Evidence that Radon is a Metalloid Element: Ion-Exchange Reactions of Cationic Radon', *Journal of the Chemical Society, Chemical Communications,* vol. 22, pp. 1631–2, doi:10.1039/C39850001631

- Stein L 1987, 'Chemical Properties of Radon' in PK Hopke (ed.) 1987, *Radon and its Decay products: Occurrence, Properties, and Health Effects,* American Chemical Society, Washington DC, pp. 240–51, ISBN 0-8412-1015-2

- Steudel R 1977, *Chemistry of the Non-metals: With an Introduction to atomic Structure and Chemical Bonding,* Walter de Gruyter, Berlin, ISBN 3-11-004882-5

- Steurer W 2007, 'Crystal Structures of the Elements' in JW Marin (ed.), *Concise Encyclopedia of the Structure of Materials,* Elsevier, Oxford, pp. 127–45, ISBN 0-08-045127-6

- Stevens SD & Klarner A 1990, *Deadly Doses: A Writer's Guide to Poisons,* Writer's Digest Books, Cincinnati, Ohio, ISBN 0-89879-371-8

- Stoker HS 2010, *General, Organic, and Biological Chemistry,* 5th ed., *Brooks/Cole, Cengage Learning, Belmont California, ISBN 0-495-83146-8*

- Stott RW 1956, *A Companion to Physical and Inorganic Chemistry,* Longmans, Green and Co., London

- Stuke J 1974, 'Optical and Electrical Properties of Selenium', in RA Zingaro & WC Cooper (eds), *Selenium,* Van Nostrand Reinhold, New York, pp. 174–297, ISBN 0-442-29575-8

- Swalin RA 1962, *Thermodynamics of Solids,* John Wiley & Sons, New York

- Swift EH & Schaefer WP 1962, *Qualitative Elemental Analysis,* WH Freeman, San Francisco

- Swink LN & Carpenter GB 1966, 'The Crystal Structure of Basic Tellurium Nitrate, $Te_2O_4 \cdot HNO_3$', *Acta Crystallographica,* vol. 21, no. 4, pp. 578–83, doi:10.1107/S0365110X66003487

- Szpunar J, Bouyssiere B & Lobinski R 2004, 'Advances in Analytical Methods for Speciation of Trace Elements in the Environment', in AV Hirner & H Emons (eds), *Organic Metal and Metalloid Species in the Environment: Analysis, Distribution Processes and Toxicological Evaluation,* Springer-Verlag, Berlin, pp. 17–40, ISBN 3-540-20829-1

- Taguena-Martinez J, Barrio RA & Chambouleyron I 1991, 'Study of Tin in Amorphous Germanium', in JA Blackman & J Tagüeña (eds), *Disorder in Condensed Matter Physics: A Volume in Honour of Roger Elliott,* Clarendon Press, Oxford, ISBN 0-19-853938-X, pp. 139–44

- Taniguchi M, Suga S, Seki M, Sakamoto H, Kanzaki H, Akahama Y, Endo S, Terada S & Narita S 1984, 'Core-Exciton Induced Resonant Photoemission in the Covalent Semiconductor Black Phosphorus', *Solid State Communications,* vol. 49, no. 9, pp. 867–70

- Tao SH & Bolger PM 1997, 'Hazard Assessment of Germanium Supplements', *Regulatory Toxicology and Pharmacology,* vol. 25, no. 3, pp. 211–19, doi:10.1006/rtph.1997.1098

- Taylor MD 1960, *First Principles of Chemistry,* D. Van Nostrand, Princeton, New Jersey

- Thayer JS 1977, 'Teaching Bio-Organometal Chemistry. I. The Metalloids', *Journal of Chemical Education,* vol. 54, no. 10, pp. 604–6, doi:10.1021/ed054p604

- *The Economist* 2012, 'Phase-Change Memory: Altered States', Technology Quarterly, September 1

- *The American Heritage Science Dictionary 2005*, Houghton Mifflin Harcourt, Boston, ISBN 0-618-45504-3

- *The Chemical News* 1897, 'Notices of Books: A Manual of Chemistry, Theoretical and Practical, by WA Tilden', vol. 75, no. 1951, p. 189

- Thomas F, Bialek B & Hensel R 2013, 'Medical Use of Bismuth: The Two Sides of the Coin', *Journal of Clinical Toxicology,* special issue 3, article 4, doi:10.4172/2161-0495

- Thomas S & Visakh PM 2012, *Handbook of Engineering and Speciality Thermoplastics: Volume 3: Polyethers and Polyesters,* John Wiley & Sons, Hoboken, New Jersey, ISBN 0470639261

- Tilden WA 1876, *Introduction to the Study of Chemical Philosophy,* D. Appleton and Co., New York

- Timm JA 1944, *General Chemistry,* McGraw-Hill, New York

- Tyler Miller G 1987, *Chemistry: A Basic Introduction,* 4th ed., Wadsworth Publishing Company, Belmont, California, ISBN 0-534-06912-6

- Togaya M 2000, 'Electrical Resistivity of Liquid Carbon at High Pressure', in MH Manghnani, W Nellis & MF.Nicol (eds), *Science and Technology of High Pressure*, proceedings of AIRAPT-17, Honolulu, Hawaii, 25–30 July 1999, vol. 2, Universities Press, Hyderabad, pp. 871–4, ISBN 81-7371-339-1

- Tom LWC, Elden LM & Marsh RR 2004, 'Topical antifungals', in PS Roland & JA Rutka, *Ototoxicity,* BC Decker, Hamilton, Ontario, pp. 134–9, ISBN 1-55009-263-4

- Tominaga J 2006, 'Application of Ge–Sb–Te Glasses for Ultrahigh Density Optical Storage', in AV Kolobov (ed.), *Photo-Induced Metastability in Amorphous Semiconductors,* Wiley-VCH, pp. 327–7, ISBN 3-527-60866-4

- Toy AD 1975, *The Chemistry of Phosphorus,* Pergamon, Oxford, ISBN 0-08-018780-3

- Träger F 2007, *Springer Handbook of Lasers and Optics,* Springer, New York, ISBN 978-0-387-95579-7

- Traynham JG 1989, 'Carbonium Ion: Waxing and Waning of a Name', *Journal of Chemical Education,* vol. 63, no. 11, pp. 930–3, doi:10.1021/ed063p930

- Trivedi Y, Yung E & Katz DS 2013, 'Imaging in Fever of Unknown Origin', in BA Cunha (ed.), *Fever of Unknown Origin,* Informa Healthcare USA, New York, pp. 209–228, ISBN 0-8493-3615-5

- Turner M 2011, 'German *E. Coli* Outbreak Caused by Previously Unknown Strain', *Nature News,* 2 Jun, doi:10.1038/n45

- Turova N 2011, *Inorganic Chemistry in Tables,* Springer, Heidelberg, ISBN 978-3-642-20486-9

- Tuthill G 2011, 'Faculty profile: Elements of Great Teaching', *The Iolani School Bulletin,* Winter, viewed 29 October 2011

- Tyler PM 1948, *From the Ground Up: Facts and Figures of the Mineral Industries of the United States,* McGraw-Hill, New York

- UCR Today 2011, 'Research Performed in Guy Bertrand's Lab Offers Vast Family of New Catalysts for use in Drug Discovery, Biotechnology', University of California, Riverside, July 28

- Uden PC 2005, 'Speciation of Selenium,' in R Cornelis, J Caruso, H Crews & K Heumann (eds), *Handbook of Elemental Speciation II: Species in the Environment, Food, Medicine and Occupational Health,* John Wiley & Sons, Chichester, pp. 346–65, ISBN 0-470-85598-3

- United Nuclear Scientific 2014, 'Disk Sources, Standard', viewed 5 April 2014

- US Bureau of Naval Personnel 1965, *Shipfitter 3 & 2,* US Government Printing Office, Washington

- US Environmental Protection Agency 1988, *Ambient Aquatic Life Water Quality Criteria for Antimony (III),* draft, Office of Research and Development, Environmental Research Laboratories, Washington

- University of Utah 2014, *New 'Topological Insulator' Could Lead to Superfast Computers,* Phys.org, viewed 15 December 2014

- Van Muylder J & Pourbaix M 1974, 'Arsenic', in M Pourbaix (ed.), *Atlas of Electrochemical Equilibria in Aqueous Solutions,* 2nd ed., National Association of Corrosion Engineers, Houston

- Van der Put PJ 1998, *The Inorganic Chemistry of Materials: How to Make Things Out of Elements,* Plenum, New York, ISBN 0-306-45731-8

- Van Setten MJ, Uijttewaal MA, de Wijs GA & Groot RA 2007, 'Thermodynamic Stability of Boron: The Role of Defects and Zero Point Motion', *Journal of the American Chemical Society,* vol. 129, no. 9, pp. 2458–65, doi:10.1021/ja0631246

- Vasáros L & Berei K 1985, 'General Properties of Astatine', pp. 107–28, in Kugler & Keller

- Vernon RE 2013, 'Which Elements Are Metalloids?', *Journal of Chemical Education,* vol. 90, no. 12, pp. 1703–1707, doi:10.1021/ed3008457

- Walker P & Tarn WH 1996, *CRC Handbook of Metal Etchants,* Boca Raton, FL, ISBN 0849336236

- Walters D 1982, *Chemistry,* Franklin Watts Science World series, Franklin Watts, London, ISBN 0-531-04581-1

- Wang Y & Robinson GH 2011, 'Building a Lewis Base with Boron', *Science,* vol. 333, no. 6042, pp. 530–531, doi:10.1126/science.1209588

- Wanga WH, Dongb C & Shek CH 2004, 'Bulk Metallic Glasses', *Materials Science and Engineering Reports,* vol. 44, nos 2–3, pp. 45–89, doi:10.1016/j.mser.2004.03.001

- Warren J & Geballe T 1981, 'Research Opportunities in New Energy-Related Materials', *Materials Science and Engineering,* vol. 50, no. 2, pp. 149–98, doi:10.1016/0025-5416(81)90177-4

- Weingart GW 1947, *Pyrotechnics,* 2nd ed., Chemical Publishing Company, New York

- Wells AF 1984, *Structural Inorganic Chemistry,* 5th ed., Clarendon, Oxford, ISBN 0-19-855370-6

- Whitten KW, Davis RE, Peck LM & Stanley GG 2007, *Chemistry,* 8th ed., Thomson Brooks/Cole, Belmont, California, ISBN 0-495-01449-4

- Wiberg N 2001, *Inorganic Chemistry,* Academic Press, San Diego, ISBN 0-12-352651-5

- Wilkie CA & Morgan AB 2009, *Fire Retardancy of Polymeric Materials,* CRC Press, Boca Raton, Florida, ISBN 1-4200-8399-6

- Witt AF & Gatos HC 1968, 'Germanium', in CA Hampel (ed.), *The Encyclopedia of the Chemical Elements,* Reinhold, New York, pp. 237–44

- Wogan T 2014, "First experimental evidence of a boron fullerene", Chemistry World, 14 July

- Woodward WE 1948, *Engineering Metallurgy,* Constable, London

- WPI-AIM (World Premier Institute – Advanced Institute for Materials Research) 2012, 'Bulk Metallic Glasses: An Unexpected Hybrid', AIMResearch, Tohoku University, Sendai, Japan, 30 April

- Wulfsberg G 2000, *Inorganic Chemistry, University Science Books, Sausalito California, ISBN 1-891389-01-7*

- Xu Y, Miotkowski I, Liu C, Tian J, Nam H, Alidoust N, Hu J, Shih C-K, Hasan M & Chen YP 2014, 'Observation of Topological Surface State Quantum Hall Effect in an Intrinsic Three-dimensional Topological Insulator,' *Nature Physics,* vol, 10, pp. 956–963, doi:10.1038/nphys3140

- Yacobi BG & Holt DB 1990, *Cathodoluminescence Microscopy of Inorganic Solids,* Plenum, New York, ISBN 0-306-43314-1

- Yang K, Setyawan W, Wang S, Nardelli MB & Curtarolo S 2012, 'A Search Model for Topological Insulators with High-throughput Robustness Descriptors,' *Nature Materials,* vol. 11, pp. 614–619, {{doi:10.1038/nmat3332}}

- Yasuda E, Inagaki M, Kaneko K, Endo M, Oya A & Tanabe Y 2003, *Carbon Alloys: Novel Concepts to Develop Carbon Science and Technology,* Elsevier Science, Oxford, pp. 3–11 et seq, ISBN 0-08-044163-7

- Yetter RA 2012, *Nanoengineered Reactive Materials and their Combustion and Synthesis,* course notes, Princeton-CEFRC Summer School On Combustion, June 25–29, 2012, Penn State University

- Young RV & Sessine S (eds) 2000, *World of Chemistry,* Gale Group, Farmington Hills, Michigan, ISBN 0-7876-3650-9

- Young TF, Finley K, Adams WF, Besser J, Hopkins WD, Jolley D, McNaughton E, Presser TS, Shaw DP & Unrine J 2010, 'What You Need to Know About Selenium', in PM Chapman, WJ Adams, M Brooks, CJ Delos, SN Luoma, WA Maher, H Ohlendorf, TS Presser & P Shaw (eds), *Ecological Assessment of Selenium in the Aquatic Environment,* CRC, Boca Raton, Florida, pp. 7–45, ISBN 1-4398-2677-3

- Zalutsky MR & Pruszynski M 2011, 'Astatine-211: Production and Availability', *Current Radiopharmaceuticals,* vol. 4, no. 3, pp. 177–185, doi:10.2174/10177

- Zhang GX 2002, 'Dissolution and Structures of Silicon Surface', in MJ Deen, D Misra & J Ruzyllo (eds), *Integrated Optoelectronics: Proceedings of the First International Symposium,* Philadelphia, PA, The Electrochemical Society, Pennington, NJ, pp. 63–78, ISBN 1-56677-370-9

- Zhang TC, Lai KCK & Surampalli AY 2008, 'Pesticides', in A Bhandari, RY Surampalli, CD Adams, P Champagne, SK Ong, RD Tyagi & TC Zhang (eds), *Contaminants of Emerging Environmental Concern,* American Society of Civil Engineers, Reston, Virginia, ISBN 978-0-7844-1014-1, pp. 343–415

- Zhdanov GS 1965, *Crystal Physics,* translated from the Russian publication of 1961 by AF Brown (ed.), Oliver & Boyd, Edinburgh

- Zingaro RA 1994, 'Arsenic: Inorganic Chemistry', in RB King (ed.) 1994, *Encyclopedia of Inorganic Chemistry,* John Wiley & Sons, Chichester, pp. 192–218, ISBN 0-471-93620-0

22.13   Further reading

- Brady JE, Humiston GE & Heikkinen H 1980, 'Chemistry of the Representative Elements: Part II, The Metalloids and Nonmetals', in *General Chemistry: Principles and Structure,* 2nd ed., SI version, John Wiley & Sons, New York, pp. 537–591, ISBN 0-471-06315-0

- Chedd G 1969, *Half-way Elements: The Technology of Metalloids,* Doubleday, New York

- Choppin GR & Johnsen RH 1972, 'Group IV and the Metalloids,' in *Introductory Chemistry,* Addison-Wesley, Reading, Massachusetts, pp. 341–357

- Dunstan S 1968, 'The Metalloids', in *Principles of Chemistry,* D. Van Nostrand Company, London, pp. 407–39

- Goldsmith RH1982, 'Metalloids',*Journal of Chemical Education,*vol.59,no.6,pp.526–527,doi:10.1021/ed0526

- Hawkes SJ2001, 'Semimetallicity',*Journal of Chemical Education,*vol.78,no.12,pp.1686–7,doi:10.1021/ed686

- Metcalfe HC, Williams JE & Castka JF 1974, 'Aluminum and the Metalloids', in *Modern Chemistry,* Holt, Rinehart and Winston, New York, pp. 538–57, ISBN 0-03-089450-6

- Moeller T, Bailar JC, Kleinberg J, Guss CO, Castellion ME & Metz C 1989, 'Carbon and the Semiconducting Elements', in *Chemistry, with Inorganic Qualitative Analysis,* 3rd ed., Harcourt Brace Jovanovich, San Diego, pp. 751–75, ISBN 0-15-506492-4

- Rieske M 1998, 'Metalloids', in *Encyclopedia of Earth and Physical Sciences,* Marshall Cavendish, New York, vol. 6, pp. 758–9, ISBN 0-7614-0551-8 (set)

- Rochow EG 1966, *The Metalloids,* DC Heath and Company, Boston

- Vernon RE 2013, 'Which Elements are Metalloids?', *Journal of Chemical Education,* vol. 90, no. 12, pp. 1703–7, doi:10.1021/ed3008457

Chapter 23

Transition metal

In chemistry, the term **transition metal** (or **transition element**) has two possible meanings:

- The IUPAC definition[1] defines a transition metal as "an element whose atom has a partially filled d sub-shell, or which can give rise to cations with an incomplete d sub-shell".

- Most scientists describe a "transition metal" as any element in the d-block of the periodic table, which includes groups 3 to 12 on the periodic table.[2][3] In actual practice, the f-block lanthanide and actinide series are also considered transition metals and are called "inner transition metals".

English chemist Charles Bury first used the word *transition* in 1921 when he referred to a transition series of elements during the change of an inner layer of electrons (for example n=3 in the 4th row of the periodic table) from a stable group of 8 to one of 18, or from 18 to 32.[4][5] These elements are now known as the d-block.

23.1 Classification

In the d-block the atoms of the elements have between 1 and 10 d electrons.

The typical electronic structure of transition metal atoms can be written as [ ]$ns^2(n-1)d^m$, following the Madelung rule where the inner d orbital is predicted to be filled after the valence-shell s orbital. This is actually not the case; the 4s electrons are higher in energy than the 3d as shown spectroscopically. An ion such as Fe2+
has no $4s$ electrons: it has the electronic configuration [Ar]$3d^6$ as compared with the configuration of the atom, [Ar]$4s^23d^6$.

The elements of groups 3–12 are now generally recognized as transition metals, although the elements La-Lu and Ac-Lr and Group 12 attract different definitions from different authors.

1. Many chemistry textbooks and printed periodic tables classify La and Ac as Group 3 elements and transition metals, since their atomic ground-state configurations are s^2d^1 like Sc and Y. The elements Ce-Lu are considered as the "lanthanide" series (or "lanthanoid" according to IUPAC) and Th-Lr as the "actinide" series.[6][7] The two series together are classified as f-block elements, or (in older sources) as "inner transition elements".

2. Some inorganic chemistry textbooks include La with the lanthanides and Ac with the actinides.[8][9][10] This classification is based on similarities in chemical behaviour, and defines 15 elements in each of the two series even though they correspond to the filling of an f subshell which can only contain 14 electrons.

3. A third classification defines the f-block elements as La-Yb and Ac-No, while placing Lu and Lr in Group 3.[4] This is based on the aufbau principle (or Madelung rule) for filling electron subshells, in which 4f is filled before 5d (and 5f before 6d), so that the f subshell is actually full at Yb (and No) while Lu (and Lr) has an [ ]$s^2f^{14}d^1$ configuration. However La and Ac are exceptions to the Aufbau principle with electron configuration [ ]s^2d^1 (not

[ ]s^2f^1 as the aufbau principle predicts) so it is not clear from atomic electron configurations whether La or Lu (Ac or Lr) should be considered as transition metals. Eric Scerri has proposed placing Lu and Lr in group 3 on the grounds of continuous sequences of atomic numbers in an expanded or long-form periodic table.[11]

Zinc, cadmium, and mercury are sometimes excluded from the transition metals[4] as they have the electronic configuration [ ]$d^{10}s^2$, with no incomplete d shell.[12] In the oxidation state +2 the ions have the electronic configuration [ ] d^{10}. However, these elements can exist in other oxidation states, including the +1 oxidation state, as in the diatomic ion Hg2+
2. The group 12 elements Zn, Cd and Hg may be classed as post-transition metals in this case, because of the formation of a covalent bond between the two atoms of the dimer. However, it is often convenient to include these elements in a discussion of the transition elements. For example, when discussing the crystal field stabilization energy of first-row transition elements, it is convenient to also include the elements calcium and zinc, as both Ca2+
and Zn2+
have a value of zero against which the value for other transition metal ions may be compared. Another example occurs in the Irving-Williams series of stability constants of complexes.

The recent synthesis of mercury(IV) fluoride (HgF
4) has been taken by some to reinforce the view that the group 12 elements should be considered transition metals,[13] but some authors still consider this compound to be exceptional.[14]

23.2   Position in the periodic table

The d-block as stated earlier, is present in the centre of the long form of periodic table. These are flanked or surrounded by elements belonging to s and p-blocks on both sides. These are called transition elements since they represent a transition i.e., there is a change from metallic character of s-block elements to non-metallic character of p-block elements through d-block elements which are also metals. As pointed above there are four transition series in this block. Since the filling of electrons takes place in $(n-1)d$ orbitals, the periods to which these series belong, is actually one more than the actual series. For example, the elements included in $3d$ series belong to fourth period ; the elements included in $4d$ series belong to the fifth period and so on.

23.3   Electronic configuration

The general electronic configuration of the d-block elements is [Inert gas] $(n-1)d^{1-10}n\,s^{1-2}$. The period 6 and 7 transition metals also add $(n-2)f^{14}$ electrons, which are omitted from the tables below.

The d-sub-shell is the penultimate (last but one) sub-shell and is denoted as $(n-1)$ d-sub-shell. The number of s electrons may vary from one to two. The s-sub-shell in the valence shell is represented as the ns sub-shell. However, palladium (Pd) is an exception with no electron in the s-sub shell. In the periodic table, the transition metals are present in ten groups (3 to 12). Group-2 belongs to the s- block with an ns^2 configuration.

The elements in group-3 have an $ns^2(n-1)d^1$ configuration. The first transition series is present in the 4thperiod, and starts after Ca (Z=20) of group-2 which has configuration [Ar]$4s^2$. The electronic configuration of scandium (Sc), the first element of group-3 with atomic number Z=21 is[Ar]$4s^23d^1$. As we move from left to right, electrons are added to the same d-sub-shell till it is complete. The element of group-12 in the first transition series is zinc (Zn) with configuration [Ar]$4s^23d^{10}$. Since the electrons added fill the $(n-1)d$ orbitals, the properties of the d-block elements are quite different from those of s and p block elements in which the filling occurs either in s or in p-orbitals of the valence shell. The electronic configuration of the individual elements present in all the transition series are given below:

First ($3d$) Transition Series (Sc-Zn)

Second ($4d$) Transition Series (Y-Cd)

Third ($5d$) Transition Series (Lu-Hg)[15]

Fourth ($6d$) Transition Series (Lr-Cn)

A careful look at the electronic configuration of the elements reveals that there are certain exceptions shown by Pt, Au and Hg.. These are either because of the symmetry or nuclear-electron and electron-electron force.

The *(n-1)d* orbitals that are involved in the transition metals are very significant because they influence such properties as magnetic character, variable oxidation states, formation of colored compounds etc. The valence *s(ns)* and *p(np)* orbitals have very little contribution in this regard since they hardly change in the moving from left to the right in a transition series. In transition metals, there is a greater horizontal similarities in the properties of the elements in a period in comparison to the periods in which the *d*-orbitals are not involved. This is because in a transition series, the valence shell electronic configuration of the elements do not change. However, there are some group similarities as well.

23.4   Characteristic properties

There are a number of properties shared by the transition elements that are not found in other elements, which results from the partially filled *d* shell. These include

- the formation of compounds whose colour is due to *d–d* electronic transitions

- the formation of compounds in many oxidation states, due to the relatively low reactivity of unpaired *d* electrons.[16]

- the formation of many paramagnetic compounds due to the presence of unpaired *d* electrons. A few compounds of main group elements are also paramagnetic (e.g. nitric oxide, oxygen)

23.4.1   Coloured compounds

From left to right, aqueous solutions of: Co(NO
3)
2 (red); K
2Cr
2O
7 (orange); K
2CrO
4 (yellow); NiCl
2 (turquoise); CuSO
4 (blue); KMnO
4 (purple).

Colour in transition-series metal compounds is generally due to electronic transitions of two principal types.

- charge transfer transitions. An electron may jump from a predominantly ligand orbital to a predominantly metal orbital, giving rise to a ligand-to-metal charge-transfer (LMCT) transition. These can most easily occur when the metal is in a high oxidation state. For example, the colour of chromate, dichromate and permanganate ions is due to LMCT transitions. Another example is that mercuric iodide, HgI_2, is red because of a LMCT transition.

A metal-to-ligand charge transfer (MLCT) transition will be most likely when the metal is in a low oxidation state and the ligand is easily reduced.

- *d-d* transitions. An electron jumps from one d-orbital to another. In complexes of the transition metals the *d* orbitals do not all have the same energy. The pattern of splitting of the *d* orbitals can be calculated using crystal field theory. The extent of the splitting depends on the particular metal, its oxidation state and the nature of the ligands. The actual energy levels are shown on Tanabe-Sugano diagrams.

In centrosymmetric complexes, such as octahedral complexes, *d-d* transitions are forbidden by the Laporte rule and only occur because of vibronic coupling in which a molecular vibration occurs together with a *d-d* transition. Tetrahedral complexes have somewhat more intense colour because mixing *d* and *p* orbitals is possible when there is no centre of symmetry, so transitions are not pure *d-d* transitions. The molar absorptivity (ε) of bands caused by *d-d* transitions are relatively low, roughly in the range 5-500 $M^{-1}cm^{-1}$ (where M = mol dm^{-3}).[17] Some *d-d* transitions are spin forbidden. An example occurs in octahedral, high-spin complexes of manganese(II), which has a d^5 configuration in which all five electron has parallel spins; the colour of such complexes is much weaker than in complexes with spin-allowed transitions. Many compounds of manganese(II) appear almost colourless. The spectrum of $[Mn(H_2O)_6]^{2+}$ shows a maximum molar absorptivity of about 0.04 $M^{-1}cm^{-1}$ in the visible spectrum.

23.4.2 Oxidation states

A characteristic of transition metals is that they exhibit two or more oxidation states, usually differing by one. For example, compounds of vanadium are known in all oxidation states between −1, such as $[V(CO)_6]^-$, and +5, such as VO_4^{3-}.

Main group elements in groups 13 to 17 also exhibit multiple oxidation states. The "common" oxidation states of these elements typically differ by two. For example, compounds of gallium in oxidation states +1 and +3 exist in which there is a single gallium atom. No compound of Ga(II) is known: any such compound would have an unpaired electron and would behave as a free radical and be destroyed rapidly. The only compounds in which gallium has a formal oxidation state of +2 are dimeric compounds, such as $[Ga_2Cl_6]^{2-}$, which contain a Ga-Ga bond formed from the unpaired electron on each Ga atom.[18] Thus the main difference in oxidation states, between transition elements and other elements is that oxidation states are known in which there is a single atom of the element and one or more unpaired electrons.

The maximum oxidation state in the first row transition metals is equal to the number of valence electrons from titanium (+4) up to manganese (+7), but decreases in the later elements. In the second row the maximum occurs with ruthenium (+8), and in the third row, the maximum occurs with iridium (+9). In compounds such as $[MnO_4]^-$ and OsO_4 the elements achieve a stable octet by forming four covalent bonds.

The lowest oxidation states are exhibited in metal carbonyl complexes such as $Cr(CO)_6$ (oxidation state zero) and $[Fe(CO)_4]^{2-}$ (oxidation state −2) in which the 18-electron rule is obeyed. These complexes are also covalent.

Ionic compounds are mostly formed with oxidation states +2 and +3. In aqueous solution the ions are hydrated by (usually) six water molecules arranged octahedrally.

23.4.3 Magnetism

Main article: magnetochemistry

Transition metal compounds are paramagnetic when they have one or more unpaired d electrons.[19] In octahedral complexes with between four and seven d electrons both high spin and low spin states are possible. Tetrahedral transition metal complexes such as [FeCl
4]2−
are high spin because the crystal field splitting is small so that the energy to be gained by virtue of the electrons being in lower energy orbitals is always less than the energy needed to pair up the spins. Some compounds are diamagnetic. These include octahedral, low-spin, d^6 and square-planar d^8 *complexes*. In these cases, crystal field splitting is such that all the electrons are paired up.

Ferromagnetism occurs when individual atoms are paramagnetic and the spin vectors are aligned parallel to each other in a crystalline material. Metallic iron and the alloy alnico are examples of ferromagnetic materials involving transition metals. Anti-ferromagnetism is another example of a magnetic property arising from a particular alignment of individual spins in the solid state.

23.4.4 Catalytic properties

The transition metals and their compounds are known for their homogeneous and heterogeneous catalytic activity. This activity is ascribed to their ability to adopt multiple oxidation states and to form complexes. Vanadium(V) oxide (in the contact process), finely divided iron (in the Haber process), and nickel (in catalytic hydrogenation) are some of the examples. Catalysts at a solid surface (nanomaterial-based catalysts) involve the formation of bonds between reactant molecules and atoms of the surface of the catalyst (first row transition metals utilize 3d and 4s electrons for bonding). This has the effect of increasing the concentration of the reactants at the catalyst surface and also weakening of the bonds in the reacting molecules (the activation energy is lowered). Also because the transition metal ions can change their oxidation states, they become more effective as catalysts.

23.4.5 Other properties

As implied by the name, all transition metals are metals and conductors of electricity.

In general, transition metals possess a high density and high melting points and boiling points. These properties are due to metallic bonding by delocalized d electrons, leading to cohesion which increases with the number of shared electrons. However the group 12 metals have much lower melting and boiling points since their full d subshells prevent d–d bonding. Mercury has a melting point of −38.83 °C (−37.89 °F) and is a liquid at room temperature.

Many transition metals can be bound to a variety of ligands.[20]

23.5 See also

- Inner transition element, a name given to any member of the f-block

- Main group element, an element other than a transition metal.

- Ligand field theory a development of crystal field theory taking covalency into account.

- Crystal field theory a model that describes the breaking of degeneracies of electronic orbital states.

- Post-transition metal, a metallic element to the right of the transition metals in the periodic table.

23.6 References

[1] IUPAC, *Compendium of Chemical Terminology*, 2nd ed. (the "Gold Book") (1997). Online corrected version: (2006–) "transition element".

[2] Petrucci, R.H.; Harwood, W.S. and Herring, F.G. (2002) *General Chemistry*, 8th ed, Prentice-Hall, pp. 341-2

[3] Housecroft, C.E. and Sharpe, A.G. (2005) *Inorganic Chemistry*, 2nd ed, Pearson Prentice-Hall, pp. 20-21

[4] Jensen, William B. (2003). "The Place of Zinc, Cadmium, and Mercury in the Periodic Table" (PDF). *Journal of Chemical Education* **80** (8): 952–961. Bibcode:2003JChEd..80..952J. doi:10.1021/ed080p952.

[5] Bury, C. R. (1921). "Langmuir's theory of the arrangement of electrons in atoms and molecules". *J. Amer. Chem. Soc.* **43** (7): 1602–1609. doi:10.1021/ja01440a023.

[6] Petrucci, R. H. *et al.* (2002), "General Chemistry", 8th edn, Prentice-Hall, pp. 49–50, 951

[7] Miessler, G. L. and Tarr, D. A. (1999) *Inorganic Chemistry*, 2nd edn, Prentice-Hall, p. 16

[8] Greenwood, Norman N.; Earnshaw, Alan (1997). *Chemistry of the Elements* (2nd ed.). Butterworth-Heinemann. ISBN 0080379419.

[9] Cotton, F.A. and Wilkinson, G. (1988) *Inorganic Chemistry*, 5th ed., Wiley , pp. 626–7

[10] Housecroft, C. E. and Sharpe, A. G. (2005) *Inorganic Chemistry*, 2nd ed., Pearson Prentice-Hall, p. 741

[11] Scerri, E.R. (2011) *A Very Short Introduction to the Periodic Table*, Oxford University Press.

[12] Cotton, F. Albert; Wilkinson, G.; Murillo, C. A. (1999). *Advanced Inorganic Chemistry* (6th ed.). New York: Wiley, ISBN 0471199575.

[13] Wang, Xuefang; Andrews, Lester; Riedel, Sebastian; Kaupp, Martin (2007). "Mercury Is a Transition Metal: The First Experimental Evidence for HgF$_4$". *Angew. Chem. Int. Ed.* **46** (44): 8371–8375. doi:10.1002/anie.200703710. PMID 17899620.

[14] Jensen, William B. (2008). "Is Mercury Now a Transition Element?".*J.Chem.Educ.***85**(9):1182–1183.Bibcode:2008JChE1182J. doi:10.1021/ed085p1182.

[15] This table follows the proposal of Scerri (see above) and considers Lu rather than La to be a transition metal in the 5d series. As noted in the section on Classification, not all chemists agree.

[16] Matsumoto, Paul S (2005). "Trends in Ionization Energy of Transition-Metal Elements". *Journal of Chemical Education* **82** (11): 1660. Bibcode:2005JChEd..82.1660M. doi:10.1021/ed082p1660.

[17] Orgel, L.E. (1966). *An Introduction to Transition-Metal Chemistry, Ligand field theory* (2nd. ed.). London: Methuen.

[18] Greenwood, Norman N.; Earnshaw, Alan (1997). *Chemistry of the Elements* (2nd ed.). Butterworth-Heinemann. ISBN 0080379419. p. 240

[19] Figgis, B.N.; Lewis, J. (1960). Lewis, J.; Wilkins, R.G., eds. *The Magnetochemistry of Complex Compounds.* Modern Coordination Chemistry. New York: Wiley Interscience. pp. 400–454.

[20] Hogan, C. Michael (2010). "Heavy metal" in *Encyclopedia of Earth*. National Council for Science and the Environment. E. Monosson and C. Cleveland (eds.) Washington DC.

Chapter 24

Metal

This article is about metallic materials. For other uses, see Metal (disambiguation).

A **metal** (from Greek μέταλλον *métallon*, "mine, quarry, metal"[1][2]) is a material (an element, compound, or alloy) that is typically hard, opaque, shiny, and has good electrical and thermal conductivity. Metals are generally malleable — that is, they can be hammered or pressed permanently out of shape without breaking or cracking — as well as fusible (able to be fused or melted) and ductile (able to be drawn out into a thin wire).[3] About 91 of the 118 elements in the periodic table are metals (some elements appear in both metallic and non-metallic forms).

Astrophysicists use the term "metal" to collectively describe all elements other than hydrogen and helium. Thus, the metallicity of an object is the proportion of its matter made up of chemical elements other than hydrogen and helium.[4]

Many elements and compounds that are not normally classified as metals become metallic under high pressures; these are formed as metallic allotropes of non-metals.

24.1 Structure and bonding

The atoms of metallic substances are closely positioned to neighboring atoms in one of two common arrangements. The first arrangement is known as body-centered cubic. In this arrangement, each atom is positioned at the center of eight others. The other is known as face-centered cubic. In this arrangement, each atom is positioned in the center of six others. The ongoing arrangement of atoms in these structures forms a crystal. Some metals adopt both structures depending on the temperature.[5]

Atoms of metals readily lose their outer shell electrons, resulting in a free flowing cloud of electrons within their otherwise solid arrangement. This provides the ability of metallic substances to easily transmit heat and electricity. While this flow of electrons occurs, the solid characteristic of the metal is produced by electrostatic interactions between each atom and the electron cloud. This type of bond is called a metallic bond.[6]

24.2 Properties

24.2.1 Chemical

Metals are usually inclined to form cations through electron loss,[6] reacting with oxygen in the air to form oxides over various timescales (iron rusts over years, while potassium burns in seconds). Examples:

$4 \, Na + O_2 \rightarrow 2 \, Na_2O$ (sodium oxide)

$2 \, Ca + O_2 \rightarrow 2 \, CaO$ (calcium oxide)

218

4 Al + 3 O$_2$ → 2 Al$_2$O$_3$ (aluminium oxide).

The transition metals (such as iron, copper, zinc, and nickel) are slower to oxidize because they form a passivating layer of oxide that protects the interior. Others, like palladium, platinum and gold, do not react with the atmosphere at all. Some metals form a barrier layer of oxide on their surface which cannot be penetrated by further oxygen molecules and thus retain their shiny appearance and good conductivity for many decades (like aluminium, magnesium, some steels, and titanium). The oxides of metals are generally basic, as opposed to those of nonmetals, which are acidic. Exceptions are largely oxides with very high oxidation states such as CrO$_3$, Mn$_2$O$_7$, and OsO$_4$, which have strictly acidic reactions.

Painting, anodizing or plating metals are good ways to prevent their corrosion. However, a more reactive metal in the electrochemical series must be chosen for coating, especially when chipping of the coating is expected. Water and the two metals form an electrochemical cell, and if the coating is less reactive than the coatee, the coating actually *promotes* corrosion.

24.2.2 Physical

Metals in general have high electrical conductivity, high thermal conductivity, and high density. Typically they are malleable and ductile, deforming under stress without cleaving.[6] In terms of optical properties, metals are shiny and lustrous. Sheets of metal beyond a few micrometres in thickness appear opaque, but gold leaf transmits green light.

Although most metals have higher densities than most nonmetals,[6] there is wide variation in their densities, Lithium being the least dense solid element and osmium the densest. The alkali and alkaline earth metals in groups I A and II A are referred to as the light metals because they have low density, low hardness, and low melting points.[6] The high density of most metals is due to the tightly packed crystal lattice of the metallic structure. The strength of metallic bonds for different metals reaches a maximum around the center of the transition metal series, as those elements have large amounts of delocalized electrons in tight binding type metallic bonds. However, other factors (such as atomic radius, nuclear charge, number of bonds orbitals, overlap of orbital energies and crystal form) are involved as well.[6]

24.2.3 Electrical

The electrical and thermal conductivities of metals originate from the fact that their outer electrons are delocalized. This situation can be visualized by seeing the atomic structure of a metal as a collection of atoms embedded in a sea of highly mobile electrons. The electrical conductivity, as well as the electrons' contribution to the heat capacity and heat conductivity of metals can be calculated from the free electron model, which does not take into account the detailed structure of the ion lattice.

When considering the electronic band structure and binding energy of a metal, it is necessary to take into account the positive potential caused by the specific arrangement of the ion cores – which is periodic in crystals. The most important consequence of the periodic potential is the formation of a small band gap at the boundary of the Brillouin zone. Mathematically, the potential of the ion cores can be treated by various models, the simplest being the nearly free electron model.

24.2.4 Mechanical

Mechanical properties of metals include ductility, i.e. their capacity for plastic deformation. Reversible elastic deformation in metals can be described by Hooke's Law for restoring forces, where the stress is linearly proportional to the strain. Forces larger than the elastic limit, or heat, may cause a permanent (irreversible) deformation of the object, known as plastic deformation or plasticity. This irreversible change in atomic arrangement may occur as a result of:

- The action of an applied force (or work). An applied force may be tensile (pulling) force, compressive (pushing) force, shear, bending or torsion (twisting) forces.

- A change in temperature (heat). A temperature change may affect the mobility of the structural defects such as grain boundaries, point vacancies, line and screw dislocations, stacking faults and twins in both crystalline and non-crystalline solids. The movement or displacement of such mobile defects is thermally activated, and thus limited by the rate of atomic diffusion.

Viscous flow near grain boundaries, for example, can give rise to internal slip, creep and fatigue in metals. It can also contribute to significant changes in the microstructure like grain growth and localized densification due to the elimination of intergranular porosity. Screw dislocations may slip in the direction of any lattice plane containing the dislocation, while the principal driving force for "dislocation climb" is the movement or diffusion of vacancies through a crystal lattice.

In addition, the nondirectional nature of metallic bonding is also thought to contribute significantly to the ductility of most metallic solids. When the planes of an ionic bond slide past one another, the resultant change in location shifts ions of the same charge into close proximity, resulting in the cleavage of the crystal; such shift is not observed in covalently bonded crystals where fracture and crystal fragmentation occurs.[7]

24.3 Alloys

Main article: Alloy

An alloy is a mixture of two or more elements in which the main component is a metal. Most pure metals are either too soft, brittle or chemically reactive for practical use. Combining different ratios of metals as alloys modifies the properties of pure metals to produce desirable characteristics. The aim of making alloys is generally to make them less brittle, harder, resistant to corrosion, or have a more desirable color and luster. Of all the metallic alloys in use today, the alloys of iron (steel, stainless steel, cast iron, tool steel, alloy steel) make up the largest proportion both by quantity and commercial value. Iron alloyed with various proportions of carbon gives low, mid and high carbon steels, with increasing carbon levels reducing ductility and toughness. The addition of silicon will produce cast irons, while the addition of chromium, nickel and molybdenum to carbon steels (more than 10%) results in stainless steels.

Other significant metallic alloys are those of aluminium, titanium, copper and magnesium. Copper alloys have been known since prehistory—bronze gave the Bronze Age its name—and have many applications today, most importantly in electrical wiring. The alloys of the other three metals have been developed relatively recently; due to their chemical reactivity they require electrolytic extraction processes. The alloys of aluminium, titanium and magnesium are valued for their high strength-to-weight ratios; magnesium can also provide electromagnetic shielding. These materials are ideal for situations where high strength-to-weight ratio is more important than material cost, such as in aerospace and some automotive applications.

Alloys specially designed for highly demanding applications, such as jet engines, may contain more than ten elements.

24.4 Categories

24.4.1 Base metal

Main article: Base metal
In chemistry, the term *base metal* is used informally to refer to a metal that oxidizes or corrodes relatively easily, and reacts variably with dilute hydrochloric acid (HCl) to form hydrogen. Examples include iron, nickel, lead and zinc. Copper is considered a base metal as it oxidizes relatively easily, although it does not react with HCl. It is commonly used in opposition to noble metal.

In alchemy, a *base metal* was a common and inexpensive metal, as opposed to precious metals, mainly gold and silver. A longtime goal of the alchemists was the transmutation of base metals into precious metals.

In numismatics, coins in the past derived their value primarily from the precious metal content. Most modern currencies are fiat currency, allowing the coins to be made of *base metal*.

24.4.2 Ferrous metal

Main article: Ferrous metallurgy
See also: Non-ferrous metals

The term "ferrous" is derived from the Latin word meaning "containing iron". This can include pure iron, such as wrought iron, or an alloy such as steel. Ferrous metals are often magnetic, but not exclusively.

24.4.3 Noble metal

Main article: Noble metal

Noble metals are metals that are resistant to corrosion or oxidation, unlike most base metals. They tend to be precious metals, often due to perceived rarity. Examples include gold, platinum, silver, rhodium, Iridium and palladium.

24.4.4 Precious metal

Main article: Precious metal

A *precious metal* is a rare metallic chemical element of high economic value.

Chemically, the precious metals are less reactive than most elements, have high luster and high electrical conductivity. Historically, precious metals were important as currency, but are now regarded mainly as investment and industrial commodities. Gold, silver, platinum and palladium each have an ISO 4217 currency code. The best-known precious metals are gold and silver. While both have industrial uses, they are better known for their uses in art, jewelry, and coinage. Other precious metals include the platinum group metals: ruthenium, rhodium, palladium, osmium, iridium, and platinum, of which platinum is the most widely traded.

The demand for precious metals is driven not only by their practical use, but also by their role as investments and a store of value. Palladium was, as of summer 2006, valued at a little under half the price of gold, and platinum at around twice that of gold. Silver is substantially less expensive than these metals, but is often traditionally considered a precious metal for its role in coinage and jewelry.

24.4.5 Heavy metal

Main article: Heavy metal (chemical element)

A heavy metal is any relatively dense metal or metalloid. More specific definitions have been proposed, but none have obtained widespread acceptance. Some heavy metals have niche uses, or are notably toxic; some are essential in trace amounts.

24.5 Extraction

Main articles: Ore, Mining and Extractive metallurgy

Metals are often extracted from the Earth by means of mining ores that are rich sources of the requisite elements, such as bauxite. Ore is located by prospecting techniques, followed by the exploration and examination of deposits. Mineral sources are generally divided into surface mines, which are mined by excavation using heavy equipment, and subsurface mines.

Once the ore is mined, the metals must be extracted, usually by chemical or electrolytic reduction. Pyrometallurgy uses high temperatures to convert ore into raw metals, while hydrometallurgy employs aqueous chemistry for the same purpose. The methods used depend on the metal and their contaminants.

When a metal ore is an ionic compound of that metal and a non-metal, the ore must usually be smelted — heated with a reducing agent — to extract the pure metal. Many common metals, such as iron, are smelted using carbon as a reducing agent. Some metals, such as aluminium and sodium, have no commercially practical reducing agent, and are extracted using electrolysis instead.[8][9]

Sulfide ores are not reduced directly to the metal but are roasted in air to convert them to oxides.

24.6   Recycling of metals

Demand for metals is closely linked to economic growth. During the 20th century, the variety of metals uses in society grew rapidly. Today, the development of major nations, such as China and India, and advances in technologies, are fuelling ever more demand. The result is that mining activities are expanding, and more and more of the world's metal stocks are above ground in use, rather than below ground as unused reserves. An example is the in-use stock of copper. Between 1932 and 1999, copper in use in the USA rose from 73g to 238g per person.[10]

Metals are inherently recyclable, so in principle, can be used over and over again, minimizing these negative environmental impacts and saving energy at the same time. For example, 95% of the energy used to make aluminium from bauxite ore is saved by using recycled material.[11] However, levels of metals recycling are generally low. In 2010, the International Resource Panel, hosted by the United Nations Environment Programme (UNEP) published reports on metal stocks that exist within society[12] and their recycling rates.[10]

The report authors observed that the metal stocks in society can serve as huge mines above ground. However, they warned that the recycling rates of some rare metals used in applications such as mobile phones, battery packs for hybrid cars and fuel cells are so low that unless future end-of-life recycling rates are dramatically stepped up these critical metals will become unavailable for use in modern technology.

24.7   Metallurgy

Main article: Metallurgy

Metallurgy is a domain of materials science that studies the physical and chemical behavior of metallic elements, their intermetallic compounds, and their mixtures, which are called alloys.

24.8   Applications

Some metals and metal alloys possess high structural strength per unit mass, making them useful materials for carrying large loads or resisting impact damage. Metal alloys can be engineered to have high resistance to shear, torque and deformation. However the same metal can also be vulnerable to fatigue damage through repeated use or from sudden stress failure when a load capacity is exceeded. The strength and resilience of metals has led to their frequent use in high-rise building and bridge construction, as well as most vehicles, many appliances, tools, pipes, non-illuminated signs and railroad tracks.

The two most commonly used structural metals, iron and aluminium, are also the most abundant metals in the Earth's crust.[13]

Metals are good conductors, making them valuable in electrical appliances and for carrying an electric current over a distance with little energy lost. Electrical power grids rely on metal cables to distribute electricity. Home electrical systems, for the most part, are wired with copper wire for its good conducting properties.

The thermal conductivity of metal is useful for containers to heat materials over a flame. Metal is also used for heat sinks to protect sensitive equipment from overheating.

The high reflectivity of some metals is important in the construction of mirrors, including precision astronomical instruments. This last property can also make metallic jewelry aesthetically appealing.

Some metals have specialized uses; radioactive metals such as uranium and plutonium are used in nuclear power plants to produce energy via nuclear fission. Mercury is a liquid at room temperature and is used in switches to complete a circuit when it flows over the switch contacts. Shape memory alloy is used for applications such as pipes, fasteners and vascular stents.

24.9 Trade

The World Bank reports that China was the top importer of ores and metals in 2005 followed by the United States and Japan.[14]

24.10 History

The nature of metals has fascinated mankind for many centuries, because these materials provided people with tools of unsurpassed properties both in war and in their preparation and processing. Sterling gold and silver were known to man since the Stone Age. Lead and silver were fused from their ores as early as the fourth millennium BC.[15]

Ancient Latin and Greek writers such as Theophrastus, Pliny the Elder in his *Natural History*, or Pedanius Dioscorides, did not try to classify metals. The ancients never attained the concept "metal" as a distinct elementary substance of fixed, characteristic chemical and physical properties. Following Empedocles, all substances within the sublunary sphere were assumed to vary in their constituent classical elements of earth, water, air and fire. Following the Pythagoreans, Plato assumed that these elements could be further reduced to plane geometrical shapes (triangles and squares) bounding space and relating to the regular polyhedra in the sequence earth:cube, water:icosahedron, air:octahedron, fire:tetrahedron. However, this philosophical extension did not become as popular as the simple four elements, after it was rejected by Aristotle. Aristotle also rejected the atomic theory of Democritus, since he classified the implied existence of a vacuum necessary for motion as a contradiction (a vacuum implies nonexistence, therefore cannot exist). Aristotle did, however, introduce underlying antagonistic qualities (or forces) of dry vs. wet and cold vs. heat into the composition of each of the four elements. The word "metal" originally meant "mines" and only later gained the general meaning of products from materials obtained in mines. In the first centuries A.D. a relation between the planets and the existing metals was assumed as Gold:Sun, Silver:Moon, Electrum:Jupiter, Iron:Mars, Copper:Venus, Tin:Mercury, Lead: Saturn. After electrum was determined to be a combination of silver and gold, the relations Tin:Jupiter and Mercury:Mercury were substituted into the previous sequence.[16]

Arabic and medieval alchemists believed that all metals, and in fact, all sublunar matter, were composed of the principle of sulfur, carrying the combustible property, and the principle of mercury, the mother of all metals and carrier of the liquidity or fusibility, and the volatility properties. These principles were not necessarily the common substances sulfur and mercury found in most laboratories. This theory reinforced the belief that the all metals were destined to become gold in the bowels of the earth through the proper combinations of heat, digestion, time, and elimination of contaminants, all of which could be developed and hastened through the knowledge and methods of alchemy. Paracelsus added the third principle of salt, carrying the nonvolatile and incombustible properties, in his *tria prima* doctrine. These theories retained the four classical elements as underlying the composition of sulfur, mercury and salt.

The first systematic text on the arts of mining and metallurgy was *De la Pirotechnia* by Vannoccio Biringuccio, which treats the examination, fusion, and working of metals. Sixteen years later, Georgius Agricola published *De Re Metallica* in 1555, a clear and complete account of the profession of mining, metallurgy, and the accessory arts and sciences, as well as qualifying as the greatest treatise on the chemical industry through the sixteenth century. He gave the following description of a metal in his *De Natura Fossilium* (1546).

> Metal is a mineral body, by nature either liquid or somewhat hard. The latter may be melted by the

heat of the fire, but when it has cooled down again and lost all heat, it becomes hard again and resumes its proper form. In this respect it differs from the stone which melts in the fire, for although the latter regain its hardness, yet it loses its pristine form and properties. Traditionally there are six different kinds of metals, namely gold, silver, copper, iron, tin and lead. There are really others, for quicksilver is a metal, although the Alchemists disagree with us on this subject, and bismuth is also. The ancient Greek writers seem to have been ignorant of bismuth, wherefore Ammonius rightly states that there are many species of metals, animals, and plants which are unknown to us. Stibium when smelted in the crucible and refined has as much right to be regarded as a proper metal as is accorded to lead by writers. If when smelted, a certain portion be added to tin, a bookseller's alloy is produced from which the type is made that is used by those who print books on paper. Each metal has its own form which it preserves when separated from those metals which were mixed with it. Therefore neither electrum nor Stannum [not meaning our tin] is of itself a real metal, but rather an alloy of two metals. Electrum is an alloy of gold and silver, Stannum of lead and silver. And yet if silver be parted from the electrum, then gold remains and not electrum; if silver be taken away from Stannum, then lead remains and not Stannum. Whether brass, however, is found as a native metal or not, cannot be ascertained with any surety. We only know of the artificial brass, which consists of copper tinted with the colour of the mineral calamine. And yet if any should be dug up, it would be a proper metal. Black and white copper seem to be different from the red kind. Metal, therefore, is by nature either solid, as I have stated, or fluid, as in the unique case of quicksilver. But enough now concerning the simple kinds.[17]

24.11   See also

- Amorphous metal

- ASM International (society)

- Ductility

- Electric field screening

- Metal theft

- Metalworking

- Properties of metals, metalloids and nonmetals

- Properties and uses of metals

- Solid

- Spin transition

- Steel

- Structural steel

- Transition metal

24.12   References

[1] μέταλλον Henry George Liddell, Robert Scott, *A Greek-English Lexicon*, on Perseus Digital Library

[2] metal, on Oxford Dictionaries

[3] metal. *Encyclopædia Britannica*

[4] John C. Martin. "What we learn from a star's metal content". *New Analysis RR Lyrae Kinematics in the Solar Neighborhood*. Retrieved September 7, 2005.

[5] Holleman, A. F.; Wiberg, E. "Inorganic Chemistry" Academic Press: San Diego, 2001. ISBN 0-12-352651-5.

[6] Mortimer, Charles E. (1975). *Chemistry: A Conceptual Approach* (3rd ed.). New York:: D. Van Nostrad Company.

[7] Ductility – strength of materials

[8] "Los Alamos National Laboratory – Sodium". Retrieved 2007-06-08.

[9] "Los Alamos National Laboratory – Aluminum". Retrieved 2007-06-08.

[10] *The Recycling Rates of Metals: A Status Report* 2010, International Resource Panel, United Nations Environment Programme

[11] *Tread lightly: Aluminium attack* Carolyn Fry, Guardian.co.uk, 22 February 2008.

[12] *Metal Stocks in Society: Scientific Synthesis* 2010, International Resource Panel, United Nations Environment Programme

[13] Frank Kreith and Yogi Goswami, eds. (2004). *The CRC Handbook of Mechanical Engineering, 2nd edition. Boca Raton. p. 12-2.*

[14] Structure of merchandise imports

[15] *Der Große Brockhaus* (in German). 7: L-MIJ (Sixteenth, altogether newly prepared ed.). Wiesbaden: Bibliographisches Institut & F. A. Brockhaus. 1955. p. 715.

[16] John Maxson Stillman, *The Story of Early Chemistry* D. Appleton (1924)

[17] Georgius Agricola, *De Re Metallica* (1556) Tr. Herbert Clark Hoover & Lou Henry Hoover (1912); Footnote quoting *De Natura Fossilium* (1546), p. 180

24.13 External links

- The dictionary definition of metal at Wiktionary

- Media related to Metals at Wikimedia Commons

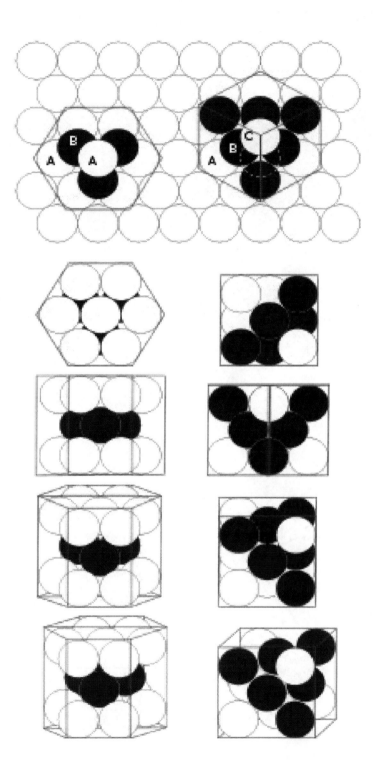

hcp and fcc close-packing of spheres

Gallium crystals

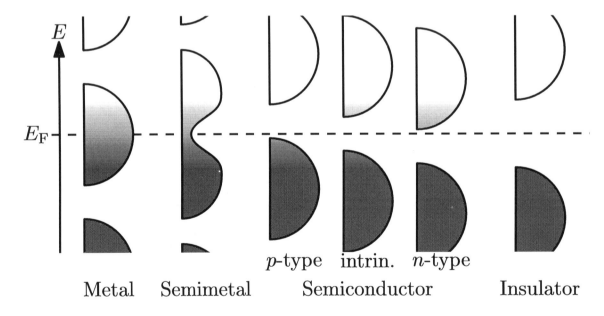

Filling of the electronic Density of states in various types of materials at equilibrium. Here the vertical axis is energy while the horizontal axis is the Density of states for a particular band in the material listed. In metals and semimetals the Fermi level E_F lies inside at least one band. In insulators and semiconductors the Fermi level is inside a band gap; however, in semiconductors the bands are near enough to the Fermi level to be thermally populated with electrons or holes.

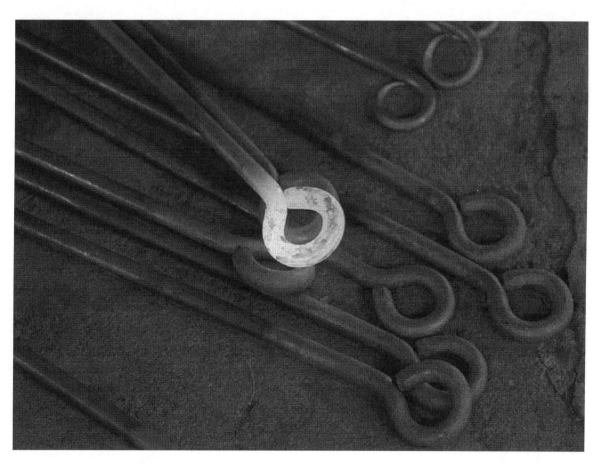

Hot metal work from a blacksmith.

Zinc, a base metal, reacting with an acid

A gold nugget

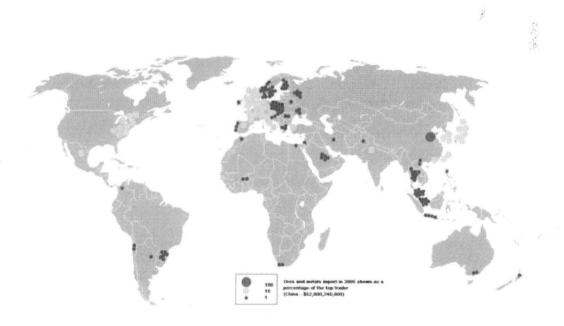

Metal and ore imports in 2005

Chapter 25

Nonmetal

Nonmetals in the periodic table:
Polyatomic nonmetal
Diatomic nonmetal
Noble gas
Apart from hydrogen, nonmetals are located in the p-block. Helium, although an s-block element, is normally placed above neon (in the p-block) on account of its noble gas properties.

In chemistry, a **nonmetal** (or **non-metal**) is a chemical element that mostly lacks metallic attributes. Physically, nonmetals tend to be highly volatile (easily vaporised), have low elasticity, and are good insulators of heat and electricity; chemically, they tend to have high ionization energy and electronegativity values, and gain or share electrons when they react with other elements or compounds. Seventeen elements are generally classified as nonmetals; most are gases (hydrogen, helium, nitrogen, oxygen, fluorine, neon, chlorine, argon, krypton, xenon and radon); one is a liquid (bromine); and a few are solids (carbon, phosphorus, sulfur, selenium, and iodine).

Moving rightward across the standard form of the periodic table, nonmetals adopt structures that have progressively fewer nearest neighbours. Polyatomic nonmetals have structures with either three nearest neighbours, as is the case (for example) with carbon (in its standard state[n 1] of graphite), or two nearest neighbours (for example) in the case of sulfur. Diatomic nonmetals, such as hydrogen, have one nearest neighbour, and the monatomic noble gases, such as helium, have none. This gradual fall in the number of nearest neighbours is associated with a reduction in metallic character and an increase in nonmetallic character. The distinction between the three categories of nonmetals, in terms of receding metallicity is not absolute. Boundary overlaps occur as outlying elements in each category show (or begin to show) less-distinct, hybrid-like or atypical properties.

Although five times more elements are metals than nonmetals, two of the nonmetals—hydrogen and helium—make up over 99 per cent of the observable Universe,[4] and one—oxygen—makes up close to half of the Earth's crust, oceans and atmosphere.[5] Living organisms are also composed almost entirely of nonmetals,[6] and nonmetals form many more compounds than metals.[7]

25.1   Definition and properties

The marvelous variety and infinite subtlety of the non-metallic elements, their compounds, structures and reactions, is not sufficiently acknowledged in the current teaching of chemistry.

"

"

JJ Zuckerman and FC Nachod
In Steudel's *Chemistry of the non-metals* (1977, preface)

There is no rigorous definition of a nonmetal. They show more variability in their properties than do metals.[8] The following are some of the chief characteristics of nonmetals.[9] Physically, they largely exist as monatomic gases, with a few having more substantial (but still open-packed) diatomic or polyatomic forms, unlike metals which are nearly all solid and close-packed; if solid, they generally have a submetallic or dull appearance and are brittle, as opposed to metals, which are lustrous, ductile or malleable; they usually have lower densities than metals; are poor conductors of heat and electricity when compared to metals; and have significantly lower melting points and boiling points than those of metals (with the exception of carbon). Chemically, the nonmetals have relatively high ionisation energy and high electronegativity; they usually exist as anions or oxyanions in aqueous solution; generally form ionic or interstitial compounds when mixed with metals, unlike metals which form alloys; and have acidic oxides whereas the common oxides of the metals are basic.

25.2   Applicable elements

The elements generally classified as nonmetals include one element in group 1 (hydrogen); one in group 14 (carbon); two in group 15 (nitrogen and phosphorus); three in group 16 (oxygen, sulfur and selenium); most of group 17 (fluorine, chlorine, bromine and iodine); and all of group 18 (with the possible exception of ununoctium).

The distinction between nonmetals and metals is by no means clear.[13] The result is that a few borderline elements lacking a preponderance of either nonmetallic or metallic properties are classified as metalloids;[14] and some elements classified as nonmetals are instead sometimes classified as metalloids, or vice versa. For example, selenium (Se), a nonmetal, is sometimes classified instead as a metalloid, particularly in environmental chemistry;[15] and astatine (At), which is a metalloid and a halogen, is sometimes classified instead as a nonmetal.[16]

25.3   Categories

Nonmetals have structures in which each atom usually forms $(8 - N)$ bonds with $(8 - N)$ nearest neighbours, where N is the number of valence electrons. Each atom is thereby able to complete its valence shell and attain a stable noble gas configuration. Exceptions to the $(8 - N)$ rule occur with hydrogen (which only needs one bond to complete its valence shell), carbon, nitrogen and oxygen. Atoms of the latter three elements are sufficiently small such that they are able

Arsenic (here sealed in a container to prevent tarnishing) is commonly classified as a metalloid,[10] or at other times as a nonmetal,[11] or as a metal.[12]

to form alternative (more stable) bonding structures, with fewer nearest neighbours.[17] Thus, carbon is able to form its layered graphite structure, and nitrogen and oxygen are able to form diatomic molecules having triple and double bonds, respectively. The larger size of the remaining non-noble nonmetals weakens their capacity to form multiple bonds and they instead form two or more single bonds to two or more different atoms.[18] Sulfur, for example, forms an eight-membered molecule in which the atoms are arranged in a ring, with each atom forming two single bonds to different atoms.

From left to right across the standard form of periodic table, as metallic character decreases,[21] nonmetals therefore adopt structures that show a gradual reduction in the numbers of nearest neighbours—three or two for the polyatomic nonmetals, through one for the diatomic nonmetals, to zero for the monatomic noble gases. A similar pattern occurs more generally, at the level of the entire periodic table, in comparing metals and nonmetals. There is a transition from metallic bonding amongst the metals on the left of the table through to covalent or Van der Waals (electrostatic) bonding amongst the nonmetals on the right of the table.[22] Metallic bonding tends to involve close-packed centrosymmetric structures with a high number of nearest neighbours.[23] Post-transition metals and metalloids, sandwiched between the true metals[n 3] and the nonmetals, tend to have more complex structures with an intermediate number of nearest neighbours.[n 4] Nonmetallic bonding, towards the right of the table, features open-packed directional (or disordered) structures with fewer or zero nearest neighbours.[26] As noted, this steady reduction in the number of nearest neighbours, as metallic character decreases and nonmetallic character increases, is mirrored among the nonmetals, the structures of which gradually change from polyatomic, to diatomic, to monatomic.

As is the case with the major categories of metals, metalloids and nonmetals,[27] there is some variation and overlapping of properties within and across each category of nonmetal. Among the polyatomic nonmetals, carbon, phosphorus and selenium—which border the metalloids—begin to show some metallic character. Sulfur (which borders the diatomic nonmetals), is the least metallic of the polyatomic nonmetals but even here shows some discernible metal-like character (discussed below). Of the diatomic nonmetals, iodine is the most metallic. Its number of nearest neighbours is sometimes described as 1+2 hence it is almost a polyatomic nonmetal.[28] Within the iodine molecule, significant electronic interactions occur with the two next nearest neighbours of each atom, and these interactions give rise, in bulk iodine, to a shiny appearance and semiconducting properties.[29] Of the monatomic nometals, radon is the most metallic and begins to show some cationic behaviour, which is unusual for a nonmetal.[30]

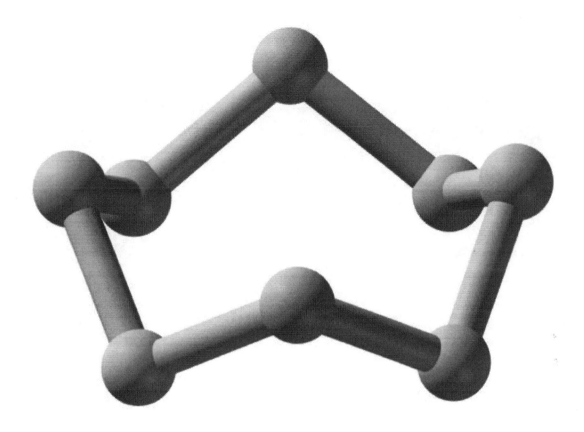

The structure of the sulfur molecule (S_8)

25.3.1 Polyatomic nonmetals

Four nonmetals are distinguished by polyatomic bonding in their standard states, in either discrete or extended molecular forms: carbon (C, as graphite sheets); phosphorus (as P_4 molecules); sulfur (as S_8 molecules); and selenium (Se, as helical chains).[32] Consistent with their higher coordination numbers (2 or 3), the polyatomic nonmetals show more metallic character than the neighbouring diatomic nonmetals; they are all solid, mostly semi-lustrous semiconductors with electronegativity values that are intermediate to moderately high (2.19–2.58). Sulfur is the least metallic of the polyatomic nonmetals given its dull appearance, brittle comportment, and low conductivity—attributes common to all sulfur allotropes. It nevertheless shows some metallic character, either intrinsically or in its compounds with other nonmetals. Examples include the malleability of plastic sulfur[33] and the lustrous-bronze appearance and metallic conductivity of polysulfur nitride (SN_x).[34][n 5]

The polyatomic nonmetals are distinguished from the diatomic nonmetals by virtue of having higher coordination numbers, higher melting points (in their thermodynamically most stable forms), and higher boiling points; and having wider liquid ranges and lower room temperature volatility.[55] More generally they show a marked tendency to exist in allotropic forms, and a stronger inclination to catenate;[56] and have a weaker ability to form hydrogen bonds.[57] The ability of carbon to catenate, in particular, is fundamental to the field of organic chemistry and life on Earth.[58][n 6] All of the polyatomic nonmetals are solids, and all are known in either malleable, pliable or ductile forms; most also have lower ionisation energies and electronegativities than those of the diatomic nonmetals.

The triple-bonded diatomic nitrogen molecule N_2, with its bond length shown in picometres (1 pm = one trillionth of a metre or 10^{-12})[n 2]

25.3.2 Diatomic nonmetals

Seven nonmetals exist as diatomic molecules in their standard states: hydrogen (H_2); nitrogen (N_2); oxygen (O_2); fluorine (F_2); chlorine (Cl_2); bromine (Br_2); and iodine (I_2).[60] They are generally highly insulating, highly electronegative, non-reflective gases, noting that bromine, a liquid, and iodine, a solid, are both volatile at room temperature.[61][62] Exceptions to this generalised description occur at the boundaries of the category: hydrogen has a comparatively low electronegativity due to its unique atomic structure;[n 7] iodine, in crystalline form, is semi-lustrous, and a semiconductor in the direction of its layers,[64][n 8] both of these attributes being consistent with incipient metallic character.

The diatomic nonmetals are distinguished from the polyatomic nonmetals by virtue of having lower coordination numbers, lower melting points (compared to the polyatomic nonmetals in their thermodynamically most stable forms), and lower boiling points; and having narrower liquid ranges[n 9] and greater room temperature volatility. More generally, they show less inclination to exist in allotropic forms, and to catenate; and have a stronger ability to form hydrogen bonds. Most are also gases, and have higher ionisation energies and higher electronegativities than those of the polyatomic nonmetals.

25.3.3 Noble gases

Main article: Noble gas

 Six nonmetals occur naturally as monatomic noble gases: helium (He), neon (Ne), argon (Ar), krypton (Kr), xenon (Xe), and the radioactive radon (Rn). They comprise a group of chemical elements with very similar properties. In their standard states they are all colorless, odourless, nonflammable gases with characteristically very low chemical reactivity.

With their closed valence shells, the noble gases have the highest first ionization potentials in each of their periods, and

Iodine crystals, showing a metallic lustre

feeble interatomic forces of attraction, with the latter property resulting in very low melting and boiling points.[67] That is why they are all gases under standard conditions, even those with atomic masses larger than many normally solid elements.[68]

The status of the period 7 congener of the noble gases, element 118 (temporary name ununoctium) is not known—it may or may not be a noble gas. It was originally predicted to be a noble gas[69] but may instead be a fairly reactive solid with an anomalously low first ionisation potential, due to relativistic effects.[70] On the other hand, if relativistic effects peak in period 7 at element 112, copernicium (as is thought to be the case), element 118 may turn out to be a noble gas after all,[71] albeit more reactive than either xenon or radon.

25.3.4   Elemental gases

Hydrogen, nitrogen, oxygen, fluorine, chlorine, plus the noble gases are collectively referred to as the *elemental gases.* These elements are gaseous at standard temperature and pressure (STP). They are also distinguished by having the lowest densities, lowest melting and boiling points, strongest insulating properties, and highest electronegativity and ionization energy values in the periodic table.

It is not known if any synthetic elements with atomic number above 99 are gases. If it transpires that copernicium and flerovium are gaseous metals at or near room temperature, as some calculations have suggested,[72] the category of elemental gases may need to be sub-divided into metallic and nonmetallic gases.

25.3.5   Organogens, CHONPS and biogens

Carbon, hydrogen, oxygen, nitrogen, phosphorus and sulfur are sometimes referred to or categorised as *organogens,*[73] *CHONPS* elements[74] or *biogens.*[75][n 10] Collectively these six nonmetals are required for all life on Earth.[76][n 11] They are further distinguished—in comparison to the halogens (F, Cl, Br, I, At) and noble gases—by their general capacity (or potential) to form allotropes;[n 12] high atomisation energies;[83] intermediate electron affinities;[n 13] reactivity combined with low toxicity;[n 14] ability to form alloys with metals;[86] and the weak or neutral acid-base character of their group hydrides.[87]

25.3.6 Other nonmetals

Selenium, and possibly boron, silicon, arsenic and tellurium, plus the organogen elements are sometimes categorized together as *other nonmetals*.[88] The first five of these (Se; B, Si, As, Te) differ from the organogens: none are universally required for life; arsenic is notoriously poisonous; and tellurium hydride is a fairly strong, rather than weak, acidic hydride.[89]

25.4 Comparison of properties

Characteristic and other properties of polyatomic nonmetals, diatomic nonmetals, and the monatomic noble gases are summarized in the following table. Physical properties are listed in loose order of ease of determination; chemical properties run from general to specific, and then to descriptive.

25.5 Allotropes

Main article: Allotropes of nonmetals

Many nonmetals have less stable allotropes, with either nonmetallic or metallic properties. Graphite, the standard state of carbon, has a lustrous appearance and is a fairly good electrical conductor. The diamond allotrope of carbon is clearly nonmetallic, however, being translucent and having a relatively poor electrical conductivity. Carbon is also known in several other allotropic forms, including semiconducting buckminsterfullerene (C_{60}). Nitrogen can form gaseous tetranitrogen (N_4), an unstable polyatomic molecule with a lifetime of about one microsecond.[103] Oxygen is a diatomic molecule in its standard state; it also exists as ozone (O_3), an unstable polyatomic nonmetallic allotrope with a half-life of around half an hour.[104] Phosphorus, uniquely, exists in several allotropic forms that are more stable than that of its standard state as white phosphorus (P_4).[n 16] The red and black allotropes are probably the best known; both are semiconductors; black phosphorus, in addition, has a lustrous appearance. Phosphorus is also known as diphosphorus (P_2), an unstable diatomic allotrope.[105] Sulfur has more allotropes than any other element;[106] all of these, except plastic sulfur (a metastable ductile mixture of allotropes)[107] have nonmetallic properties. Selenium has several nonmetallic allotropes, all of which are much less electrically conducting than its standard state of grey "metallic" selenium.[108] Iodine is also known in a semiconducting amorphous form.[109] Under sufficiently high pressures, just over half of the nonmetals, starting with phosphorus at 1.7 GPa,[110] have been observed to form metallic allotropes.

25.6 Abundance and extraction

Hydrogen and helium are estimated to make up approximately 99 per cent of all ordinary matter in the universe. Less than five per cent of the Universe is believed to be made of ordinary matter, represented by stars, planets and living beings. The balance is made of dark energy and dark matter, both of which are poorly understood at present.[111]

Hydrogen, carbon, nitrogen, oxygen and constitute the great bulk of the Earth's atmosphere, oceans, crust, and biosphere; the remaining nonmetals have abundances of 0.5 per cent or less. In comparison, 35 per cent of the crust is made up of the metals sodium, magnesium, aluminium, potassium and iron; together with a metalloid, silicon. All other metals and metalloids have abundances within the crust, oceans or biosphere of 0.2 per cent or less.[112]

Nonmetals, in their elemental forms, are extracted from:[113] *brine:* Cl, Br, I; *liquid air:* N, O, Ne, Ar, Kr, Xe; *minerals:* C (coal; diamond; graphite); F (fluorite); P (phosphates); I (in sodium iodate $NaIO_3$ and sodium iodide NaI); *natural gas:* H, He, S; and from *ores,* as processing byproducts: Se (especially copper ores); and Rn (uranium bearing ores).

25.7   Applications in common

For prevalent and speciality applications of individual nonmetals see the main article for each element.

Nonmetals do not have any universal or near-universal applications. This is not the case with metals, most of which have structural uses; nor the metalloids, the typical uses of which extend to (for example) oxide glasses, alloying components, and semiconductors.

Shared applications of different subsets of the nonmetals instead encompass their presence in, or specific uses in the fields of: *cryogenics and refrigerants:* H, He, N, O, F and Ne; *fertilisers:* H, N, P, S, Cl (as a micronutrient) and Se; *household accoutrements:* H (primary constituent of water), He (party balloons), C (in pencils, as graphite), N (beer widgets), O (as peroxide, in detergents), F (as fluoride, in toothpaste), Ne (lighting), P (matches), S (garden treatments), Cl (bleach constituent), Ar (insulated windows), Se (glass; solar cells), Br (as bromide, for purification of spa water), Kr (energy saving fluorescent lamps), I (in antiseptic solutions), Xe (in plasma TV display cells) and Rn (as an unwanted, potentially hazardous indoor pollutant);[115] *industrial acids:* C, N, F, P, S and Cl; *inert air replacements:* N, Ne, S (in sulfur hexafluoride SF_6), Ar, Kr and Xe; *lasers and lighting:* He, C (in carbon dioxide lasers, CO_2), N, O (in a chemical oxygen iodine laser), F (in a hydrogen fluoride laser, HF), Ne, S (in a sulfur lamp), Ar, Kr and Xe; and *medicine and pharmaceuticals:* He, O, F, Cl, Br, I, Xe and Rn.

The number of compounds formed by nonmetals is vast.[116] The first nine places in a "top 20" table of elements most frequently encountered in 8,427,300 compounds, as listed in the Chemical Abstracts Service register for July 1987, were occupied by nonmetals. Hydrogen, carbon, oxygen and nitrogen were found in the majority (greater than 64 per cent) of compounds. The highest rated metal, with an occurrence frequency of 2.3 per cent, was iron, in 11th place.[117]

25.8   Discovery

25.8.1   Antiquity: C, S

Sulfur and carbon were known in antiquity. The earliest known use of charcoal dates to around 3750 BCE. The Egyptians and Sumerians employed it for the reduction of copper, zinc, and tin ores in the manufacture of bronze. Diamonds were probably known from as early as 2500 BCE. The first true chemical analyses were made in the 18th century; Lavoisier recognized carbon as an element in 1789. Sulfur usage dates from before 2500 BCE; it was recognized as an element by Antoine Lavoisier in 1777.

25.8.2   17th century: P

Phosphorus was prepared from urine, by Hennig Brand, in 1669. It was the first element to be chemically discovered.

25.8.3   18th century: H, O, N, Cl

Cavendish, in 1766, was the first to distinguish hydrogen from other gases, although Paracelsus around 1500, Robert Boyle (1670), and Joseph Priestley {?} had observed its production by reacting strong acids with metals. Lavoisier named it in 1793. Carl Wilhelm Scheele obtained oxygen by heating mercuric oxide and nitrates in 1771, but did not publish his findings until 1777. Priestley also prepared this new "air" by 1774, but only Lavoisier recognized it as a true element; he named it in 1777. Rutherford discovered nitrogen while he was studying at the University of Edinburgh. He showed that the air in which animals breathed, after removal of exhaled carbon dioxide, was no longer able to burn a candle. Scheele, Henry Cavendish, and Priestley also studied this element at about the same time; Lavoisier named it in 1775-6. Scheele obtained chlorine from hydrochloric acid, but thought it was an oxide. Only in 1808 did Humphry Davy recognize it as an element

25.8.4   Early 19th century: I, Se, Br

Courtois, in 1811, discovered iodine in the ashes of seaweed. In 1817, when Berzelius and Johan Gottlieb Gahn were working with lead they discovered a substance that reminded of tellurium. After more investigation Berzelius concluded that it was a new element, related to sulfur and tellurium. Because of tellurium has been named for the Earth, Berzelius named the new element "selenium", after the moon. Balard and Gmelin both discovered bromine in the autumn of 1825 and published their results in the following year.

25.8.5   Late 19th century: He, F, Ar, Kr, Ne, Xe, Rn

In 1868, Janssen and Lockyer independently observed a yellow line in the solar spectrum that did not match that of any other element. In 1895, in each case at around the same time, Ramsay, Cleve, and Langlet independently observed helium trapped in cleveite. André-Marie Ampère predicted an element analogous to chlorine obtainable from hydrofluoric acid, and between 1812 and 1886 many researchers tried to obtain it. Fluorine was eventually isolated by Moissan, in 1886. In 1894, Lord Rayleigh and Ramsay discovered argon by comparing the molecular weights of nitrogen prepared by liquefaction from air and nitrogen prepared by chemical means. It was the first noble gas to be isolated. In 1898, within a period of three weeks, Ramsay and Travers successively separated krypton, neon and xenon from liquid argon by their differences in boiling points. In 1898, Dorn discovered a radioactive gas resulting from the radioactive decay of radium; Ramsay and Robert Whytlaw-Gray subsequently isolated radon in 1910.

25.9   Notes

[1] The standard state of an element (with one exception) is the most thermodynamically stable form of the element at ambient conditions. The exception is phosphorus, the standard state of which is the white allotrope, the most thermodynamically unstable, as well as the most volatile and reactive form.[11] It is also the most common, industrially important,[12] and easily reproducible allotrope. For those reasons white phosphorus (rather than the black allotrope, which is the most thermodynamically stable form) is the standard state of the element.[3]

[2] Triple bonding is thought to represent the limit for main-group elements.[19] Of the diatomic nonmetals only hydrogen has a shorter bond length, at 74 pm.[20]

[3] True metals are the alkali metals, alkaline earth metals, lanthanides, actinides, and d-block metals up to group 11.[24]

[4] Structural complexity in the post-transition metals arises due to the influence of partially covalent bonding, the directionality of which dictates fewer nearest neighbours.[25]

[5] Metallic or metalloidal character is further shown by sulfur, as follows. It is a solid, as are nearly all metals, and all metalloids, whereas the large majority of nonmetals are not. It has the highest superconducting transition phase critical temperature (10 K at 93 GPa) amongst the non-metallic elements.[35] Its most stable oxidation state is +6.[36] It has a well-established cationic chemistry in superacidic media, extending to the isolation of sulfur salts such as $[S_4]^{2+}[SbF_6]^-_2$, a pale-yellow solid that is stable at room temperature.[37][38] The lower electronegativity of sulfur compared to its lighter congener, oxygen, largely means that transition metal sulfides are more likely to be alloy-like semiconductors or metallic conductors than the corresponding oxides,[39] a trend also evident in lanthanide sulfides. Thus, cerium sulfide (CeS) has a metallic bronze lustre, exhibits high-level metallic electrical conductivity, and can be machined like a metal,[40][41] and samarium sulfide (SmS), a black semiconductor, can reportedly adopt a golden metallic phase by way of pressure, polishing or simply scratching on single crystals.[42] The diminished nonmetallic character of sulfur (and conversely its increased metallic character) is further shown by the fact that zinc sulfide does not dissolve in alkaline solution whereas zinc oxide does.[43] Sulfur can be combined with carbon to form a disulfide (CS_2) which, above 50 GPa, undergoes an insulator-to-metal transition. As yet this is puzzling behaviour given the most "metallic" organic polymers, other than compressed CS_2, exhibit barely metallic conductivity.[44] When sulfur is introduced as a doping agent, it causes silicon to exhibit metal-like conduction and associated enhanced light absorption characteristics (from low-frequency visible light through near- and short-wave-infrared wavelengths that would normally pass right through regular silicon).[45][46] Sulfur is a photoconductor (sometimes described as a semiconductor),[47] which means that its electrical conductivity increases by up to a million-fold when illuminated.[48] It becomes a (liquid) semiconductor at 900°C, with an electrical conductivity of 5×10^{-5} S•cm^{-1} (about a trillion times that of its room temperature conductivity, and twice that of boron, a metalloid).[49][50] In 2013 researchers reported metallic conductivity in linear chains of sulfur atoms, isolated inside

carbon nanotubes, at ambient conditions.[51] Sulfur trioxide is a glass-former,[52] as are oxides of phosphorus and selenium[53] and, at 40 GPa, carbon dioxide.[54]

[6] After carbon, phosphorus shows the next strongest ability to catenate.[59]

[7] Hydrogen's single electron is not shielded from the single proton in its nucleus, resulting in hydrogen having an ionisation energy on par with that of oxygen. However, this configuration also means that a hydrogen atom's ability to attract another electron to itself is compromised by its single proton not being able to fully stabilise the electron-electron repulsion forces that arise between two valence shell electrons (resulting in the instability of the hydride H$^-$ ion).[63]

[8] Iodine is an insulator in the direction perpendicular to its crystalline layers.[65]

[9] The liquid range of an element is the difference between its melting point and boiling point.[66]

[10] Selenium is counted as a metalloid in the last case, rather than a nonmetal.[75]

[11] The same elements comprise the bulk of terrestrial life; [77] a rough estimate[78] of the composition of the biosphere is
$$C_{1450}H_{3000}O_{1450}N_{15}P_1S_1$$

[12] Carbon, oxygen, phosphorus and sulfur exist in well known allotropic forms. Hydrogen is known in metastable monatomic and unstable triatomic forms, in addition to its stable diatomic form. Monatomic hydrogen has a lifetime of a few tenths of a second to hours depending on the production technique (heating, electrical arc or discharge, UV or microwave irradiation, electron bombardment) and containment method.[79] It can be used, via a Langmuir torch, to weld or melt high melting point metals or compounds.[80] Triatomic hydrogen is unstable and breaks up in under a millionth of a second. Its fleeting lifetime makes it rare, but it is quite commonly formed and destroyed in the universe.[81] For nitrogen, the as yet unsynthesized allotrope octaazacubane (N_8) is predicted to be metastable.[82]

[13] Electron affinities of the CHNOPS elements are −0.07 to +2.08 eV; halogens +2.8 to +3.4; noble gases −0.08 to −0.42[84]

[14] Phosphorus in its most thermodynamically stable black form is generally inert (although still reactive compared to the noble gases) and has a low toxicity.[85] White phosphorus, the most commonly known and encountered form, is metastable, highly reactive, flammable and poisonous.

[15] Carbon as exfoliated (expanded) graphite,[90] and as metre-long carbon nanotube wire;[91] phosphorus as white phosphorus (soft as wax, pliable and can be cut with a knife, at room temperature);[92] sulfur as plastic sulfur;[33] and selenium as selenium wires.[93]

[16] White phosphorus is the most common, industrially important,[2] and easily reproducible allotrope. For those reasons it is the standard state of the element.[3] Paradoxically, it is also thermodynamically the least stable, as well as the most volatile and reactive form.[1]

25.10   Citations

[1] Greenwood & Earnshaw 2002, pp. 479, 482

[2] Eagleson 1994, p. 820

[3] Oxtoby, Gillis & Campion 2008, p. 508

[4] Sukys 1999, p. 60

[5] Bettelheim et al. 2010, p. 37

[6] Schulze-Makuch & Irwin 2008, p. 89

[7] Steurer 2007, p. 7

[8] Brown & Rogers 1987, p. 40

[9] Kneen, Rogers & Simpson, 1972, p. 263

[10] Stwertka 2012, p. 104

[11] Patten 1989, p. 192

[12] Russell & Lee 2005, p. 419

[13] Cracolice & Peters 2011, p. 335

[14] Cracolice & Peters 2011, p. 336

[15] Meyer et al. 2005, p. 284; Manahan 2001, p. 911; Szpunar et al. 2004, p. 17

[16] Emsley 1971, p. 1

[17] Addison 1964, pp. 41–2, 51, 61–3

[18] Brady & Senese 2009, pp. 858–63

[19] Ball 2013

[20] Aylward & Findlay, p. 124

[21] Shipman, Wilson & Todd 2009, p. 297

[22] Russell & Lee 2005, p. 5; Zumdahl & DeCoste 2013, pp. 35, 784

[23] Borg & Dienes 1992, p. 15

[24] Wells 1984, pp. 1275, 77

[25] Russell & Lee 2005, p. 5

[26] Patterson, Kuper & Nanney 1967, p. 388; King 2004, p. 197; DeKock & Gray 1989, p. 426

[27] Ashford 1967, p. 329

[28] Townes 1952

[29] Steudel 1977, p. 240; Greenwood & Earnshaw 2002, p. 803; Wiberg 2001, p. 416

[30] Stein 1983, p. 165

[31] Wiberg 2001, p. 680

[32] Taylor 1960, p. 377; Miller 1987, p. 62; Irving 2005, p. 131

[33] Partington 1944, p. 405

[34] Labes et al. 1979

[35] Steudel & Eckert 2003, p. 60

[36] Mitchell 2006, p. 24

[37] Wiberg 2001, p. 517

[38] Greenwood & Earnshaw 2002, pp. 664–5

[39] Phillips & Williams 1965, pp. 577, 580–1, 583–4

[40] Eastman et al. 1950, p. 2250

[41] Krikorian & Curtis 1988

[42] Cotton 2006, pp. 29–30, 32

[43] Martin & Lander 1946, p. 195

[44] Dias et al. 2011

[45] Winkler 2009, pp. 16, 139

[46] Winkler et al. 2011

[47] Yu & Cardona 2010, p. 1

[48] Moss 1952, pp. 180–84

[49] Steudel 2003, pp. 106–7

[50] Schaefer 1968, p. 76

[51] Fujimori et al. 2013

[52] Phifer 2000, p. 1

[53] Rao 2002, p. 22

[54] McMillan 2006

[55] Lide 2003

[56] Steudel & Strauss 1984, p. 135

[57] Novak 1979, p. 281

[58] Chapman & Jarvis 2003, p. 23

[59] Wiberg 2001, p. 686

[60] Shipman, Wilson & Todd 2009, p. 293

[61] Cairns 2012, p. 147

[62] Wiberg 2001, p. 416

[63] Rogers 2012, p. pp. 267–269; Murray 1976, p. 103–4

[64] Greenwood & Earnshaw 2002, pp. 800,804

[65] Nelson 1998, p. 25

[66] Rayner-Canham & Overton 2006, p. 353

[67] Jolly 1966, p. 20

[68] Clugston & Flemming 2000, pp. 100–1, 104–5, 302

[69] Seaborg 1969, p. 626

[70] Nash 2005

[71] Scerri 2013, pp. 204–8

[72] Kratz, J. V. (5 September 2011). *The Impact of Superheavy Elements on the Chemical and Physical Sciences* (PDF). 4th International Conference on the Chemistry and Physics of the Transactinide Elements. Retrieved 27 August 2013.

[73] Ivanenko et al. 2011, p. 784

[74] Catling 2013, p. 12

[75] Crawford 1968, p. 540

[76] Merchant & Helmann 2012, pp. 94–96

[77] Berkowitz 2012, p. 293

[78] Jørgensen & Mitsch 1983, p. 59

[79] Wiberg 2001, p. 245–246; Silvera & Walraven 1981, pp. 204–207

[80] Wiberg 2001, p. 246

[81]   Oka 2006; McCall & Oka 2003; Mitchell & McGowan 1983, 310

[82]   Patil, Dhumal & Gejji 2004

[83]   Aylward & Findlay, p. 127–129

[84]   Atkins & Paula p. 355; Raju 2005, p. 495; Bunge & Bunge 1979

[85]   Bryson 1989, p. 511

[86]   Desch 1914, p. 86

[87]   Sherwin & Weston 1966, p. 100, 123, 152; Smith 2011, p. 754; Finney 2015, p. 88

[88]   Challoner 2014, p. 5; Myers, Oldham & Tocci 2004, pp. 120–121: The latter authors categorize nonmetals as hydrogen; semiconductors "(also known as metalloids)"; other nonmetals (C, N, O, P, S, Se); halogens; or noble gases; Government of Canada 2015; Gargaud et al. 2006, p. 447

[89]   Jorgensen 2012, p. 56

[90]   Chung 1987; Godfrin & Lauter 1995

[91]   Cambridge Enterprise 2013

[92]   Faraday 1853, p. 42; Holderness & Berry 1979, p. 255

[93]   Regnault 1853, p. 208

[94]   Siebring & Schaff 1980, p. 276

[95]   Conroy 1968, p. 672

[96]   Jenkins & Kawamura 1976, p. 88

[97]   Bogoroditskii & Pasynkov 1967, p. 77

[98]   Greenwood & Earnshaw 2002, p. 804

[99]   Stein 1969; Pitzer 1975; Schrobilgen 2011

[100] Arunan et al. 2011, p. 1623–4

[101] Henderson 2000, p. 134

[102] Ritter 2011, p. 10

[103] Cacace, de Petris & Troiani 2002

[104] Koziel 2002, p. 18

[105] Piro et al. 2006

[106] Steudel & Eckert 2003, p. 1

[107] Greenwood & Earnshaw 2002, pp. 659–660

[108] Moss 1952, p. 192; Greenwood & Earnshaw 2002, p. 751

[109] Shanabrook, Lannin & Hisatsune 1981

[110] Yousuf 1998, p. 425

[111] Ostriker & Steinhardt 2001

[112] Nelson 1987, p. 732

[113] Emsley 2001, p. 428

[114] Bolin 2012, p. 2-1

[115] Maroni 1995

[116] King & Caldwell 1954, p. 17; Brady & Senese 2009, p. 69

[117] Nelson 1987, p. 735

25.11   References

- Addison WE 1964, *The allotropy of the elements*, Oldbourne Press, London

- Arunan E, Desiraju GR, Klein RA, Sadlej J, Scheiner S, Alkorta I, Clary DC, Crabtree RH, Dannenberg JJ, Hobza P, Kjaergaard HG, Legon AC, Mennucci B & Nesbitt DJ 2011, 'Defining the hydrogen bond: An account (IUPAC Technical Report)', *Pure and Applied Chemistry,* vol. 83, no. 8, pp. 1619–36, doi:10.1351/PAC-REP-10-01-01

- Ashford TA 1967, *The physical sciences: From atoms to stars,* 2nd ed., Holt, Rinehart and Winston, New York

- Atkins P & de Paula J 2011, *Physical chemistry for the life sciences*, 2nd ed., Oxford University Press, Oxford, ISBN 978-1429231145

- Aylward G & Findlay T 2008, *SI chemical data,* 6th ed., John Wiley & Sons Australia, Milton, Queensland

- Ball P 2013, 'The name's bond', *Chemistry World,* vol. 10, no. 6, p. 41

- Berkowitz J 2012, *The stardust revolution: the new story of our origin in the stars,* Prometheus Books, Amherst, New York, ISBN 978-1-61614-549-1

- Bettelheim FA, Brown WH, Campbell MK, Farrell SO 2010, *Introduction to general, organic, and biochemistry,* 9th ed., Brooks/Cole, Belmont California, ISBN 9780495391128

- Bogoroditskii NP & Pasynkov VV 1967, *Radio and electronic materials,* Iliffe Books, London

- Bolin P 2000, 'Gas-insulated substations', in JD McDonald (ed.), *Electric power substations engineering,* 3rd, ed., CRC Press, Boca Raton, FL, pp. 2–1–2-19, ISBN 9781439856383

- Borg RJ & Dienes GJ 1992, *The physical chemistry of solids,* Academic Press, San Diego, California, ISBN 9780121184209

- Brady JE & Senese F 2009, *Chemistry: The study of matter and its changes,* 5th ed., John Wiley & Sons, New York, ISBN 9780470576427

- Brown WH & Rogers EP 1987, *General, organic and biochemistry,* 3rd ed., Brooks/Cole, Monterey, California, ISBN 0534068707

- Bryson PD 1989, *Comprehensive review in toxicology,* Aspen Publishers, Rockville, Maryland, ISBN 0871897776

- Bunge AV & Bunge CF 1979, 'Electron affinity of helium $(1s2s)^3S$ ', *Physical Review A,* vol. 19, no. 2, pp. 452–456, doi:10.1103/PhysRevA.19.452

- Cacace F, de Petris G & Troiani A 2002, 'Experimental detection of tetranitrogen', *Science,* vol. 295, no. 5554, pp. 480–81, doi:10.1126/science.1067681

- Cairns D2012,Essentials of pharmaceutical chemistry,4th ed.,Pharmaceutical Press,London,ISBN97808536997

- Cambridge Enterprise 2013, 'Carbon 'candy floss' could help prevent energy blackouts', Cambridge University, viewed 28 August 2013

- Catling DC 2013, *Astrobiology: A very short introduction,* Oxford University Press, Oxford, ISBN 978-0-19-958645-5

- Challoner J 2014, *The elements: The new guide to the building blocks of our universe,* Carlton Publishing Group, ISBN 978-0-233-00436-5

- Chapman B & Jarvis A 2003, *Organic chemistry, kinetics and equilibrium,* rev. ed., Nelson Thornes, Cheltenham, ISBN 978-0-7487-7656-6

- Chung DD1987, 'Review of exfoliated graphite',*Journal of Materials Science,*vol.22,pp.4190–98,doi:10.1007/BF08

- Clugston MJ & Flemming R 2000, *Advanced chemistry*, Oxford University Press, Oxford, ISBN 9780199146338

- Conroy EH 1968, 'Sulfur', in CA Hampel (ed.), *The encyclopedia of the chemical elements,* Reinhold, New York, pp. 665–680

- Cotton S 2006, *Lanthanide and actinide chemistry,* 2nd ed., John Wiley & Sons, New York, ISBN 9780470010068

- Cracolice MS & Peters EI 2011, *Basics of introductory chemistry: An active learning approach,* 2nd ed., Brooks/Cole, Belmont California, ISBN 9780495558507

- Crawford FH 1968, *Introduction to the science of physics,* Harcourt, Brace & World, New York

- DeKock RL & Gray HB 1989, *Chemical structure and bonding,* 2nd ed., University Science Books, Mill Valley, California, ISBN 093570261X

- Desch CH 1914, *Intermetallic Compounds,* Longmans, Green and Co., New York

- Dias RP, Yoo C, Kim M & Tse JS 2011, 'Insulator-metal transition of highly compressed carbon disulfide,' *Physical Review B*, vol. 84, pp. 144104–1–6, doi:10.1103/PhysRevB.84.144104

- Eagleson M 1994, *Concise encyclopedia chemistry,* Walter de Gruyter, Berlin, ISBN 3110114518

- Eastman ED, Brewer L, Bromley LA, Gilles PW, Lofgren NL 1950, 'Preparation and properties of refractory cerium sulfides', *Journal of the American Chemical Society,* vol. 72, no. 5, pp. 2248–50, doi:10.1021/ja01161a102

- Emsley J 1971, *The inorganic chemistry of the non-metals,* Methuen Educational, London, ISBN 0423861204

- Emsley J 2001, *Nature's building blocks: An A–Z guide to the elements, Oxford University Press, Oxford, ISBN 0198503415*

- Faraday M 1853, *The subject matter of a course of six lectures on the non-metallic elements,* (arranged by J Scoffern), Longman, Brown, Green, and Longmans, London

- Finney J 2015, *Water: A Very Short Introduction,* Oxford University Press, Oxford, ISBN 978-0198708728,

- Fujimori T, Morelos-Gómez A, Zhu Z, Muramatsu H, Futamura R, Urita K, Terrones M, Hayashi T, Endo M, Hong SY, Choi YC, Tománek D & Kaneko K 2013, 'Conducting linear chains of sulphur inside carbon nanotubes,' *Nature Communications,* vol. 4, article no. 2162, doi:10.1038/ncomms3162

- Gargaud M, Barbier B, Martin H & Reisse J (eds) 2006, *Lectures in astrobiology, vol. 1, part 1: The early Earth and other cosmic habitats for life,* Springer, Berlin, ISBN 3-540-29005-2

- Government of Canada 2015, *Periodic table of the elements*, accessed 30 August 2015

- Godfrin H & Lauter HJ 1995, 'Experimental properties of [3]He adsorbed on graphite', in WP Halperin (ed.), *Progress in low temperature physics, volume 14*, pp. 213–320 (216–8), Elsevier Science B.V., Amsterdam, ISBN 9780080539935

- Greenwood NN & Earnshaw A 2002,*Chemistry of the elements,*2nd ed., Butterworth-Heinemann,ISBN 075063364

- Henderson W 2000, *Main group chemistry,* Royal Society of Chemistry, Cambridge, ISBN 9780854046171

- Holderness A & Berry M 1979, *Advanced level inorganic chemistry,* 3rd ed., Heinemann Educational Books, London, ISBN 9780435654351

- Irving KE 2005, 'Using chime simulations to visualize molecules', in RL Bell & J Garofalo (eds), *Science units for Grades 9–12,* International Society for Technology in Education, Eugene, Oregon, ISBN 9781564842176

- Ivanenko NB, Ganeev AA, Solovyev ND & Moskvin LN 2011, "Determination of trace elements in biological fluids", *Journal of Analytical Chemistry,* vol. 66, no. 9, pp. 784–799 (784), doi:10.1134/S1061934811090036

- Jenkins GM & Kawamura K 1976, *Polymeric carbons—carbon fibre, glass and char,* Cambridge University Press, Cambridge, ISBN 0521206936

- Jolly WL 1966, *The chemistry of the non-metals,* Prentice-Hall, Englewood Cliffs, New Jersey

- Jorgensen CK 2012, *Oxidation numbers and oxidation states,* Springer-Verlag, Berlin, ISBN 978-3-642-87760-5

- Jørgensen SE & Mitsch WJ (eds) 1983, *Application of ecological modelling in environmental management, part A,* Elsevier Science Publishing, Amsterdam, ISBN 0-444-42155-6

- King RB 2004, 'The metallurgist's periodic table and the Zintl-Klemm concept', in DH Rouvray & BR King (eds), *The periodic table: into the 21st century,* Research Studies Press, Philadelphia, pp. 189–206, ISBN 0863802923

- King GB & Caldwell WE 1954, *The fundamentals of college chemistry,* American Book Company, New York

- Kneen WR, Rogers MJW & Simpson P 1972, *Chemistry: Facts, patterns, and principles,* Addison-Wesley, London, ISBN 0201037793

- Koziel JA 2002, 'Sampling and sample preparation for indoor air analysis', in J Pawliszyn (ed.), *Comprehensive analytical chemistry,* vol. 37, Elsevier Science B.V., Amsterdam, pp. 1–32, ISBN 0444505105

- Krikorian OH & Curtis PG 1988, 'Synthesis of CeS and interactions with molten metals,' *High Temperatures-High Pressures,* vol. 20, pp. 9–17, ISSN 0018-1544

- Labes MM, Love P & Nichols LF 1979, 'Polysulfur nitride—a metallic, superconducting polymer', *Chemical Review,* vol. 79, no. 1, pp. 1–15, doi:10.1021/cr60317a002

- Lide DR (ed.) 2003, *CRC handbook of chemistry and physics,* 84th ed., CRC Press, Boca Raton, Florida, Section 6, Fluid properties; Vapor pressure, ISBN 0849304849

- Manahan SE 2001, *Fundamentals of environmental chemistry,* 2nd ed., CRC Press, Boca Raton, Florida, ISBN 156670491X

- Maroni M, Seifert B & Lindvall T (eds) 1995, 'Physical pollutants', in *Indoor air quality: A comprehensive reference book,* Elsevier, Amsterdam, pp. 108–123, ISBN 0444816429

- Martin RM & Lander GD 1946, *Systematic inorganic chemistry: From the standpoint of the periodic law,* 6th ed., Blackie & Son, London

- McCall BJ & Oka T 2003, 'Enigma of H_3^+ in diffuse interstellar clouds', in SL Guberman (ed.), *Dissociative recombination of molecular ions with electrons,* Springer Science+Business Media, New York, ISBN 978-1-4613-4915-0

- McMillan PF 2006, 'Solid-state chemistry:A glass of carbon dioxide,'*Nature,*vol.441,p.823,doi:10.1038/44182

- Merchant SS & Helmann JD 2012, 'Elemental economy: Microbial strategies for optimizing growth in the face of nutrient limitation', in Poole RK (ed), *Advances in Microbial Physiology,* vol. 60, pp. 91–210, doi:10.1016/B978-0-12-398264-3.00002-4

- Meyer JS, Adams WJ, Brix KV, Luoma SM, Mount DR, Stubblefield WA & Wood CM (eds) 2005, *Toxicity of dietborne metals to aquatic organisms,* Proceedings from the Pellston Workshop on Toxicity of Dietborne Metals to Aquatic Organisms, 27 July–1 August 2002, Fairmont Hot Springs, British Columbia, Canada, Society of Environmental Toxicology and Chemistry, Pensacola, Florida, ISBN 1880611708

- Miller T 1987, *Chemistry: a basic introduction,* 4th ed., Wadsworth, Belmont, California, ISBN 0534069126

- Mitchell JBA & McGowan JW 1983, 'Experimental studies of electron-ion combination', *Physics of ion-ion and electron-ion collisions,* F Brouillard F & JW McGowan (eds), Plenum Press, ISBN 978-1-4613-3547-4

- Mitchell SC 2006, 'Biology of sulfur', in SC Mitchell (ed.), *Biological interactions of sulfur compounds,* Taylor & Francis, London, pp. 20–41, ISBN 0203375122

- Moss T 1952, *Photoconductivity in the elements,* Butterworths Scientific Publications, London

- Murray PRS & Dawson PR 1976, *Structural and comparative inorganic chemistry: A modern approach for schools and colleges,* Heinemann Educational Book, London, ISBN 9780435656447

- Myers RT, Oldham KB & Tocci S 2004, "Holt Chemistry," teacher ed., Holt, Rinehart & Winston, Orlando, ISBN 0-03-066463-2

- Nash CS 2005, 'Atomic and molecular properties of elements 112, 114, and 118', *Journal of Physical Chemistry A,* vol. 109, pp. 3493–500, doi:10.1021/jp050736o

- Nelson PG 1987, 'Important elements', *Journal of Chemical Education,* vol.68,no.9,pp.732–737,doi:10.1021/ed068

- Nelson PG 1998, 'Classifying substances by electrical character: An alternative to classifying by bond type', *Journal of Chemical Education,* vol. 71, no. 1, pp. 24–6, doi:10.1021/ed071p24

- Novak A 1979, 'Vibrational spectroscopy of hydrogen bonded systems', in TM Theophanides (ed.), *Infrared and Raman spectroscopy of biological molecules,* proceedings of the NATO Advanced Study Institute held at Athens, Greece, August 22–31, 1978, D. Reidel Publishing Company, Dordrecht, Holland, pp. 279–304, ISBN 9027709661

- Oka T 2006, 'Interstellar H+
  3', *PNAS,* vol. 103, no. 33, doi:10.1073_pnas.0601242103

- Ostriker JP & Steinhardt PJ 2001, 'The quintessential universe', *Scientific American,* January, pp. 46–53

- Oxtoby DW, Gillis HP & Campion A 2008, *Principles of modern chemistry, 6th ed., Thomson Brooks/Cole, Belmont, California, ISBN 0534493661*

- Partington JR 1944, *A text-book of inorganic chemistry,* 5th ed., Macmillan & Co., London

- Patil UN, Dhumal NR & Gejji SP 2004, 'Theoretical studies on the molecular electron densities and electrostatic potentials in azacubanes', *Theoretica Chimica Acta,* vol. 112, no. 1, pp 27–32, doi:10.1007/s00214-004-0551-2

- Patten MN 1989, *Other metals and some related materials*, in MN Patten (ed.), Information sources in metallic materials, Bowker-Saur, London, ISBN 0408014911

- Patterson CS, Kuper HS & Nanney TR 1967, *Principles of chemistry,* Appleton Century Crofts, New York

- Phifer C 2000, 'Ceramics, glass structure and properties', in *Kirk-Othmer Encyclopedia of Chemical Technology,* doi:10.1002/0471238961.0712011916080906.a01

- Phillips CSG & Williams RJP 1965, *Inorganic chemistry, I: Principles and non-metals,* Clarendon Press, Oxford

- Piro NA, Figueroa JS, McKellar JT & Troiani CC 2006, 'Triple-bond reactivity of diphosphorus molecules', *Science,* vol. 313, no. 5791, pp. 1276–9, doi:10.1126/science.1129630

- Pitzer K 1975, 'Fluorides of radon and elements 118', *Journal of the Chemical Society, Chemical Communications,* no. 18, pp. 760–1, doi:10.1039/C3975000760B

- Raju GG 2005, *Gaseous Electronics: Theory and Practice,* CRC Press, Boca Raton, Florida, ISBN 978-0-203-02526-0

- Rao KY 2002, *Structural chemistry of glasses, Elsevier, Oxford, ISBN 0080439586*

- Rayner-Canham G & Overton T 2006, *Descriptive inorganic chemistry,* 4th ed., WH Freeman, New York, ISBN 0716789639

- Regnault MV 1853, *Elements of chemistry,* vol. 1, 2nd ed., Clark & Hesser, Philadelphia

- Ritter SK 2011, 'The case of the missing xenon', *Chemical & Engineering News,* vol. 89, no. 9, ISSN 0009-2347

- Rodgers GE 2012, *Descriptive inorganic, coordination, & solid-state chemistry,* 3rd ed., Brooks/Cole, Belmont, California, ISBN 9780840068460

- Russell AM & Lee KL 2005, *Structure-property relations in nonferrous metals*, Wiley-Interscience, New York, ISBN 047164952X

- Scerri E 2013, *A tale of seven elements*, Oxford University Press, Oxford, ISBN 9780195391312

- Schaefer JC 1968, 'Boron' in CA Hampel (ed.), *The encyclopedia of the chemical elements*, Reinhold, New York, pp. 73–81

- Schrobilgen GJ 2011, 'radon (Rn)', in *Encyclopædia Britannica*, accessed 7 Aug 2011

- Schulze-Makuch D & Irwin LN 2008, *Life in the Universe: Expectations and constraints*, 2nd ed., Springer-Verlag, Berlin, ISBN 9783540768166

- Seaborg GT 1969, 'Prospects for further considerable extension of the periodic table', *Journal of Chemical Education*, vol. 46, no. 10, pp. 626–34, doi:10.1021/ed046p626

- Shanabrook BV, Lannin JS & Hisatsune IC 1981, 'Inelastic light scattering in a onefold-coordinated amorphous semiconductor', *Physical Review Letters*, vol. 46, no. 2, 12 January, pp. 130–133

- Sherwin E & Weston GJ 1966, *Chemistry of the non-metallic elements*, Pergamon Press, Oxford

- Shipman JT, Wilson JD & Todd AW 2009, *An introduction to physical science*, 12th ed., Houghton Mifflin Company, Boston, ISBN 9780618935963

- Siebring BR & Schaff ME 1980, *General chemistry*, Wadsworth Publishing, Belmont, California

- Silvera I & Walraven JTM 1981, 'Monatomic hydrogen – a new stable gas', *New Scientist*, 22 January

- Smith MB 2011, *Organic Chemistry: An Acid—Base Approach*, CRC Press, Boca Raton, Florida, ISBN 978-1-4200-7921-0

- Stein L 1969, 'Oxidized radon in halogen fluoride solutions', *Journal of the American Chemical Society*, vol. 19, no. 19, pp. 5396–7, doi:10.1021/ja01047a042

- Stein L 1983, 'The chemistry of radon', *Radiochimica Acta*, vol. 32, pp. 163–71

- Steudel R 1977, *Chemistry of the non-metals: With an introduction to atomic structure and chemical bonding*, Walter de Gruyter, Berlin, ISBN 3110048825

- Steudel R 2003, 'Liquid sulfur', in R Steudel (ed.), *Elemental sulfur and sulfur-rich compounds I*, Springer-Verlag, Berlin, pp. 81–116, ISBN 9783540401919

- Steudel R & Eckert B 2003, 'Solid sulfur allotropes', in R Steudel (ed.), *Elemental sulfur and sulfur-rich compounds I*, Springer-Verlag, Berlin, pp. 1–80, ISBN 9783540401919

- Steudel R & Strauss E 1984, 'Homcyclic selenium molecules and related cations', in HJ Emeleus (ed.), *Advances in inorganic chemistry and radiochemistry*, vol. 28, Academic Press, Orlando, Florida, pp. 135–167, ISBN 9780080578774

- Steurer W 2007, 'Crystal structures of the elements' in JW Marin (ed.), *Concise encyclopedia of the structure of materials*, Elsevier, Oxford, pp. 127–45, ISBN 0080451276

- Stwertka A 2012, *A guide to the elements*, 3rd ed., Oxford University Press, Oxford, ISBN 9780199832521

- Sukys P 1999, *Lifting the scientific veil: Science appreciation for the nonscientist*, Rowman & Littlefield, Oxford, ISBN 0847696006

- Szpunar J, Bouyssiere B & Lobinski R 2004, 'Advances in analytical methods for speciation of trace elements in the environment', in AV Hirner & H Emons (eds), *Organic metal and metalloid species in the environment: Analysis, distribution processes and toxicological evaluation*, Springer-Verlag, Berlin, pp. 17–40, ISBN 3540208291

- Taylor MD 1960, *First principles of chemistry*, Van Nostrand, Princeton, New Jersey

- Townes CH & Dailey BP 1952, 'Nuclear quadrupole effects and electronic structure of molecules in the solid state', *Journal of Chemical Physics,* vol. 20, pp. 35–40, doi:10.1063/1.1700192

- Wells AF 1984, *Structural inorganic chemistry,* 5th ed., Clarendon Press, Oxfordshire, ISBN 0198553706

- Wiberg N 2001, *Inorganic chemistry, Academic Press, San Diego, ISBN 0123526515*

- Winkler MT 2009, 'Non-equilibrium chalcogen concentrations in silicon: Physical structure, electronic transport, and photovoltaic potential,' PhD thesis, Harvard University, Cambridge, Massachusetts

- Winkler MT, Recht D, Sher M, Said AJ, Mazur E & Aziz MJ 2011, 'Insulator-to-metal transition in sulfur-doped silicon', *Physical Review Letters,* vol. 106, pp. 178701–4

- Yousuf M 1998, 'Diamond anvil cells in high-pressure studies of semiconductors', in T Suski & W Paul (eds), *High pressure in semiconductor physics II,* Semiconductors and semimetals, vol. 55, Academic Press, San Diego, pp. 382–436, ISBN 9780080864532

- Yu PY & Cardona M 2010, *Fundamentals of semiconductors: Physics and materials properties,* 4th ed., Springer, Heidelberg, ISBN 9783642007101

- Zumdahl SS&DeCoste DJ2013,*Chemical principles,*7th ed.,Brooks/Cole,Belmont,California,ISBN978111158

25.12 Monographs

- Emsley J 1971, *The inorganic chemistry of the non-metals,* Methuen Educational, London, ISBN 0423861204

- Johnson RC 1966, *Introductory descriptive chemistry: selected nonmetals, their properties, and behavior,* WA Benjamin, New York

- Jolly WL 1966, *The chemistry of the non-metals,* Prentice-Hall, Englewood Cliffs, New Jersey

- Powell P & Timms PL 1974, *The chemistry of the non-metals,* Chapman & Hall, London, ISBN 0470695706

- Sherwin E & Weston GJ 1966, *Chemistry of the non-metallic elements,* Pergamon Press, Oxford

- Steudel R 1977, *Chemistry of the non-metals: with an introduction to atomic structure and chemical bonding,* English edition by FC Nachod & JJ Zuckerman, Berlin, Walter de Gruyter, ISBN 3110048825

White phosphorus, stored under water to prevent its oxidation[31]

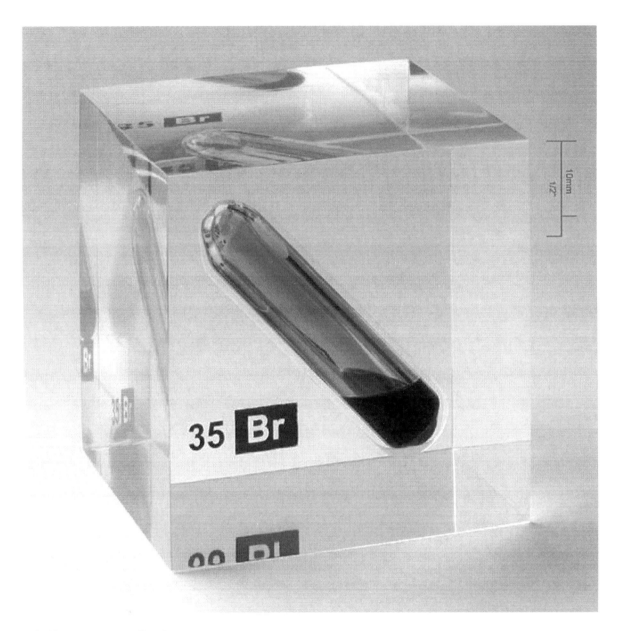

A vial of bromine in an acrylic cube

A small piece of melting argon ice

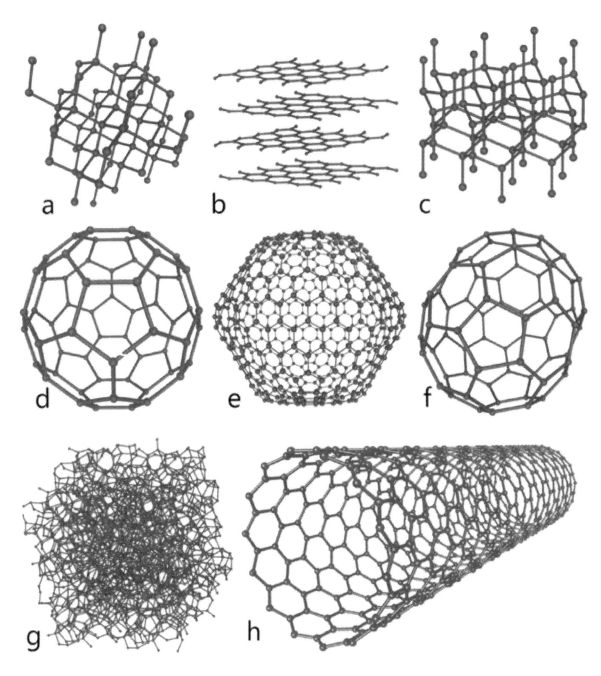

Some allotropes of carbon

Fluorite, a source of fluorine, in the form of an isolated crystal

A high-voltage circuit-breaker employing sulfur hexafluoride SF$_6$ as its inert (air replacement) interrupting medium[114]

Chapter 26

Refractory metals

Refractory metals are a class of metals that are extraordinarily resistant to heat and wear. The expression is mostly used in the context of materials science, metallurgy and engineering. The definition of which elements belong to this group differs. The most common definition includes five elements: two of the fifth period (niobium and molybdenum) and three of the sixth period (tantalum, tungsten, and rhenium). They all share some properties, including a melting point above 2000 °C and high hardness at room temperature. They are chemically inert and have a relatively high density. Their high melting points make powder metallurgy the method of choice for fabricating components from these metals. Some of their applications include tools to work metals at high temperatures, wire filaments, casting molds, and chemical reaction vessels in corrosive environments. Partly due to the high melting point, refractory metals are stable against creep deformation to very high temperatures.

26.1 Definition

Most definitions of the term 'refractory metals' list the extraordinarily high melting point as a key requirement for inclusion. By one definition, a melting point above 4,000 °F (2,200 °C) is necessary to qualify.[2] The five elements niobium, molybdenum, tantalum, tungsten and rhenium are included in all definitions,[3] while the wider definition, including all elements with a melting point above 2,123 K (1,850 °C), includes a varying number of nine additional elements, titanium, vanadium, chromium, zirconium, hafnium, ruthenium, rhodium, osmium and iridium. Transuranium elements (those above uranium, which are all unstable and not found naturally on earth) and technetium (melting point 2430 K or 2157 °C) are never considered to be part of the refractory metals.[4]

26.2 Properties

26.2.1 Physical

The melting point of the refractory metals are the highest for all elements except carbon, osmium and iridium. This high melting point defines most of their applications. All the metals are body-centered cubic except rhenium which is hexagonal close-packed. Most physical properties of the elements in this group vary significantly because they are members of different groups.[5][6]

Creep resistance is a key property of the refractory metals. In metals, the starting of creep correlates with the melting point of the material; the creep in aluminium alloys starts at 200 °C, while for refractory metals temperatures above 1500 °C are necessary. This resistance against deformation at high temperatures makes the refractory metals suitable against strong forces at high temperature, for example in jet engines, or tools used during forging.[7][8]

26.2.2 Chemical

The refractory metals show a wide variety of chemical properties because they are members of three distinct groups in the periodic table. They are easily oxidized, but this reaction is slowed down in the bulk metal by the formation of stable oxide layers on the surface. Especially the oxide of rhenium is more volatile than the metal, and therefore at high temperature the stabilization against the attack of oxygen is lost, because the oxide layer evaporates. They all are relatively stable against acids.[5]

26.3 Applications

Refractory metals are used in lighting, tools, lubricants, nuclear reaction control rods, as catalysts, and for their chemical or electrical properties. Because of their high melting point, refractory metal components are never fabricated by casting. The process of powder metallurgy is used. Powders of the pure metal are compacted, heated using electric current, and further fabricated by cold working with annealing steps. Refractory metals can be worked into wire, ingots, rebars, sheets or foil.

26.3.1 Molybdenum alloys

Main articles: Molybdenum and Molybdenum § Applications

Molybdenum based alloys are widely used, because they are cheaper than superior tungsten alloys. The most widely used alloy of molybdenum is the Titanium-Zirconium-Molybdenum alloy TZM, composed of 0.5% titanium and 0.08% of zirconium (with molybdenum being the rest). The alloy exhibits a higher creep resistance and strength at high temperatures, making service temperatures of above 1060 °C possible for the material. The high resistivity of Mo-30W an alloy of 70% molybdenum and 30 tungsten against the attack of molten zinc makes it the ideal material for casting zinc. It is also used to construct valves for molten zinc.[9]

Molybdenum is used in mercury wetted reed relays, because molybdenum does not form amalgams and is therefore resistant to corrosion by liquid mercury.[10][11]

Molybdenum is the most commonly used of the refractory metals. Its most important use is as a strengthening alloy of steel. Structural tubing and piping often contains molybdenum, as do many stainless steels. Its strength at high temperatures, resistance to wear and low coefficient of friction are all properties which make it invaluable as an alloying compound. Its excellent anti-friction properties lead to its incorporation in greases and oils where reliability and performance are critical. Automotive constant-velocity joints use grease containing molybdenum. The compound sticks readily to metal and forms a very hard, friction resistant coating. Most of the world's molybdenum ore can be found in China, the USA, Chile and Canada.[12][13][14][15]

26.3.2 Tungsten and its alloys

Main articles: Tungsten and Tungsten § Applications

Tungsten was discovered in 1781 by the Swedish chemist, Carl Wilhelm Scheele. Tungsten has the highest melting point of all metals, at 3,410 °C (6,170 °F).

Up to 22% rhenium is alloyed with tungsten to improve its high temperature strength and corrosion resistance. Thorium as an alloying compound is used when electric arcs have to be established. The ignition is easier and the arc burns more stably than without the addition of thorium. For powder metallurgy applications, binders have to be used for the sintering process. For the production of the tungsten heavy alloy, binder mixtures of nickel and iron or nickel and copper are widely used. The tungsten content of the alloy is normally above 90%. The diffusion of the binder elements into the tungsten grains is low even at the sintering temperatures and therefore the interior of the grains is pure tungsten.[16]

Filament of a 200 watt incandescent lightbulb highly magnified

Tungsten and its alloys are often used in applications where high temperatures are present but still a high strength is necessary and the high density is not troublesome.[17] Tungsten wire filaments provide the vast majority of household incandescent lighting, but are also common in industrial lighting as electrodes in arc lamps. Lamps get more efficient in the conversion of electric energy to light with higher temperatures and therefore a high melting point is essential for the application as filament in incandescent light.[18] In the Gas tungsten arc welding (GTAW, also known as tungsten inert gas (TIG) welding) equipment uses a permanent, non-melting electrode. The high melting point and the wear resistance against the electric arc makes tungsten a suitable material for the electrode.[19][20]

Tungsten's high density and strength is also a key property for its use in weapon projectiles, for example as an alternative to depleted Uranium for tank guns.[21] Its high melting point makes tungsten a good material for applications like rocket nozzles, for example in the UGM-27 Polaris.[22] Some of the applications of tungsten are not related to its refractory properties but simply to its density. For example, it is used in balance weights for planes and helicopters or for heads of golf clubs.[23][24] In this applications similar dense materials like the more expensive osmium can also be used.

26.3.3   Niobium alloys

Main articles: Niobium § Applications and Niobium_alloy
 Niobium is nearly always found together with tantalum, and was named after Niobe, the daughter of the mythical Greek king Tantalus for whom tantalum was named. Niobium has many uses, some of which it shares with other refractory metals. It is unique in that it can be worked through annealing to achieve a wide range of strength and elasticity, and is the least dense of the refractory metals. It can also be found in electrolytic capacitors and in the most practical superconducting alloys. Niobium can be found in aircraft gas turbines, vacuum tubes and nuclear reactors.

An alloy used for liquid rocket thruster nozzles, such as in the main engine of the Apollo Lunar Modules, is C103,

Apollo CSM with the dark rocket nozzle made from niobium-titanium alloy

which consists of 89% niobium, 10% hafnium and 1% titanium.[25] Another niobium alloy was used for the nozzle of the Apollo Service Module. As niobium is oxidized at temperatures above 400 °C, a protective coating is necessary for these applications to prevent the alloy from becoming brittle.[25]

26.3.4 Tantalum and its alloys

Main articles: Tantalum and Tantalum § Applications

Tantalum is one of the most corrosion resistant substances available.

Many important uses have been found for tantalum owing to this property, particularly in the medical and surgical fields, and also in harsh acidic environments. It is also used to make superior electrolytic capacitors. Tantalum films provide the second most capacitance per volume of any substance after Aerogel, and allow miniaturization of electronic components and circuitry. Many cellular phones and computers contain tantalum capacitors.

26.3.5 Rhenium alloys

Main article: Rhenium

Rhenium is the most recently discovered refractory metal. It is found in low concentrations with many other metals, in the ores of other refractory metals, platinum or copper ores. It is useful as an alloy to other refractory metals, where it adds ductility and tensile strength. Rhenium alloys are being used in electronic components, gyroscopes and nuclear reactors. Rhenium finds its most important use as a catalyst. It is used as a catalyst in reactions such as alkylation, dealkylation, hydrogenation and oxidation. However its rarity makes it the most expensive of the refractory metals.[26]

26.4   Advantages and shortfalls

Refractory metals and alloys attract the attention of investigators because of their remarkable properties and promising practical usefulness.

Physical properties of refractory metals, such as molybdenum, tantalum and tungsten, their strength, and high-temperature stability make them suitable material for hot metalworking applications and for vacuum furnace technology. Many special applications exploit these properties: for example, tungsten lamp filaments operate at temperatures up to 3073 K, and molybdenum furnace windings withstand to 2273 K.

However, poor low-temperature fabricability and extreme oxidability at high temperatures are shortcomings of most refractory metals. Interactions with the environment can significantly influence their high-temperature creep strength. Application of these metals requires a protective atmosphere or coating.

The refractory metal alloys of molybdenum, niobium, tantalum, and tungsten have been applied to space nuclear power systems. These systems were designed to operate at temperatures from 1350 K to approximately 1900 K. An environment must not interact with the material in question. Liquid alkali metals as the heat transfer fluids are used as well as the ultra-high vacuum.

The high-temperature creep strain of alloys must be limited for them to be used. The creep strain should not exceed 1–2%. An additional complication in studying creep behavior of the refractory metals is interactions with environment, which can significantly influence the creep behavior.

26.5   See also

- Refractory

26.6   References

[1] "International Journal of Refractory Metals and Hard Materials". Elsevier. Retrieved 2010-02-07.

[2] Bauccio, Michael; American Society for Metals (1993). "Refractory metals". *ASM metals reference book*. ASM International. pp. 120–122. ISBN 978-0-87170-478-8.

[3] Metals, Behavior Of; Wilson, J. W (1965-06-01). "General Behaviour of Refractory Metals". *Behavior and Properties of Refractory Metals*. pp. 1–28. ISBN 978-0-8047-0162-4.

[4] Davis, Joseph R (2001). *Alloying: understanding the basics*. pp. 308–333. ISBN 978-0-87170-744-4.

[5] Borisenko, V. A. (1963). "Investigation of the temperature dependence of the hardness of molybdenum in the range of 20–2500°C". *Soviet Powder Metallurgy and Metal Ceramics* **1** (3): 182. doi:10.1007/BF00775076.

[6] Fathi, Habashi (2001). "Historical Introduction to Refractory Metals". *Mineral Processing and Extractive Metallurgy Review* **22** (1): 25–53. doi:10.1080/08827509808962488.

[7] Schmid, Kalpakjian (2006). "Creep". *Manufacturing engineering and technology*. Pearson Prentice Hall. pp. 86–93. ISBN 978-7-302-12535-8.

[8] Weroński, Andrzej; Hejwowski, Tadeusz (1991). "Creep-Resisting Materials". *Thermal fatigue of metals*. CRC Press. pp. 81–93. ISBN 978-0-8247-7726-5.

[9] Smallwood, Robert E. (1984). "TZM Moly Alloy". *ASTM special technical publication 849: Refractory metals and their industrial applications: a symposium*. ASTM International. p. 9. ISBN 978-0-8031-0203-3.

[10] Kozbagarova, G. A.; Musina, A. S.; Mikhaleva, V. A. (2003). "Corrosion Resistance of Molybdenum in Mercury". *Protection of Metals* **39** (4): 374–376. doi:10.1023/A:1024903616630.

[11] Gupta, C. K. (1992). "Electric and Electronic Industry". *Extractive Metallurgy of Molybdenum*. CRC Press. pp. 48–49. ISBN 978-0-8493-4758-0.

[12] Magyar, Michael J. "Commodity Summary 2009:Molybdenum" (PDF). United States Geological Survey. Retrieved 2010-04-01.

[13] Ervin, D.R.; Bourell, D.L.; Persad, C.; Rabenberg, L. (1988). "Structure and properties of high energy, high rate consolidated molybdenum alloy TZM". *Materials Science and Engineering: A* **102**: 25. doi:10.1016/0025-5416(88)90529-0.

[14] Oleg D., Neikov (2009). "Properties of Molybdenum and Molybdenum Alloys powder". *Handbook of Non-Ferrous Metal Powders: Technologies and Applications*. Elsevier. pp. 464–466. ISBN 978-1-85617-422-0.

[15] Davis, Joseph R. (1997). "Refractory Metalls and Alloys". *ASM specialty handbook: Heat-resistant materials*. pp. 361–382. ISBN 978-0-87170-596-9.

[16] Lassner, Erik; Schubert, Wolf-Dieter (1999). *Tungsten: properties, chemistry, technology of the element, alloys, and chemical compounds*. Springer. pp. 255–282. ISBN 978-0-306-45053-2.

[17] National Research Council (U.S.), Panel on Tungsten, Committee on Technical Aspects of Critical and Strategic Material (1973). *Trends in Usage of Tungsten: Report*. National Research Council, National Academy of Sciences-National Academy of Engineering. pp. 1–3.

[18] Lassner, Erik; Schubert, Wolf-Dieter (1999). *Tungsten: properties, chemistry, technology of the element, alloys, and chemical compounds*. Springer. ISBN 978-0-306-45053-2.

[19] Harris, Michael K. (2002). "Welding Health and Safety". *Welding health and safety: a field guide for OEHS professionals*. AIHA. p. 28. ISBN 978-1-931504-28-7.

[20] Galvery, William L.; Marlow, Frank M. (2001). *Welding essentials: questions & answers*. Industrial Press Inc. p. 185. ISBN 978-0-8311-3151-7.

[21] Lanz, W.; Odermatt, W.; Weihrauch3, G. (7–11 May 2001). *KINETIC ENERGY PROJECTILES: DEVELOPMENT HISTORY, STATE OF THE ART, TRENDS* (PDF). 19th International Symposium of Ballistics. Interlaken, Switzerland.

[22] Ramakrishnan, P. (2007-01-01). "Powder metallurgyfor Aerospace Applications". *Powder metallurgy : processing for automotive, electrical/electronic and engineering industry*. New Age International. p. 38. ISBN 81-224-2030-3.

[23] Arora, Arran (2004). "Tungsten Heavy Alloy For Defence Applications". *Materials Technology* **19** (4): 210–216.

[24] Moxson, V. S.; (sam) Froes, F. H. (2001). "Fabricating sports equipment components via powder metallurgy". *JOM* **53** (4): 39. Bibcode:2001JOM....53d..39M. doi:10.1007/s11837-001-0147-z.

[25] Hebda, John (2001-05-02). "Niobium alloys and high Temperature Applications" (pdf). *Niobium Science & Technology: Proceedings of the International Symposium Niobium 2001 (Orlando, Florida, USA)* (Companhia Brasileira de Metalurgia e Mineração).

[26] Wilson, J. W. (1965). "Rhenium". *Behavior and Properties of Refractory Metals*. Stanford University Press. ISBN 978-0-8047-0162-4.

26.7 Further reading

- Levitin, Valim (2006). *High Temperature Strain of Metals and Alloys: Physical Fundamentals*. WILEY-VCH. ISBN 978-3-527-31338-9.

- Brunner, T (2000). "Chemical and structural analyses of aerosol and fly-ash particles from fixed-bed biomass combustion plants by electron microscopy". *1st World Conference on Biomass for Energy and Industry: proceedings of the conference held in Sevilla, Spain, 5–9 June 2000* (London: James & James Ltd). ISBN 1-902916-15-8.

- Spink, Donald (1961). "Reactive Metals. Zirconium, Hafnium, and Titanium". *Industrial & Engineering Chemistry* **53** (2): 97–104. doi:10.1021/ie50614a019.

- Hayes,Earl(1961). "Chromium and Vanadium".*Industrial&Engineering Chemistry***53**(2):105–107.doi:10.1021/0.

Chapter 27

Noble metal

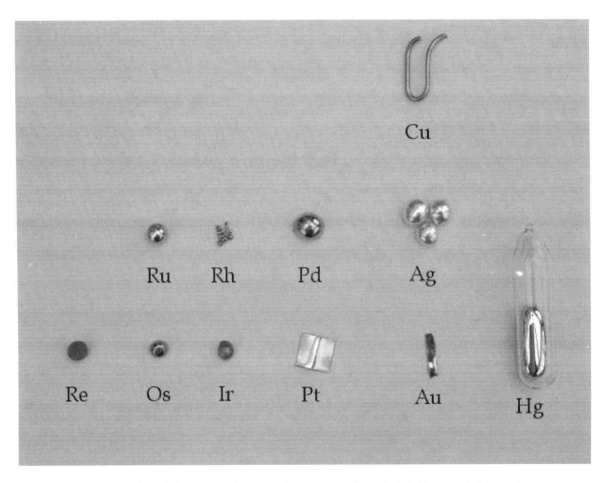

A collection of the noble metals, including copper, rhenium and mercury, which are included by some definitions. These are arranged according to their position in the periodic table.

In chemistry, the **noble metals** are metals that are resistant to corrosion and oxidation in moist air (unlike most base metals). The short list of chemically noble metals (those elements upon which almost all chemists agree) comprises ruthenium, rhodium, palladium, silver, osmium, iridium, platinum, and gold.[1]

More inclusive lists include one or more of mercury,[2][3][4] rhenium[5] or copper as noble metals. On the other hand, titanium, niobium, and tantalum are not included as noble metals although they are very resistant to corrosion.

While the noble metals tend to be valuable – due to both their rarity in the Earth's crust and their usefulness in areas like metallurgy, high technology, and ornamentation (jewelry, art, sacred objects, etc.) – the terms "noble metal" and "precious metal" are not synonymous.

The term *noble metal* can be traced back to at least the late 14th century[6] and has slightly different meanings in different fields of study and application. Only in atomic physics is there a strict definition. For this reason there are many quite different lists of "noble metals".

In addition to this term's function as a compound noun, there are circumstances where "noble" is used as an adjective for the noun "metal". A "galvanic series" is a hierarchy of metals (or other electrically conductive materials, including composites and semimetals) that runs from *noble* to *active*, and allows one to predict how materials will interact in the environment used to generate the series. In this sense of the word, graphite is more noble than silver and the relative nobility of many materials is highly dependent upon context, as for aluminium and stainless steel in conditions of varying pH.[7]

27.1   Properties

Palladium, platinum, gold and mercury can be dissolved in aqua regia, a highly concentrated mixture of hydrochloric acid and nitric acid, but iridium and silver cannot. Silver is, however, soluble in pure nitric acid. Ruthenium can be dissolved in aqua regia only when in the presence of oxygen, while rhodium must be in a fine pulverized form. Niobium and tantalum are resistant to all acids, including aqua regia. [8]

27.2   Physics

In physics, the definition of a noble metal is most strict. It requires that the d-bands of the electronic structure are filled. From this perspective, only copper, silver and gold are noble metals, as all d-like bands are filled and do not cross the Fermi level.[9] However, d-hybridized bands do cross the Fermi level to a minimal extent. For platinum, two d-bands cross the Fermi level, changing its chemical behaviour such that it can function as a catalyst. The difference in reactivity can easily be seen during the preparation of clean metal surfaces in an ultra-high vacuum: surfaces of "physically defined" noble metals (e.g., gold) are easy to clean and keep clean for a long time, while those of platinum or palladium, for example, are covered by carbon monoxide very quickly.[10]

27.3   Electrochemistry

Metallic elements, including noble and several non-noble metals (noble metals bolded):[11]

The columns *group* and *period* denote its position in the periodic table, hence electronic configuration. The simplified *reaction*s, listed in the next column, can also be read in detail from the Pourbaix diagrams of the considered element in water. Finally the column *potential* indicates the electric potential of the element measured against a Standard hydrogen electrode. All missing elements in this table are either not metals or have a negative standard potential.

Antimony is considered to be a metalloid and thus cannot be a noble metal. Also chemists and metallurgists consider copper and bismuth not noble metals because they easily oxidize due to the reaction O
2 + 2 H
2O + 4 e
- ⇌ 4 OH–
(aq) + 0.40 V which is possible in moist air.

The film of silver is due to its high sensitivity to hydrogen sulfide. Chemically patina is caused by an attack of oxygen in wet air and by CO
2 afterward.[8] On the other hand, rhenium coated mirrors are said to be very durable,[8] although rhenium and technetium are said to tarnish slowly in moist atmosphere.[13]

27.4 See also

- Base metal

- Minor metals

- Precious metal

27.5 References

[1] A. Holleman, N. Wiberg, "Lehrbuch der Anorganischen Chemie", de Gruyter, 1985, 33. edition, p. 1486

[2] Die Adresse für Ausbildung, Studium und Beruf

[3] "Dictionary of Mining, Mineral, and Related Terms", Compiled by the American Geological Institute, 2nd edition, 1997

[4] Scoullos, M.J., Vonkeman, G.H., Thornton, I., Makuch, Z., "Mercury - Cadmium - Lead: Handbook for Sustainable Heavy Metals Policy and Regulation",Series: Environment & Policy, Vol. 31, Springer-Verlag, 2002

[5] The New Encyclopædia Britannica, 15th edition, Vol. VII, 1976

[6] http://dictionary.reference.com/browse/noble+metal

[7] Everett Collier, "The Boatowner's Guide to Corrosion", International Marine Publishing, 2001, p. 21

[8] A. Holleman, N. Wiberg, "Inorganic Chemistry", Academic Press, 2001

[9] Hüger, E.; Osuch, K. (2005). "Making a noble metal of Pd". *EPL (Europhysics Letters)* **71** (2): 276. Bibcode:2005EL......71..276H. doi:10.1209/epl/i2005-10075-5.

[10] S. Fuchs, T.Hahn, H.G. Lintz, "The oxidation of carbon monoxide by oxygen over platinum, palladium and rhodium catalysts from 10^{-10} to 1 bar", Chemical engineering and processing, 1994, V 33(5), pp. 363-369

[11] D. R. Lidle editor, "CRC Handbook of Chemistry and Physics", 86th edition, 2005

[12] A. J. Bard, "Encyclopedia of the Electrochemistry of the Elements", Vol. IV, Marcel Dekker Inc., 1975

[13] R. D. Peack, "The Chemistry of Technetium and Rhenium", Elsevier, 1966

Notes

- R. R. Brooks, "Noble metals and biological systems: their role in Medicine, Mineral Exploration, and the Environment", CRC Press, 1992

27.6 External links

- noble metal - chemistry Encyclopædia Britannica, online edition

- To see which bands cross the Fermi level, the Fermi surfaces of almost all the metals can be found at the Fermi Surface Database

- The following article might also clarify the correlation between *band structure* and the term *noble metal*: Hüger, E.; Osuch, K. (2005). "Making a noble metal of Pd". *EPL (Europhysics Letters)* **71** (2): 276. Bibcode:2005EL..71276H. doi:10.1209/epl/i2005-10075-5.

Chapter 28

Periodic trends

This article refers to periodic trends in chemistry, for other periodic trends see trend (disambiguation).

When comparing the properties of the chemical elements, recurring ('periodic') trends are apparent. This led to the

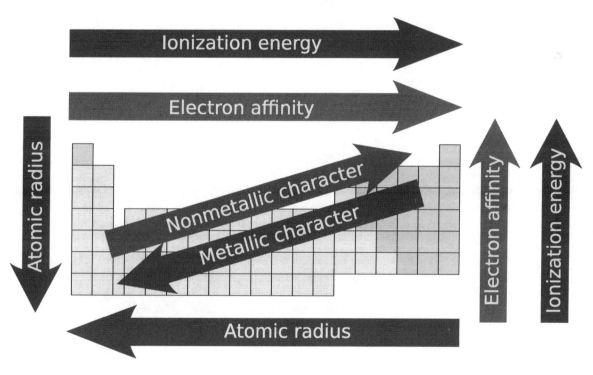

The Periodic Trends

creation of the periodic table as a useful way to display the elements and rationalize their behavior. When laid out in tabular form, many trends in properties can be observed to increase or decrease as one progresses along a row or column.

These period trends can be explained by theories of atomic structure. The elements are laid out in order of increasing atomic number, which represents increasing positive charge in the atomic nucleus. Negative electrons are arranged in orbitals around the nucleus; recurring properties are due to recurring configurations of these electrons.

These periodic trends are distributed among 3 different properties namely, physical properties, chemical properties and on the basis of chemical reactivity. In chemical properties, it is classified on the basis of two i.e.periodicity of valence or oxidation states, anomalous properties of second period elements. Here, we are going to discuss about the periodic trends with respect to their physical properties.

28.1 Atomic radius

Main article: Atomic radius

The atomic radius is the distance from the atomic nucleus to the outermost stable electron orbital in an atom that is at equilibrium. The atomic radii tend to decrease across a period from left to right. The atomic radius usually increases while going down a group due to the addition of a new energy level (shell). However, atomic radii tend to increase diagonally, since the number of electrons has a larger effect than the sizeable nucleus. For example, lithium (145 picometer) has a smaller atomic radius than magnesium (150 picometer).

Atomic radius can be further specified as:

- **Covalent radius:** half the distance between two atoms of a diatomic compound, singly bonded.

- **Van der Waals radius:** half the distance between the nuclei of atoms of different molecules in a lattice of covalent molecules.

- **Metallic radius:** half the distance between two adjacent nuclei of atoms in a metallic lattice.

- **Ionic radius:** half the distance between two nuclei

28.2 Ionization energy

Main article: Ionization energy

The ionization potential is the minimum amount of energy required to remove one electron from each atom in a mole of atoms in the gaseous state. The *first ionization energy* is the energy required to remove two, the *ionization energy* is the energy required to remove the atom's nth electron, after the $(n-1)$ electrons before it have been removed. Trend-wise, ionization energy tends to increase while one progresses across a period because the greater number of protons (higher nuclear charge) attract the orbiting electrons more strongly, thereby increasing the energy required to remove one of the electrons. Ionization energy and ionization potentials are completely different. The potential is an intensive property and it is measured by "volt" ; whereas the energy is an extensive property expressed by "eV" or "kJ/mole".

As one progresses down a group on the periodic table, the ionization energy will likely decrease since the valence electrons are farther away from the nucleus and experience a weaker attraction to the nucleus's positive charge. There will be an increase of ionization energy from left to right of a given period and a decrease from top to bottom. As a rule, it requires far less energy to remove an outer-shell electron than an inner-shell electron. As a result the ionization energies for a given element will increase steadily within a given shell, and when starting on the next shell down will show a drastic jump in ionization energy. Simply put, the lower the principal quantum number, the higher the ionization energy for the electrons within that shell. The exceptions are the elements in the boron and oxygen family, which require slightly less energy than the general trend.

28.3 Electron affinity

Main article: Electron affinity

The electron affinity of an atom can be described either as the energy gained by an atom when an electron is added to it, or conversely as the energy required to detach an electron from a singly charged anion. The sign of the electron affinity can be quite confusing, as atoms that become more stable with the addition of an electron (and so are considered to have a higher electron affinity) show a decrease in potential energy; i.e. the energy gained by the atom appears to be negative. For atoms that become less stable upon gaining an electron, potential energy increases, which implies that the atom gains energy. In such a case, the atom's electron affinity value is positive.[1] Consequently, atoms with a more negative electron affinity value are considered to have a higher electron affinity (they are more receptive to gaining electrons), and vice versa. However in the reverse scenario where electron affinity is defined as the energy required to detach an electron from an anion, the energy value obtained will be of the same magnitude but have the opposite sign. This is because those atoms with a high electron affinity are less inclined to give up an electron, and so take more energy to remove the electron from the atom. In this case, the atom with the more positive energy value has the higher electron affinity. As one progresses from left to right across a period, the electron affinity will increase.

Although it may seem that Fluorine should have the greatest electron affinity, the small size of fluorine generates enough repulsion that Chlorine has the greatest electron affinity.

28.4   Electronegativity

Main article: Electronegativity

Electronegativity is a measure of the ability of an atom or molecule to attract pairs of electrons in the context of a chemical bond. The type of bond formed is largely determined by the difference in electronegativity between the atoms involved, using the Pauling scale. Trend-wise, as one moves from left to right across a period in the periodic table, the electronegativity increases due to the stronger attraction that the atoms obtain as the nuclear charge increases. Moving down in a group, the electronegativity decreases due to the longer distance between the nucleus and the valence electron shell, thereby decreasing the attraction, making the atom have less of an attraction for electrons or protons.

However, in the group 13 elements electronegativity decreases from aluminium to thallium, and in group 14 electronegativity of lead is lower than that of tin.

28.5   Valence Electrons

Main article: Valence (chemistry)

Valence electrons [2] are the electrons in the outermost electron shell of an isolated atom of an element. Sometimes, it is also regarded as the basis of Modern Periodic Table. In a period, the number of valence electrons increases (mostly for light metal/elements) as we move from left to right side. However, in a group this periodic trend is constant, that is the number of valence electrons remains the same.

However, this periodic trend is sparsely followed for heavier elements (elements with atomic number greater than 20), especially for lanthanide and actinide series.

It is also important to consider the core electrons when speaking about the valence electrons.

28.6   Summary

Moving Left → Right

• Atomic Radius Decreases

• Ionization Energy Increases

• Electronegativity Increases

Moving Top → Bottom

• Atomic Radius Increases

• Ionization Energy Decreases

• Electronegativity Decreases

28.7 Metallic properties

Metallic properties increase down groups as decreasing attraction between the nuclei and the outermost electrons causes the outermost electrons to be loosely bound and thus able to conduct heat and electricity. Across the period, increasing attraction between the nuclei and the outermost electrons causes metallic character to decrease.

28.8 Non-metallic properties

Non-metallic property increases across a period and decreases down the group due to the same reason.

28.9 See also

- List of elements by atomic properties

28.10 References

[1] http://www.sparknotes.com/chemistry/fundamentals/atomicstructure/section3.rhtml

[2] "Valence Electrons". *chemed.chem.purdue.edu*. Retrieved 2015-09-29.

http://www.jstage.jst.go.jp/article/jlve/33/2/33_67/_article

Chapter 29

Aufbau principle

"Atomic build-up" redirects here. For the spread of nuclear weapons, see Nuclear proliferation.

The **Aufbau principle** states that, hypothetically, electrons orbiting one or more atoms fill the lowest available energy levels before filling higher levels (e.g., 1s before 2s). In this way, the electrons of an atom, molecule, or ion harmonize into the most stable electron configuration possible.

Aufbau is a German noun that means "construction". The Aufbau principle is sometimes called the **building-up principle** or the **Aufbau rule**.

The details of this "building-up" tendency are described mathematically by atomic orbital functions. Electron behavior is elaborated by other principles of atomic physics, such as Hund's rule and the Pauli exclusion principle. Hund's rule asserts that even if multiple orbitals of the same energy are available, electrons fill unoccupied orbitals first, before reusing orbitals occupied by other electrons. But, according to the Pauli exclusion principle, in order for electrons to occupy the same orbital, they must have different spins.

A version of the Aufbau principle known as the nuclear shell model is used to predict the configuration of protons and neutrons in an atomic nucleus.[1]

29.1 Madelung energy ordering rule

The order in which these orbitals are filled is given by the $n + \ell$ **rule**, also known as the **Madelung rule** (after Erwin Madelung), or the **Janet rule** or the **Klechkowski rule** (after Charles Janet or Vsevolod Klechkovsky in some, mostly French and Russian-speaking, countries), or the **diagonal rule**.[2] Orbitals with a lower $n + \ell$ value are filled before those with higher $n + \ell$ values. In this context, n represents the principal quantum number and ℓ the azimuthal quantum number; the values $\ell = 0, 1, 2, 3$ correspond to the *s*, *p*, *d*, and *f* labels, respectively.

The rule is based on the total number of nodes in the atomic orbital, $n + \ell$, which is related to the energy.[3] In the case of equal $n + \ell$ values, the orbital with a lower n value is filled first. The fact that most of the ground state configurations of neutral atoms fill orbitals following this $n + \ell$, n pattern was obtained experimentally, by reference to the spectroscopic characteristics of the elements.[4]

The Madelung energy ordering rule applies only to neutral atoms in their ground state, and even in that case, there are several elements for which it predicts configurations that differ from those determined experimentally.[5] Copper, chromium, and palladium are common examples of this property. According to the Madelung rule, the 4s orbital ($n + \ell = 4 + 0 = 4$) is occupied before the 3d orbital ($n + \ell = 3 + 2 = 5$). The rule then predicts the configuration of $_{29}$Cu to be $1s^2 2s^2 2p^6 3s^2\, 3p^6 4s^2 3d^9$, abbreviated [Ar]$4s^2 3d^9$ where [Ar] denotes the configuration of Ar (the preceding noble gas). However the experimental electronic configuration of the copper atom is [Ar]$4s^1 3d^{10}$. By filling the 3d orbital, copper can be in a lower energy state. Similarly, chromium takes the electronic configuration of [Ar]$4s^1 3d^5$ instead of [Ar]$4s^2 3d^4$. In this case, chromium has a half-full 3d shell. For palladium, the Madelung rule predicts [Kr]$5s^2 4d^8$, but

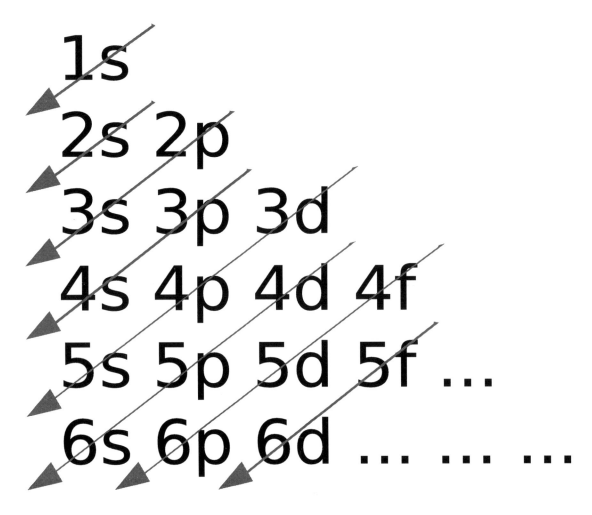

Order in which orbitals are arranged by increasing energy according to the Madelung rule. Each diagonal red arrow corresponds to a different value of $n + \ell$.

the experimental configuration $[Kr]4d^{10}$ differs in the placement of two electrons.

29.2 History

29.2.1 The Aufbau principle in the new quantum theory

The principle takes its name from the German, *Aufbauprinzip*, "building-up principle", rather than being named for a scientist. In fact, it was formulated by Niels Bohr and Wolfgang Pauli in the early 1920s, and states that:

This was an early application of quantum mechanics to the properties of electrons, and explained chemical properties in physical terms. Each added electron is subject to the electric field created by the positive charge of the atomic nucleus *and* the negative charge of other electrons that are bound to the nucleus. Although in hydrogen there is no energy difference between orbitals with the same principal quantum number n, this is not true for the outer electrons of other atoms.

In the old quantum theory prior to quantum mechanics, electrons were supposed to occupy classical elliptical orbits. The orbits with the highest angular momentum are 'circular orbits' outside the inner electrons, but orbits with low angular momentum (*s*- and *p*-orbitals) have high orbital eccentricity, so that they get closer to the nucleus and feel on average a less strongly screened nuclear charge.

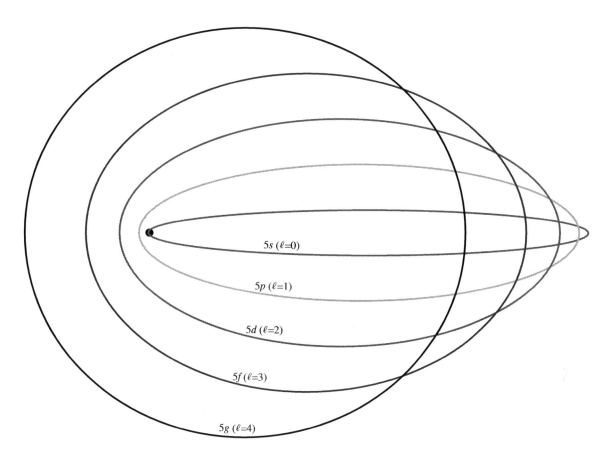

5s ($\ell=0$)

5p ($\ell=1$)

5d ($\ell=2$)

5f ($\ell=3$)

5g ($\ell=4$)

In the old quantum theory, orbits with low angular momentum (s- and p-orbitals) get closer to the nucleus.

29.2.2   The $n + \ell$ energy ordering rule

A periodic table in which each row corresponds to one value of $n + \ell$ was suggested by Charles Janet in 1927. In 1936, the German physicist Erwin Madelung proposed his empirical rules for the order of filling atomic subshells, based on knowledge of atomic ground states determined by the analysis of atomic spectra, and most English-language sources therefore refer to the Madelung rule. Madelung may have been aware of this pattern as early as 1926.[6] In 1962 the Russian agricultural chemist V.M. Klechkowski proposed the first theoretical explanation for the importance of the sum $n + \ell$, based on the statistical Thomas–Fermi model of the atom.[7] Many French- and Russian-language sources therefore refer to the Klechkowski rule. In recent years some authors have challenged the validity of Madelung's rule in predicting the order of filling of atomic orbitals. For example, it has been claimed, not for the first time, that in the case of the scandium atom a 3d orbital is occupied 'before' the occupation of the 4s orbital. In addition to there being ample experimental evidence to support this view, it makes the explanation of the order of ionization of electrons in this and other transition metals far more intelligible, given that 4s electrons are invariably preferentially ionized.[8]

29.3   See also

- Electron configuration

- Valence electrons

- Wiswesser's rule

29.4 References

[1] Cottingham W.N. and Greenwood D.A., *An introduction to nuclear physics* (Cambridge University Press 1986) ISBN 0 521 31960 9. chap.5 *Ground state properties of nuclei: the shell model*

[2] Electron Configuration from WyzAnt

[3] Weinhold, Frank; Landis, Clark R. (2005). *Valency and bonding: A Natural Bond Orbital Donor-Acceptor Perspective*. Cambridge: Cambridge University Press. pp. 715–16. ISBN 0-521-83128-8.

[4] Scerri, Eric R. (1998). "How Good is the Quantum Mechanical Explanation of the Periodic System?" (PDF). *J. Chem. Ed.* **75** (11): 1384–85. Bibcode:1998JChEd..75.1384S. doi:10.1021/ed075p1384.

[5] Meek, Terry L.; Allen, Leland C. (2002). "Configuration irregularities: deviations from the Madelung rule and inversion of orbital energy levels". *Chem. Phys. Lett.* **362** (5–6): 362–64. Bibcode:2002CPL...362..362M. doi:10.1016/S0009-2614(02)00919-3.

[6] Goudsmit, S. A.; Richards, Paul I. (1964). "The Order of Electron Shells in Ionized Atoms" (PDF). *Proc. Natl. Acad. Sci.* **51** (4): 664–671 (with correction on p 906). Bibcode:1964PNAS...51..664G. doi:10.1073/pnas.51.4.664.

[7] Wong, D. Pan (1979). "Theoretical justification of Madelung's rule" (PDF). *J. Chem. Ed.* **56** (11): 714–18. Bibcode:1979JChEd..5W. doi:10.1021/ed056p714.

[8] Scerri, Eric (2013). "The Trouble With the Aufbau Principle". *Education in Chemistry* **50** (11): 24–26.

29.5 Further reading

- Image: Understanding order of shell filling

- Boeyens, J. C. A.: *Chemistry from First Principles*. Berlin: Springer Science 2008, ISBN 978-1-4020-8546-8

- Ostrovsky, V.N. (2005). "On Recent Discussion Concerning Quantum Justification of the Periodic Table of the Elements" (PDF). *Foundations of Chemistry* **7** (3): 235–39. doi:10.1007/s10698-005-2141-y. Abstract.

- Kitagawara, Y.; Barut, A.O. (1984). "On the dynamical symmetry of the periodic table. II. Modified Demkov-Ostrovsky atomic model" (PDF). *J. Phys. B: At. Mol. Phys.* **17** (21): 4251–59. Bibcode:1984JPhB...17.4251K. doi:10.1088/0022-3700/17/21/013.

- Scerri, E.R. (2013). "The Trouble with the Aufbau Principle". *Education in Chemistry* (November): 24–26.

- Vanquickenborne, L. G. (1994). "Transition Metals and the Aufbau Principle" (PDF). *Journal of Chemical Education* **71** (6): 469–471. Bibcode:1994JChEd..71..469V. doi:10.1021/ed071p469.

29.6 External links

- Electron Configurations, the Aufbau Principle, Degenerate Orbitals, and Hund's Rule from Purdue University

Chapter 30

History of the periodic table

The periodic table is an arrangement of the chemical elements, organized on the basis of their atomic numbers, electron configurations and recurring chemical properties. Elements are presented in order of increasing atomic number. The standard form of the table consists of a grid of elements, with rows called periods and columns called groups.

The **history of the periodic table** reflects over a century of growth in the understanding of chemical properties. The most important event in its history occurred in 1869, when the table was published by Dmitri Mendeleev,[1] who built upon earlier discoveries by scientists such as Antoine-Laurent de Lavoisier and John Newlands, but who is nevertheless generally given sole credit for its development.

30.1 Ancient times

A number of physical elements (such as gold, silver and copper) have been known from antiquity, as they are found in their native form and are relatively simple to mine with primitive tools.[2] However, the notion that there were a limited number of elements from which everything was composed originated in around 330 BCE, when the Greek philosopher Aristotle proposed that everything is made up of a mixture of one or more *roots*, an idea that had originally been suggested by the Sicilian philosopher Empedocles. The four roots, which were later renamed as *elements* by Plato, were *earth*, *water*, *air* and *fire*. While Aristotle and Plato introduced the concept of an element, their ideas did nothing to advance the understanding of the nature of matter.

30.2 Age of Enlightenment

30.2.1 Hennig Brand

The history of the periodic table is also a history of the discovery of the chemical elements. The first person in history to discover a new element was Hennig Brand, a bankrupt German merchant. Brand tried to discover the Philosopher's Stone — a mythical object that was supposed to turn inexpensive base metals into gold. In 1649, his experiments with distilled human urine resulted in the production of a glowing white substance, which he named phosphorus.[3] He kept his discovery secret until 1680, when Robert Boyle rediscovered phosphorus and published his findings. The discovery of phosphorus helped to raise the question of what it meant for a substance to be an element.

In 1661, Boyle defined an element as "a substance that cannot be broken down into a simpler substance by a chemical reaction". This simple definition served for three centuries and lasted until the discovery of subatomic particles.

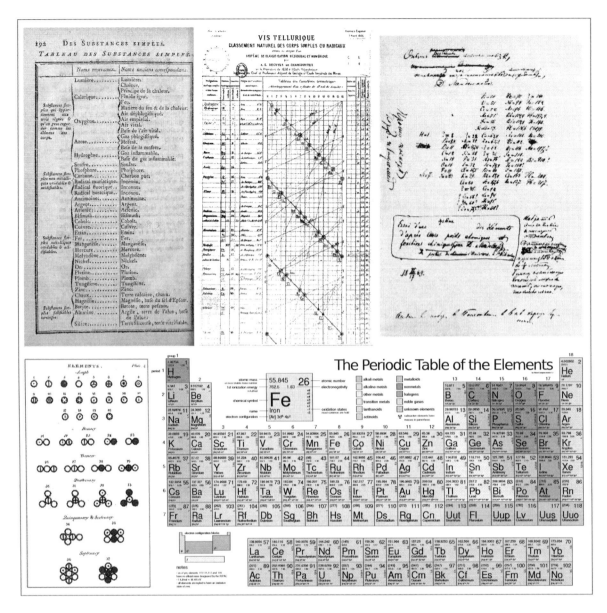

A collection of historic documents that led to the development of the modern periodic table (clockwise from top left) - Lavoisier's 'Table of Simple substances'; de Chancourtois' 'Vis Tellurique'; Mendeleev's hand-written periodic table; a modern periodic table; John Dalton's list of atomic weights & symbols.

30.2.2   Antoine-Laurent de Lavoisier

Lavoisier's *Traité Élémentaire de Chimie* (*Elementary Treatise of Chemistry*), which was written in 1789 and first translated into English by the writer Robert Kerr, is considered to be the first modern textbook about chemistry. It contained a list of "simple substances" that Lavoisier believed could not be broken down further, which included oxygen, nitrogen, hydrogen, phosphorus, mercury, zinc and sulfur, which formed the basis for the modern list of elements. Lavoisier's list also included 'light' and 'caloric', which at the time were believed to be material substances. He has classified these substances into metals and non metals. While many leading chemists refused to believe Lavoisier's new revelations, the *Elementary Treatise* was written well enough to convince the younger generation. However, Lavoisier's descriptions of his elements lack completeness, as he only classified them as metals and non-metals.

Hennig Brand, as shown in The Alchemist Discovering Phosphorus

30.3   19th century

30.3.1   Johann Wolfgang Döbereiner

In 1817, Johann Wolfgang Döbereiner began to formulate one of the earliest attempts to classify the elements. In 1829, he found that he could form some of the elements into groups of three, with the members of each group having related properties. He termed these groups *triads*. Some of the triads that were classified by Döbereiner are:

1. chlorine, bromine, and iodine

2. calcium, strontium, and barium

Antoine Laurent de Lavoisier

3. sulfur, selenium, and tellurium

4. lithium, sodium, and potassium

In all of the triads, the atomic weight of the middle element was almost exactly the average of the atomic weights of the other two elements.[4]

30.3.2   Alexandre-Emile Béguyer de Chancourtois

Alexandre-Emile Béguyer de Chancourtois, a French geologist, was the first person to notice the periodicity of the elements — similar elements occurring at regular intervals when they are ordered by their atomic weights. In 1862 he devised an early form of periodic table, which he named *Vis tellurique* (the 'telluric helix'), after the element tellurium, which fell near the center of his diagram.[5] With the elements arranged in a spiral on a cylinder by order of increasing atomic weight, de Chancourtois saw that elements with similar properties lined up vertically. His 1863 publication included a chart (which contained ions and compounds, in addition to elements), but his original paper in the *Comptes Rendus Academie des Scéances* used geological rather than chemical terms and did not include a diagram. As a result, de Chancourtois' ideas received little attention until after the work of Dmitri Mendeleev had been publicised.[6]

30.3.3   John Newlands

No.		No.		No.		No.		No.		No.		No.		No.	
H	1	F	8	Cl	15	Co & Ni	22	Br	29	Pd	36	I	42	Pt & Ir	50
Li	2	Na	9	K	16	Cu	23	Rb	30	Ag	37	Cs	44	Os	51
G	3	Mg	10	Ca	17	Zn	24	Sr	31	Cd	38	Ba & V	45	Hg	52
Bo	4	Al	11	Cr	19	Y	25	Ce & La	33	U	40	Ta	46	Tl	53
C	5	Si	12	Ti	18	In	26	Zr	32	Sn	39	W	47	Pb	54
N	6	P	13	Mn	20	As	27	Di & Mo	34	Sb	41	Nb	48	Bi	55
O	7	S	14	Fe	21	Se	28	Ro & Ru	35	Te	43	Au	49	Th	56

Newlands' law of octaves

In 1864, the English chemist John Newlands classified the sixty-two known elements into seven groups, based on their physical properties.[7][8]

Newlands noted that many pairs of similar elements existed, which differed by some multiple of eight in mass number, and was the first to assign them an atomic number.[9] When his 'law of octaves' was printed in *Chemistry News*, likening this periodicity of eights to the musical scale, it was ridiculed by some of his contemporaries. His lecture to the Chemistry Society on 1 March 1866 was not published, the Society defending their decision by saying that such 'theoretical' topics might be controversial.

The importance of Newlands' analysis was eventually recognised by the Chemistry Society with a Gold Medal five years after they recognised Mendeleev's work. It was not until the following century, with Gilbert N. Lewis' valence bond theory (1916) and Irving Langmuir's octet theory of chemical bonding (1919), that the importance of the periodicity of eight would be accepted.[10][11] The Royal Chemistry Society acknowledged Newlands' contribution to science in 2008, when they put a Blue Plaque on the house where he was born, which described him as the "discoverer of the Periodic Law for the chemical elements".[9] He contributed the word 'periodic' in chemistry.

30.3.4   Dmitri Mendeleev

The Russian chemist Dmitri Mendeleev was the first scientist to make a periodic table similar to the one used today. Mendeleev arranged the elements by atomic mass, corresponding to relative molar mass. It is sometimes said that he played 'chemical solitaire' on long train journeys, using cards with various facts about the known elements.[12] On March 6, 1869, a formal presentation was made to the Russian Chemical Society, entitled *The Dependence Between the Properties*

Dmitri Ivanovich Mendeleev

of the Atomic Weights of the Elements. In 1869, the table was published in an obscure Russian journal and then republished in a German journal, *Zeitschrift für Chemie*.[13] In it, Mendeleev stated that:

1. The elements, if arranged according to their atomic mass, exhibit an apparent periodicity of properties.

2. Elements which are similar as regards to their chemical properties have atomic weights which are either of nearly the same value (e.g., Pt, Ir, Os) or which increase regularly (e.g., K, Rb, Cs).

3. The arrangement of the elements, or of groups of elements in the order of their atomic masses, corresponds to their so-called valencies, as well as, to some extent, to their distinctive chemical properties; as is apparent among other series in that of Li, Be, B, C, N, O, and F.

4. The elements which are the most widely diffused have small atomic weights.

5. The magnitude of the atomic weight determines the character of the element, just as the magnitude of the molecule determines the character of a compound body.

6. We must expect the discovery of many yet unknown elements – for example, elements analogous to aluminium and silicon – whose atomic weight would be between 65 and 75.

7. The atomic weight of an element may sometimes be amended by a knowledge of those of its contiguous elements. Thus the atomic weight of tellurium must lie between 123 and 126, and cannot be 128.

8. Certain characteristic properties of elements can be foretold from their atomic masses.

Scientific benefits of Mendeleev's table

- It enabled Mendeleev to predict the discovery of new elements and left spaces for them, namely eka-silicon (germanium), eka-aluminium (gallium), and eka-boron (scandium). Thus, there was no disturbance in the periodic table.

- It could be used by Mendeleev to point out that some of the atomic weights being used at the time were incorrect.

- It provided for variance from atomic weight order.

Shortcomings of Mendeleev's table

- The table was not able to predict the existence of the noble gases. However, when this entire family of elements was discovered, Sir William Ramsay was able to add them to the table as Group 0, without the basic concept of the periodic table being disturbed.

- A single position could not be assigned to hydrogen, which could be placed either in the alkali metals group, the halogens group or separately above the table between boron and carbon.[14]

30.3.5   Lothar Meyer

Unknown to Mendeleev, the German chemist Lothar Meyer was also working on a periodic table. Although his work was published in 1864, and was done independently of Mendeleev, few historians regard him as an equal co-creator of the periodic table. Meyer's table only included twenty-eight elements, which were not classified by atomic weight, but by valence and he never reached the idea of predicting new elements and correcting atomic weights. A few months after Mendeleev published his periodic table of the known elements, predicted new elements to help complete his table and corrected the atomic weights of some of the elements, Meyer published a virtually identical periodic table.

Meyer and Mendeleev are considered by some historians of science to be the co-creators of the periodic table, but Mendeleev's accurate prediction of the qualities of undiscovered elements enables him to have the larger share of the credit.

30.3.6   William Odling

In 1864, the English chemist William Odling also drew up a table that was remarkably similar to the table produced by Mendeleev. Odling overcame the tellurium-iodine problem and even managed to get thallium, lead, mercury and platinum into the right groups, which is something that Mendeleev failed to do at his first attempt. Odling failed to achieve recognition, however, since it is suspected that he, as Secretary of the Chemical Society of London, was instrumental in discrediting Newlands' earlier work on the periodic table.

30.4   20th century

30.4.1   Henry Moseley

In 1914, a year before he was killed in action at Gallipoli, the English physicist Henry Moseley found a relationship between the X-ray wavelength of an element and its atomic number. He was then able to re-sequence the periodic table by nuclear charge, rather than by atomic weight. Before this discovery, atomic numbers were sequential numbers based on an element's atomic weight. Moseley's discovery showed that atomic numbers were in fact based upon experimental measurements.

Using information about their X-ray wavelengths, Moseley placed argon (with an atomic number Z=18) before potassium (Z=19), despite the fact that argon's atomic weight of 39.9 is greater than the atomic weight of potassium (39.1). The new order was in agreement with the chemical properties of these elements, since argon is a noble gas and potassium is an alkali metal. Similarly, Moseley placed cobalt before nickel and was able to explain that tellurium occurs before iodine, without revising the experimental atomic weight of tellurium, as had been proposed by Mendeleev.

Moseley's research showed that there were gaps in the periodic table at atomic numbers 43 and 61, which are now known to be occupied by technetium and promethium respectively.

30.4.2 Glenn T. Seaborg

During his Manhattan Project research in 1943, Glenn T. Seaborg experienced unexpected difficulties in isolating the elements americium and curium. Seaborg wondered if these elements belonged to a different series, which would explain why their chemical properties were different from what was expected. In 1945, against the advice of colleagues, he proposed a significant change to Mendeleev's table: the actinide series.

Seaborg's actinide concept of heavy element electronic structure, predicting that the actinides form a transition series analogous to the rare earth series of lanthanide elements, is now well accepted and included in the periodic table. The actinide series is the second row of the f-block (5f series). In both the actinide and lanthanide series, an inner electron shell is being filled. The actinide series comprises the elements from actinium to lawrencium. Seaborg's subsequent elaborations of the actinide concept theorized a series of superheavy elements in a transactinide series comprising elements from 104 to 121 and a superactinide series of elements from 122 to 153.

30.5 See also

- Alternative periodic tables
- History of chemistry
- Periodic Systems of Small Molecules
- Prout's hypothesis
- The Mystery of Matter: Search for the Elements (PBS film)
- Timeline of chemical element discoveries

30.6 References

[1] IUPAC article on periodic table

[2] Scerri, E. R. (2006). *The Periodic Table: Its Story ad Its Significance*; New York City, New York; Oxford University Press.

[3] "A Brief History of the Development of Periodic Table".

[4] Leicester, Henry M. (1971). *The Historical Background of Chemistry*; New York City, New York; Dover Publications.

[5] Chancourtois, *Comptes rendus Academie des sciences*, volume 55, p. 600.

[6] Annales des Mines history page.

[7] in a letter published in *Chemistry News* in February 1863, according to the Notable Names Data Base

[8] Newlands on classification of elements

[9] John Newlands, Chemistry Review, November 2003, pp15-16

[10] Irving Langmuir, "The Structure of Atoms and the Octet Theory of Valence", Proceedings of the National Academy of Science, Vol. V, 252, Letters (1919) – online at

[11] Irving Langmuir, "The Arrangement of Electrons in Atoms and Molecules", Journal of the American Chemical Society, Vol. 41, No, 6, pg. 868 (June 1919) – beginning and ending of the paper are transcribed online at ; the middle is missing

[12] *Physical Science*, Holt Rinehart & Winston (January 2004), page 302 ISBN 0-03-073168-2

[13] Mendeleev, Dmitri (1869). "Ueber die Beziehungen der Eigenschaften zu den Atomgewichten der Elemente". *Zeitschrift für Chemie* **12**: 405–406. Retrieved 29 November 2013.

[14] http://www.reed.edu/reed_magazine/summer2009/columns/NoAA/from_the_archives.html

30.7   External links

- Development of the periodic table (part of a collection of pages that explores the periodic table and the elements) by the Royal Society of Chemistry

- The path to the periodic table by the Chemical Heritage Foundation

- Dr. Eric Scerri's web page, which has links to interviews, lectures and articles on various aspects of the periodic system, including the history of the periodic table.

- The Internet Database of Periodic Tables - a large collection of periodic tables and periodic system formulations.

- History of Mendeleev periodic table of elements as a data visualization at CrossValidated Stack Exchange

Ueber die Beziehungen der Eigenschaften zu den Atomgewichten der Elemente. Von D. Mendelejeff. — Ordnet man Elemente nach zunehmenden Atomgewichten in verticale Reihen so, dass die Horizontalreihen analoge Elemente enthalten, wieder nach zunehmendem Atomgewicht geordnet, so erhält man folgende Zusammenstellung, aus der sich einige allgemeinere Folgerungen ableiten lassen.

			Ti = 50	Zr = 90	? = 180
			V = 51	Nb = 94	Ta = 182
			Cr = 52	Mo = 96	W = 186
			Mn = 55	Rh = 104,4	Pt = 197,4
			Fe = 56	Ru = 104,4	Ir = 198
		Ni = Co = 59		Pd = 106,6	Os = 199
H = 1			Cu = 63,4	Ag = 108	Hg = 200
	Be = 9,4	Mg = 24	Zn = 65,2	Cd = 112	
	B = 11	Al = 27,4	? = 68	Ur = 116	Au = 197 ?
	C = 12	Si = 28	? = 70	Sn = 118	
	N = 14	P = 31	As = 75	Sb = 122	Bi = 210 ?
	O = 16	S = 32	Se = 79,4	Te = 128 ?	
	F = 19	Cl = 35,5	Br = 80	J = 127	
Li = 7	Na = 23	K = 39	Rb = 85,4	Cs = 133	Tl = 204
		Ca = 40	Sr = 87,6	Ba = 137	Pb = 207
		? = 45	Ce = 92		
		?Er = 56	La = 94		
		?Yt = 60	Di = 95		
		?In = 75,6	Th = 118 ?		

1. Die nach der Grösse des Atomgewichts geordneten Elemente zeigen eine stufenweise Abänderung in den Eigenschaften.

2. Chemisch-analoge Elemente haben entweder übereinstimmende Atomgewichte (Pt, Ir, Os), oder letztere nehmen gleichviel zu (K, Rb, Cs).

3. Das Anordnen nach den Atomgewichten entspricht der *Werthigkeit* der Elemente und bis zu einem gewissen Grade der Verschiedenheit im chemischen Verhalten, z. B. Li, Be, B, C, N, O, F.

4. Die in der Natur verbreitetsten Elemente haben *kleine* Atomgewichte und alle solche Elemente zeichnen sich durch Schärfe des Verhaltens aus. Es sind also *typische* Elemente und mit Recht wird daher das leichteste Element H als typischer Massstab gewählt.

5. Die *Grösse* des Atomgewichtes bedingt die Eigenschaften des Elementes, weshalb beim Studium von Verbindungen nicht nur auf Anzahl und Eigenschaften der Elemente und deren gegenseitiges Verhalten Rücksicht zu nehmen ist, sondern auf die *Atomgewichte* der Elemente. Daher zeigen bei mancher Analogie die Verbindungen von S und Te, Cl und J, doch auffallende Verschiedenheiten.

6. Es lässt sich die Entdeckung noch vieler *neuen* Elemente vorhersehen, z. B. Analoge des Si und Al mit Atomgewichten von 65   75.

7. Einige Atomgewichte werden voraussichtlich eine Correction erfahren, z. B. Te kann nicht das Atomgewicht 128 haben, sondern 123—126.

8. Aus obiger Tabelle ergeben sich neue Analogien zwischen Elementen. So erscheint Bo (?) als ein Analoges von Bo und Al, was bekanntlich schon längst experimentell festgesetzt ist.   (Russ. chem. Ges. 1, 60.)

Zeitschrift für Chemie (*1869, pages 405-6*), *in which Mendeleev's periodic table is first published outside Russia.*

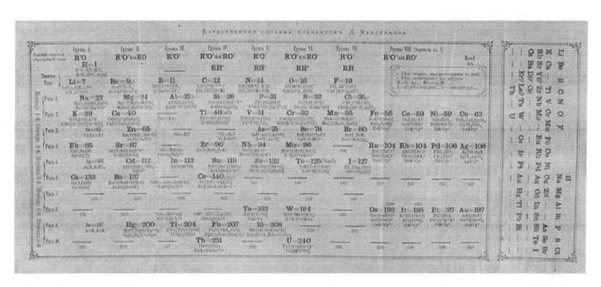

Mendeleev's 1871 periodic table. Dashes: unknown elements. Group I-VII: modern group 1–2 and 3–7 with transition metals added; some of these extend into a group VIII. Noble gasses unknown (and unpredicted).

Henry Moseley

Chapter 31

Alternative periodic tables

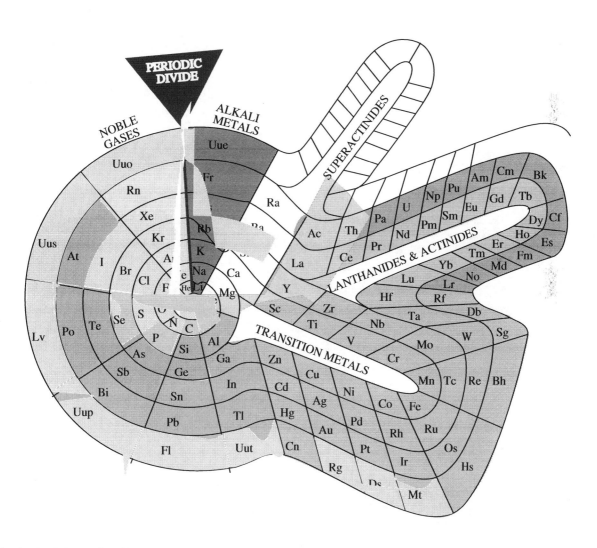

Theodor Benfey's periodic table (1964)

Alternative periodic tables are tabulations of chemical elements differing significantly in their organization from the traditional depiction of the Periodic System.[1][2] Several have been devised, often purely for didactic reasons, as not all correlations between the chemical elements are effectively captured by the standard periodic table.

Alternative periodic tables are developed often to highlight or emphasize different chemical or physical properties of the elements which are not as apparent in traditional periodic tables. Some tables aim to emphasize both the nucleon and electronic structure of atoms. This can be done by changing the spatial relationship or representation each element has with respect to another element in the table. Other tables aim to emphasize the chemical element isolations by humans over time.

31.1 Major alternative structures

31.1.1 Left step periodic table (Janet, 1928)

Charles Janet's **Left Step periodic table** (1928)[3] is considered to be the most significant alternative to the traditional depiction of the periodic system. It organizes elements according to orbital filling (instead of valence) and is widely used by physicists.[4]

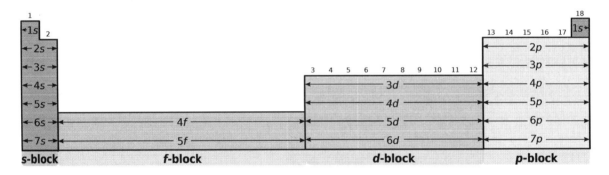

Left to right: s-, f-, d-, p-block in the common periodic table presentation; for sufficiently high principal quantum numbers, these blocks fill out in the order of s, p, d, and f. The left-step periodic table is organized according to a reversal of this order, so that the true order is maintained through a proper reading.

Compared to the common layout, the Left step table has these changes:

- Helium is placed in group 2,

- Groups 1 and 2 (the s-block), including elements 119 and 120 in extended period 8, are moved to the right side of the table

- The s-block is shifted upwards one row, now predicted elements 119 and 120 appear in row 7.

31.1.2 ADOMAH (2006)

A modern version of the Left Step is constructed by Valery Tsimmerman, known as the **ADOMAH periodic table** (2006).[5] Its structure is based on the four quantum numbers of the electron configurations.[6]

31.1.3 Two-dimensional spiral (Benfey, 1964)

In Theodor Benfey's periodic table (1964), the elements form a **two-dimensional spiral**, starting from hydrogen, and folding their way around two peninsulars, the transition metals, and lanthanides and actinides. A superactinide island is already slotted in.[7] The Chemical Galaxy (2004) is organized in a similar way.

31.1.4   Three-dimensional, physicist's (Timmothy Stowe)

Timmothy Stowe's physicist's periodic table is three-dimensional with the **three axes** representing the **principal quantum number**, orbital quantum number, and orbital magnetic quantum number. Helium is again a group 2 element.

31.1.5   Three-dimensional, flower-like (Paul Giguère, 1966)

Paul Giguère's 3-D periodic table consists of **4 connected billboards** with the elements written on the front and the back. The first billboard has the group 1 elements on the front and the group 2 elements at the back, with hydrogen and helium omitted altogether. At a 90° angle the second billboard contains the groups 13 to 18 front and back. Two more billboard each making 90° angles contain the other elements.[8][9]

31.1.6   Elements repeating (Ronald L. Rich, 2005)

Ronald L. Rich has proposed a periodic table where **elements appear more than once** when appropriate.[10] He notes that hydrogen shares properties with group 1 elements based on valency, with group 17 elements because hydrogen is a non-metal but also with the carbon group based on similarities in chemical bonding to transition metals and a similar electronegativity. In this rendition of the periodic table carbon and silicon also appear in the same group as titanium and zirconium.

31.1.7   Other

A chemists' table ("Newlands Revisited") with an alternative positioning of hydrogen, helium and the lanthanides was published by EG Marks and JA Marks in 2010.[11]

31.2   Variants of the classical layout

From Mendeleev's original periodic table, elements have been basically arranged by valence (groups in columns) and the repetition therein (periods in rows). Over the years and with discoveries in atomic structure, this schema has been adjusted and expanded, but not changed as a principle.

The oldest periodic table is the **short form** table (columns I–VIII) by Dmitri Mendeleev, which shows secondary chemical kinships. For example, the alkali metals and the coinage metals (copper, silver, gold) are in the same column because both groups tend to have a valence of one. This format is still used by many, as shown by this contemporary Russian short form table which includes all elements and element names until roentgenium.

H.G. Deming used the so-called **long periodic table** (18 columns) in his textbook General Chemistry, which appeared in the USA for the first time in 1923 (Wiley), and was the first to designate the first two and the last five Main Groups with the notation "A", and the intervening Transition Groups with the notation "B".

The numeration was chosen so that the characteristic oxides of the B groups would correspond to those of the A groups. The iron, cobalt, and nickel groups were designated neither A nor B. The Noble Gas Group was originally attached (by Deming) to the left side of the periodic table. The group was later switched to the right side and usually labeled as Group VIIIA.

31.3   Extension of the periodic table

In the **extended periodic table**, suggested by Glenn T. Seaborg in 1969, yet unknown elements are included up to atomic number 218. Theoretical periods above regular period 7 are added.

In the research field of superatoms, clusters of atoms have properties of single atoms of another element. It is suggested to extend the periodic table with a second layer to be occupied with these cluster compounds. The latest addition to this multi-story table is the aluminium cluster ion Al–
7, which behaves like a multivalent germanium atom.[12]

31.4 Gallery

- Spiral periodic table (Robert W Harrison)

- *The Ring Of Periodic Elements* (TROPE)

- Circular periodic table

- Alternative circular periodic table

- Spiral periodic table (Jan Scholten)

- Mendeleev's Flower (Flower periodic table)

- Binary electron shells periodic table

- "Stowe" periodic table

- "Zmaczynski & Bayley" periodic table

- ADOMAH Periodic Table (V.Tsimmerman)

- Newlands revisited

- Pyramidal periodic table

- Stowe-Janet-Scerri with 3D electron orbitals

- 4D Stowe-Janet-Scerri Periodic Table

31.5 References

[1] E.R. Scerri. *The Periodic Table, Its Story and Its Significance.* Oxford University Press, New York, 2006, ISBN 0195345673.

[2] Henry Bent. *New Ideas in Chemistry from Fresh Energy for the Periodic Law.* AuthorHouse, 2006, ISBN 978-1-4259-4862-7.

[3] "Left Step Periodic Table". 1928. Retrieved 2014-02-15.

[4] Stewart, Philip J. (2009). "Charles Janet: Unrecognized genius of the periodic system". *Foundations of Chemistry* **12**: 5. doi:10.1007/s10698-008-9062-5.

[5] Tsimmerman, Valery (2006). "ADOMAH Periodic Table". Retrieved 2014-02-16.

[6] Tsimmerman, Valery (2008). "Periodic Law can be understood in terms of the Tetrahedral Sphere Packing!". Retrieved 2014-02-16. creation of the first man, Adam, from the dust of the earth, in Hebrew, Adomah

[7] Benfey's table appears in an article by Glenn Seaborg, "Plutonium: The Ornery Element", *Chemistry*, June 1964, 37 (6), 12–17, on p. 14.

[8] Mazurs, E.G. (1974). *Graphical Representations of the Periodic System During One Hundred Years.* Alabama: University of Alabama Press. p. 111. ISBN 978-0-8173-3200-6.

[9] The animated depiction of Giguère's periodic table that is widely available on the internet (including from here) is erroneous, as it does not include hydrogen and helium. Giguère included hydrogen, above lithium, and helium, above beryllium. See: Giguère P.A. (1966). "The "new look" for the periodic system". *Chemistry in Canada* vol. **18** (12): 36–39 (see p.37).

[10] Rich, Ronald L. (2005). "Are Some Elements More Equal Than Others?". *J. Chem. Educ.* **82** (12): 1761. doi:10.1021/ed082p1761.

[11] Marks, E. G.; Marks, J. A. (2010). "Newlands revisited: A display of the periodicity of the chemical elements for chemists". *Foundations of Chemistry* **12**: 85. doi:10.1007/s10698-010-9083-8.

[12] Amato, Ivan (November 21, 2006). "Beyond The Periodic Table Metal clusters mimic chemical properties of atoms". *Chemical & Engineering News.*

31.6   Further reading

- A 1974 review of the tables then known is considered a definitive work on the topic: Mazurs, E. G. Graphical Representations of the Periodic System During One Hundred Years. Alabama; University of Alabama Press, 1974, ISBN 0-8173-3200-6.

- Hjørland, Birger (2011). The periodic table and the philosophy of classification. Knowledge Organization, 38(1), 9–21.

31.7   External links

- Representing the Periodic Table in Different Ways a site curated by the Michigan State University Alumni Association Knowledge Network

- Robert Harrison's modern spiral periodic table

- Janet's Left Step Periodic Table

- Correction to Physicist periodic table offered by Jeries Rihani as Meitnerium occupies the position that Hassium should have.

- A Wired Article on Alternate Periodic Tables

- A Selection of Periodic Tables

- http://periodicspiral.com/ arranges the periodic table in a (hexagonal) spiral.

- *Rotaperiod.com* A new periodic table.

- Note on the T-shirt topology of the Z-spiral.

- New Periodic Table of the Elements this is in a square-triangular periodic arrangement.

- Periodic Table based on **electron configurations**

- Database of Periodic Tables

- Periodic Fractal of the Elements

- Bob Doyle Periodic Table of the Elements A regrouping by properties used to better explain electron grouping

- Earth Scientist's Periodic Table

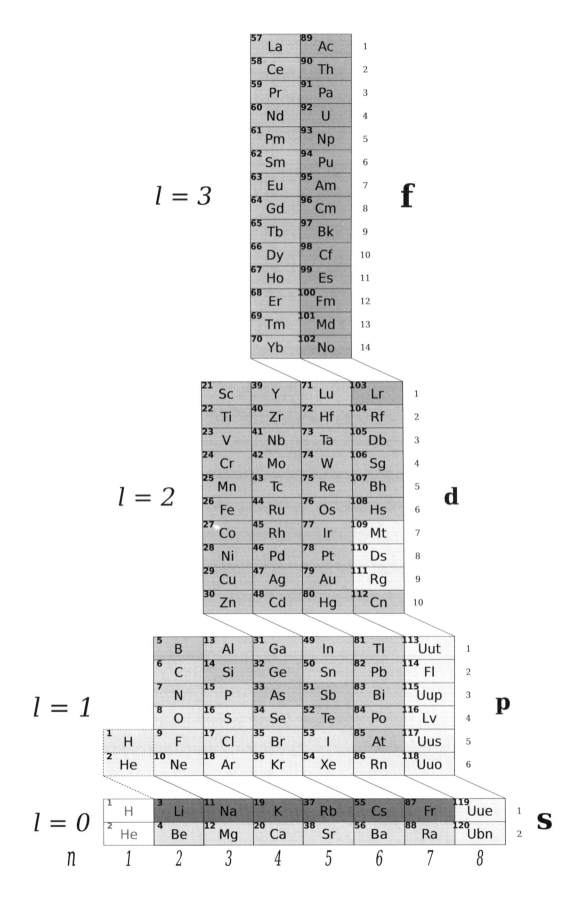

The ADOMAH periodic table is based on the electron's quantum numbers

Reihen	Gruppe I. — R^2O	Gruppe II. — RO	Gruppe III. — R^2O^3	Gruppe IV. RH^4 RO^2	Gruppe V. RH^3 R^2O^5	Gruppe VI. RH^2 RO^3	Gruppe VII. RH R^2O^7	Gruppe VIII. — RO^4
1	H=1							
2	Li=7	Be=9,4	B=11	C=12	N=14	O=16	F=19	
3	Na=23	Mg=24	Al=27,8	Si=28	P=31	S=32	Cl=35,5	
4	K=39	Ca=40	—=44	Ti=48	V=51	Cr=52	Mn=55	Fe=56, Co=59, Ni=59, Cu=63.
5	(Cu=63)	Zn=65	—=68	—=72	As=75	Se=78	Br=80	
6	Rb=85	Sr=87	?Yt=88	Zr=90	Nb=94	Mo=96	—=100	Ru=104, Rh=104, Pd=106, Ag=108.
7	(Ag=108)	Cd=112	In=113	Sn=118	Sb=122	Te=125	J=127	
8	Cs=133	Ba=137	?Di=138	?Ce=140	—	—	—	— — — —
9	(—)	—	—	—	—	—	—	
10	—	—	?Er=178	?La=180	Ta=182	W=184	—	Os=195, Ir=197, Pt=198, Au=199.
11	(Au=199)	Hg=200	Tl=204	Pb=207	Bi=208	—	—	
12	—	—	—	Th=231	—	U=240	—	— — — —

Mendeleev's 1871 periodic table in VIII columns. Nowadays, roughly spoken, pairs of Reihen *are shown as grouplabels A, B (for example: Reihen 4, 5 are written as period 3 and groups (columns) IA–VIIIA, IB–VIIIB).*

Chapter 32

Extended periodic table

An **extended periodic table** theorizes about elements beyond element 118 (beyond period 7, or row 7). Currently seven periods in the periodic table of chemical elements are known and proven, culminating with atomic number 118. If further elements with higher atomic numbers than this are discovered, they will be placed in additional periods, laid out (as with the existing periods) to illustrate periodically recurring trends in the properties of the elements concerned. Any additional periods are expected to contain a larger number of elements than the seventh period, as they are calculated to have an additional so-called **g-block**, containing at least 18 elements with partially filled g-orbitals in each period. An **eight-period table** containing this block was suggested by Glenn T. Seaborg in 1969.[1][2] IUPAC defines an element to exist if its lifetime is longer than 10^{-14} seconds, which is the time it takes for the nucleus to form an electronic cloud.[3]

No elements in this region have been synthesized or discovered in nature.[4] The first element of the g-block may have atomic number 121, and thus would have the systematic name unbiunium. Elements in this region are likely to be highly unstable with respect to radioactive decay, and have extremely short half lives, although element 126 is hypothesized to be within an island of stability that is resistant to fission but not to alpha decay. It is not clear how many elements beyond the expected island of stability are physically possible, if period 8 is complete, or if there is a period 9.

According to the orbital approximation in quantum mechanical descriptions of atomic structure, the g-block would correspond to elements with partially filled g-orbitals, but spin-orbit coupling effects reduce the validity of the orbital approximation substantially for elements of high atomic number. While Seaborg's version of the extended period had the heavier elements following the pattern set by lighter elements, as it did not take into account relativistic effects, models that take relativistic effects into account do not. Pekka Pyykkö and B. Fricke used computer modeling to calculate the positions of elements up to $Z = 184$ (comprising periods 8, 9, and the beginning of 10), and found that several were displaced from the Madelung rule.[5][6]

32.1   History

It is unknown how far the periodic table might extend beyond the known 118 elements. Glenn T. Seaborg suggested that the highest possible element may be under $Z = 130$,[7] while Walter Greiner predicted that there may not be a highest possible element.[8]

All of these hypothetical undiscovered elements are named by the International Union of Pure and Applied Chemistry (IUPAC) systematic element name standard which creates a generic name for use until the element has been discovered, confirmed, and an official name approved. These names are typically not used in the literature, and are referred to by their atomic numbers; hence, element 164 would usually not be called "unhexquadium" (the IUPAC systematic name), but rather "element 164" with symbol "164", "(164)", or "E164".

As of April 2011, synthesis has been attempted for only ununennium, unbinilium, unbibium, unbiquadium, unbihexium, and unbiseptium. ($Z = 119$, 120, 122, 124, 126, and 127)

At element 118, the orbitals 1s, 2s, 2p, 3s, 3p, 3d, 4s, 4p, 4d, 4f, 5s, 5p, 5d, 5f, 6s, 6p, 6d, 7s and 7p are assumed to be filled, with the remaining orbitals unfilled. A simple extrapolation from the Aufbau principle would predict the eighth row to fill orbitals in the order 8s, 5g, 6f, 7d, 8p; but after element 120, the proximity of the electron shells makes placement in a simple table problematic.

Not all models show the higher elements following the pattern established by lighter elements. Pekka Pyykkö, for example, used computer modeling to calculate the positions of elements up to Z=172, and found that several were displaced from the Madelung energy-ordering rule.[6] He predicts that the orbital shells will fill up in this order:

- 8s,

- 5g,

- the first two spaces of 8p,

- 6f,

- 7d,

- 9s,

- the first two spaces of 9p,

- the rest of 8p.

He also suggests that period 8 be split into three parts:

- 8a, containing 8s,

- 8b, containing the first two elements of 8p,

- 8c, containing 7d and the rest of 8p.[9]

Fricke *et al.* also predicted the extended periodic table up to 184.[5] This model has been more widely used among scientists and is shown above as the main form of the extended periodic table.

32.2   Predicted properties of undiscovered elements

Element 118 is the last element that has been claimed to have been synthesized. The next two elements, elements 119 and 120, should form an 8s series and be an alkali and alkaline earth metal respectively. Beyond element 120, the superactinide series is expected to begin, when the 8s electrons and the filling $8p_{1/2}$, $7d_{3/2}$, $6f_{5/2}$, and $5g_{7/2}$ subshells determine the chemistry of these elements. Complete and accurate CCSD calculations are not available for elements beyond 122 because of the extreme complexity of the situation: the 5g, 6f, and 7d orbitals should have about the same energy level, and in the region of element 160 the 9s, $8p_{3/2}$, and $9p_{1/2}$ orbitals should also be about equal in energy. This will cause the electron shells to mix so that the block concept no longer applies very well, and will also result in novel chemical properties that will make positioning these elements in a periodic table very difficult. For example, element 164 is expected to mix characteristics of the elements of group 10, 12, 14, and 18.[10]

32.2.1   Chemical and physical properties

8s elements

Main articles: Ununennium and Unbinilium

The first two elements of period 8 are expected to be ununennium and unbinilium, elements 119 and 120. Their electron configurations should have the 8s orbital being filled. This orbital is relativistically stabilized and contracted and thus, elements 119 and 120 should be more like caesium and barium than their immediate neighbours above, francium and radium. Another effect of the relativistic contraction of the 8s orbital is that the atomic radii of these two elements should be about the same of those of francium and radium. They should behave like normal alkali and alkaline earth metals, normally forming +1 and +2 oxidation states respectively, but the relativistic destabilization of the $7p_{3/2}$ subshell and the relatively low ionization energies of the $7p_{3/2}$ electrons should make higher oxidation states like +3 and +4 (respectively) possible as well.[5][10]

Superactinides

The superactinide series is expected to contain elements 121 to 155. In the superactinide series, the $7d_{3/2}$, $8p_{1/2}$, $6f_{5/2}$ and $5g_{7/2}$ shells should all fill simultaneously:[11] this creates very complicated situations, so much so that complete and accurate CCSD calculations have been done only for elements 121 and 122.[10] The first superactinide, unbiunium (element 121), should be a congener of lanthanum and actinium and should have similar properties to them:[12] its main oxidation state should be +3, although the closeness of the valence subshells' energy levels may permit higher oxidation states, just like in elements 119 and 120.[10] Relativistic stabilization of the 8p subshell should result in a ground-state $8s^2 8p^1$ valence electron configuration for element 121, in contrast to the ds^2 configurations of lanthanum and actinium.[10] Its first ionization energy is predicted to be 429.4 kJ/mol, which would be lower than those of all known elements except for the alkali metals potassium, rubidium, caesium, and francium: this value is even lower than that of the period 8 alkali metal ununennium (463.1 kJ/mol). Similarly, the next superactinide, unbibium (element 122), may be a congener of cerium and thorium, with a main oxidation state of +4, but would have a ground-state $7d^1 8s^2 8p^1$ valence electron configuration, unlike thorium's $6d^2 7s^2$ configuration. Hence, its first ionization energy would be smaller than thorium's (Th: 6.54 eV; Ubb: 5.6 eV) because of the greater ease of ionizing unbibium's $8p_{1/2}$ electron than thorium's 7s electron.[10]

In the first few superactinides, the binding energies of the added electrons are predicted to be small enough that they can lose all their valence electrons; for example, unbihexium (element 126) could easily form a +8 oxidation state, and even higher oxidation states for the next few elements may be possible. Unbihexium is also predicted to display a variety of other oxidation states: recent calculations have suggested a stable monofluoride UbhF may be possible, resulting from a bonding interaction between the 5g orbital on unbihexium and the 2p orbital on fluorine.[13] Other predicted oxidation states include +2, +4, and +6; +4 is expected to be the most usual oxidation state of unbihexium.[11] The presence of electrons in g-orbitals, which do not exist in the ground state electron configuration of any currently known element, should allow presently unknown hybrid orbitals to form and influence the chemistry of the superactinides in new ways, although the absence of *g* electrons in known elements makes predicting their chemistry more difficult.[5]

In the later superactinides, the oxidation states should become lower. By element 132, the predominant most stable oxidation state will be only +6; this is further reduced to +3 and +4 by element 144, and at the end of the superactinide series it will be only +2 (and possibly even 0) because the 6f shell, which is being filled at that point, is deep inside the electron cloud and the 8s and $8p_{1/2}$ electrons are bound too strongly to be chemically active. The 5g shell should be filled at element 144 and the 6f shell at around element 154, and at this region of the superactinides the $8p_{1/2}$ electrons are bound so strongly that they are no longer active chemically, so that only a few electrons can participate in chemical reactions. Calculations by Fricke *et al.* predict that at element 154, the 6f shell is full and there are no d- or other electron wave functions outside the chemically inactive 8s and $8p_{1/2}$ shells. This would cause element 154 to be very unreactive, so that it may exhibit properties similar to those of the noble gases.[5][10]

Similarly to the lanthanide and actinide contractions, there should be a superactinide contraction in the superactinide series where the ionic radii of the superactinides are smaller than expected. In the lanthanides, the contraction is about 4.4 pm per element; in the actinides, it is about 3 pm per element. The contraction is larger in the lanthanides than in the actinides due to the greater localization of the 4f wave function as compared to the 5f wave function. Comparisons with the wave functions of the outer electrons of the lanthanides, actinides, and superactinides lead to a prediction of a contraction of about 2 pm per element in the superactinides; although this is smaller than the contractions in the lanthanides and actinides, its total effect is larger due to the fact that 32 electrons are filled in the deeply buried 5g and 6f shells, instead of just 14 electrons being filled in the 4f and 5f shells in the lanthanides and actinides respectively.[5]

Pekka Pyykkö divides these superactinides into three series: a 5g series (elements 121 to 138), an $8p_{1/2}$ series (elements 139 to 140), and a 6f series (elements 141 to 155), although noting that there would be a great deal of overlapping between energy levels and that the 6f, 7d, or $8p_{1/2}$ orbitals could well also be occupied in the early superactinide atoms or ions. He also expects that they would behave more like "superlanthanides", in the sense that the 5g electrons would mostly be chemically inactive, similarly to how only one or two 4f electrons in each lanthanide are ever ionized in chemical compounds. He also predicted that the possible oxidation states of the superactinides might rise very high in the 6f series, to values such as +12 in element 148.[9]

7d transition metals

The transition metals in period 8 are expected to be elements 156 to 164. Although the 8s and $8p_{1/2}$ electrons are bound so strongly in these elements that they should not be able to take part in any chemical reactions, the 9s and $9p_{1/2}$ levels are expected to be readily available for hybridization such that these elements will still behave chemically like their lighter homologues in the periodic table, showing the same oxidation states as they do, in contrast to earlier predictions which predicted the period 8 transition metals to have main oxidation states two less than those of their lighter congeners.[5][10]

The noble metals of this series of transition metals are not expected to be as noble as their lighter homologues, due to the absence of an outer *s* shell for shielding and also because the 7d shell is strongly split into two subshells due to relativistic effects. This causes the first ionization energies of the 7d transition metals to be smaller than those of their lighter congeners.[5][10][11]

Calculations predict that the 7d electrons of element 164 (unhexquadium) should participate very readily in chemical reactions, so that unhexquadium should be able to show stable +6 and +4 oxidation states in addition to the normal +2 state in aqueous solutions with strong ligands. Unhexquadium should thus be able to form compounds like $Uhq(CO)_4$, $Uhq(PF_3)_4$ (both tetrahedral), and Uhq(CN)2–
2 (linear), which is very different behavior from that of lead, which unhexquadium would be a heavier homologue of if not for relativistic effects. Nevertheless, the divalent state would be the main one in aqueous solution, and unhexquadium(II) should behave more similarly to lead than unhexquadium(IV) and unhexquadium(VI).[10][11]

Unhexquadium should be a soft metal like mercury, and metallic unhexquadium should have a high melting point as it is predicted to bond covalently. It is also expected to be a soft Lewis acid and have Ahrlands softness parameter close to 4 eV. It should also have some similarities to ununoctium as well as to the other group 12 elements.[10] Unhexquadium should be at most moderately reactive, having a first ionization energy that should be around 685 kJ/mol, comparable to that of molybdenum.[5][11] Due to the lanthanide, actinide, and superactinide contractions, unhexquadium should have an metallic radius of only 158 pm, very close to that of the much lighter magnesium, despite its being expected to have an atomic weight of around 474 u, about 19.5 times as much as that of magnesium.[5] This small radius and high weight cause it to be expected to have an extremely high density of around 46 $g \cdot cm^{-3}$, over twice that of osmium, currently the most dense element known, at 22.61 $g \cdot cm^{-3}$; unhexquadium should be the second most dense element in the first 172 elements in the periodic table, with only its neighbour unhextrium (element 163) being more dense (at 47 $g \cdot cm^{-3}$).[5] Metallic unhexquadium should be quite stable, as the 8s and $8p_{1/2}$ electrons are very deeply buried in the electron core and only the 7d electrons are available for bonding. Metallic unhexquadium should have a very large cohesive energy (enthalpy of crystallization) due to its covalent bonds, most probably resulting in a high melting point.[11]

Theoretical interest in the chemistry of unhexquadium is largely motivated by theoretical predictions that it, especially the isotope ^{482}Uhq (with 164 protons and 318 neutrons), would be at the center of a hypothetical second island of stability (the first being centered on ^{306}Ubb).[15][16][17]

Elements 165 to 172

Elements 165 (unhexpentium) and 166 (unhexhexium) should behave as normal alkali and alkaline earth metals when in the +1 and +2 oxidation states respectively. The 9s electrons should have ionization energies comparable to those of the 3s electrons of sodium and magnesium, due to relativistic effects causing the 9s electrons to be much more strongly bound than non-relativistic calculations would predict. Elements 165 and 166 should normally exhibit the +1 and +2 oxidation states respectively, although the ionization energies of the 7d electrons are low enough to allow higher oxidation states like +3 and +4 to occur quite commonly.[5][10]

In elements 167 to 172, the $9p_1/2$ and $8p_3/2$ shells will be filled. Their energy eigenvalues are so close together that they behaves as one combined *p* shell, similar to the non-relativistic 2p and 3p shells. Thus, the inert pair effect does not occur and the most common oxidation states of elements 167 to 170 should be +3, +4, +5, and +6 respectively. Element 171 (unseptunium) is expected to be a halogen, showing various oxidation states ranging from –1 to +7. Its electron affinity should be 3.0 eV, allowing it to form a hydrogen halide, HUsu. The Usu⁻ ion is expected to be a soft base, comparable to iodide (I⁻). Element 172 (unseptbium) should be a noble gas with chemical behaviour similar to that of xenon, as their ionization energies should be very similar (Xe, 1170.4 kJ/mol; Usb, 1090.3 kJ/mol). The only main difference between them is that element 172, unlike xenon, is expected to be a liquid or a solid at standard temperature and pressure due to its much higher atomic weight.[5] Unseptbium should be a strong Lewis acid, forming fluorides and oxides, similarly to its lighter congener xenon.[11] Because of this analogy of elements 165–172 to periods 2 and 3, Fricke *et al.* considered them to form a ninth period of the periodic table, while the eighth period was taken by them to end at the noble metal element 164. This ninth period would be similar to the second and third period in that it should have no transition metals.[11]

"Eka-superactinides"

Immediately after element 172 (unseptbium), the first noble gas after element 118 (the last period 7 element), another long transition series like the superactinides should begin, filling the 6g, 7f, 8d, and perhaps 6h shells. These electrons would be very loosely bound, rendering extremely high oxidation states possibly easy to reach.[11] This series may be termed the eka-superactinides, as it is the next long transition series in the periodic table after the superactinides.

The ground-state electron configuration of element 184 (unoctquadium) is expected to be $[Usb]6g^57f^48d^3$: only the 8d and 7f electrons should be chemically active, with possible reasons being small radial extension and large binding energy. The absence of $6h_{11}/2$, 10s, and $10p_1/2$ electrons from this ground-state electron configuration suggests that it would behave chemically simpler than the early superactinides, and more similar to uranium or neptunium. As more electrons are ionized, the number of 6g electrons in the unoctquadium ion will increase: these are buried in the electron core and would not participate in chemical reactions, but the 7f electrons could. Extrapolation from uranium suggests that the +4 state would be the most stable in aqueous solution, with +5 and +6 readily obtainable in solid compounds. Higher states would necessitate the ionization of the deeply buried 6g electrons and are probably unlikely: furthermore, their binding energy becomes much higher as more electrons are removed. This effect is so important that the 9s and $9p_1/2$ electrons, part of the closed [Usb] electron core, would enter the 6g subshell in the +8 oxidation state and higher. This suggests that the multitude of simultaneously-filling outer electron shells as one proceeds down a long transition series might not lead to exceptionally high or exotic oxidation states, nor should it lead to anomalously low increases in ionization energy.[5][11] This contradicts preliminary extrapolations (without calculation) that expected that unoctquadium would have many oxidation states ranging from +4 (with 8 6g electrons) to +12 (with no 6g electrons).[18]

32.2.2 Nuclear properties

The first island of stability is expected to be centered on unbibium−306 (with 122 protons and 184 neutrons),[15] and the second is expected to be centered on unhexquadium−482 (with 164 protons and 318 neutrons).[16][17] This second island of stability should confer additional stability on elements 152–168.[11]

Calculations according to the Hartree–Fock–Bogoliubov Method using the non-relativistic Skyrme interaction have proposed Z=126 as a closed proton shell. In this region of the periodic table, N=184 and N=196 have been suggested as closed neutron shells. Therefore the isotopes of most interest are ³¹⁰Ubh and ³²²Ubh, for these might be considerably longer-lived than other isotopes. Unbihexium, having a magic number of protons, is predicted to be more stable than other elements in this region, and may have nuclear isomers with very long half-lives.[19]

32.2.3   Electron configurations

The following are the expected electron configurations of elements 119–172 and 184.[11]

32.3   Attempts to synthesize still undiscovered elements

The only period 8 elements that have had synthesis attempts were elements 119, 120, 122, 124, 126, and 127. So far, none of these synthesis attempts were successful.

32.3.1   Ununennium

The synthesis of ununennium was attempted in 1985 by bombarding a target of einsteinium−254 with calcium−48 ions at the superHILAC accelerator at Berkeley, California:

$$^{254}_{99}\text{Es} + {}^{48}_{20}\text{Ca} \rightarrow {}^{302}_{119}\text{Uue}^*$$

No atoms were identified, leading to a limiting yield of 300 nb.[20] As of May 2012, plans are under way to attempt to synthesize the isotopes ^{295}Uue and ^{296}Uue by bombarding a target of berkelium with titanium at the GSI Helmholtz Centre for Heavy Ion Research in Darmstadt, Germany:[21][22]

$$^{249}_{97}\text{Bk} + {}^{50}_{22}\text{Ti} \rightarrow {}^{296}_{119}\text{Uue} + 3\,{}^{1}_{0}\text{n}$$

$$^{249}_{97}\text{Bk} + {}^{50}_{22}\text{Ti} \rightarrow {}^{295}_{119}\text{Uue} + 4\,{}^{1}_{0}\text{n}$$

32.3.2   Unbinilium

Attempts to date to synthesize the element using fusion reactions at low excitation energy have met with failure, although there are reports that the fission of nuclei of unbinilium at very high excitation has been successfully measured, indicating a strong shell effect at Z=120. In March–April 2007, the synthesis of unbinilium was attempted at the Flerov Laboratory of Nuclear Reactions in Dubna by bombarding a plutonium−244 target with iron−58 ions.[23] Initial analysis revealed that no atoms of element 120 were produced providing a limit of 400 fb for the cross section at the energy studied.[24]

$$^{244}_{94}\text{Pu} + {}^{58}_{26}\text{Fe} \rightarrow {}^{302}_{120}\text{Ubn}^* \rightarrow \textit{fission only}$$

The Russian team are planning to upgrade their facilities before attempting the reaction again.[22]

In April 2007, the team at GSI attempted to create unbinilium using uranium−238 and nickel−64:[22]

$$^{238}_{92}\text{U} + {}^{64}_{28}\text{Ni} \rightarrow {}^{302}_{120}\text{Ubn}^* \rightarrow \textit{fission only}$$

No atoms were detected providing a limit of 1.6 pb on the cross section at the energy provided. The GSI repeated the experiment with higher sensitivity in three separate runs from April–May 2007, Jan–March 2008, and Sept–Oct 2008, all with negative results and providing a cross section limit of 90 fb.[22]

In June–July 2010, scientists at the GSI attempted the fusion reaction:[22]

$$^{248}_{96}\text{Cm} + ^{54}_{24}\text{Cr} \rightarrow ^{302}_{120}\text{Ubn}^*$$

They were unable to detect any atoms but exact details are not currently available.[22]

In August–October 2011, a different team at the GSI using the TASCA facility tried the new reaction:[22]

$$^{249}_{98}\text{Cf} + ^{50}_{22}\text{Ti} \rightarrow ^{299}_{120}\text{Ubn}^*$$

Results from this experiment are not yet available.[22] In 2008, the team at GANIL, France, described the results from a new technique which attempts to measure the fission half-life of a compound nucleus at high excitation energy, since the yields are significantly higher than from neutron evaporation channels. It is also a useful method for probing the effects of shell closures on the survivability of compound nuclei in the super-heavy region, which can indicate the exact position of the next proton shell (Z=114, 120, 124, or 126). The team studied the nuclear fusion reaction between uranium ions and a target of natural nickel:

$$^{238}_{92}\text{U} + ^{nat}_{28}\text{Ni} \rightarrow ^{296,298,299,300,302}\text{Ubn}^* \rightarrow \textit{fission}.$$

The results indicated that nuclei of unbinilium were produced at high (~70 MeV) excitation energy which underwent fission with measurable half-lives $> 10^{-18}$ s. Although very short, the ability to measure such a process indicates a strong shell effect at Z=120. At lower excitation energy (see neutron evaporation), the effect of the shell will be enhanced and ground-state nuclei can be expected to have relatively long half-lives. This result could partially explain the relatively long half-life of ^{294}Uuo measured in experiments at Dubna. Similar experiments have indicated a similar phenomenon at Z=124 (see unbiquadium) but not for flerovium, suggesting that the next proton shell does in fact lie at Z>120.[25][26] The team at RIKEN have begun a program utilizing ^{248}Cm targets and have indicated future experiments to probe the possibility of Z=120 being the next magic number using the aforementioned nuclear reactions to form ^{302}Ubn.[27]

32.3.3 Unbibium

The first attempt to synthesize unbibium was performed in 1972 by Flerov *et al.* at JINR, using the hot fusion reaction:[28]

$$^{238}_{92}\text{U} + ^{66}_{30}\text{Zn} \rightarrow ^{304}_{122}\text{Ubb}^* \rightarrow \text{no atoms}.$$

No atoms were detected and a yield limit of 5 mb (5,000,000,000 pb) was measured. Current results (see flerovium) have shown that the sensitivity of this experiment was too low by at least 6 orders of magnitude.

In 2000, the Gesellschaft für Schwerionenforschung (GSI) performed a very similar experiment with much higher sensitivity:

$$^{238}_{92}\text{U} + ^{70}_{30}\text{Zn} \rightarrow ^{308}_{122}\text{Ubb}^* \rightarrow \text{no atoms}.$$

These results indicate that the synthesis of such heavier elements remains a significant challenge and further improvements of beam intensity and experimental efficiency is required. The sensitivity should be increased to 1 fb.

Another unsuccessful attempt to synthesize unbibium was carried out in 1978 at the GSI, where a natural erbium target was bombarded with xenon-136 ions:[28]

$$^{nat}_{68}\text{Er} + ^{136}_{54}\text{Xe} \rightarrow ^{298,300,302,303,304,306}\text{Ubb}^* \rightarrow \text{no atoms}.$$

The two attempts in the 1970s to synthesize unbibium were caused by research investigating whether superheavy elements could potentially be naturally occurring.[28] Several experiments have been performed between 2000-2004 at the Flerov laboratory of Nuclear Reactions studying the fission characteristics of the compound nucleus ^{306}Ubb. Two nuclear reactions have been used, namely ^{248}Cm + ^{58}Fe and ^{242}Pu + ^{64}Ni.[28] The results have revealed how nuclei such as this fission predominantly by expelling closed shell nuclei such as ^{132}Sn (Z=50, N=82). It was also found that the yield for the fusion-fission pathway was similar between ^{48}Ca and ^{58}Fe projectiles, indicating a possible future use of ^{58}Fe projectiles in superheavy element formation.[29]

32.3.4   Unbiquadium

In a series of experiments, scientists at GANIL have attempted to measure the direct and delayed fission of compound nuclei of elements with Z=114, 120, and 124 in order to probe shell effects in this region and to pinpoint the next spherical proton shell. This is because having complete nuclear shells (or, equivalently, having a magic number of protons or neutrons) would confer more stability on the nuclei of such superheavy elements, thus moving closer to the island of stability. In 2006, with full results published in 2008, the team provided results from a reaction involving the bombardment of a natural germanium target with uranium ions:

$$^{238}_{92}\text{U} + {}^{nat}_{32}\text{Ge} \rightarrow {}^{308,310,311,312,314}\text{Ubq}^* \rightarrow fission.$$

The team reported that they had been able to identify compound nuclei fissioning with half-lives > 10^{-18} s. This result suggests a strong stabilizing effect at Z=124 and points to the next proton shell at Z>120, not at Z=114 as previously thought. A compound nucleus is a loose combination of nucleons that have not arranged themselves into nuclear shells yet. It has no internal structure and is held together only by the collision forces between the target and projectile nuclei. It is estimated that it requires around 10^{-14} s for the nucleons to arrange themselves into nuclear shells, at which point the compound nucleus becomes an nuclide, and this number is used by IUPAC as the minimum half-life a claimed isotope must have to potentially be recognised as being discovered. Thus, the GANIL experiments do not count as a discovery of element 124.[28]

32.3.5   Unbihexium

The first and only attempt to synthesize unbihexium, which was unsuccessful, was performed in 1971 at CERN by René Bimbot and John M. Alexander using the hot fusion reaction:[28]

232
90Th + 84
36Kr → 316
126Ubh* → *no atoms*

A high energy alpha particle was observed and taken as possible evidence for the synthesis of unbihexium. Recent research suggests that this is highly unlikely as the sensitivity of experiments performed in 1971 would have been several orders of magnitude too low according to current understanding.

32.3.6   Unbiseptium

Unbiseptium has had one failed attempt at synthesis in 1978 at the Darmstadt UNILAC accelerator by bombarding a natural tantalum target with xenon ions:[28]

$$^{nat}_{73}\text{Ta} + {}^{136}_{54}\text{Xe} \rightarrow {}^{316,317}\text{Ubs}^* \rightarrow \text{no atoms.}$$

32.4 Possible natural occurrence

On April 24, 2008, a group led by Amnon Marinov at the Hebrew University of Jerusalem claimed to have found single atoms of unbibium-292 in naturally occurring thorium deposits at an abundance of between 10^{-11} and 10^{-12}, relative to thorium.[30] The claim of Marinov *et al.* was criticized by a part of the scientific community, and Marinov says he has submitted the article to the journals *Nature* and *Nature Physics* but both turned it down without sending it for peer review.[31] The unbibium-292 atoms were claimed to be superdeformed or hyperdeformed isomers, with a half-life of at least 100 million years.[28]

A criticism of the technique, previously used in purportedly identifying lighter thorium isotopes by mass spectrometry,[32] was published in Physical Review C in 2008.[33] A rebuttal by the Marinov group was published in Physical Review C after the published comment.[34]

A repeat of the thorium-experiment using the superior method of Accelerator Mass Spectrometry (AMS) failed to confirm the results, despite a 100-fold better sensitivity.[35] This result throws considerable doubt on the results of the Marinov collaboration with regards to their claims of long-lived isotopes of thorium,[32] roentgenium[36] and unbibium.[30] It is still possible that traces of unbibium might only exist in some thorium samples, although this is unlikely.[28]

It was suggested in 1976 that primordial superheavy elements (mainly livermorium, unbiquadium, unbihexium, and unbiseptium) could be a cause of unexplained radiation damage in minerals. This prompted many researchers to search for it in nature from 1976 to 1983. Some claimed that they had detected alpha particles with the right energies to cause the damage observed, supporting the presence of unbihexium, while some claimed that no unbihexium had been detected. The possible extent of primordial unbihexium on Earth is uncertain; it might now only exist in traces, or could even have completely decayed by now after having caused the radiation damage long ago.[19]

32.5 End of the periodic table

The number of physically possible elements is unknown. A low estimate is that the periodic table may end soon after the island of stability,[7] which is expected to center on $Z = 126$, as the extension of the periodic and nuclides tables is restricted by the proton and the neutron drip lines;[37] some, such as Walter Greiner, predict that there may not be an end to the periodic table.[8] Other predictions of an end to the periodic table include $Z = 128$ (John Emsley) and $Z = 155$ (Albert Khazan).[28]

32.5.1 Feynmanium and elements above the atomic number 137

Richard Feynman noted[38] that a simplistic interpretation of the relativistic Dirac equation runs into problems with electron orbitals at $Z > 1/\alpha \approx 137$ as described in the sections below, suggesting that neutral atoms cannot exist beyond untriseptium, and that a periodic table of elements based on electron orbitals therefore breaks down at this point. On the other hand, a more rigorous analysis calculates the limit to be $Z \approx 173$, and also that this limit would not actually spell the end of the periodic table.

Bohr model

The Bohr model exhibits difficulty for atoms with atomic number greater than 137, for the speed of an electron in a 1s electron orbital, v, is given by

$$v = Z\alpha c \approx \frac{Zc}{137.036}$$

where Z is the atomic number, and α is the fine structure constant, a measure of the strength of electromagnetic interactions. Under this approximation, any element with an atomic number of greater than 137 would require 1s electrons to be traveling faster than c, the speed of light. Hence the non-relativistic Bohr model is clearly inaccurate when applied to such an element.

Relativistic Dirac equation

The relativistic Dirac equation gives the ground state energy as

$$E = \frac{mc^2}{\sqrt{1 - \frac{v^2}{c^2}}} = \frac{mc^2}{\sqrt{1 - Z^2\alpha^2}},$$

where m is the rest mass of the electron. For $Z > 137$, the wave function of the Dirac ground state is oscillatory, rather than bound, and there is no gap between the positive and negative energy spectra, as in the Klein paradox.[40] More accurate calculations taking into account the effects of the finite size of the nucleus indicate that the binding energy first exceeds $2mc^2$ for $Z > Z_{\text{cr}} \approx 173$. For $Z > Z_{\text{cr}}$, if the innermost orbital (1s) is not filled, the electric field of the nucleus will pull an electron out of the vacuum, resulting in the spontaneous emission of a positron.[41] This does not happen if the innermost orbital is filled, so that $Z = 173$ does not constitute a limit to the periodic table, only a limit to fully ionized nuclei.[8]

32.6 See also

- Table of nuclides (combined)

- Hypernucleus

32.7 References

[1] Seaborg, Glenn T. (August 26, 1996). "An Early History of LBNL".

[2] Frazier, K. (1978). "Superheavy Elements". *Science News* **113** (15): 236–238. doi:10.2307/3963006. JSTOR 3963006.

[3] Kernchemie

[4] Element 122 was claimed to exist naturally in April 2008, but this claim was widely believed to be erroneous. "Heaviest element claim criticised". Rsc.org. 2008-05-02. Retrieved 2010-03-16.

[5] Fricke, B.; Greiner, W.; Waber, J. T. (1971). "The continuation of the periodic table up to $Z = 172$. The chemistry of superheavy elements". *Theoretica chimica acta* (Springer-Verlag) **21** (3): 235–260. doi:10.1007/BF01172015. Retrieved 28 November 2012.

[6] "Extended elements: new periodic table". 2010.

[7] Seaborg, Glenn T. (c. 2006). "transuranium element (chemical element)". Encyclopædia Britannica. Retrieved 2010-03-16.

[8] Philip Ball (November 2010). "Would element 137 really spell the end of the periodic table? Philip Ball examines the evidence". *Chemistry World*. Royal Society of Chemistry. Retrieved 2012-09-30.

[9] Pyykkö, Pekka (2011). "A suggested periodic table up to $Z \le 172$, based on Dirac–Fock calculations on atoms and ions". *Physical Chemistry Chemical Physics* **13** (1): 161–8. Bibcode:2011PCCP...13..161P. doi:10.1039/c0cp01575j. PMID 20967377.

[10] Hoffman, Darleane C.; Lee, Diana M.; Pershina, Valeria (2006). "Transactinides and the future elements". In Morss; Edelstein, Norman M.; Fuger, Jean. *The Chemistry of the Actinide and Transactinide Elements* (3rd ed.). Dordrecht, The Netherlands: Springer Science+Business Media. ISBN 1-4020-3555-1.

[11] Fricke, Burkhard (1975). "Superheavy elements: a prediction of their chemical and physical properties". *Recent Impact of Physics on Inorganic Chemistry* **21**: 89–144. doi:10.1007/BFb0116498. Retrieved 4 October 2013.

[12] Waber, J. T. (1969). "SCF Dirac–Slater Calculations of the Translawrencium Elements". *The Journal of Chemical Physics* **51** (2): 664–661. Bibcode:1969JChPh..51..664W. doi:10.1063/1.1672054.

[13] Jacoby, Mitch (2006). "As-yet-unsynthesized superheavy atom should form a stable diatomic molecule with fluorine". *Chemical & Engineering News* **84** (10): 19. doi:10.1021/cen-v084n010.p019a. Retrieved 2008-01-14.

[14] Makhyoun, M. A. (October 1988). "On the electronic structure of $5g^1$ complexes of element 125: a quasi-relativistic MS-Xα study". *Journal de Chimie Physique et de Physico-Chimie Biologique* (EDP Sciences, Les Ulis, FRANCE (1903-2000)) **85** (10): 917–24.

[15] Kratz, J. V. (5 September 2011). *The Impact of Superheavy Elements on the Chemical and Physical Sciences* (PDF). 4th International Conference on the Chemistry and Physics of the Transactinide Elements. Retrieved 27 August 2013.

[16] http://www.eurekalert.org/pub_releases/2008-04/acs-nse031108.php

[17] http://link.springer.com/article/10.1007%2FBF01406719/lookinside/000.png

[18] Penneman, R. A.; Mann, J. B.; Jørgensen, C. K. (February 1971). "Speculations on the chemistry of superheavy elements such as Z = 164". *Chemical Physics Letters* **8** (4): 321–326. Bibcode:1971CPL.....8..321P. doi:10.1016/0009-2614(71)80054-4.

[19] Emsley, John (2011). *Nature's Building Blocks: An A-Z Guide to the Elements* (New ed.). New York, NY: Oxford University Press. p. 592. ISBN 978-0-19-960563-7.

[20] R. W. Lougheed, J. H. Landrum, E. K. Hulet, J. F. Wild, R. J. Dougan, A. D. Dougan, H. Gäggeler, M. Schädel, K. J. Moody, K. E. Gregorich, and G. T. Seaborg (1985). "Search for superheavy elements using ^{48}Ca + ^{254}Es^g reaction". *Physical Reviews C* **32** (5): 1760–1763. Bibcode:1985PhRvC..32.1760L. doi:10.1103/PhysRevC.32.1760.

[21] Modern alchemy: Turning a line, The Economist

[22] http://fias.uni-frankfurt.de/kollo/Duellmann_FIAS-Kolloquium.pdf

[23] THEME03-5-1004-94/2009

[24] Oganessian, Yu. Ts.; Utyonkov, V.; Lobanov, Yu.; Abdullin, F.; Polyakov, A.; Sagaidak, R.; Shirokovsky, I.; Tsyganov, Yu.; Voinov, A.; et al. (2009). "Attempt to produce element 120 in the ^{244}Pu+^{58}Fe reaction". *Phys. Rev. C* **79** (2): 024603. Bibcode:2009PhRvC..79b4603O. doi:10.1103/PhysRevC.79.024603.

[25] Natowitz, Joseph (2008). "How stable are the heaviest nuclei?". *Physics* **1**: 12. Bibcode:2008PhyOJ...1...12N. doi:10.1103/Physics.

[26] Morjean, M.; et al. (2008). "Fission Time Measurements: A New Probe into Superheavy Element Stability". *Phys. Rev. Lett.* **101** (7): 072701. Bibcode:2008PhRvL.101g2701M. doi:10.1103/PhysRevLett.101.072701. PMID 18764526.

[27] see slide 11 in Future Plan of the Experimental Program on Synthesizing the Heaviest Element at RIKEN

[28] Emsley, John (2011). *Nature's Building Blocks: An A-Z Guide to the Elements* (New ed.). New York, NY: Oxford University Press. p. 588. ISBN 978-0-19-960563-7.

[29] see Flerov lab annual reports 2000–2004 inclusive http://www1.jinr.ru/Reports/Reports_eng_arh.html

[30] Marinov, A.; Rodushkin, I.; Kolb, D.; Pape, A.; Kashiv, Y.; Brandt, R.; Gentry, R. V.; Miller, H. W. (2008). "Evidence for a long-lived superheavy nucleus with atomic mass number A=292 and atomic number Z=~122 in natural Th". *International Journal of Modern Physics E* **19**: 131. arXiv:0804.3869. Bibcode:2010IJMPE..19..131M. doi:10.1142/S0218301310014662.

[31] Royal Society of Chemistry, "Heaviest element claim criticised", Chemical World.

[32] Marinov, A.; Rodushkin, I.; Kashiv, Y.; Halicz, L.; Segal, I.; Pape, A.; Gentry, R. V.; Miller, H. W.; Kolb, D.; Brandt, R. (2007). "Existence of long-lived isomeric states in naturally-occurring neutron-deficient Th isotopes". *Phys. Rev. C* **76** (2): 021303(R). arXiv:nucl-ex/0605008. Bibcode:2007PhRvC..76b1303M. doi:10.1103/PhysRevC.76.021303.

[33] R. C. Barber; J. R. De Laeter (2009). "Comment on "Existence of long-lived isomeric states in naturally-occurring neutron-deficient Th isotopes"". *Phys. Rev. C* **79** (4): 049801. Bibcode:2009PhRvC..79d9801B. doi:10.1103/PhysRevC.79.049801.

[34] A. Marinov; I. Rodushkin; Y. Kashiv; L. Halicz; I. Segal; A. Pape; R. V. Gentry; H. W. Miller; D. Kolb; R. Brandt (2009). "Reply to "Comment on 'Existence of long-lived isomeric states in naturally-occurring neutron-deficient Th isotopes'"". *Phys. Rev. C* **79** (4): 049802. Bibcode:2009PhRvC..79d9802M. doi:10.1103/PhysRevC.79.049802.

[35] J. Lachner; I. Dillmann; T. Faestermann; G. Korschinek; G. Poutivtsev; G. Rugel (2008). "Search for long-lived isomeric states in neutron-deficient thorium isotopes". *Phys. Rev. C* **78** (6): 064313. arXiv:0907.0126. Bibcode:2008PhRvC..78f4313L. doi:10.1103/PhysRevC.78.064313.

[36] Marinov, A.; Rodushkin, I.; Pape, A.; Kashiv, Y.; Kolb, D.; Brandt, R.; Gentry, R. V.; Miller, H. W.; Halicz, L.; Segal, I. (2009). "Existence of Long-Lived Isotopes of a Superheavy Element in Natural Au" (PDF). *International Journal of Modern Physics E* (World Scientific Publishing Company) **18** (3): 621–629. arXiv:nucl-ex/0702051. Bibcode:2009IJMPE..18..621M. doi:10.1142/S021830130901280X. Retrieved February 12, 2012.

[37] Cwiok, S.; Heenen, P.-H.; Nazarewicz, W. (2005). "Shape coexistence and triaxiality in the superheavy nuclei". *Nature* **433** (7027): 705–9. Bibcode:2005Natur.433..705C. doi:10.1038/nature03336. PMID 15716943.

[38] Elert, G. "Atomic Models". *The Physics Hypertextbook*. Retrieved 2009-10-09.

[39] Eisberg, R.; Resnick, R. (1985). *Quantum Physics of Atoms, Molecules, Solids, Nuclei and Particles*. Wiley.

[40] Bjorken, J. D.; Drell, S. D. (1964). *Relativistic Quantum Mechanics*. McGraw-Hill.

[41] Greiner, W.; Schramm, S. (2008). "American Journal of Physics" **76**. p. 509., and references therein

32.8　Further reading

- Kaldor, U. (2005). "Superheavy Elements—Chemistry and Spectroscopy". *Encyclopedia of Computational Chemistry*. doi:10.1002/0470845015.cu0044. ISBN 0470845015.

- Seaborg, G. T. (1968). "Elements Beyond 100, Present Status and Future Prospects". *Annual Review of Nuclear Science* **18**: 53–15. Bibcode:1968ARNPS..18...53S. doi:10.1146/annurev.ns.18.120168.000413.

- Scerri, Eric. (2011). *A Very Short Introduction to the Periodic Table, Oxford University Press, Oxford*. ISBN 978-0-19-958249-5.

32.9　External links

- Holler, Jim. "Images of g-orbitals". University of Kentucky.

- Rihani, Jeries A. "The extended periodic table of the elements".

- Scerri, Eric. "Eric Scerri's website for the elements and the periodic table".

Chapter 33

Rare earth element

Rare earth ore *(shown with a penny coin for size comparison)*

These rare-earth oxides are used as tracers to determine which parts of a drainage basin are eroding.[1]
Legend:
gadolinium · praseodymium · cerium
samarium · lanthanum
neodymium

A **rare earth element** (**REE**) or **rare earth metal** (**REM**), as defined by IUPAC, is one of a set of seventeen chemical elements in the periodic table, specifically the fifteen lanthanides, as well as scandium and yttrium.[2] Scandium and yttrium are considered rare earth elements because they tend to occur in the same ore deposits as the lanthanides and exhibit similar chemical properties.

Despite their name, rare earth elements are – with the exception of the radioactive promethium – relatively plentiful in Earth's crust, with cerium being the 25th most abundant element at 68 parts per million, or as abundant as copper. However, because of their geochemical properties, rare earth elements are typically dispersed and not often found concentrated as rare earth minerals in economically exploitable ore deposits.[3] It was the very scarcity of these minerals (previously called "earths") that led to the term "rare earth". The first such mineral discovered was gadolinite, a mineral composed of cerium, yttrium, iron, silicon and other elements. This mineral was extracted from a mine in the village of

Ytterby in Sweden; four of the rare earth elements bear names derived from this single location.

33.1 List

A table listing the seventeen rare earth elements, their atomic number and symbol, the etymology of their names, and their main usages (see also Applications of lanthanides) is provided here. Some of the rare earth elements are named after the scientists who discovered or elucidated their elemental properties, and some after their geographical discovery.

A mnemonic for the names of the sixth-row elements in order is "Lately college parties never produce sexy European girls that drink heavily even though you look".[6]

33.2 Abbreviations

The following abbreviations are often used:

- RE = rare earth

- REM = rare-earth metals

- REE = rare-earth elements

- REO = rare-earth oxides

- REY = rare-earth elements and yttrium

- LREE = light rare earth elements (Sc, La, Ce, Pr, Nd, Pm, Sm, Eu, and Gd; also known as the cerium group)[7][8]

- HREE = heavy rare earth elements (Y, Tb, Dy, Ho, Er, Tm, Yb, and Lu; also known as the yttrium group)[7][8]

The densities of the LREEs (as pure elements) range from 2.989 (scandium) to 7.9 g/cc (gadolinium), whereas those of the HREEs are from 8.2 to 9.8, except for yttrium (4.47) and ytterbium (between 6.9 and 7). The distinction between the groups is more to do with atomic volume and geological behavior (see lower down).

33.3 Discovery and early history

Rare earth elements became known to the world with the discovery of the black mineral "Ytterbite" (renamed to Gadolinite in 1800) by Lieutenant Carl Axel Arrhenius in 1787, at a quarry in the village of Ytterby, Sweden.[9]

Arrhenius's "ytterbite" reached Johan Gadolin, a Royal Academy of Turku professor, and his analysis yielded an unknown oxide (earth) that he called yttria. Anders Gustav Ekeberg isolated beryllium from the gadolinite but failed to recognize other elements that the ore contained. After this discovery in 1794 a mineral from Bastnäs near Riddarhyttan, Sweden, which was believed to be an iron–tungsten mineral, was re-examined by Jöns Jacob Berzelius and Wilhelm Hisinger. In 1803 they obtained a white oxide and called it ceria. Martin Heinrich Klaproth independently discovered the same oxide and called it *ochroia*.

Thus by 1803 there were two known rare earth elements, *yttrium* and *cerium*, although it took another 30 years for researchers to determine that other elements were contained in the two ores ceria and yttria (the similarity of the rare earth metals' chemical properties made their separation difficult).

In 1839 Carl Gustav Mosander, an assistant of Berzelius, separated ceria by heating the nitrate and dissolving the product in nitric acid. He called the oxide of the soluble salt *lanthana*. It took him three more years to separate the lanthana further into *didymia* and pure lanthana. Didymia, although not further separable by Mosander's techniques, was a mixture of oxides.

In 1842 Mosander also separated the yttria into three oxides: pure yttria, terbia and erbia (all the names are derived from the town name "Ytterby"). The earth giving pink salts he called *terbium*; the one that yielded yellow peroxide he called *erbium*.

So in 1842 the number of known rare earth elements had reached six: *yttrium, cerium, lanthanum, didymium, erbium* and *terbium*.

Nils Johan Berlin and Marc Delafontaine tried also to separate the crude yttria and found the same substances that Mosander obtained, but Berlin named (1860) the substance giving pink salts *erbium* and Delafontaine named the substance with the yellow peroxide *terbium*. This confusion led to several false claims of new elements, such as the *mosandrium* of J. Lawrence Smith, or the *philippium* and *decipium* of Delafontaine.

33.3.1 Spectroscopy

There were no further discoveries for 30 years, and the element didymium was listed in the periodic table of elements with a molecular mass of 138. In 1879 Delafontaine used the new physical process of optical-flame spectroscopy, and he found several new spectral lines in didymia. Also in 1879, the new element *samarium* was isolated by Paul Émile Lecoq de Boisbaudran from the mineral samarskite.

The samaria earth was further separated by Lecoq de Boisbaudran in 1886 and a similar result was obtained by Jean Charles Galissard de Marignac by direct isolation from samarskite. They named the element *gadolinium* after Johan Gadolin, and its oxide was named "gadolinia".

Further spectroscopic analysis between 1886 and 1901 of samaria, yttria, and samarskite by William Crookes, Lecoq de Boisbaudran and Eugène-Anatole Demarçay yielded several new spectroscopic lines that indicated the existence of an unknown element. The fractional crystallization of the oxides then yielded *europium* in 1901.

In 1839 the third source for rare earths became available. This is a mineral similar to gadolinite, *uranotantalum* (now called "samarskite"). This mineral from Miass in the southern Ural Mountains was documented by Gustave Rose. The Russian chemist R. Harmann proposed that a new element he called "ilmenium" should be present in this mineral, but later, Christian Wilhelm Blomstrand, Galissard de Marignac, and Heinrich Rose found only tantalum and niobium (columbium) in it.

The exact number of rare earth elements that existed was highly unclear, and a maximum number of 25 was estimated. The use of X-ray spectra (obtained by X-ray crystallography) by Henry Gwyn Jeffreys Moseley made it possible to assign atomic numbers to the elements. Moseley found that the exact number of lanthanides had to be 15 and that element 61 had yet to be discovered.

Using these facts about atomic numbers from X-ray crystallography, Moseley also showed that *hafnium* (element 72) would not be a rare earth element. Moseley was killed in World War I in 1915, years before hafnium was discovered. Hence, the claim of Georges Urbain that he had discovered element 72 was untrue. Hafnium is an element that lies in the periodic table immediately below zirconium, and hafnium and zirconium are very similar in their chemical and physical properties.

During the 1940s, Frank Spedding and others in the United States (during the Manhattan Project) developed the chemical ion exchange procedures for separating and purifying the rare earth elements. This method was first applied to the actinides for separating plutonium−239 and neptunium, from uranium, thorium, actinium, and the other actinide rare earths in the materials produced in nuclear reactors. The plutonium-239 was very desirable because it is a fissile material.

The principal sources of rare earth elements are the minerals bastnäsite, monazite, and loparite and the lateritic ion-adsorption clays. Despite their high relative abundance, rare earth minerals are more difficult to mine and extract than equivalent sources of transition metals (due in part to their similar chemical properties), making the rare earth elements relatively expensive. Their industrial use was very limited until efficient separation techniques were developed, such as ion exchange, fractional crystallization and liquid-liquid extraction during the late 1950s and early 1960s.[10]

33.3.2   Early classification

Before the time that ion exchange methods and elution were available, the separation of the rare earths was primarily achieved by repeated precipitation or crystallisation. In those days, the first separation was into two main groups, the cerium group earths (scandium, lanthanum, cerium, praseodymium, neodymium, and samarium) and the yttrium group earths (yttrium, dysprosium, holmium, erbium, thulium, ytterbium, and lutetium). Europium, gadolinium, and terbium were either considered as a separate group of rare earth elements (the terbium group), or europium was included in the cerium group, and gadolinium and terbium were included in the yttrium group. The reason for this division arose from the difference in solubility of rare earth double sulfates with sodium and potassium. The sodium double sulfates of the cerium group are difficultly soluble, those of the terbium group slightly, and those of the yttrium group are very soluble.[11]

33.4   Origin

Rare earth elements, except scandium, are heavier than iron and thus are produced by supernova nucleosynthesis or the s-process in asymptotic giant branch stars. In nature, spontaneous fission of uranium-238 produces trace amounts of radioactive promethium, but most promethium is synthetically produced in nuclear reactors.

Due to their chemical similarity, the concentrations of rare earths in rocks are only slowly changed by geochemical processes, making their proportions useful for geochronology and dating fossils.

33.5   Geological distribution

Rare earth cerium is actually the 25th most abundant element in Earth's crust, having 68 parts per million (about as common as copper). Only the highly unstable and radioactive promethium "rare earth" is quite scarce.

The rare earth elements are often found together. The longest-lived isotope of promethium has a half life of 17.7 years, so the element exists in nature in only negligible amounts (approximately 572 g in the entire Earth's crust).[12] Promethium is one of the two elements that do not have stable (non-radioactive) isotopes and are followed by (i.e. with higher atomic number) stable elements (the other being technetium).

Due to lanthanide contraction, yttrium, which is trivalent, is of similar ionic size as dysprosium and its lanthanide neighbors. Due to the relatively gradual decrease in ionic size with increasing atomic number, the rare earth elements have always been difficult to separate. Even with eons of geological time, geochemical separation of the lanthanides has only rarely progressed much farther than a broad separation between light versus heavy lanthanides, otherwise known as the cerium and yttrium earths. This geochemical divide is reflected in the first two rare earths that were discovered, yttria in 1794 and ceria in 1803. As originally found, each comprised the entire mixture of the associated earths. Rare earth minerals, as found, usually are dominated by one group or the other, depending on which size range best fits the structural lattice.

Thus, among the anhydrous rare earth phosphates, it is the tetragonal mineral xenotime that incorporates yttrium and the yttrium earths, whereas the monoclinic monazite phase incorporates cerium and the cerium earths preferentially. The smaller size of the yttrium group allows it a greater solid solubility in the rock-forming minerals that comprise Earth's mantle, and thus yttrium and the yttrium earths show less enrichment in Earth's crust relative to chondritic abundance, than does cerium and the cerium earths. This has economic consequences: large ore bodies of the cerium earths are known around the world, and are being exploited. Corresponding orebodies for yttrium tend to be rarer, smaller, and less concentrated. Most of the current supply of yttrium originates in the "ion absorption clay" ores of Southern China. Some versions provide concentrates containing about 65% yttrium oxide, with the heavy lanthanides being present in ratios reflecting the Oddo-Harkins rule: even-numbered heavy lanthanides at abundances of about 5% each, and odd-numbered lanthanides at abundances of about 1% each. Similar compositions are found in xenotime or gadolinite.

Well-known minerals containing yttrium include gadolinite, xenotime, samarskite, euxenite, fergusonite, yttrotantalite, yttrotungstite, yttrofluorite (a variety of fluorite), thalenite, yttrialite. Small amounts occur in zircon, which derives its typical yellow fluorescence from some of the accompanying heavy lanthanides. The zirconium mineral eudialyte, such as is found in southern Greenland, contains small but potentially useful amounts of yttrium. Of the above yttrium minerals,

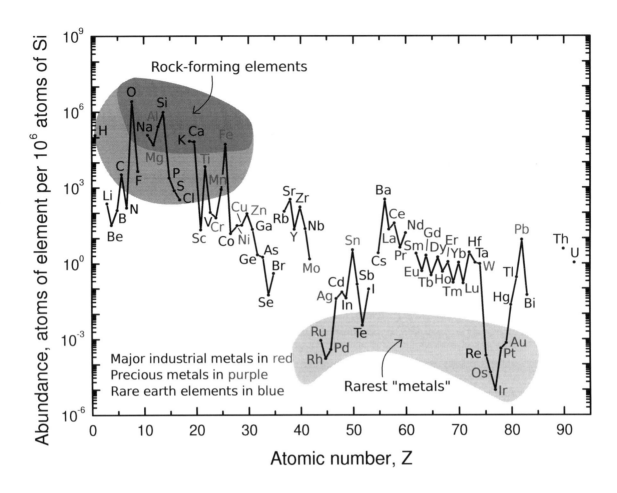

Abundance of elements in Earth's crust per million of Si atoms

most played a part in providing research quantities of lanthanides during the discovery days. Xenotime is occasionally recovered as a byproduct of heavy sand processing, but is not as abundant as the similarly recovered monazite (which typically contains a few percent of yttrium). Uranium ores from Ontario have occasionally yielded yttrium as a byproduct.

Well-known minerals containing cerium and the light lanthanides include bastnäsite, monazite, allanite, loparite, ancylite, parisite, lanthanite, chevkinite, cerite, stillwellite, britholite, fluocerite, and cerianite. Monazite (marine sands from Brazil, India, or Australia; rock from South Africa), bastnäsite (from Mountain Pass, California, or several localities in China), and loparite (Kola Peninsula, Russia) have been the principal ores of cerium and the light lanthanides.

In 2011, Yasuhiro Kato, a geologist at the University of Tokyo who led a study of Pacific Ocean seabed mud, published results indicating the mud could hold rich concentrations of rare earth minerals. The deposits, studied at 78 sites, came from "[h]ot plumes from hydrothermal vents pull[ing] these materials out of seawater and deposit[ing] them on the seafloor, bit by bit, over tens of millions of years. One square patch of metal-rich mud 2.3 kilometers wide might contain enough rare earths to meet most of the global demand for a year, Japanese geologists report July 3 in *Nature Geoscience*." "I believe that rare earth resources undersea are much more promising than on-land resources," said Kato. "[C]oncentrations of rare earths were comparable to those found in clays mined in China. Some deposits contained twice as much heavy rare earths such as dysprosium, a component of magnets in hybrid car motors."[13]

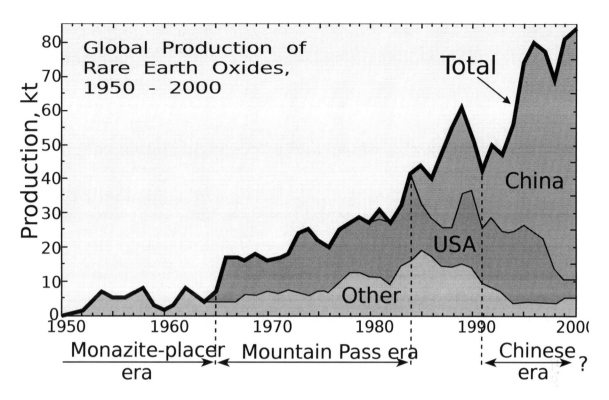

Global production 1950–2000

33.6   Global rare earth production

Until 1948, most of the world's rare earths were sourced from placer sand deposits in India and Brazil.[14] Through the 1950s, South Africa took the status as the world's rare earth source, after large veins of rare earth bearing monazite were discovered there.[14] Through the 1960s until the 1980s, the Mountain Pass rare earth mine in California was the leading producer. Today, the Indian and South African deposits still produce some rare earth concentrates, but they are dwarfed by the scale of Chinese production. In 2010, China produced over 95% of the world's rare earth supply, mostly in Inner Mongolia,[3][15] although it had only 37% of proven reserves;[16] the latter number has been reported to be only 23% in 2012.[17] All of the world's heavy rare earths (such as dysprosium) come from Chinese rare earth sources such as the polymetallic Bayan Obo deposit.[15][18] In 2010, the United States Geological Survey (USGS) released a study that found that the United States had 13 million metric tons of rare earth elements.[19]

New demand has recently strained supply, and there is growing concern that the world may soon face a shortage of the rare earths.[20] In several years from 2009 worldwide demand for rare earth elements is expected to exceed supply by 40,000 tonnes annually unless major new sources are developed.[21]

33.6.1   China

See also: Rare Earths Trade Dispute

These concerns have intensified due to the actions of China, the predominant supplier.[22] Specifically, China has announced regulations on exports and a crackdown on smuggling.[23] On September 1, 2009, China announced plans to reduce its export quota to 35,000 tons per year in 2010–2015 to conserve scarce resources and protect the environment.[24] On October 19, 2010, *China Daily*, citing an unnamed Ministry of Commerce official, reported that China will "further reduce quotas for rare earth exports by 30 percent at most next year to protect the precious metals from over-exploitation".[25] The government in Beijing further increased its control by forcing smaller, independent miners to merge into state-owned corporations or face closure. At the end of 2010, China announced that the first round of export quotas in 2011 for rare earths would be 14,446 tons, which was a 35% decrease from the previous first round of quotas in 2010.[26] China announced further export quotas on 14 July 2011 for the second half of the year with total allocation at 30,184 tons with total production capped at 93,800 tonnes.[27] In September 2011, China announced the halt in production of three of its eight major rare earth mines, responsible for almost 40% of China's total rare earth production.[28] In March 2012, the US, EU, and Japan confronted China at WTO about these export and production restrictions. China responded with claims that the restrictions had environmental protection in mind.[29] In August 2012, China announced a further 20% reduction in production.[30] These restrictions have damaged industries in other countries and forced producers of rare earth products to relocate their operations to China.[29] The Chinese restrictions on supply failed in 2012 as prices dropped in response to the opening of other sources.[31] The price of dysprosium oxide was $994/kg in 2011, but dropped to $265/kg by 2014.[32]

On August 29, 2014, the WTO ruled that China had broken free trade agreements, and the WTO said in the summary of key findings that

> the Panel concluded that the overall effect of the foreign and domestic restrictions is to encourage domestic extraction and secure preferential use of those materials by Chinese manufacturers.[33]

China declared it would implement the ruling on September 26, 2014, but would need some time to do so. By January 5, 2015, China had lifted all quotas from the export of rare earths, however export licences will still be required.[32]

33.6.2 Outside of China

As a result of the increased demand and tightening restrictions on exports of the metals from China, some countries are stockpiling rare earth resources.[34] Searches for alternative sources in Australia, Brazil, Canada, South Africa, Tanzania, Greenland, and the United States are ongoing.[35] Mines in these countries were closed when China undercut world prices in the 1990s, and it will take a few years to restart production as there are many barriers to entry.[23] One example is the Mountain Pass mine in California, which announced its resumption of operations on a start-up basis on August 27, 2012.[15][36] Other significant sites under development outside of China include the Nolans Project in Central Australia, the remote Hoidas Lake project in northern Canada,[37] and the Mount Weld project in Australia.[15][36][38] The Hoidas Lake project has the potential to supply about 10% of the $1 billion of REE consumption that occurs in North America every year.[39] Vietnam signed an agreement in October 2010 to supply Japan with rare earths[40] from its northwestern Lai Châu Province.[41]

Also under consideration for mining are sites such as Thor Lake in the Northwest Territories, various locations in Vietnam,[15][21][42] and a site in southeast Nebraska in the US, where Quantum Rare Earth Development, a Canadian company, is currently conducting test drilling and economic feasibility studies toward opening a niobium mine.[43] Additionally, a large deposit of rare earth minerals was recently discovered in Kvanefjeld in southern Greenland.[44] Pre-feasibility drilling at this site has confirmed significant quantities of black lujavrite, which contains about 1% rare-earth oxides (REO).[45] The European Union has urged Greenland to restrict Chinese development of rare-earth projects there, but as of early 2013, the government of Greenland has said that it has no plans to impose such restrictions.[46] Many Danish politicians have expressed concerns that other nations, including China, could gain influence in thinly populated Greenland, given the number of foreign workers and investment that could come from Chinese companies in the near future because of the law passed December 2012.[47]

Adding to potential mine sites, ASX listed Peak Resources announced in February 2012, that their Tanzanian-based Ngualla project contained not only the 6th largest deposit by tonnage outside of China, but also the highest grade of rare earth elements of the 6.[48]

North Korea has been reported to have sold rare earth metals to China. During May and June 2014, North Korea sold over US$1.88 million worth of rare earth metals to China.[49] Other sources suggest that North Korea has the world's second largest reserve of rare earth metals, with potentially over 20 million tons in total.[50]

33.6.3   Other sources

Significant quantities of rare earth oxides are found in tailings accumulated from 50 years of uranium ore, shale and loparite mining at Sillamäe, Estonia.[51] Due to the rising prices of rare earths, extraction of these oxides has become economically viable. The country currently exports around 3,000 tonnes per year, representing around 2% of world production.[52] Similar resources are suspected in the western United States, where gold rush-era mines are believed to have discarded large amounts of rare earths, because they had no value at the time.[53]

Nuclear reprocessing is another potential source of rare earth or any other elements. Nuclear fission of uranium or plutonium produces a full range of elements, including all their isotopes. However, due to the radioactivity of many of these isotopes, it is unlikely that extracting them from the mixture can be done safely and economically.

In May 2012, researchers from two prevalent universities in Japan announced that they had discovered rare earths in Ehime Prefecture, Japan.[54][55] In 2012, Japanese scientists discovered about 6.8 million tons of rare earth elements near the island of Minami-Tori-Shima, enough to supply Japan's current consumption for over 200 years. Around 90% of the world's production of REE comes from China, and Japan imports 60% of that.[56]

Recycling

Another recently developed source of rare earths is electronic waste and other wastes that have significant rare earth components. New advances in recycling technology have made extraction of rare earths from these materials more feasible,[57] and recycling plants are currently operating in Japan, where there is an estimated 300,000 tons of rare earths stored in unused electronics.[58] In France, the Rhodia group is setting up two factories, in La Rochelle and Saint-Fons, that will produce 200 tons of rare earths a year from used fluorescent lamps, magnets and batteries.[59][60]

33.6.4   Malaysian refining plans

In early 2011, Australian mining company, Lynas, was reported to be "hurrying to finish" a US$230 million rare earth refinery on the eastern coast of Peninsular Malaysia's industrial port of Kuantan. The plant would refine ore— lanthanides concentrate from the Mount Weld mine in Australia. The ore would be trucked to Fremantle and transported by container ship to Kuantan. However, the Malaysian authorities confirmed that as of October 2011, Lynas was not given any permit to import any rare earth ore into Malaysia. On February 2, 2012, the Malaysian AELB (Atomic Energy Licensing Board) recommended that Lynas be issued a Temporary Operating License (TOL) subject to completion of a number of conditions. On April 3, 2012, Lynas announced to the Malaysian media that these conditions had been met, and was now waiting on the issuance of the licence. Within two years, Lynas was said to expect the refinery to be able to meet nearly a third of the world's demand for rare earth materials, not counting China."[61] The Kuantan development brought renewed attention to the Malaysian town of Bukit Merah in Perak, where a rare-earth mine operated by a Mitsubishi Chemical subsidiary, Asian Rare Earth, closed in 1992 and left continuing environmental and health concerns.[62] In mid-2011, after protests, Malaysian government restrictions on the Lynas plant were announced. At that time, citing subscription-only Dow Jones Newswire reports, a *Barrons* report said the Lynas investment was $730 million, and the projected share of the global market it would fill put at "about a sixth."[63] An independent review was initiated by Malaysian Government and United Nations and conducted by the International Atomic Energy Agency (IAEA) between 29 May and 3 June 2011 to address concerns of radioactive hazards. The IAEA team was not able to identify any non-compliance with international radiation safety standards.[64]

On 2 September 2014, Lynas was issued a 2-year Full Operating Stage License (FOSL) by the Malaysian Atomic Energy Licensing Board (AELB).[65]

33.6.5   Environmental considerations

Mining, refining, and recycling of rare earths have serious environmental consequences if not properly managed. A particular hazard is mildly radioactive slurry tailings resulting from the common occurrence of thorium and uranium in rare earth element ores.[66] Additionally, toxic acids are required during the refining process.[16] Improper handling of these substances can result in extensive environmental damage. In May 2010, China announced a major, five-month crackdown

Satellite image of the Bayan Obo Mining District, 2006.

on illegal mining in order to protect the environment and its resources. This campaign is expected to be concentrated in the South,[67] where mines – commonly small, rural, and illegal operations – are particularly prone to releasing toxic wastes into the general water supply.[15][68] However, even the major operation in Baotou, in Inner Mongolia, where much of the world's rare earth supply is refined, has caused major environmental damage.[16]

Residents blamed a rare earth refinery at Bukit Merah for birth defects and eight leukemia cases within five years in a community of 11,000 — after many years with no leukemia cases. Seven of the leukemia victims died. Osamu Shimizu, a director of Asian Rare Earth, said, "the company might have sold a few bags of calcium phosphate fertilizer on a trial basis as it sought to market byproducts; calcium phosphate is not radioactive or dangerous," in reply to a former resident of Bukit Merah who said, "The cows that ate the grass [grown with the fertilizer] all died."[69] Malaysia's Supreme Court ruled on 23 December 1993 that there was no evidence that the local chemical joint venture Asian Rare Earth was contaminating the local environment.[70]

The Bukit Merah mine in Malaysia has been the focus of a US$100 million cleanup that is proceeding in 2011. After having accomplished the hilltop entombment of 11,000 truckloads of radioactively contaminated material, the project is expected to entail in summer, 2011, the removal of "more than 80,000 steel barrels of radioactive waste to the hilltop repository."[62]

In May 2011, after the Fukushima Daiichi nuclear disaster, widespread protests took place in Kuantan over the Lynas refinery and radioactive waste from it. The ore to be processed has very low levels of thorium, and Lynas founder and chief executive Nicholas Curtis said "There is absolutely no risk to public health." T. Jayabalan, a doctor who says he has been monitoring and treating patients affected by the Mitsubishi plant, "is wary of Lynas's assurances. The argument that low levels of thorium in the ore make it safer doesn't make sense, he says, because radiation exposure is cumulative."[69] Construction of the facility has been halted until an independent United Nations IAEA panel investigation is completed, which is expected by the end of June 2011.[71] New restrictions were announced by the Malaysian government in late June.[63]

IAEA panel investigation is completed and no construction has been halted. Lynas is on budget and on schedule to start producing 2011. The IAEA report has concluded in a report issued on Thursday June 2011 said it did not find any instance of "any non-compliance with international radiation safety standards" in the project.[72]

33.6.6   Geo-political considerations

China has officially cited resource depletion and environmental concerns as the reasons for a nationwide crackdown on its rare earth mineral production sector.[28] However, non-environmental motives have also been imputed to China's rare earth policy.[16] According to *The Economist*, "Slashing their exports of rare-earth metals...is all about moving Chinese manufacturers up the supply chain, so they can sell valuable finished goods to the world rather than lowly raw materials."[73] One possible example is the division of General Motors that deals with miniaturized magnet research, which shut down its US office and moved its entire staff to China in 2006[74] (it should be noted that China's export quota only applies to the metal but not products made from these metals such as magnets).

It was reported,[75] but officially denied,[76] that China instituted an export ban on shipments of rare earth oxides (but not alloys) to Japan on 22 September 2010, in response to the detainment of a Chinese fishing boat captain by the Japanese Coast Guard.[77] On September 2, 2010, a few days before the fishing boat incident, *The Economist* reported that "China...in July announced the latest in a series of annual export reductions, this time by 40% to precisely 30,258 tonnes."[78]

The United States Department of Energy in its 2010 Critical Materials Strategy report identified dysprosium as the element that was most critical in terms of import reliance.[79]

A 2011 report issued by the US Geological Survey and US Department of the Interior, "China's Rare-Earth Industry," outlines industry trends within China and examines national policies that may guide the future of the country's production. The report notes that China's lead in the production of rare-earth minerals has accelerated over the past two decades. In 1990, China accounted for only 27% of such minerals. In 2009, world production was 132,000 metric tons; China produced 129,000 of those tons. According to the report, recent patterns suggest that China will slow the export of such materials to the world: "Owing to the increase in domestic demand, the Government has gradually reduced the export quota during the past several years." In 2006, China allowed 47 domestic rare-earth producers and traders and 12 Sino-foreign rare-earth producers to export. Controls have since tightened annually; by 2011, only 22 domestic rare-earth producers and traders and 9 Sino-foreign rare-earth producers were authorized. The government's future policies will likely keep in place strict controls: "According to China's draft rare-earth development plan, annual rare-earth production may be limited to between 130,000 and 140,000 [metric tons] during the period from 2009 to 2015. The export quota for rare-earth products may be about 35,000 [metric tons] and the Government may allow 20 domestic rare-earth producers and traders to export rare earths."[80]

The United States Geological Survey is actively surveying southern Afghanistan for rare earth deposits under the protection of United States military forces. Since 2009 the USGS has conducted remote sensing surveys as well as fieldwork to verify Soviet claims that volcanic rocks containing rare earth metals exist in Helmand province near the village of Khanneshin. The USGS study team has located a sizable area of rocks in the center of an extinct volcano containing light rare earth elements including cerium and neodymium. It has mapped 1.3 million metric tons of desirable rock, or about 10 years of supply at current demand levels. The Pentagon has estimated its value at about $7.4 billion.[81]

33.6.7   Rare earth pricing

Rare earth elements are not exchange-traded in the same way that precious (for instance, gold and silver) or non-ferrous metals (such as nickel, tin, copper, and aluminium) are. Instead they are sold on the private market, which makes their prices difficult to monitor and track. The 17 elements are not usually sold in their pure form, but instead are distributed in mixtures of varying purity, e.g. "Neodymium metal ≥ 99%". As such, pricing can vary based on the quantity and quality required by the end user's application.

33.7   See also

- Group 3 element

- KREEP

- Rare earth mineral

33.8 References

[1] "News and events". US Department of Agriculture. Retrieved 2012-03-13.

[2] Edited by N G Connelly and T Damhus (with R M Hartshorn and A T Hutton), ed. (2005). *Nomenclature of Inorganic Chemistry: IUPAC Recommendations 2005* (PDF). Cambridge: RSC Publ. ISBN 0-85404-438-8. Archived from the original (PDF) on 2008-05-27. Retrieved 2012-03-13.

[3] "Haxel G, Hedrick J, Orris J. 2006. Rare earth elements critical resources for high technology. Reston (VA): United States Geological Survey. USGS Fact Sheet: 087-02." (PDF). Retrieved 2012-03-13.

[4] C. R. Hammond, "Section 4; The Elements", in *CRC Handbook of Chemistry and Physics, 89th Edition (Internet Version 2009), David R. Lide, ed., CRC Press/Taylor and Francis, Boca Raton, FL.*

[5] Energy-efficient light bulbs containing yttrium

[6] Mentioned by Prof. Andrea Sella on a BBC Business Daily programme, March 19, 2014 . Unfortunately this mnemonic doesn't distinguish very well between terbium and thulium.

[7] Gschneidner, Karl A., Jr. 1966. |title=Rare Earths-The Fraternal Fifteen. Washington, DC, US atomic Energy Commission, Divisions of Technical Information, 42 pages.

[8] Hedrick, James B. "REE Handbook -- The ultimate guide to Rare Earth Elements,". *Rare Metal Blog.* Toronto, Canada.

[9] Gschneidner KA, Cappellen, ed. (1987). "1787–1987 Two hundred Years of Rare Earths". *Rare Earth Information Center, IPRT, North-Holland.* IS-RIC 10.

[10] Spedding F, Daane AH: "The Rare Earths", John Wiley & Sons, Inc., 1961

[11] B. Smith Hopkins: "Chemistry of the rarer elements", D. C. Heath & Company, 1923

[12] P. Belli, R. Bernabei, F. Cappella, R. Cerulli, C.J. Dai, F.A. Danevich, A. d'Angelo, A. Incicchitti, V.V. Kobychev, S.S. Nagorny, S. Nisi, F. Nozzoli, D. Prosperi, V.I. Tretyak, S.S. Yurchenko (2007). "Search for α decay of natural Europium". *Nuclear Physics A* **789** (1–4): 15–29. Bibcode:2007NuPhA.789...15B. doi:10.1016/j.nuclphysa.2007.03.001.

[13] Powell, Devin, "Rare earth elements plentiful in ocean sediments", *ScienceNews*, July 3rd, 2011. Via Kurt Brouwer's Fundmastery Blog, *MarketWatch*, 2011-07-05.. Retrieved 2011-07-05.

[14] ER, Rose. Rare Earths of the Grenville Sub-Province Ontatio and Quebec. GSC Report Number 59-10. Ottawa: Geological Survey of Canada Department of Mines and Technical Surveys, 1960.

[15] China's Rare Earth Dominance, Wikinvest. Retrieved on 11 Aug 2010.

[16] Bradsher, Keith (October 29, 2010). "After China's Rare Earth Embargo, a New Calculus". *The New York Times.* Retrieved October 30, 2010.

[17] China Warns its Rare Earth Reserves are Declining BBC News June 20, 2012. Retrieved June 20, 2012

[18] Chao ECT, Back JM, Minkin J, Tatsumoto M, Junwen W, Conrad JE, McKee EH, Zonglin H, Qingrun M. "Sedimentary carbonate-hosted giant Bayan Obo REE-Fe-Nb ore deposit of Inner Mongolia, China; a cornerstone example for giant polymetallic ore deposits of hydrothermal origin." 1997. United States Geological Survey Publications Warehouse. 29 February 2008.

[19] USGS. Rare Earth Elements in US Not So Rare: Significant Deposits Found in 14 States. US Department of the Interior. Full Report: The Principal Rare Earth Elements Deposits of the United States—A Summary of Domestic Deposits and a Global Perspective.

[20] "Cox C. 2008. Rare earth innovation. Herndon (VA): The Anchor House Inc;". Retrieved 2008-04-19.

[21] "As hybrid cars gobble rare metals, shortage looms". Reuters. August 31, 2009. Retrieved Aug 31, 2009

[22] How Beijing Cornered the Rare Earths Market April 25, 2012 Foreign Affairs

[23] Livergood R. (2010). "Rare Earth Elements: A Wrench in the Supply Chain" (PDF). Center for Strategic and International Studies. Retrieved 2012-03-13.

[24] China To Limit Rare Earths Exports, Manufacturing.net, 1 September 2009. Retrieved 2010-08-30.

[25] China to cut exports of rare earth minerals vital to energy tech" thehill.com, 19 Oct. 2009. Retrieved 2010-10-19.

[26] China's Rare Earth Exports Surge in Value" thechinaperspective.com, January 19. 2011

[27] Zhang, Ding, Fu, Qi, Qingfen, Jing. "Rare earths export quota unchanged". ChinaDaily.com.cn. Retrieved 2011-07-15.

[28] *China halts rare earth production at three mines*, Reuters, 2011-09-06, retrieved 2011-09-07

[29] "WRAPUP 4-US, EU, Japan take on China at WTO over rare earths" - Reuters, March 13, 2012

[30] CNN *China cuts mines vital to tech industry*

[31] "El Reg man: Too bad, China - I was RIGHT about hoarding rare earths."

[32] "China scraps quotas on rare earths after WTO complaint". *China scraps quotas on rare earths after WTO complaint*. The Guardian. Jan 5, 2015. Retrieved Jan 5, 2015.

[33] "DISPUTE SETTLEMENT: DISPUTE DS431 China — Measures Related to the Exportation of Rare Earths, Tungsten and Molybdenum". *DISPUTE SETTLEMENT: DISPUTE DS431 China — Measures Related to the Exportation of Rare Earths, Tungsten and Molybdenum*. World Trade Organisation. Retrieved 5/1/14. Check date values in: |access-date= (help)

[34] *EU stockpiles rare earths as tensions with china rise*, Financial Post, retrieved 2011-09-07

[35] "Canadian Firms Step Up Search for Rare-Earth Metals". *NYTimes.com* (Reuters). 2009-09-09. Retrieved 2009-09-15.

[36] Leifert, H. Restarting US rare earth production?. Earth magazine. June 2010. Pgs 20–21.

[37] "Lunn J. 2006. Great western minerals. London: Insigner Beaufort Equity Research" (PDF). Retrieved 2008-04-19.

[38] Gorman, Steve (2009-08-31). "California mine digs in for 'green' gold rush.". *Reuters*. Retrieved 2010-03-22.

[39] "Hoidas Lake Project". Retrieved 2008-09-24.

[40] "Rare earths supply deal between Japan and Vietnam". BBC News. 31 October 2010.

[41] "Vietnam signs major nuclear pacts". AlJazeera. 31 October 2010. Retrieved 31 October 2010.

[42] "Federal minister approves N.W.T. rare earth mine". CBC News. 2013-11-04. It follows the recommendation from the Mackenzie Valley Environmental Review Board in July, and marks a major milestone in the company's effort to turn the project into an operating mine. Avalon claims Nechalacho is "the most advanced large heavy rare earth development project in the world."

[43] "High-tech buried treasure.". Retrieved 2010-05-05.

[44] Greenland "Rare Earth Elements at Kvanefjeld, Greenland", Retrieved on 2010-11-10.

[45] Greenland "New Multi-Element Targets and Overall Resource Potential", Retrieved on 2010-11-10.

[46] *Chinese Workers—in Greenland?* February 10, 2013 *BusinessWeek*

[47] *Greenland Votes to Get Tough on Investors; New Ruling Party Campaigned to Backtrack on Country's Recent Opening to Investment From Foreign Mining Ventures* March 13, 2013 WSJ

[48] Peak Resources – Maiden Resource, Ngualla Rare Earth Project, ASX Announcement, 29th February 2012

[49] [Korean only http://www.voakorea.com/content/north-korea-rare-earth/1966603.html]

[50] [Korean only http://www.ohmynews.com/NWS_Web/View/at_pg.aspx?CNTN_CD=A0001938381]

[51] Rofer, Cheryl K.; Tõnis Kaasik (2000). *Turning a Problem Into a Resource: Remediation and Waste Management at the Sillamäe Site, Estonia*. Volume 28 of NATO science series: Disarmament technologies. Springer. p. 229. ISBN 978-0-7923-6187-9.

[52] Anneli Reigas (2010-11-30). "Estonia's rare earth break China's market grip". *AFP*. Retrieved 2010-12-01.

[53] Cone, Tracie (July 21, 2013). "Gold Rush Trash is Information Age Treasure". *USA Today*. Retrieved July 21, 2013.

[54] "Japan Discovers Domestic Rare Earths Reserve". BrightWire.

[55] http://www.brightwire.com/news/search?utf8=%E2%9C%93&fuzzy_path=%2Ffuzzy_search%2Fentity&q=rare+earth+

[56] Westlake, Adam. "Scientists in Japan discover rare earths in Pacific Ocean east of Tokyo" *Japan Daily Press*, 29 June 2012. Retrieved: 29 June 2012.

[57] Recycling of neodymium and samarium

[58] Tabuchi, Hiroko. "Japan Recycles Minerals From Used Electronics". *New York Times*. October 5, 2010.

[59] Rhodia press release http://www.rhodia.com/en/news_center/news_releases/Recycle_rare_earths_031011.tcm

[60]

[61] Bradsher, Keith, "Taking a Risk for Rare Earths", *The New York Times*, March 8, 2011 (March 9, 2011 p. B1 NY ed.). Retrieved 2011-03-09.

[62] Bradsher, Keith, "Mitsubishi Quietly Cleans Up Its Former Refinery", *The New York Times*, March 8, 2011 (March 9, 2011 p. B4 NY ed.). Retrieved 2011-03-09.

[63] Coleman, Murray, "Rare Earth ETF Jumps As Plans To Break China's Hold Suffer Setback", *Barrons* blog, June 30, 2011 1:52 PM ET. Retrieved 2011=−6-30.

[64] Report of the International Review Mission on the Radiation Safety Aspects of a Proposed Rare Earths Processing Facility (Lynas Project). (PDF) . Retrieved on 2011-09-27.

[65] Ng, Eileen (2 September 2014). "Lynas gets full operating licence before TOL expiry date". The Malaysian Insider. Retrieved 3 September 2014.

[66] Bourzac, Katherine. "Can the US Rare-Earth Industry Rebound?" *Technology Review*. October 29, 2010.

[67] Govt cracks whip on rare earth mining. *China Daily,* May 21, 2010. Accessed June 3rd, 2010.

[68] Y, Lee. "South China Villagers Slam Pollution From Rare Earth Mine." 22 February 2008. RFA English Website. 16 March 2008

[69] Lee, Yoolim, "Malaysia Rare Earths in Largest Would-Be Refinery Incite Protest", *Bloomberg Markets Magazine*, May 31, 2011 5:00 PM ET.

[70] "Malaysia court rejects pollution suit against ARE", World Information Service on Energy, February 11, 1994.

[71] "UN investigation into Malaysia rare-earth plant safety", BBC, 30 May 2011 05:52 ET.

[72] IAEA Submits Lynas Report to Malaysian Government. Iaea.org (2011-06-29). Retrieved on 2011-09-27.

[73] "The Difference Engine: More precious than gold". *The Economist* September 17, 2010.

[74] C, Cox. "Rare earth innovation: the silent shift to China". 16 November 2006. The Anchor House: Research on Rare Earth Elements Accessed 29 February 2008

[75] Bradsher, Keith (2010-09-22). "Amid Tension, China Blocks Vital Exports to Japan". The New York Times Company. Retrieved 22 September 2010.

[76] James T. Areddy, David Fickling And Norihiko Shirouzu (2010-09-23). "China Denies Halting Rare-Earth Exports to Japan". Wall Street Journal. Retrieved 22 September 2010.

[77] Backlash over the alleged China curb on metal exports, *Daily Telegraph*, London, 29 Aug 2010. Retrieved 2010-08-30.

[78] "Rare earths: Digging in" *The Economist* September 2, 2010

[79] Mills, Mark P. "Tech's Mineral Infrastructure – Time to Emulate China's Rare Earth Policies." *Forbes*, 1 January 2010.

[80] "US Geological Survey: China's Rare-Earth Industry". Journalist's Resource.org.

[81] Simpson, S.: Afghanistan's Buried Riches, "Scientific American", October 2011

33.9 External links

- Tabuchi, Hiroko (5 October 2010). "Japan Recycles Rare Earth Minerals From Used Electronics". *The New York Times*.

- Kan, Michael (7 October 2010). "Common gadgets may be affected by shortage of rare earths". New Zealand PC World Magazine. Retrieved 6 October 2010.

- Auslin, Michael (13 October 2010). "Japan's Rare-Earth Jolt". Wall Street Journal. Retrieved 13 October 2010.

- Aston, Adam (15 October 2010). "China's Rare-Earth Monopoly". Technology Review (MIT). Retrieved 17 October 2010.

- Hurst, Cindy (March 2010). "China's Rare Earth Elements Industry: What Can the West Learn?" (PDF). Institute for the Analysis of Global Security (IAGS). Retrieved 18 October 2010.

- Rare earths mining: China's 21st Century gold rush, BBC News June 2010 infographic examining China's role in the rare earths market.

- Rare Earth Elements in National Defense: Background, Oversight Issues, and Options for Congress Congressional Research Service, March 31, 2011.

- Digging for rare earths: The mines where iPhones are born | Apple - CNET News, September 26, 2012

- Khan, Malek; Lundmark, Martin; Hellström, Jerker "Rare Earth Elements and Europe's Dependence on China" in *Strategic Outlook 2013*, Swedish Defence Research Agency (FOI), June 2013, pp. 93–98.

- Terra Rara: The strange story of some political elements Prof Andrea Sella, Royal Institution filmed event, 31 May 2013

- BBC feature

- Rare-earth Metals

- Moon exploration will reduce the shortage of rare earth metals

- The dystopian lake filled by the world's tech lust (April 2015), *BBC Future*

Chapter 34

Abundance of the chemical elements

The **abundance** of a chemical element measures how common is the element relative to all other elements in a given environment. Abundance is measured in one of three ways: by the mass-fraction (the same as weight fraction); by the mole-fraction (fraction of atoms by numerical count, or sometimes fraction of molecules in gases); or by the volume-fraction. Volume-fraction is a common abundance measure in mixed gases such as planetary atmospheres, and is similar in value to molecular mole-fraction for gas mixtures at relatively low densities and pressures, and ideal gas mixtures. Most abundance values in this article are given as mass-fractions.

For example, the abundance of oxygen in pure water can be measured in two ways: the *mass fraction* is about 89%, because that is the fraction of water's mass which is oxygen. However, the *mole-fraction* is 33.3333...% because only 1 atom of 3 in water, H_2O, is oxygen.

As another example, looking at the *mass-fraction* abundance of hydrogen and helium in both the Universe as a whole and in the atmospheres of gas-giant planets such as Jupiter, it is 74% for hydrogen and 23-25% for helium; while the *(atomic) mole-fraction* for hydrogen is 92%, and for helium is 8%, in these environments. Changing the given environment to Jupiter's outer atmosphere, where hydrogen is diatomic while helium is not, changes the *molecular* mole-fraction (fraction of total gas molecules), as well as the fraction of atmosphere by volume, of hydrogen to about 86%, and of helium to 13%.[Note 1]

34.1   Abundance of elements in the Universe

See also: Stellar population, Cosmochemistry and Astrochemistry

The elements – that is, ordinary (baryonic) matter made of protons, neutrons, and electrons, are only a small part of the content of the Universe. Cosmological observations suggest that only 4.6% of the universe's energy (including the mass contributed by energy, $E = mc^2 \leftrightarrow m = E / c^2$) comprises the visible baryonic matter that constitutes stars, planets, and living beings. The rest is made up of dark energy (72%) and dark matter (23%).[2] These are forms of matter and energy believed to exist on the basis of scientific theory and observational deductions, but they have not been directly observed and their nature is not well understood.

Most standard (baryonic) matter is found in stars and interstellar clouds, in the form of atoms or ions (plasma), although it can be found in degenerate forms in extreme astrophysical settings, such as the high densities inside white dwarfs and neutron stars.

Hydrogen is the most abundant element in the Universe; helium is second. However, after this, the rank of abundance does not continue to correspond to the atomic number; oxygen has abundance rank 3, but atomic number 8. All others are substantially less common.

The abundance of the lightest elements is well predicted by the standard cosmological model, since they were mostly produced shortly (i.e., within a few hundred seconds) after the Big Bang, in a process known as Big Bang nucleosynthesis.

Heavier elements were mostly produced much later, inside of stars.

Hydrogen and helium are estimated to make up roughly 74% and 24% of all baryonic matter in the universe respectively. Despite comprising only a very small fraction of the universe, the remaining "heavy elements" can greatly influence astronomical phenomena. Only about 2% (by mass) of the Milky Way galaxy's disk is composed of heavy elements.

These other elements are generated by stellar processes.[3][4][5] In astronomy, a "metal" is any element other than hydrogen or helium. This distinction is significant because hydrogen and helium are the only elements that were produced in significant quantities in the Big Bang. Thus, the metallicity of a galaxy or other object is an indication of stellar activity, after the Big Bang.

The following graph (note log scale) shows abundance of elements in our solar system. The table shows the twelve most common elements in our galaxy (estimated spectroscopically), as measured in parts per million, by mass.[1] Nearby galaxies that have evolved along similar lines have a corresponding enrichment of elements heavier than hydrogen and helium. The more distant galaxies are being viewed as they appeared in the past, so their abundances of elements appear closer to the primordial mixture. Since physical laws and processes are uniform throughout the universe, however, it is expected that these galaxies will likewise have evolved similar abundances of elements.

The abundance of elements in the Solar System (see graph) is in keeping with their origin from the Big Bang and nucleosynthesis in a number of progenitor supernova stars. Very abundant hydrogen and helium are products of the Big Bang, while the next three elements are rare since they had little time to form in the Big Bang and are not made in stars (they are, however, produced in small quantities by breakup of heavier elements in interstellar dust, as a result of impact by cosmic rays).

Beginning with carbon, elements have been produced in stars by buildup from alpha particles (helium nuclei), resulting in an alternatingly larger abundance of elements with even atomic numbers (these are also more stable). The effect of odd-numbered chemical elements generally being more rare in the universe was empirically noticed in 1914, and is known as the Oddo-Harkins rule.

Cosmogenesis: In general, such elements up to iron are made in large stars in the process of becoming supernovae. Iron-56 is particularly common, since it is the most stable element that can easily be made from alpha particles (being a product of decay of radioactive nickel-56, ultimately made from 14 helium nuclei). Elements heavier than iron are made in energy-absorbing processes in large stars, and their abundance in the universe (and on Earth) generally decreases with increasing atomic number.

34.1.1   Elemental abundance and nuclear binding energy

Loose correlations have been observed between estimated elemental abundances in the universe and the nuclear binding energy curve. Roughly speaking, the relative stability of various atomic isotopes has exerted a strong influence on the relative abundance of elements formed in the Big Bang, and during the development of the universe thereafter. [7] See the article about nucleosynthesis for the explanation on how certain nuclear fusion processes in stars (such as carbon burning, etc.) create the elements heavier than hydrogen and helium.

A further observed peculiarity is the jagged alternation between relative abundance and scarcity of adjacent atomic numbers in the elemental abundance curve, and a similar pattern of energy levels in the nuclear binding energy curve. This alternation is caused by the higher relative binding energy (corresponding to relative stability) of even atomic numbers compared to odd atomic numbers, and is explained by the Pauli Exclusion Principle.[8] The semi-empirical mass formula (SEMF), also called **Weizsäcker's formula** or the **Bethe-Weizsäcker mass formula**, gives a theoretical explanation of the overall shape of the curve of nuclear binding energy.[9]

34.2   Abundance of elements in the Earth

See also: Earth § Chemical composition

The Earth formed from the same cloud of matter that formed the Sun, but the planets acquired different compositions during the formation and evolution of the solar system. In turn, the natural history of the Earth caused parts of this planet to have differing concentrations of the elements.

The mass of the Earth is approximately 5.98×10^{24} kg. In bulk, by mass, it is composed mostly of iron (32.1%), oxygen (30.1%), silicon (15.1%), magnesium (13.9%), sulfur (2.9%), nickel (1.8%), calcium (1.5%), and aluminium (1.4%); with the remaining 1.2% consisting of trace amounts of other elements.[10]

The bulk composition of the Earth by elemental-mass is roughly similar to the gross composition of the solar system, with the major differences being that Earth is missing a great deal of the volatile elements hydrogen, helium, neon, and nitrogen, as well as carbon which has been lost as volatile hydrocarbons. The remaining elemental composition is roughly typical of the "rocky" inner planets, which formed in the thermal zone where solar heat drove volatile compounds into space. The Earth retains oxygen as the second-largest component of its mass (and largest atomic-fraction), mainly from this element being retained in silicate minerals which have a very high melting point and low vapor pressure.

34.2.1 Earth's detailed bulk (total) elemental abundance in table form

Click "show" at right, to show more numerical values in a full table. Note that these are ordered by atom-fraction abundance (right-most column), not mass-abundance.

An estimate[11] of the elemental abundances in the total mass of the Earth. Note that numbers are estimates, and they will vary depending on source and method of estimation. Order of magnitude of data can roughly be relied upon. ppb (atoms) is parts per billion, meaning that is the number of atoms of a given element in every billion atoms in the Earth.

34.2.2 Earth's crustal elemental abundance

Main article: Abundance of elements in Earth's crust
The mass-abundance of the nine most abundant elements in the Earth's crust (see main article above) is approximately: oxygen 46%, silicon 28%, aluminum 8.2%, iron 5.6%, calcium 4.2%, sodium 2.5%, magnesium 2.4%, potassium, 2.0%, and titanium 0.61%. Other elements occur at less than 0.15%.

The graph at left illustrates the relative atomic-abundance of the chemical elements in Earth's upper continental crust, which is relatively accessible for measurements and estimation. Many of the elements shown in the graph are classified into (partially overlapping) categories:

1. rock-forming elements (major elements in green field, and minor elements in light green field);

2. rare earth elements (lanthanides, La-Lu, and Y; labeled in blue);

3. major industrial metals (global production $> \sim 3 \times 10^7$ kg/year; labeled in red);

4. precious metals (labeled in purple);

5. the nine rarest "metals" — the six platinum group elements plus Au, Re, and Te (a metalloid) — in the yellow field.

Note that there are two breaks where the unstable elements technetium (atomic number: 43) and promethium (atomic number: 61) would be. These are both extremely rare, since on Earth they are only produced through the spontaneous fission of very heavy radioactive elements (for example, uranium, thorium, or the trace amounts of plutonium that exist in uranium ores), or by the interaction of certain other elements with cosmic rays. Both of the first two of these elements have been identified spectroscopically in the atmospheres of stars, where they are produced by ongoing nucleosynthetic processes. There are also breaks where the six noble gases would be, since they are not chemically bound in the Earth's crust, and they are only generated by decay chains from radioactive elements and are therefore extremely rare there. The twelve naturally occurring very rare, highly radioactive elements (polonium, astatine, francium, radium, actinium, protactinium, neptunium, plutonium, americium, curium, berkelium, and californium) are not included, since any of these elements that were present at the formation of the Earth have decayed away eons ago, and their quantity today is negligible and is only produced from the radioactive decay of uranium and thorium.

Oxygen and silicon are notably quite common elements in the crust. They have frequently combined with each other to form common silicate minerals.

Crustal rare-earth elemental abundance

"Rare" earth elements is a historical misnomer. The persistence of the term reflects unfamiliarity rather than true rarity. The more abundant rare earth elements are each similar in crustal concentration to commonplace industrial metals such as chromium, nickel, copper, zinc, molybdenum, tin, tungsten, or lead. The two least abundant rare earth elements (thulium and lutetium) are nearly 200 times more common than gold. However, in contrast to the ordinary base and precious metals, rare earth elements have very little tendency to become concentrated in exploitable ore deposits. Consequently, most of the world's supply of rare earth elements comes from only a handful of sources. Furthermore, the rare earth metals are all quite chemically similar to each other, and they are thus quite difficult to separate into quantities of the pure elements.

Differences in abundances of individual rare earth elements in the upper continental crust of the Earth represent the superposition of two effects, one nuclear and one geochemical. First, the rare earth elements with even atomic numbers ($_{58}$Ce, $_{60}$Nd, ...) have greater cosmic and terrestrial abundances than the adjacent rare earth elements with odd atomic numbers ($_{57}$La, $_{59}$Pr, ...). Second, the lighter rare earth elements are more incompatible (because they have larger ionic radii) and therefore more strongly concentrated in the continental crust than the heavier rare earth elements. In most rare earth ore deposits, the first four rare earth elements – lanthanum, cerium, praseodymium, and neodymium – constitute 80% to 99% of the total amount of rare earth metal that can be found in the ore.

34.2.3 Earth's mantle elemental abundance

Main article: Mantle (geology)

The mass-abundance of the eight most abundant elements in the Earth's crust (see main article above) is approximately: oxygen 45%, magnesium 23%, silicon 22%, iron 5.8%, calcium 2.3%, aluminum 2.2%, sodium 0.3%, potassium 0.3%.

The mantle differs in elemental composition from the crust in having a great deal more magnesium and significantly more iron, while having much less aluminum and sodium.

34.2.4 Earth's core elemental abundance

Due to mass segregation, the core of the Earth is believed to be primarily composed of iron (88.8%), with smaller amounts of nickel (5.8%), sulfur (4.5%), and less than 1% trace elements.[10]

34.2.5 Oceanic elemental abundance

For a complete list of the abundance of elements in the ocean, see Abundances of the elements (data page)#Sea water.

34.2.6 Atmospheric elemental abundance

The order of elements by volume-fraction (which is approximately molecular mole-fraction) in the atmosphere is nitrogen (78.1%), oxygen (20.9%),[12] argon (0.96%), followed by (in uncertain order) carbon and hydrogen because water vapor and carbon dioxide, which represent most of these two elements in the air, are variable components. Sulfur, phosphorus, and all other elements are present in significantly lower proportions.

According to the abundance curve graph (above right), argon, a significant if not major component of the atmosphere, does not appear in the crust at all. This is because the atmosphere has a far smaller mass than the crust, so argon remaining in the crust contributes little to mass-fraction there, while at the same time buildup of argon in the atmosphere has become large enough to be significant.

34.2.7 Abundances of elements in urban soils

For a complete list of the abundance of elements in urban soils, see Abundances of the elements (data page)#Urban soils.

Reasons for establishing

In the time of life existence, or at least in the time of the existence of human beings, the abundances of chemical elements within the Earth's crust have not been changed dramatically due to migration and concentration processes except the radioactive elements and their decay products and also noble gases. However, significant changes took place in the distribution of chemical elements. But within the biosphere not only the distribution, but also the abundances of elements have changed during the last centuries.

The rate of a number of geochemical changes taking place during the last decades in the biosphere has become catastrophically high. Such changes are often connected with human activities. To study these changes and to make better informed decisions on diminishing their adverse impact on living organisms, and especially on people, it is necessary to estimate the contemporary abundances of chemical elements in geochemical systems susceptible to the highest anthropogenic impact and having a significant effect on the development and existence of living organisms. One of such systems is the soil of urban landscapes. Settlements occupy less than 10% of the land area, but virtually the entire population of the planet lives within them. The main deposing medium in cities is soil, which ecological and geochemical conditions largely determine the life safety of citizens. So that, one of the priority tasks of the environmental geochemistry is to establish the average contents (abundances) of chemical elements in the soils of settlements.

Methods and results

The geochemical properties of urban soils from more than 300 cities in Europe, Asia, Africa, Australia, and America were evaluated.[13] In each settlement samples were collected uniformly throughout the territory, covering residential, industrial, recreational and other urban areas. The sampling was carried out directly from the soil surface and specifically traversed pits, ditches and wells from the upper soil horizon. The number of samples in each locality ranged from 30 to 1000. The published data and the materials kindly provided by a number of geochemists were also incorporated into the research. Considering the great importance of the defined contents, quantitative and quantitative emission spectral, gravimetric, X-ray fluorescence, and partly neutron activation analyses were carried out in parallel approximately in the samples. In a volume of 3–5% of the total number of samples, sampling and analyses of the inner and external controls were conducted. Calculation of random errors and systematic errors allowed to consider the sampling and analytical laboratory work as good.

For every city the average concentrations of elements in soils were determined. To avoid the errors related to unequal number of samples, each city was then represented by only one "averaged" sample. The statistical processing of this data allowed to calculate the average concentrations, which can be considered as the abundances of chemical elements in urban soils.

This graph illustrates the relative abundance of the chemical elements in urban soils, irregularly decreasing in proportion with the increasing atomic masses. Therefore, the evolution of organisms in this system occurs in the conditions of light elements' prevalence. It corresponds to the conditions of the evolutional development of the living matter on the Earth. The irregularity of element decreasing may be somewhat connected, as stated above, with the technogenic influence. The Oddo-Harkins rule, which holds that elements with an even atomic number are more common than elements with an odd atomic number, is saved in the urban soils but with some technogenic complications. Among the considered abundances the even-atomic elements make 91.48% of the urban soils mass. As it is in the Earth's crust, elements with the 4-divisible atomic masses of leading isotope (oxygen — 16, silicon — 28, calcium — 40, carbon — 12, iron — 56) are sharply prevailing in urban soils.

In spite of significant differences between abundances of several elements in urban soils and those values calculated for the Earth's crust, the general patterns of element abundances in urban soils repeat those in the Earth's crust in a great measure. The established abundances of chemical elements in urban soils can be considered as their geochemical (ecological and geochemical) characteristic, reflecting the combined impact of technogenic and natural processes occurring during certain time period (the end of the 20th century–beginning of the 21st century). With the development of science and technology the abundances may gradually change. The rate of these changes is still poorly predictable. The abundances of chemical elements may be used during various ecological and geochemical studies.

34.3   Human body elemental abundance

Main article: Chemical makeup of the human body

By mass, human cells consist of 65–90% water (H_2O), and a significant portion of the remainder is composed of carbon-containing organic molecules. Oxygen therefore contributes a majority of a human body's mass, followed by carbon. Almost 99% of the mass of the human body is made up of six elements: oxygen, carbon, hydrogen, nitrogen, calcium, and phosphorus. The next 0.75% is made up of the next five elements: potassium, sulfur, chlorine, sodium, and magnesium. Only 17 elements are known for certain to be necessary to human life, with one additional element (fluorine) thought to be helpful for tooth enamel strength. A few more trace elements may play some role in the health of mammals. Boron and silicon are notably necessary for plants but have uncertain roles in animals. The elements aluminium and silicon, although very common in the earth's crust, are conspicuously rare in the human body.[14]

Periodic table highlighting nutritional elements[15]

Periodic table highlighting dietary elements

34.4   See also

- Abundances of the elements (data page)

- Natural abundance (isotopic abundance)

- Primordial nuclide

34.5   References

34.5.1   Footnotes

[1] Croswell, Ken (February 1996). *Alchemy of the Heavens*. Anchor. ISBN 0-385-47214-5.

[2] WMAP- Content of the Universe

[3] Suess, Hans; Urey, Harold (1956). "Abundances of the Elements". *Reviews of Modern Physics* **28**:53. Bibcode:1956RvMP...28. doi:10.1103/RevModPhys.28.53.

[4] Cameron, A.G.W. (1973). "Abundances of the elements in the solar system". *Space Science Reviews* **15**:121. Bibcode:1973SSRv.. doi:10.1007/BF00172440.

[5] Anders, E; Ebihara, M (1982). "Solar-system abundances of the elements". *Geochimica et Cosmochimica Acta* **46** (11): 2363. Bibcode:1982GeCoA..46.2363A. doi:10.1016/0016-7037(82)90208-3.

[6] Arnett, David (1996). *Supernovae and Nucleosynthesis* (First ed.). Princeton, New Jersey: Princeton University Press. ISBN 0-691-01147-8. OCLC 33162440.

[7] Bell, Jerry A.; GenChem Editorial/Writing Team (2005). "Chapter 3: Origin of Atoms". *Chemistry: a project of the American Chemical Society*. New York [u.a.]: Freeman. pp. 191–193. ISBN 978-0-7167-3126-9. Correlations between abundance and nuclear binding energy [Subsection title]

[8] Bell, Jerry A.; GenChem Editorial/Writing Team (2005). "Chapter 3: Origin of Atoms". *Chemistry: a project of the American Chemical Society*. New York [u.a.]: Freeman. p. 192. ISBN 978-0-7167-3126-9. The higher abundance of elements with even atomic numbers [Subsection title]

[9] Bailey, David. "Semi-empirical Nuclear Mass Formula". *PHY357: Strings & Binding Energy*. University of Toronto. Retrieved 2011-03-31.

[10] Morgan, J. W.; Anders, E. (1980). "Chemical composition of Earth, Venus, and Mercury". *Proceedings of the National Academy of Sciences* **77** (12): 6973–6977. Bibcode:1980PNAS...77.6973M. doi:10.1073/pnas.77.12.6973. PMC 350422. PMID 16592930.

[11] William F McDonough The composition of the Earth. quake.mit.edu

[12] Zimmer, Carl (3 October 2013). "Earth's Oxygen: A Mystery Easy to Take for Granted". *New York Times*. Retrieved 3 October 2013.

[13] Vladimir Alekseenko; Alexey Alekseenko (2014). "The abundances of chemical elements in urban soils". *Journal of Geochemical Exploration* (Elsevier B.V.) **147**: 245–249. doi:10.1016/j.gexplo.2014.08.003. ISSN 0375-6742.

[14] Table data from Chang, Raymond (2007). *Chemistry, Ninth Edition*. McGraw-Hill. p. 52. ISBN 0-07-110595-6.

[15] Ultratrace minerals. Authors: Nielsen, Forrest H. USDA, ARS Source: Modern nutrition in health and disease / editors, Maurice E. Shils ... et al.. Baltimore : Williams & Wilkins, c1999., p. 283-303. Issue Date: 1999 URI:

34.5.2 Notes

[1] Below Jupiter's outer atmosphere, volume fractions are significantly different from mole fractions due to high temperatures (ionization and disproportionation) and high density where the Ideal Gas Law is inapplicable.

34.5.3 Notations

- http://geopubs.wr.usgs.gov/fact-sheet/fs087-02/

- http://imagine.gsfc.nasa.gov/docs/dict_ei.html

34.6 External links

- List of elements in order of abundance in the Earth's crust (only correct for the twenty most common elements)

- Cosmic abundance of the elements and nucleosynthesis

- webelements.com Lists of elemental abundances for the Universe, Sun, meteorites, Earth, ocean, streamwater

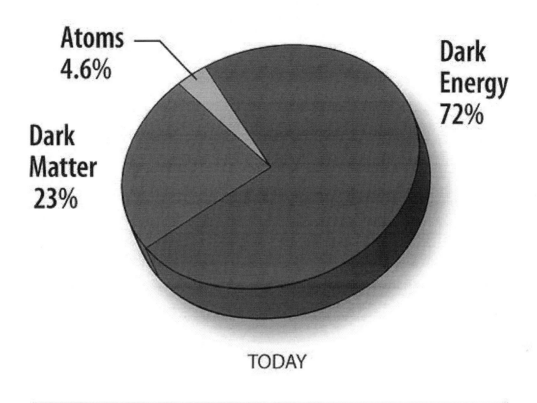

TODAY

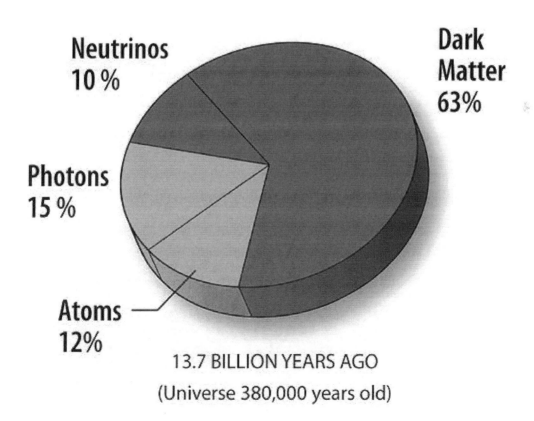

13.7 BILLION YEARS AGO
(Universe 380,000 years old)

Estimated proportions of matter, dark matter and dark energy in the universe. Only the fraction of the mass and energy in the universe labeled "atoms" is composed of chemical elements.

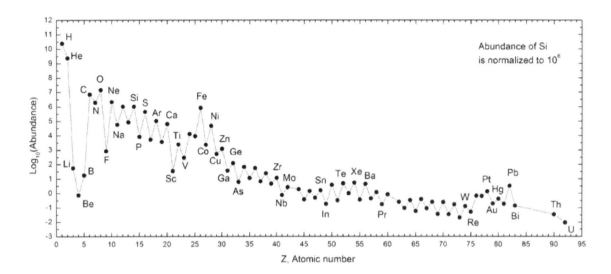

*Estimated abundances of the chemical **elements in the Solar system**. Hydrogen and helium are most common, from the Big Bang. The next three elements (Li, Be, B) are rare because they are poorly synthesized in the Big Bang and also in stars. The two general trends in the remaining stellar-produced elements are: (1) an alternation of abundance in elements as they have even or odd atomic numbers (the Oddo-Harkins rule), and (2) a general decrease in abundance, as elements become heavier. Iron is especially common because it represents the minimum energy nuclide that can be made by fusion of helium in supernovae.*

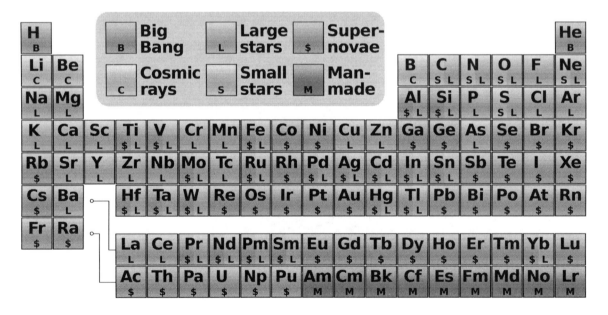

Periodic table showing the cosmogenic origin of each element

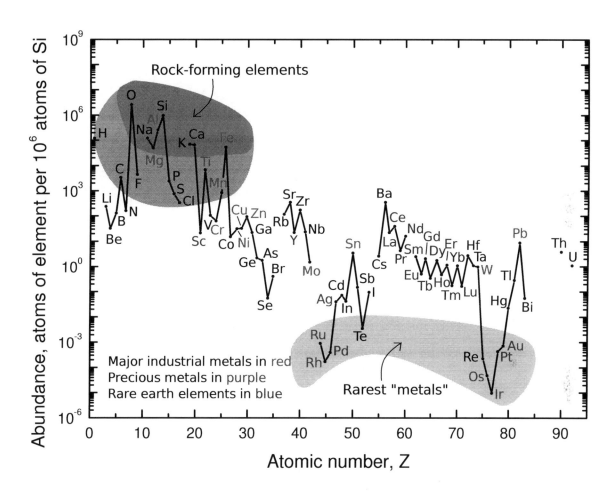

Abundance (atom fraction) of the chemical elements in Earth's upper continental crust as a function of atomic number. The rarest elements in the crust (shown in yellow) are the most dense. They were further rarefied in the crust by being siderophile (iron-loving) elements, in the Goldschmidt classification of elements. Siderophiles were depleted by being relocated into the Earth's core. Their abundance in meteoroid materials is relatively higher. Additionally, tellurium and selenium have been depleted from the crust due to formation of volatile hydrides.

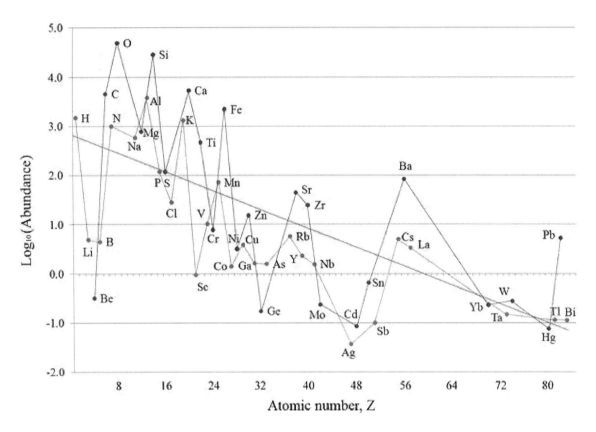

The half-logarithm graph of the abundances of chemical elements in urban soils. *(Alekseenko and Alekseenko, 2014) Chemical elements are distributed extremely irregularly in urban soils, what is also typical for the Earth's crust. Nine elements (O, Si, Ca, C, Al, Fe, H, K, N) make the 97.68% of the considering geochemical system (urban soils). These elements and also Zn, Sr, Zr, Ba, and Pb essentially prevail over the trend line. Part of them could be considered as "inherited" from the concentrations in the Earth's crust; another part is explained as a result of intensive technogenic activity in the cities.*

Chapter 35

List of elements

The following is a list of the 118 identified chemical elements.

35.1   List

35.2   Notes

- ^1 The element does not have any stable nuclides, and a value in brackets, e.g. [209], indicates the mass number of the longest-lived isotope of the element. However, four such elements, bismuth, thorium, protactinium, and uranium, have characteristic terrestrial isotopic compositions, and thus their standard atomic weights are given.

- ^2 The isotopic composition of this element varies in some geological specimens, and the variation may exceed the uncertainty stated in the table.

- ^3 The isotopic composition of the element can vary in commercial materials, which can cause the atomic weight to deviate significantly from the given value.

- ^4 The isotopic composition varies in terrestrial material such that a more precise atomic weight can not be given.

- ^5 The atomic weight of commercial lithium can vary between 6.939 and 6.996—analysis of the specific material is necessary to find a more accurate value.

- ^6 This element does not solidify at a pressure of one atmosphere. The value listed above, 0.95 K, is the temperature at which helium does solidify at a pressure of 25 atmospheres.

- ^7 This element sublimes at one atmosphere of pressure

- ^8 The transuranic elements 95 and above do not occur naturally, but some of them can be produced artificially.

- ^9 The value listed is the conventional atomic-weight value suitable for trade and commerce. The actual value may differ depending on the isotopic composition of the sample. Since 2009, IUPAC provides the standard atomic-weight values for these elements using the interval notation. The corresponding standard atomic weights are:

  - Hydrogen: [1.00784, 1.00811]
  - Lithium: [6.938, 6.997]
  - Boron: [10.806, 10.821]
  - Carbon: [12.0096, 12.0116]
  - Nitrogen: [14.00643, 14.00728]
  - Oxygen: [15.99903, 15.99977]

- Magnesium: [24.304, 24.307]
- Silicon: [28.084, 28.086]
- Sulfur: [32.059, 32.076]
- Chlorine: [35.446, 35.457]
- Bromine: [79.901, 79.907]
- Thallium: [204.382, 204.385]

- **^10** Electronegativity on the Pauling scale. Standard symbol: χ

- **^11** The value has not been precisely measured, usually because of the element's short half-life; the value given in parentheses is a prediction.

- **^12** With error bars: 357+112 −108 K.

- **^13** This predicted value is for liquid ununoctium, not gaseous ununoctium.

35.3 References

[1] Royal Society of Chemistry – *Visual Element Periodic Table*

[2] – Online Etymological Dictionary

[3]

[4] Holman, Lawrence and Barr

- Wieser, Michael E.; et al. (2013). "Atomic weights of the elements 2011 (IUPAC Technical Report)". *Pure Appl. Chem.* (IUPAC) **85** (5): 1047–1078. doi:10.1351/PAC-REP-13-03-02. (for standard atomic weights of elements)

- Sonzogni, Alejandro. "Interactive Chart of Nuclides". National Nuclear Data Center: Brookhaven National Laboratory. Retrieved 2008-06-06. (for atomic weights of elements with atomic numbers 103–118)

35.4 External links

- Atoms made thinkable, an interactive visualisation of the elements allowing physical and chemical properties to be compared

Chapter 36

Table of nuclides

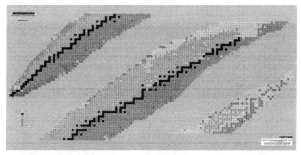

A chart of nuclides. Above, cut into three parts for better presentation; below, combined.

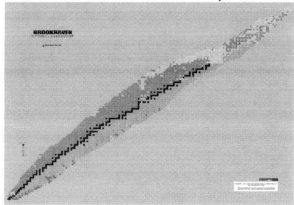

Decay modes

- Positron emission or Electron capture

A **table of nuclides** or **chart of nuclides** is a two-dimensional graph in which one axis represents the number of neutrons and the other represents the number of protons in an atomic nucleus. Each point plotted on the graph thus represents the nuclide of a real or hypothetical chemical element. This system of ordering nuclides can offer a greater insight into the characteristics of isotopes than the better-known periodic table, which shows only elements instead of each of their isotopes.

36.1 Description and utility

A chart or table of nuclides is a simple map to the nuclear, or radioactive, behaviour of nuclides, as it distinguishes the isotopes of an element. It contrasts with a periodic table, which only maps their chemical behavior, since isotopes of the same element do not differ chemically. Nuclide charts organize isotopes along the X axis by their numbers of neutrons and along the Y axis by their numbers of protons, out to the limits of the neutron and proton drip lines. This representation was first published by Giorgio Fea in 1935,[1] and expanded by Emilio Segrè in 1945 or G. Seaborg. In 1958, Walter Seelmann-Eggebert and Gerda Pfennig published the first edition of the Karlsruhe Nuclide Chart. Its 7th edition was made available in 2006. Today, one finds several nuclide charts, four of them have a wide distribution: the Karlsruhe Nuclide Chart, the Strasbourg Universal Nuclide Chart, the Chart of the Nuclides from the JAEA and the Nuclide Chart from Knolls Atomic Power Laboratory.[2] It has become a basic tool of the nuclear community.

36.2 Trends in the chart of nuclides

Fragment of table of nuclides for Polonium, Radium, Copernicium and Curium, as seen on a monument in front of University of Warsaw's Centre of New Technologies

- **Isotopes** are nuclides with the same number of protons but differing numbers of neutrons; that is, they have the same atomic number and are therefore the same chemical element. Isotopes neighbor each other vertically, e.g., carbon-12, carbon-13, carbon-14 or oxygen-15, oxygen-16, oxygen-17.

- **Isotones** are nuclides with the same number of neutrons but differing number of protons. Isotones neighbor each other horizontally. Example: carbon-14, nitrogen-15, oxygen-16 in the sample table above.

- **Isobars** are nuclides with the same number of nucleons, i.e. mass number, but different numbers of protons and different number of neutrons. Isobars neighbor each other diagonally from lower-left to upper-right. Example: carbon-14, nitrogen-14, oxygen-14 in the sample table above.

- **Isodiaphers** are nuclides with the same difference between neutrons and protons (N−Z). Like isobars, they follow diagonal lines, but at right angles to the isobar lines; from upper-left to lower right. Examples: boron-10, carbon-12, nitrogen-14 where N−Z=0; boron-12, carbon-14, nitrogen-16 where N−Z=2.

- Beyond the **neutron drip line** along the right, nuclides decay by neutron emission.

- Beyond the **proton drip line** along the upper left, nuclides decay by proton emission. Drip lines have only been established for some elements.

- The island of stability is a hypothetical region of the table of nuclides that contains isotopes far more stable than other transuranic elements.

- There are no stable atoms having an equal number of protons and neutrons in their nuclei with atomic number greater than 20 (i.e. calcium) as can be readily "read" from the chart. Nuclei of greater atomic number require an excess of neutrons for stability.

- The only stable atoms having an odd number of protons and an odd number of neutrons are hydrogen-2, lithium-6, boron-10, nitrogen-14 and (observationally) tantalum-180m. This is because the mass-energy of such atoms is usually higher than that of their neighbors on the same isobar, so most of them are unstable to beta decay.

- There are no stable atoms with mass numbers 5 or 8. There are stable atoms with all other mass numbers up to 208 with the exceptions of 147 and 151. (Bismuth-209 was found to be radioactive in 2003, but with a half-life of 1.9×10^{19} years.)

- With the possible exception of the pair tellurium-123 and antimony-123, odd mass numbers are never represented by more than one stable atom. This is because the mass-energy is a convex function of atomic number, so all nuclides on an odd isobar except one have a lower-energy neighbor to which they can decay by beta decay.

- There are no stable atoms having atomic number greater than Z=82 (lead),[3] although bismuth (Z=83) is stable for all practical human purposes. Atoms with atomic numbers from 1 to 82 all have stable isotopes, with the exceptions of technetium (Z=43) and promethium (Z=61).

36.3   Available representations

36.4   References

[1] Georgio Fea. Il Nuovo Cimento 2 (1935) 368

[2] "What We Do: The Chart of Nuclides". Knolls Atomic Power Laboratory. Retrieved 14 May 2009.

[3] Holden,CRC Handbook of Chemistry and Physics, 90th Edition §11

36.5   External links

- **Interactive Chart of Nuclides (Brookhaven National Laboratory)**

- *YChartElements* dynamic periodic table and chart of the nuclides, a Yoix application

- Compact Chart of Nuclides (non-standard representation with elements along a diagonal) 70x74.

- **The Live Chart of Nuclides - IAEA**

- Another example of a Chart of Nuclides from Korea Data up to Jan 1999 only

- Map of the Nuclides (Dead URL Aug 2012)

- **The Lund/LBNL Nuclear Data Search** (Dead URL Aug 2012)

- Tendency equation and curve of stable nuclides.

- Order the Karlsruhe Nuclide Chart (Nucleonica GmbH))

- The Nuclear Science Portal Nucleonica Commercial site

Chapter 37

Timeline of chemical element discoveries

The discovery of the elements known to exist today is presented here in chronological order. The elements are listed generally in the order in which each was first defined as the pure element, as the exact date of discovery of most elements cannot be accurately defined.

Given is each element's name, atomic number, year of first report, name of the discoverer, and some notes related to the discovery.

37.1   Table

37.2   Unrecorded discoveries

37.3   Recorded discoveries

37.4   Unconfirmed discoveries

37.5   Graphics

37.6   See also

- History of the periodic table
- Periodic table
- The Mystery of Matter: Search for the Elements (2015 PBS film)

37.7   References

[1] "Copper History". Rameria.com. Retrieved 2008-09-12.

[2] CSA – Discovery Guides, A Brief History of Copper

[3] "The History of Lead – Part 3". Lead.org.au. Retrieved 2008-09-12.

[4] 47 Silver

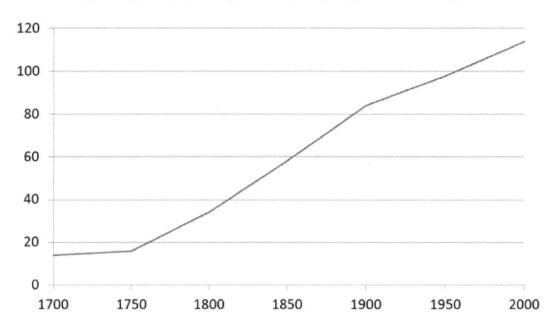

Number of known chemical elements 1700-2000

Development in discovery

[5] "Silver Facts – Periodic Table of the Elements". Chemistry.about.com. Retrieved 2008-09-12.

[6] "26 Iron". Elements.vanderkrogt.net. Retrieved 2008-09-12.

[7] Weeks, Mary Elvira; Leichester, Henry M. (1968). "Elements Known to the Ancients". *Discovery of the Elements.* Easton, PA: Journal of Chemical Education. pp. 29–40. ISBN 0-7661-3872-0. LCCCN 68-15217.

[8] "Notes on the Significance of the First Persian Empire in World History". Courses.wcupa.edu. Retrieved 2008-09-12.

[9] "History of Carbon and Carbon Materials – Center for Applied Energy Research – University of Kentucky". Caer.uky.edu. Retrieved 2008-09-12.

[10] "Chinese made first use of diamond". BBC News. 17 May 2005. Retrieved 2007-03-21.

[11] Ferchault de Réaumur, R-A (1722). *L'art de convertir le fer forgé en acier, et l'art d'adoucir le fer fondu, ou de faire des ouvrages de fer fondu aussi finis que le fer forgé (English translation from 1956).* Paris, Chicago.

[12] Senese, Fred (September 9, 2009). "Who discovered carbon?". Frostburg State University. Retrieved 2007-11-24.

[13] "50 Tin". Elements.vanderkrogt.net. Retrieved 2008-09-12.

[14] "History of Metals". Neon.mems.cmu.edu. Retrieved 2008-09-12.

[15] "Sulfur History". Georgiagulfsulfur.com. Retrieved 2008-09-12.

[16] "Mercury and the environment — Basic facts". *Environment Canada, Federal Government of Canada.* 2004. Retrieved 2008-03-27.

[17] Craddock, P. T. et al. (1983), "Zinc production in medieval India", *World Archaeology* **15** (2), Industrial Archaeology, p. 13

[18] "30 Zinc". Elements.vanderkrogt.net. Retrieved 2008-09-12.

[19] Weeks, Mary Elvira (1933). "III. Some Eighteenth-Century Metals". *The Discovery of the Elements.* Easton, PA: Journal of Chemical Education. p. 21. ISBN 0-7661-3872-0.

[20] "Arsenic". Los Alamos National Laboratory. Retrieved 3 March 2013.

[21] SHORTLAND, A. J. (2006-11-01). "APPLICATION OF LEAD ISOTOPE ANALYSIS TO A WIDE RANGE OF LATE BRONZE AGE EGYPTIAN MATERIALS". *Archaeometry* **48** (4): 657–669. doi:10.1111/j.1475-4754.2006.00279.x.

[22] "15 Phosphorus". Elements.vanderkrogt.net. Retrieved 2008-09-12.

[23] "27 Cobalt". Elements.vanderkrogt.net. Retrieved 2008-09-12.

[24] "78 Platinum". Elements.vanderkrogt.net. Retrieved 2008-09-12.

[25] "28 Nickel". Elements.vanderkrogt.net. Retrieved 2008-09-12.

[26] "Bismuth". Los Alamos National Laboratory. Retrieved 3 March 2013.

[27] "12 Magnesium". Elements.vanderkrogt.net. Retrieved 2008-09-12.

[28] "01 Hydrogen". Elements.vanderkrogt.net. Retrieved 2008-09-12.

[29] Andrews, A. C. (1968). "Oxygen". In Clifford A. Hampel. *The Encyclopedia of the Chemical Elements*. New York: Reinhold Book Corporation. p. 272. LCCN 68-29938.

[30] "08 Oxygen". Elements.vanderkrogt.net. Retrieved 2008-09-12.

[31] Cook, Gerhard A.; Lauer, Carol M. (1968). "Oxygen". In Clifford A. Hampel. *The Encyclopedia of the Chemical Elements*. New York: Reinhold Book Corporation. pp. 499–500. LCCN 68-29938.

[32] Roza, Greg (2010). *The Nitrogen Elements: Nitrogen, Phosphorus, Arsenic, Antimony, Bismuth*. p. 7. ISBN 9781435853355.

[33] "07 Nitrogen". Elements.vanderkrogt.net. Retrieved 2008-09-12.

[34] "17 Chlorine". Elements.vanderkrogt.net. Retrieved 2008-09-12.

[35] "25 Manganese". Elements.vanderkrogt.net. Retrieved 2008-09-12.

[36] "56 Barium". Elements.vanderkrogt.net. Retrieved 2008-09-12.

[37] "42 Molybdenum". Elements.vanderkrogt.net. Retrieved 2008-09-12.

[38] "52 Tellurium". Elements.vanderkrogt.net. Retrieved 2008-09-12.

[39] IUPAC. "74 Tungsten". Elements.vanderkrogt.net. Retrieved 2008-09-12.

[40] "38 Strontium". Elements.vanderkrogt.net. Retrieved 2008-09-12.

[41] "Lavoisier". Homepage.mac.com. Retrieved 2008-09-12.

[42] "Chronology – Elementymology". Elements.vanderkrogt.net. Retrieved 2008-09-12.

[43] Lide, David R., ed. (2007–2008). "CRC Handbook of Chemistry and Physics" **4**. New York: CRC Press. p. 42. 978-0-8493-0488-0. |contribution= ignored (help)

[44] M. H. Klaproth (1789). "Chemische Untersuchung des Uranits, einer neuentdeckten metallischen Substanz". *Chemische Annalen* **2**: 387–403.

[45] E.-M. Péligot (1842). "Recherches Sur L'Uranium". *Annales de chimie et de physique* **5** (5): 5–47.

[46] "Titanium". Los Alamos National Laboratory. 2004. Retrieved 2006-12-29.

[47] Barksdale, Jelks (1968). The Encyclopedia of the Chemical Elements. Skokie, Illinois: Reinhold Book Corporation. pp. 732–38 "Titanium". LCCCN 68-29938.

[48] Browning, Philip Embury (1917). "Introduction to the Rarer Elements". *Kongl. Vet. Acad. Handl.* **XV**: 137.

[49] *Crell Anal.* **I**: 313. 1796. Missing or empty |title= (help)

[50] Vauquelin, Louis Nicolas (1798). "Memoir on a New Metallic Acid which exists in the Red Lead of Sibiria". *Journal of Natural Philosophy, Chemistry, and the Art* **3**: 146.

[51] "04 Beryllium". Elements.vanderkrogt.net. Retrieved 2008-09-12.

[52] "23 Vanadium". Elements.vanderkrogt.net. Retrieved 2008-09-12.

[53] "41 Niobium". Elements.vanderkrogt.net. Retrieved 2008-09-12.

[54] "73 Tantalum". Elements.vanderkrogt.net. Retrieved 2008-09-12.

[55] "46 Palladium". Elements.vanderkrogt.net. Retrieved 2008-09-12.

[56] "58 Cerium". Elements.vanderkrogt.net. Retrieved 2008-09-12.

[57] "76 Osmium". Elements.vanderkrogt.net. Retrieved 2008-09-12.

[58] "77 Iridium". Elements.vanderkrogt.net. Retrieved 2008-09-12.

[59] "45 Rhodium". Elements.vanderkrogt.net. Retrieved 2008-09-12.

[60] "19 Potassium". Elements.vanderkrogt.net. Retrieved 2008-09-12.

[61] "11 Sodium". Elements.vanderkrogt.net. Retrieved 2008-09-12.

[62] "05 Boron". Elements.vanderkrogt.net. Retrieved 2008-09-12.

[63] "09 Fluorine". Elements.vanderkrogt.net. Retrieved 2008-09-12.

[64] "53 Iodine". Elements.vanderkrogt.net. Retrieved 2008-09-12.

[65] "03 Lithium". Elements.vanderkrogt.net. Retrieved 2008-09-12.

[66] "48 Cadmium". Elements.vanderkrogt.net. Retrieved 2008-09-12.

[67] "34 Selenium". Elements.vanderkrogt.net. Retrieved 2008-09-12.

[68] "14 Silicon". Elements.vanderkrogt.net. Retrieved 2008-09-12.

[69] "13 Aluminium". Elements.vanderkrogt.net. Retrieved 2008-09-12.

[70] "35 Bromine". Elements.vanderkrogt.net. Retrieved 2008-09-12.

[71] "90 Thorium". Elements.vanderkrogt.net. Retrieved 2008-09-12.

[72] "57 Lanthanum". Elements.vanderkrogt.net. Retrieved 2008-09-12.

[73] "68 Erbium". Elements.vanderkrogt.net. Retrieved 2008-09-12.

[74] "65 Terbium". Elements.vanderkrogt.net. Retrieved 2008-09-12.

[75] "44 Ruthenium". Elements.vanderkrogt.net. Retrieved 2008-09-12.

[76] "55 Caesium". Elements.vanderkrogt.net. Retrieved 2008-09-12.

[77] Caesium

[78] "37 Rubidium". Elements.vanderkrogt.net. Retrieved 2008-09-12.

[79] "81 Thallium". Elements.vanderkrogt.net. Retrieved 2008-09-12.

[80] "49 Indium". Elements.vanderkrogt.net. Retrieved 2008-09-12.

[81] "02 Helium". Elements.vanderkrogt.net. Retrieved 2008-09-12.

[82] "31 Gallium". Elements.vanderkrogt.net. Retrieved 2008-09-12.

[83] "70 Ytterbium". Elements.vanderkrogt.net. Retrieved 2008-09-12.

[84] "67 Holmium". Elements.vanderkrogt.net. Retrieved 2008-09-12.

[85] "69 Thulium". Elements.vanderkrogt.net. Retrieved 2008-09-12.

[86] "21 Scandium". Elements.vanderkrogt.net. Retrieved 2008-09-12.

[87] "62 Samarium". Elements.vanderkrogt.net. Retrieved 2008-09-12.

[88] "64 Gadolinium". Elements.vanderkrogt.net. Retrieved 2008-09-12.

[89] "59 Praseodymium". Elements.vanderkrogt.net. Retrieved 2008-09-12.

[90] "60 Neodymium". Elements.vanderkrogt.net. Retrieved 2008-09-12.

[91] "32 Germanium". Elements.vanderkrogt.net. Retrieved 2008-09-12.

[92] "18 Argon". Elements.vanderkrogt.net. Retrieved 2008-09-12.

[93] "10 Neon". Elements.vanderkrogt.net. Retrieved 2008-09-12.

[94] "54 Xenon". Elements.vanderkrogt.net. Retrieved 2008-09-12.

[95] "84 Polonium". Elements.vanderkrogt.net. Retrieved 2008-09-12.

[96] "88 Radium". Elements.vanderkrogt.net. Retrieved 2008-09-12.

[97] Partington, J.R. (May 1957). "Discovery of Radon". *Nature* **179** (4566): 912. Bibcode:1957Natur.179..912P. doi:10.1038.

[98] Ramsay, W.; Gray, R. W. (1910). "La densité de l'emanation du radium". *Comptes rendus hebdomadaires des séances de l'Académie des sciences* **151**: 126–128.

[99] "89 Actinium". Elements.vanderkrogt.net. Retrieved 2008-09-12.

[100] "63 Europium". Elements.vanderkrogt.net. Retrieved 2008-09-12.

[101] "71 Lutetium". Elements.vanderkrogt.net. Retrieved 2008-09-12.

[102] http://www.maik.ru/abstract/radchem/0/radchem0535_abstract.pdf

[103] "72 Hafnium". Elements.vanderkrogt.net. Retrieved 2008-09-12.

[104] Noddack, W.; Tacke, I.; Berg, O (1925). "Die Ekamangane". *Naturwissenschaften* **13** (26): 567. Bibcode:1925NW.....13..567.. doi:10.1007/BF01558746.

[105] "91 Protactinium". Elements.vanderkrogt.net. Retrieved 2008-09-12.

[106] Emsley, John (2001). *Nature's Building Blocks* ((Hardcover, First Edition) ed.). Oxford University Press. p. 347. ISBN 0-19-850340-7.

[107] "43 Technetium". Elements.vanderkrogt.net. Retrieved 2008-09-12.

[108] *History of the Origin of the Chemical Elements and Their Discoverers*, Individual Element Names and History, "Technetium"

[109] "87 Francium". Elements.vanderkrogt.net. Retrieved 2008-09-12.

[110] Adloff, Jean-Pierre; Kaufman, George B. (2005-09-25). Francium (Atomic Number 87), the Last Discovered Natural Element. *The Chemical Educator* **10** (5). [2007-03-26]

[111] "85 Astatine". Elements.vanderkrogt.net. Retrieved 2008-09-12.

[112] Close, Frank E. (2004). *Particle Physics: A Very Short Introduction*. Oxford University Press. p. 2. ISBN 978-0-19-280434-1.

[113] "93 Neptunium". Elements.vanderkrogt.net. Retrieved 2008-09-12.

[114] "94 Plutonium". Elements.vanderkrogt.net. Retrieved 2008-09-12.

[115] "95 Americium". Elements.vanderkrogt.net. Retrieved 2008-09-12.

[116] "96 Curium". Elements.vanderkrogt.net. Retrieved 2008-09-12.

[117] "97 Berkelium". Elements.vanderkrogt.net. Retrieved 2008-09-12.

[118] "98 Californium". Elements.vanderkrogt.net. Retrieved 2008-09-12.

[119] "99 Einsteinium". Elements.vanderkrogt.net. Retrieved 2008-09-12.

[120] "100 Fermium". Elements.vanderkrogt.net. Retrieved 2008-09-12.

[121] "101 Mendelevium". Elements.vanderkrogt.net. Retrieved 2008-09-12.

[122] "102 Nobelium". Elements.vanderkrogt.net. Retrieved 2008-09-12.

[123] "103 Lawrencium". Elements.vanderkrogt.net. Retrieved 2008-09-12.

[124] "104 Rutherfordium". Elements.vanderkrogt.net. Retrieved 2008-09-12.

[125] "105 Dubnium". Elements.vanderkrogt.net. Retrieved 2008-09-12.

[126] "106 Seaborgium". Elements.vanderkrogt.net. Retrieved 2008-09-12.

[127] "107 Bohrium". Elements.vanderkrogt.net. Retrieved 2008-09-12.

[128] "109 Meitnerium". Elements.vanderkrogt.net. Retrieved 2008-09-12.

[129] "108 Hassium". Elements.vanderkrogt.net. Retrieved 2008-09-12.

[130] "110 Darmstadtium". Elements.vanderkrogt.net. Retrieved 2008-09-12.

[131] "111 Roentgenium". Elements.vanderkrogt.net. Retrieved 2008-09-12.

[132] "112 Copernicium". Elements.vanderkrogt.net. Retrieved 2009-07-17.

[133] "Discovery of the Element with Atomic Number 112". www.iupac.org. 2009-06-26. Retrieved 2009-07-17.

[134] Oganessian, Yu. Ts.; Utyonkov, V. K.; Lobanov, Yu. V.; Abdullin, F. Sh.; Polyakov, A. N.; Shirokovsky, I. V.; Tsyganov, Yu. S.; Gulbekian, G. G.; Bogomolov, S. L.; Gikal, B.; Mezentsev, A.; Iliev, S.; Subbotin, V.; Sukhov, A.; Buklanov, G.; Subotic, K.; Itkis, M.; Moody, K.; Wild, J.; Stoyer, N.; Stoyer, M.; Lougheed, R. (October 1999). "Synthesis of Superheavy Nuclei in the ^{48}Ca + ^{244}Pu Reaction". *Physical Review Letters* **83** (16): 3154. Bibcode:1999PhRvL..83.3154O. doi:10.1103/PhysRevLett.83.3154.

[135] Oganessian, Yu. Ts.; Utyonkov, V. K.; Lobanov, Yu. V.; Abdullin, F. Sh.; Polyakov, A. N.; Shirokovsky, I. V.; Tsyganov, Yu. S.; Gulbekian, G. G.; Bogomolov, S. L.; Gikal, B.; Mezentsev, A.; Iliev, S.; Subbotin, V.; Sukhov, A.; Ivanov, O.; Buklanov, G.; Subotic, K.; Itkis, M.; Moody, K.; Wild, J.; Stoyer, N.; Stoyer, M.; Lougheed, R.; Laue, C.; Karelin, Ye.; Tatarinov, A. (2000). "Observation of the decay of ^{292}116". *Physical Review C* **63**: 011301. Bibcode:2001PhRvC..63a1301O. doi:10.1103/PhysRevC.63.011301.

[136] Oganessian, Yu. Ts.; Utyonkov, V. K.; Lobanov, Yu. V.; Abdullin, F. Sh.; Polyakov, A. N.; Sagaidak, R. N.; Shirokovsky, I. V.; Tsyganov, Yu. S.; Voinov, A. A.; Gulbekian, G.; Bogomolov, S.; Gikal, B.; Mezentsev, A.; Iliev, S.; Subbotin, V.; Sukhov, A.; Subotic, K.; Zagrebaev, V.; Vostokin, G.; Itkis, M.; Moody, K.; Patin, J.; Shaughnessy, D.; Stoyer, M.; Stoyer, N.; Wilk, P.; Kenneally, J.; Landrum, J.; Wild, J.; Lougheed, R. (2006). "Synthesis of the isotopes of elements 118 and 116 in the ^{249}Cf and ^{245}Cm+^{48}Ca fusion reactions". *Physical Review C* **74** (4): 044602. Bibcode:2006PhRvC..74d4602O. doi:10.1103/PhysRevC.74.044602.

[137] Oganessian, Yu. Ts.; Utyonkov, V. K.; Dmitriev, S. N.; Lobanov, Yu. V.; Itkis, M. G.; Polyakov, A. N.; Tsyganov, Yu. S.; Mezentsev, A. N.; Yeremin, A. V.; Voinov, A.; Sokol, E.; Gulbekian, G.; Bogomolov, S.; Iliev, S.; Subbotin, V.; Sukhov, A.; Buklanov, G.; Shishkin, S.; Chepygin, V.; Vostokin, G.; Aksenov, N.; Hussonnois, M.; Subotic, K.; Zagrebaev, V.; Moody, K.; Patin, J.; Wild, J.; Stoyer, M.; Stoyer, N.; et al. (2005). "Synthesis of elements 115 and 113 in the reaction ^{243}Am + ^{48}Ca". *Physical Review C* **72** (3): 034611. Bibcode:2005PhRvC..72c4611O. doi:10.1103/PhysRevC.72.034611.

[138] Oganessian, Yu. Ts.; Abdullin, F. Sh.; Bailey, P. D.; Benker, D. E.; Bennett, M. E.; Dmitriev, S. N.; Ezold, J. G.; Hamilton, J. H.; Henderson, R. A.; Itkis, M. G.; Lobanov, Yu. V.; Mezentsev, A. N.; Moody, K. J.; Nelson, S. L.; Polyakov, A. N.; Porter, C. E.; Ramayya, A. V.; Riley, F. D.; Roberto, J. B.; Ryabinin, M. A.; Rykaczewski, K. P.; Sagaidak, R. N.; Shaughnessy, D. A.; Shirokovsky, I. V.; Stoyer, M. A.; Subbotin, V. G.; Sudowe, R.; Sukhov, A. M.; Tsyganov, Yu. S.; et al. (April 2010). "Synthesis of a New Element with Atomic Number Z=117". *Physical Review Letters* **104** (14): 142502. Bibcode:2010PhRvL.104n2502O. doi:10.1103/PhysRevLett.104.142502. PMID 20481935.

37.8   External links

- History of the Origin of the Chemical Elements and Their Discoverers Last updated by Boris Pritychenko on March 30, 2004

- History of Elements of the Periodic Table

- Timeline of Element Discoveries

- Discovery of the Elements - The Movie - YouTube (1:18)

- The History Of Metals Timeline. A timeline showing the discovery of metals and the development of metallurgy.

37.9 Text and image sources, contributors, and licenses

37.9.1 Text

- **Periodic table** *Source:* https://en.wikipedia.org/wiki/Periodic_table?oldid=686835311 *Contributors:* AxelBoldt, Dreamyshade, Chuck Smith, Lee Daniel Crocker, Mav, Bryan Derksen, Timo Honkasalo, The Anome, Tarquin, DanKeshet, Rjstott, Andre Engels, XJaM, Christian List, PierreAbbat, Heron, Fonzy, Youandme, Olivier, Someone else, Bob Jonkman, Patrick, Infrogmation, Michael Hardy, Erik Zachte, TMC, Kwertii, Dan Koehl, Shellreef, Taras, Wapcaplet, Ixfd64, Dcljr, Tomi, Eric119, Kosebamse, Egil, Mdebets, Ahoerstemeier, Stan Shebs, Ronz, Jpatokal, Theresa knott, Snoyes, Suisui, Den fjättrade ankan~enwiki, Kragen, Salsa Shark, Cyan, Stefan-S, Poor Yorick, Kwekubo, Jiang, Eirik (usurped), Mxn, BRG, Smack, Schneelocke, Jengod, Okome~enwiki, Emperorbma, EL Willy, Eszett, Adam Bishop, Reddi, Stone, Piolinfax, Dtgm, Selket, Tpbradbury, Rarb, Maximus Rex, Nv8200pa, Tempshill, Bevo, Traroth, Shizhao, Stormie, Dpbsmith, Bcorr, Secretlondon, Jusjih, Just another user 2, Darthchaos, Jeffq, Lumos3, Denelson83, Jni, Nofutureuk, Gromlakh, Gentgeen, Robbot, Phisite, Juve82, Fredrik, Chris 73, WormRunner, Altenmann, Romanm, Naddy, Lowellian, WebElements, Yosri, Rfc1394, Texture, Hippietrail, Caknuck, Bkell, David Edgar, Borislav, Eliashedberg, Radagast, David Gerard, Giftlite, DocWatson42, Haeleth, Ævar Arnfjörð Bjarmason, Tom harrison, Lupin, Everyking, Bkonrad, No Guru, NeoJustin, Bensaccount, Zaphod Beeblebrox, AJim, Avsa, Yekrats, Dmmaus, Archenzo, Brockert, Darrien, SWAdair, Bobblewik, Deus Ex, Edcolins, Lucky 6.9, Peter Ellis, Gadfium, Zed0, Ran, Antandrus, Ctachme, PDH, Jossi, Exigentsky, Kesac, Vbs, Icairns, Sam Hocevar, Clemwang, Karl Dickman, Adashiel, Iwilcox, EagleOne, Mike Rosoft, Alkivar, D6, Andrew11, Poccil, Zarxos, EugeneZelenko, Felix Wan, A-giau, Noisy, Discospinster, Rich Farmbrough, KarlaQat, Cacycle, Inkypaws, Vsmith, Samboy, Joeclark, SpookyMulder, Bender235, TerraFrost, Sunborn, Klenje, RJHall, El C, Kwamikagami, Shanes, Briséis~enwiki, RoyBoy, Femto, Semper discens, Grick, Bobo192, AlHalawi, Whosyourjudas, Nyenyec, Reinyday, Clawson, Cwolfsheep, Dbchip, Giraffedata, SpeedyGonsales, Jojit fb, Nk, Eddideigel, Conget~enwiki, Jhd, Conny, Stephen G. Brown, Danski14, Honeycake, Orzetto, Alansohn, Mo0, Atlant, Keenan Pepper, Plumbago, Sl, Damnreds, AzaToth, Mac Davis, Caesura, Blobglob, Wtmitchell, ClockworkSoul, Unconventional, Helixblue, Stephan Leeds, Harej, RJFJR, Skatebiker, Computerjoe, GabrielF, Ghirlandajo, HGB, Feline1, Weyes, Lucent, Philthecow, Cimex, TigerShark, Benbest, Mpatel, Schzmo, U10ajf, Bluemoose, CharlesC, Waldir, SeventyThree, EarthmatriX, MarcoTolo, Cataclysm, V8rik, Qwertyus, Kbdank71, FreplySpang, DePiep, Dwaipayanc, Canderson7, Drbogdan, Saperaud~enwiki, Angusmclellan, Joe Decker, Koavf, Oblivious, SeanMack, Shalmanese, Sango123, Ptdecker, Yamamoto Ichiro, RobertG, Pumeleon, Nivix, Pathoschild, RexNL, Gurch, Kolbasz, Brendan Moody, Scerri, Alphachimp, Kri, Dalta~enwiki, Glenn L, Physchim62, Innotminkus, Chobot, Visor, Jared Preston, DVdm, Bgwhite, Gwernol, EamonnPKeane, Roboto de Ajvol, Mercury McKinnon, YurikBot, Wavelength, Hairy Dude, Deeptrivia, Phantomsteve, RussBot, Vlad4599, Fabartus, SpuriousQ, IanManka, Stephenb, Rintrah, Alvinrune, Schoen, Rsrikanth05, Bovineone, Wimt, Stassats, Anomalocaris, EngineerScotty, NawlinWiki, Wiki alf, E123, Test-tools~enwiki, Jaxl, Terfili, Yahya Abdal-Aziz, Mkouklis, Nick, Ragesoss, Dhollm, Cholmes75, Dmoss, Matticus78, RUL3R, AdiJapan, Ryanminier, Juanpdp, Hv, Misza13, Beanyk, Aaron Schulz, Bota47, CorbieVreccan, Derek.cashman, DRosenbach, Elkman, Phaedrus86, Smaines, Wknight94, Tetracube, FF2010, Ageekgal, Closedmouth, Jwissick, Ketsuekigata, Sean Whitton, Petri Krohn, DGaw, CWenger, Smurrayinchester, Kungfuadam, Junglecat, RG2, NeilN, DVD R W, Itub, Thecroman, SmackBot, Android 93, Bobet, Reedy, InverseHypercube, KnowledgeOfSelf, TestPilot, Melchoir, Unyoyega, KocjoBot~enwiki, Davewild, Thunderboltz, Milesnfowler, Anastrophe, Delldot, J0lt C0la, Knowhow, Elk Salmon, Edgar181, HalfShadow, Eupedia, Srnec, Gilliam, Ohnoitsjamie, Skizzik, Carbon-16, JRSP, Chris the speller, Keegan, Iskander32, RDBrown, Thumperward, Fuzzform, Lollerskates, EncMstr, MalafayaBot, OrangeDog, Roscelese, Bonaparte, Xoyorkie13, Metacomet, Dustimagic, DHN-bot~enwiki, DNAmaster, Darth Panda, Suicidalhamster, Can't sleep, clown will eat me, Onorem, Clorox, Konczewski, Squadoosh, Andy120290, Ddon, DR04, UU, Grover cleveland, Jachapo, PiMaster3, TotalSpaceshipGuy3, Savidan, Dreadstar, Pwjb, Aco47, Peterwhy, Jklin, DMacks, BrotherFlounder, Suidafrikaan, Sadi Carnot, The undertow, SashatoBot, Mchavez, Nishkid64, Tarantola, LtPowers, Archimerged, Khazar, Vitall, Scientizzle, Gobonobo, Btg2290, Anoop.m, Olin, ManiF, JohnWittle, Moop stick, Jaywubba1887, Ckatz, Dale101usa, Chrisch, Garudabd, Digger3000, Slakr, Rainwarrior, Beetstra, Mr Stephen, AxG, Arkrishna, Mets501, Ambuj.Saxena, Ryulong, RichardF, Jose77, DGtal, WOWGeek, Sifaka, Asyndeton, Ramuman, Dead3y3, Michaelbusch, Walton One, Tabfugnic, David Little, J Di, CapitalR, DavidOaks, Supertigerman, Pearson3372, Az1568, Courcelles, Túrelio, Ziusudra, Dpeters11, Tawkerbot2, Bobby131313, VinceB, Cryptic C62, Kaischwartz, Lincmad, TranClan, JForget, Betaeleven, Deon, Eli84, Van helsing, NullAshton, CBM, Rawling, DSachan, GHe, Fork me, Egmonster, Black and White, FlyingToaster, Wikiman7~enwiki, WeggeBot, Logical2u, Pi Guy 31415, Johnlogic, MrFish, Bill Sayre, Dmsc893, Rudjek, Nebular110, Reywas92, Grahamec, MC10, Rasmus vendelboe, Vanished user vjhsduheuiui4t5hjri, Rifleman 82, Corpx, GeorgeTopouria, Islander, Mycroft.Holmes, Methyl~enwiki, Fifo, Christian75, DumbBOT, Chrislk02, Shrikethestalker, Taylor4452, Ndufour, Memorymike, JodyB, Rowlaj01, Calvero JP, Satori Son, Casliber, Thijs!bot, Full On, Barticus88, ShayneRyan, Opabinia regalis, Kiwi137, Corsair18, Dagrimdialer619, Headbomb, Sobreira, Marek69, Ydoommas, John254, Racantrell, Dmitri Lytov, Philippe, Nezzington, Nemti, Escarbot, Mentifisto, Lani123, Tom dl, AntiVandalBot, Luna Santin, Michael phan, Bigtimepeace, Random user 8384993, Chill doubt, LegitimateAndEvenCompelling, Myanw, Figma, Tomertomer, JAnDbot, Barek, MER-C, Zerotjon, Sanchom, Hut 8.5, Kirrages, Kerotan, Maurakt, Magioladitis, Canjth, Bongwarrior, VoABot II, JNW, Kinston eagle, Redaktor, Aa35te, SparrowsWing, Avicennasis, Superworms, Ahecht, Nposs, Wikiak, Dirac66, Adrian J. Hunter, Allstarecho, ChrisSmol, StuFifeScotland, User A1, Musicloudball, Cpl Syx, Vssun, Just James, DerHexer, Khalid Mahmood, TheRanger, DancingPenguin, FisherQueen, Hdt83, MartinBot, HLewis, PostScript, Rettetast, Roastytoast, 1993 lol, Glrx, Kateshortforbob, CommonsDelinker, Supia, PrestonH, Thomasrive, Exodecai101, Slash, J.delanoy, Pharaoh of the Wizards, Ilovestars89, ChickenMarengo, Hans Dunkelberg, Psycho Kirby, Smartweb, I2yu, Sergeibernstein, Tempnegro, Extransit, WarthogDemon, Munkimunki, Richard777, Thom.fynn, Tdadamemd, Yvonr, Adamsbriand, EH74DK, Rescorbic, McSly, Aonrotar, Ryan Postlethwaite, Ephebi, Notapotato, Gurchzilla, Wasitgood69, Wasitgood, Paulbkirk, Monkeybutt5423, AntiSpamBot, Gffootball58, Coin945, Bigsnake 19, NewEnglandYankee, Creator58, Joka1991, SJP, C0RNF1AK35, Shoessss, Bob, Vanished user 39948282, BrianScanlan, Nat682, Darklama, Hyuuganeji0123, Arjun Rana, AnjuX, Suuperturtle, Idioma-bot, Johnnieblue, Deor, 28bytes, VolkovBot, Thedjatclubrock, Iosef, Jmocenigo, Christophenstein, Jeff G., JohnBlackburne, TheOtherJesse, RemoteCar, Barneca, Philip Trueman, Af648, Drunkenmonkey, Sweetness46, TXiKiBoT, TheVault, A4bot, Quilbert, Caster23, GDonato, Miranda, Chrisk12, Sankalpdravid, Qxz, Littlealien182, Anna Lincoln, Corvus cornix, Martin451, Jackfork, LeaveSleaves, Andrewrost3241981, DBragagnolo, Luuva, Quindraco, Gona.eu, RadiantRay, Madhero88, Jinglesmells999, Finngall, Aciddoll, Superjustinbros., Deeryh01, Synthebot, Enviroboy, Chengyq19942007, Insanity Incarnate, Everybody's Got One, Why Not A Duck, Brianga, Jaybo007, HiDrNick, LuigiManiac, Petergans, ConnTorrodon, NHRHS2010, SieBot, Coffee, Mikemoral, TJRC, Rihanij, PlanetStar, Jmwwiki, Borgdylan, Gprince007, Tiddly Tom, Scarian, WereSpielChequers, Jauerback, Jack Merridew, Gerakibot, Dawn Bard, Viskonsas, Caltas, Kragenz, The way, the truth, and the light,

Michfg, Sat84, Whiteghost.ink, Til Eulenspiegel, Purbo T, Tiptoety, Exert, Elcobbola, Nopetro, Hamilton hogs, Hiddenfromview, Segalsegal, SquirrelMonkeySpiderFace, Oxymoron83, Nuttycoconut, Lightmouse, Hwn tls, Hak-kâ-ngìn, BenoniBot~enwiki, Jack the Stripper, Sunrise, Dillard421, Werldwayd, Pappapasd, Maelgwnbot, DixonD, Nergaal, Precious Roy, Escape Orbit, Into The Fray, Jimmy Slade, Kanonkas, Georgedriver, Mr. Granger, Twinsday, ClueBot, PipepBot, Dinamik, The Thing That Should Not Be, Apastrophe, Techdawg667, Gawaxay, Syhon, Thompsontm, Drmies, Polyamorph, Elsweyn, Ryoutou, CounterVandalismBot, Ansh666, Blanchardb, LizardJr8, Dylan620, Timex987, Wifiless, Puchiko, Natasha.fielding, Dlorang, Robert Skyhawk, Excirial, Alexbot, BirgerH, Omgosh2, Jazjaz92, NuclearWarfare, Pearrari, Kaeso Dio, Jotterbot, LarryMorseDCOhio, Psinu, Realm up, DeltaQuad, Kaiba, Dekisugi, Pwntskater, NolanRichard, Thehelpfulone, La Pianista, Another Believer, Kiran the great, Viper275, Aitias, Blargblarg89, Versus22, Sch00l3r, SoxBot III, MairAW, JDT1991, BalkanFever, Vanished user uih38riiw4hjlsd, Indopug, TimothyRias, Jean-claude perez, Neuralwarp, XLinkBot, Shpakovich, Gonzonoir, Nsimya, TheSickBehemoth, Ryuken14, Nsim, WikHead, Noctibus, JinJian, ZooFari, MystBot, RyanCross, Thatguyflint, Roentgenium111, Tomilee0001, Yoenit, OmgItsTheSmartGuy, NicholasSThompson, JenR32, Vchorozopoulos, WFPM, Cst17, Download, SoSaysChappy, EconoPhysicist, Mathmarker, Glane23, Plutonium55, Debresser, LinkFA-Bot, Drova, LiveAgain, PoliteCarbide, Numbo3-bot, Tide rolls, BrianKnez, Luckas Blade, Greyhood, Arbitrarily0, Angrysockhop, Jack who built the house, Luckas-bot, Yobot, Essam Sharaf, Azylber, KamikazeBot, Jobroluver98, EnDaLeCoMpLeX, MacTire02, Tempodivalse, AnomieBOT, Zhieaanm, Helixer, Rubinbot, Jake Fuersturm, Degg444, Daniele Pugliesi, Piano non troppo, Collieuk, AdjustShift, Kingpin13, Ulric1313, Materialscientist, RadioBroadcast, The High Fin Sperm Whale, Citation bot, ArthurBot, Carturo222, Xqbot, Ziaix, Timir2, Sionus, Gopal81, Vidshow, Tfts, YBG, Grim23, Srich32977, WingedSkiCap, Pirateer, GrouchoBot, RibotBOT, Ian Fraser at Temple Newsam House, Spesh531, FaTony, AlimanRuna, A. di M., Nolimits5017, Dougofborg, Thehelpfulbot, R8R Gtrs, FrescoBot, Ylime715, Dogposter, StaticVision, Michael93555, Car132, Zach112233, Weetoddid, Strongbadmanofme, Maxus96, Grandiose, Xhaoz, Dellacomp, Citation bot 1, Pezzells, Pshent, Redrose64, ArnaudContet, Pinethicket, I dream of horses, M.pois, Hamtechperson, BTolli, Gemmi3, RedBot, Fishekad, NarSakSasLee, Kangxi emperor6868, Hardwigg, Abc518, Tjlafave, Lightlowemon, Gamewizard71, FoxBot, Double sharp, Adult Swim Addict, TobeBot, Graniggo, Jaeger Lotno, Pitcroft, Sillyboy67, Hughbert512369, Dinamik-bot, BZRatfink, Fyandcena, Kelvin35, Mattvirajrenaudbrandon, Turn off 2, Imawsome 09, Mass09, Tbhotch, Pldx1, Naughtysriram, DARTH SIDIOUS 2, Luhar1997, Twonernator, Pickweed, Dancojocari, EmausBot, Sir Arthur Williams, WikitanvirBot, GA bot, Franjklogos, Tf1321, Tommy2010, P. S. F. Freitas, Kitrkatr, 쓰쓰쓰쓰, Zainiadragon10000, JSquish, Jakers69, StringTheory11, Dffgd, Cobaltcigs, Шугуан, Joshlepaknpsa, Hzb pangus, Elly4web, Arman Cagle, KotVa, Sven nestle2, Hh73wiki, Ego White Tray, Negovori, ChuispastonBot, GermanJoe, Sunshine4921, Rmashhadi, Chaotic iak, Whoop whoop pull up, Heidslovesearl, Xanchester, ClueBot NG, Ihakeycakeyabreak, Wd930, Ozkithar Salas, Manubot, CocuBot, Lanthanum-138, Frietjes, Hazhk, Moneya, Parcly Taxel, Metaknowledge, Willzuk, Helpful Pixie Bot, RobertGustafson, ಅಭಿ ಕಾರಂತ್, Curb Chain, Bibcode Bot, BG19bot, MKar, Vagobot, Nagasturg, Sandbh, Mark Arsten, IraChesterfield, Soerfm, Zedshort, FeralOink, Razzat99, SoylentPurple, BattyBot, Justincheng12345-bot, Judiakok1985, Ziggypowe, Ushau97, Maxronnersjo, ChrisGualtieri, JYBot, Dexbot, LightandDark2000, Aditya Mahar, Mogism, Burzuchius, Carpelogos, Jtrevor99, Leitoxx, Kazim5294, AmericanLemming, WikiEditor2563, 08adamsm, Michel Djerzinski, The Herald, Shearflyer, Quenhitran, Asadwarraich, Kind Tennis Fan, Monkbot, HiYahhFriend, Deepak harshal nagle, Trackteur, Encyclopedia Lu, IiKkEe, Shane Stachwick, Mogie Bear, R. Portela F., Forscienceonly, KasparBot, Alistairgray42, Equinox, Lexi sioz and Anonymous: 1391

- **Atomic number** *Source:* https://en.wikipedia.org/wiki/Atomic_number?oldid=686410124 *Contributors:* AxelBoldt, Kpjas, Sodium, Mav, The Anome, Andre Engels, Peterlin~enwiki, Fonzy, Tim Starling, Liftarn, Fruge~enwiki, Eric119, Docu, Suisui, Александър, Julesd, Andres, The Anomebot, Paul-L~enwiki, Philopp, Skood, Donarreiskoffer, Gentgeen, Robbot, Fredrik, Jredmond, Romanm, Naddy, DHN, Hadal, Wikibot, Michael Snow, Giftlite, Guanaco, Eequor, Pne, Antandrus, Icairns, Tsemii, Mormegil, HedgeHog, Discospinster, Rich Farmbrough, Guanabot, Vsmith, Ponder, Bender235, Spearhead, RoyBoy, Femto, Bobo192, Nigelj, Brim, Sam Korn, Methegreat, Jumbuck, Alansohn, Bart133, Nuno Tavares, ^demon, Palica, Graham87, Rsg, Chun-hian, Ceinturion, Vary, SMC, Ian Dunster, Watcharakorn, ChongDae, Kolbasz, Physchim62, Chobot, DVdm, Mhking, VolatileChemical, Akke~enwiki, YurikBot, JWB, Jimp, Kirill Lokshin, Eleassar, NawlinWiki, Bota47, Zarboki, FF2010, Lt-wiki-bot, Adilch, Nemu, Dspradau, Petri Krohn, Paul D. Anderson, Katieh5584, Itub, SmackBot, F, Unyoyega, Bomac, Kilo-Lima, WookieInHeat, MelancholieBot, Gilliam, Betacommand, Rmosler2100, Persian Poet Gal, Catchpole, Bonaparte, DHN-bot~enwiki, Sbharris, Darth Panda, Suicidalhamster, Scott3, Addshore, SundarBot, Astroview120mm, DMacks, Vina-iwbot~enwiki, SashatoBot, FrozenMan, Sir Nicholas de Mimsy-Porpington, NongBot~enwiki, Cielomobile, Muadd, Hetar, Courcelles, Tawkerbot2, Ale jrb, Erik Kennedy, Jsd, Nick Y., Clovis Sangrail, Christian75, DumbBOT, Chrislk02, FastLizard4, Thijs!bot, Epbr123, Mojo Hand, Headbomb, Escarbot, Dantheman531, Seaphoto, Quintote, Pwhitwor, Spencer, JAnDbot, Leuko, Plantsurfer, Fragilityfemme, Andonic, .anacondabot, Bongwarrior, VoABot II, Loonymonkey, Allstarecho, DerHexer, JaGa, Lilac Soul, J.delanoy, Pharaoh of the Wizards, Ginsengbomb, It Is Me Here, Gman124, L'Aquatique, Skier Dude, JohnnyRush10, KylieTastic, Treisijs, TraceyR, Idioma-bot, X!, CWii, AlnoktaBOT, Philip Trueman, TXiKiBoT, Oshwah, The Original Wildbear, Java7837, Mainstream Nerd, Seraphim, Martin451, JhsBot, LeaveSleaves, Querl, Tgm1024, Roarman113, AlleborgoBot, Sssprkrao~enwiki, EmxBot, Tswei, SieBot, Coffee, YonaBot, ToePeu.bot, Winchelsea, Caltas, Keilana, Scorpion451, Lightmouse, Mygerardromance, Troy 07, ImageRemovalBot, XDanielx, Atif.t2, ClueBot, Bobathon71, The Thing That Should Not Be, Wwheaton, Polyamorph, ChandlerMapBot, DragonBot, Excirial, Alexbot, Jusdafax, Estirabot, Jotterbot, Razorflame, Ace of Spades IV, DumZiBoT, Tarheel95, Dsvyas, Feinoha, Satinchair, ElMeBot, Yes, I'm A Scientist, Thatguyflint, Ketchup krew, Addbot, Wotsinaname, Ronhjones, TutterMouse, Leszek Jańczuk, Ehrenkater, Kukers12, Tide rolls, Hunyadym, Arbitrarily0, Luckas-bot, Yobot, Washburnmav, THEN WHO WAS PHONE?, Nallimbot, KamikazeBot, AnomieBOT, Rubinbot, IRP, JackieBot, Piano non troppo, Flewis, Materialscientist, Maxis ftw, Hyperlaosboy, Vuerqex, Maniadis, Frankenpuppy, Xqbot, TinucherianBot II, Capricorn42, Jsharpminor, GrouchoBot, RibotBOT, 78.26, Ernsts, A. di M., Nixón, FrescoBot, Tobby72, Wikipe-tan, Phillyfanjd, Dnafish, PSPatel, Colubedy, Pinethicket, Mathsman91, RedBot, SpaceFlight89, RandomStringOfCharacters, Theeditor19494, ItsZippy, Vrenator, Sampathsris, Tbhotch, Reach Out to the Truth, The Utahraptor, Bhawani Gautam, Salvio giuliano, DASHBot, EmausBot, WikitanvirBot, Avenue X at Cicero, Racerx11, Razor2988, Tommy2010, K6ka, Werieth, JSquish, ZéroBot, Bollyjeff, StringTheory11, Wayne Slam, Ocaasi, Rcsprinter123, L Kensington, Donner60, Ihardlythinkso, RockMagnetist, Kowl kaz, ClueBot NG, Widr, Diyar se, Lowercase sigmabot, Will102, Greeny 12 1, Dan653, Mark Arsten, Joydeep, Vladihernando, ChrisGualtieri, EuroCarGT, 786b6364, Bashiaden, Warp Dragon, The Herald, Reginatr08, Dannyzhaofb, TheEpTic, De Riban5, Poepkop, Filedelinkerbot, Guttyut, Rstensby, Aryavat Kapoor, GeneralizationsAreBad, KasparBot, JJMC89 and Anonymous: 399

- **Electron configuration** *Source:* https://en.wikipedia.org/wiki/Electron_configuration?oldid=681720067 *Contributors:* The Anome, Andre Engels, SimonP, Robot5005, Bth, Lir, Patrick, Michael Hardy, Erik Zachte, Kku, Karada, Looxix~enwiki, Snoyes, Darkwind, GaryW, Julesd, Djnjwd, Andres, Smack, Ddoherty, Malbi, Timwi, Dysprosia, LMB, Topbanana, Fvw, Donarreiskoffer, Gentgeen, Robbot, Astronautics~enwiki, R3m0t, Yelyos, Romanm, Ashley Y, Ojigiri~enwiki, Hadal, Timemutt, Tobias Bergemann, Unfree, Mor~enwiki, Giftlite,

Anville, Mboverload, Nayuki, Lucky 6.9, Utcursch, Antandrus, Mako098765, Rdsmith4, Discospinster, FT2, Vsmith, Bender235, Donsimon~enwiki, Brian0918, Kwamikagami, Kouhoutek, Spoon!, WhiteTimberwolf, Femto, Nk, NickSchweitzer, Jumbuck, Alansohn, Anthony Appleyard, Dark Shikari, Bucephalus, Colin Kimbrell, Cburnett, Arag0rn, RainbowOfLight, Dirac1933, Linas, StradivariusTV, Kurzon, Mpatel, Marudubshinki, V8rik, DePiep, MZMcBride, Azure8472, Margosbot~enwiki, Crazycomputers, RexNL, Alexjohnc3, Fresheneesz, Mattopia, Glenn L, Physchim62, Chobot, YurikBot, Wavelength, RobotE, X42bn6, Jengelh, KSmrq, RadioFan, Pseudomonas, Nawlin-Wiki, SEWilcoBot, Dumoren, Nsmith 84, Vb, E2mb0t~enwiki, Alex43223, EEMIV, BOT-Superzerocool, Cstaffa, Djdaedalus, Tetracube, Johndburger, 2over0, Jованб6, Arthur Rubin, Josh3580, TBadger, LeonardoRob0t, Kungfuadam, RG2, Janek Kozicki, P. B. Mann, DrJolo, SmackBot, KnowledgeOfSelf, Bomac, Jacek Kendysz, Yamaguchi⁇⁇, Skizzik, Kmarinas86, Rmosler2100, Hugo-cs, Chris the speller, Jayko, Bduke, Dreg743, Robth, DHN-bot~enwiki, Sbharris, Colonies Chris, John Reaves, Dethme0w, Sergio.ballestrero, OrphanBot, Rrburke, Aldaron, Nakon, Nrcprm2026, Fuzzypeg, DMacks, Bhludzin, SashatoBot, OhioFred, John, AstroChemist, Anoop.m, Beetstra, Aeluwas, Iridescent, CmdrObot, Calmargulis, Necessary Evil, Tawkerbot4, Quibik, Steviedpeele, Cinderblock63, Thijs!bot, Epbr123, Qwyrxian, Headbomb, CharlotteWebb, Escarbot, AntiVandalBot, Luna Santin, Seaphoto, Tyco.skinner, Byrgenwulf, Res2216firestar, Deadbeef, JAnDbot, PhilKnight, Meeples, Bongwarrior, VoABot II, Marik7772003, Baccyak4H, Animum, EagleFan, Dirac66, Adventurer, JohnofPhoenix, Patstuart, MartinBot, Schmloof, Rettetast, Yannledu, AlphaEta, J.delanoy, Maurice Carbonaro, KeepItClean, Brant.merrell, Skier Dude, Matt18224, SmilesALot, Juliancolton, DavidCBryant, CardinalDan, Idioma-bot, VolkovBot, Mrh30, LokiClock, Af648, TXiKiBoT, Pjstewart, Anna Lincoln, Cerebellum, Axiosaurus, Broadbot, Cremepuff222, BotKung, Ilyushka88, Krazywrath, Falcon8765, MrChupon, HansHermans, Coffee, OMCV, Graham Beards, RJaguar3, Tiptoety, Radon210, Fzhi555, Oxymoron83, Jdaloner, IdealEric, Svick, Retireduser1111, Anchor Link Bot, Tesi1700, Pinkadelica, ClueBot, The Thing That Should Not Be, CounterVandalismBot, Ankit.sunrise, Tamariandre, Kevkevkevkev, DragonBot, Djr32, NuclearWarfare, Yuk ngan, Asaad900, Bliz1, Aitias, Ace of Spades IV, Footballfan190, RMFan1, Qbmaster, Dnvrfantj, Aritate, Addbot, Glane23, AtheWeatherman, Weekwhom, Drova, Tide rolls, Teles, Gail, Nantaskot, Legobot, आशीष भटनागर, Luckasbot, Amirobot, KarlHegbloom, عالم, محبوب, AnomieBOT, Götz, Daniele Pugliesi, Jim1138, 9258fahsflkh917fas, Chuckiesdad, Ulric1313, Materialscientist, The High Fin Sperm Whale, Citation bot, ArthurBot, Xqbot, S h i v a (Visnu), 4twenty42o, Earlypsychosis, RibotBOT, SassoBot, Shadowjams, Prari, VS6507, Chau7, Spacecadet262, Redrose64, TokioHotel93, Pinethicket, MJ94, Meddlingwithfire, Appy3, Jelson25, Double sharp, علی ویکی, Diannaa, Mttcmbs, DARTH SIDIOUS 2, Deagle AP, EmausBot, Manifolded, Wikipelli, Holyhell5050, AManWithNoPlan, RaptureBot, Donner60, NTox, DASHBotAV, Rocketrod1960, Trevorhailey1, Fucktosh, ClueBot NG, Dipanshu.sheru, A520, Wd930Bot, Lanthanum-138, Helpful Pixie Bot, HMSSolent, Bibcode Bot, BG19bot, MusikAnimal, Dan653, Glevum, Gwickwire, Virtualzx, Aisteco, Samanthaclark11, Davidwhite18, Mediran, Putodog, M Farooq 2012, Maryann gersaniva, Denden0019, Mychael23, Monkbot, Apisani82, Cautious Chemist, Wobbieftw, Alango1998, Sayambohra, Pfam32, CyberWarfare and Anonymous: 490

- **Chemical property** *Source:* https://en.wikipedia.org/wiki/Chemical_property?oldid=684705155 *Contributors:* You~enwiki, Grimm Ripper, Kku, Reddi, Taxman, Gentgeen, TomPhil, Wile E. Heresiarch, Hung yao, Slyguy, Jackol, Antandrus, Karol Langner, Trevor MacInnis, Maestrosync, Discospinster, Cacycle, Vsmith, ESkog, Bobo192, Hurricane111, Smalljim, Kjkolb, Alansohn, Andrewpmk, Plumbago, Cburnett, Sciurinæ, Morios, Isnow, V8rik, S Schaffter, AJR, Gurch, Physchim62, DVdm, YurikBot, Retaggio, Tony1, CWenger, Katieh5584, Junglecat, Paul Erik, DVD R W, ChemGardener, Itub, SmackBot, Elonka, Jab843, Edgar181, Commander Keane bot, Gilliam, Skizzik, Andy M. Wang, Tytrain, Hallenrm, Can't sleep, clown will eat me, Frap, Rrburke, Dreadstar, Astroview120mm, BinaryTed, Jklin, DMacks, Nishkid64, NongBot~enwiki, IronGargoyle, Beetstra, Sifaka, Dead3y3, Iridescent, Arkhiver, IvanLanin, Lenoxus, Muhand, Dycedarg, Reeves, Adluolas, Peripitus, Gogo Dodo, Christian75, Roberta F., Alaibot, Epbr123, Kablammo, AntiVandalBot, Spencer, Leuko, Davewho2, PhilKnight, VoABot II, JamesBWatson, Riceplaytexas, WODUP, Catgut, DerHexer, Video game fan11, Hdt83, MartinBot, ChemNerd, J.delanoy, Nigholith, Stan J Klimas, Ycdkwm, NewEnglandYankee, Scoterican, WilfriedC, Tygrrr, CardinalDan, Striker30, VolkovBot, AlnoktaBOT, Oshwah, KyleRGiggs, W4chris, Dawn Bard, Calabraxthis, Keilana, Aillema, Oxymoron83, Manway, OKBot, Vig vimarsh, Pinkadelica, Into The Fray, Atif.t2, ClueBot, PipepBot, The Thing That Should Not Be, Arakunem, Blanchardb, Xrangestrxx, Chemsmith, Excirial, Jusdafax, PixelBot, Jerry Zhang, Versus22, SoxBot III, Spitfire, Gonzonoir, Frood, Menthaxpiperita, Cäsium137~enwiki, Thatguyflint, Addbot, Myhero22, Crazysane, Ronhjones, Cst17, Glane23, Ehrenkater, Tide rolls, Legobot, Luckas-bot, Adi, Ojay123, AnomieBOT, Bsimmons666, Jim1138, Piano non troppo, Sz-iwbot, Materialscientist, The High Fin Sperm Whale, Neurolysis, Crix Madine, Xqbot, Sionus, Cureden, Elvim, Craftyminion, YBG, SassoBot, Smallman12q, Middle 8, Erik9bot, Michael93555, Jamesooders, Pinethicket, Vicenarian, Edderso, Half price, Hamtechperson, Fumitol, FoxBot, Sanchezman, Clarkcj12, MrX, Mileycamille, Reaper Eternal, Soccergamer11, El Mayimbe, Forenti, GeneralCheese, Immunize, RA0808, NiJoKeJONASROCKS, K6ka, Maxviwe, Access Denied, Aeonx, Wayne Slam, Tolly4bolly, Jay-Sebastos, ClueBot NG, Widr, Krenair, Sandbh, Hallows AG, Makerman123, Jorgeposada14, Suradnik50, Lugia2453, Maniesansdelire, Harlem Baker Hughes, Glaisher, Jianhui67, Blake532001, Amortias, THEYCALLMELOBBY, WyattAlex, Lily6978 and Anonymous: 436

- **Block (periodic table)** *Source:* https://en.wikipedia.org/wiki/Block_(periodic_table)?oldid=675584164 *Contributors:* Mav, The Anome, Tarquin, Bth, Looxix~enwiki, Ahoerstemeier, Suisui, Smack, Emperorbma, Gentgeen, Robbot, SimonMayer, Ctachme, Rich Farmbrough, Hashar-Bot~enwiki, Jumbuck, Urod, DePiep, Butros, EamonnPKeane, YurikBot, RobotE, Höyhens, ⁇⁇⁇⁇ robot, SmackBot, Gilliam, Sbharris, SundarBot, DMacks, Sombrero, IvanLanin, Wfructose, N2e, Christian75, Thijs!bot, Barticus88, Dirac66, Numbo3, CardinalDan, VolkovBot, TXiKiBoT, Pjstewart, Cosmium, Synthebot, BotMultichill, Nergaal, ClueBot, Mardetanha, ChandlerMapBot, Mikaey, Cantor, Addbot, Yoenit, WFPM, Download, RTG, Legobot, आशीष भटनागर, AnomieBOT, Materialscientist, Am13gore, Brane.Blokar, Xqbot, Лев Дубовой, Erik9bot, Haida19, Cyanos, MondalorBot, Gamewizard71, FoxBot, Double sharp, Krit-tonkla, J'88, StringTheory11, ClueBot NG, Lanthanum-138, Tholme, Sandbh, Marcocapelle, Kc kennylau, Pratyya Ghosh, Carbon6, Bsmayank11, Bergfurher, MetazoanMarek and Anonymous: 27

- **Period (periodic table)** *Source:* https://en.wikipedia.org/wiki/Period_(periodic_table)?oldid=686879565 *Contributors:* Mav, The Anome, Dcljr, Ahoerstemeier, Suisui, Александър, Xnybre, Jusjih, Gentgeen, Robbot, Yekrats, Kandar, Ctachme, 1297, Icairns, LiSrt, Mike Rosoft, Kwamikagami, Fremsley, Jumbuck, Alansohn, Wtmitchell, Velella, Feline1, Blaxthos, Woohookitty, Benbest, Tckma, M412k, DePiep, PHenry, MZMcBride, Jared999, Glenn L, Chobot, Bgwhite, Roboto de Ajvol, YurikBot, RobotE, Grafen, Ms2ger, Adilch, Josh3580, Allens, Kungfuadam, GrinBot~enwiki, SmackBot, Melchoir, Eskimbot, Skizzik, DHN-bot~enwiki, Shunpiker, VMS Mosaic, SashatoBot, Lambiam, Rigadoun, Stwalkerster, Jimbuckar00, N2e, Rifleman 82, Thijs!bot, Epbr123, Headbomb, Nick Number, Northumbrian, Tharkon, Jayron32, Legolost, .anacondabot, Easchiff, DerHexer, Kayau, MartinBot, R'n'B, Tgeairn, RockMFR, Reedy Bot, ThsTorturedSoul, VolkovBot, AlnoktaBOT, Soliloquial, TXiKiBoT, Rei-bot, Qxz, Pimemorizer, Luuva, Insanity Incarnate, SieBot, Rdx-77, Keilana, Plainandsimple, ClueBot, Wikijens, ChandlerMapBot, BOTarate, WikHead, SilvonenBot, HexaChord, Addbot, Ronhjones, Laurinavicius, LinkFA-Bot, Tide rolls, Legobot, आशीष भटनागर, Luckas-bot, Yobot, AnomieBOT, DemocraticLuntz, IRP, JackieBot, Materialscientist, Brane.Blokar, MauritsBot, Xqbot, YBG, Jakwra, GrouchoBot, Kieryh, Amaury, Erik9bot, LucienBOT, Pinethicket, Rushbugled13, RedBot, Double sharp, RjwilmsiBot, TjBot,

EmausBot, Noobatron30, Golfandme, ZéroBot, StringTheory11, Bethastrong, Isarra, Palosirkka, ClueBot NG, Lanthanum-138, Marechal Ney, Widr, Shovan Luessi, Helpful Pixie Bot, Strike Eagle, Titodutta, Sandbh, Vivek7de, ChrisGualtieri, JYBot, Tripodno1, Suradnik50, Vanamonde93, Eyesnore, George8211, Monkbot, Rider ranger47, Zschrambo, Phizrotythe the Beautiful Mustache and Anonymous: 142

- **Group (periodic table)** *Source:* https://en.wikipedia.org/wiki/Group_(periodic_table)?oldid=686844397 *Contributors:* AxelBoldt, The Anome, Css, Fonzy, Michael Hardy, Tannin, Dcljr, ArnoLagrange, Ahoerstemeier, Suisui, Александър, Caramdir~enwiki, GCarty, Hollgor, Gentgeen, Robbot, Andrew Levine, Filemon, Giftlite, Robin Patterson, Herbee, Ctachme, 1297, DragonflySixtyseven, Elektron, Icairns, Felix Wan, Discospinster, Mani1, ESkog, El C, Edwinstearns, Femto, Jumbuck, Alansohn, Arthena, Gene Nygaard, Ianweller, DePiep, Farazy, Canderson7, FlaBot, Chobot, Roboto de Ajvol, YurikBot, RobotE, Deeptrivia, Hede2000, RadioFan, Thane, Adilch, Badgettrg, Stumps, Itub, MattieTK, Narson, Unyoyega, KocjoBot~enwiki, Edgar181, Hugo-cs, The Benefactor, Can't sleep, clown will eat me, Cybercobra, Steve Pucci, DMacks, John, Rigadoun, Mathias-S, Hvn0413, Ace Frahm, Paul Foxworthy, Beve, Dlohcierekim, Hammer Raccoon, J Milburn, Laplacian, Rwflammang, Christian75, UberScienceNerd, Marek69, Escarbot, AntiVandalBot, Seaphoto, Mister Macbeth, JAnDbot, Legolost, Adams kevin, .anacondabot, VoABot II, Dirac66, Rettetast, RockMFR, J.delanoy, Pharaoh of the Wizards, (jarbarf), Cometstyles, STBotD, CardinalDan, PeaceNT, Philip Trueman, TXiKiBoT, Cosmium, JhsBot, FourteenDays, Luuva, Falcon8765, PRimaVeriTaVIV~enwiki, SieBot, Malcolmxl5, ToePeu.bot, LeadSongDog, Keilana, Flyer22 Reborn, Nergaal, ClueBot, The Thing That Should Not Be, Plastikspork, MDov, Uncle Milty, Otolemur crassicaudatus, Ktr101, Excirial, ZJB3, Aitias, Plasmic Physics, WikHead, Tmark111, Addbot, Roentgenium111, Some jerk on the Internet, Axecution, Ronhjones, CactusWriter, NjardarBot, Kj cheetham, LinkFA-Bot, Numbo3-bot, आशीष भटनागर, Luckasbot, Fraggle81, TaBOT-zerem, AnomieBOT, Jim1138, IRP, Limideen, Citation bot, Xqbot, MrsZanolini, YBG, Jezhotwells, RibotBOT, Bob323457364584649, Shadowjams, StaticVision, Louperibot, Killian441, I dream of horses, DarrenHensley, RedBot, Reconsider the static, Double sharp, J'88, HK-90, TjBot, Agent Smith (The Matrix), Slon02, EmausBot, Orphan Wiki, Rabbabodrool, ZéroBot, StringTheory11, Zap Rowsdower, Mpicciotti, ClueBot NG, Alexgx, Lanthanum-138, Frietjes, O.Koslowski, Helpful Pixie Bot, Zscheidegger, Curb Chain, Frze, YodaRULZ, Justincheng12345-bot, HueSatLum, The Illusive Man, ChrisGualtieri, Scholani, Webclient101, Suradnik50, 14GTR, Lugia2453, T42N24T, Kilabotzx987, Fred88freddy, 7Sidz, Psizzle03, Kurousagi and Anonymous: 215

- **Halogen** *Source:* https://en.wikipedia.org/wiki/Halogen?oldid=687110940 *Contributors:* Mav, Tarquin, Jeronimo, Malcolm Farmer, Andre Engels, Josh Grosse, William Avery, Caltrop, Heron, Eric119, Ahoerstemeier, Suisui, Jebba, Александър, Julesd, Glenn, Mxn, Timwi, Stone, Dysprosia, Andrewman327, Taxman, LMB, Paul-L~enwiki, Topbanana, Johnleemk, David.Monniaux, Jeffq, Lumos3, Gentgeen, Robbot, Pigsonthewing, Rfc1394, Pko, Giftlite, No Guru, Jorge Stolfi, Rchandra, OldakQuill, OverlordQ, 1297, Icairns, Guybrush, Mike Rosoft, Venu62, NathanHurst, Vsmith, Mani1, Kbh3rd, Geoking66, RJHall, El C, Dajhorn, Remember, Femto, JRM, Sole Soul, Vortexrealm, Dungodung, La goutte de pluie, Officiallyover, Eddideigel, Jumbuck, Alansohn, Atlant, SemperBlotto, Plumbago, Benjah-bmm27, Lightdarkness, Caesura, DaMoose, Docboat, Gene Nygaard, Feline1, HenryLi, Ceyockey, Woohookitty, Camw, Dolfrog, Blackcats, Wayward, Palica, Stevey7788, DePiep, Mendaliv, Canderson7, Quiddity, Yamamoto Ichiro, Titoxd, APC, FlaBot, Latka, Drumguy8800, Scerri, Physchim62, Chobot, DVdm, Skraz, YurikBot, Wavelength, Al Silonov, Jimp, Hellbus, Russoc4, Bovineone, NawlinWiki, Welsh, RL0919, Nick C, Tim Goodwyn, Lockesdonkey, BOT-Superzerocool, Gadget850, Bota47, Nlu, Deeday-UK, Lt-wiki-bot, Ninly, Wd40, Pb30, KGasso, Naught101, HereToHelp, Memodude, TrygveFlathen, Itub, FocalPoint, KnowledgeOfSelf, Blue520, Jfurr1981, Jab843, Knowhow, Flameeyes, Nscheffey, Edgar181, HeartofaDog, Gilliam, Hmains, Skizzik, Chris the speller, Bluebot, MalafayaBot, RayAYang, DHN-bot~enwiki, Gruzd, Sbharris, MyNameIsVlad, Buttered Bread, Addshore, T-borg, DMacks, Kukini, SashatoBot, Titus III, Mgiganteus1, CaptainVindaloo, Brett leigh dicks, Nong bel~enwiki, Dicklyon, Anonymous anonymous, Solipse, Tawkerbot2, Dlohcierekim, Daniel5127, Orangutan, JForget, Charvex, WeggeBot, Karenjc, Ispy1981, Myasuda, Wild Wizard, Slazenger, Abeg92, Nick Y., Rifleman 82, Gogo Dodo, Kotiwalo, Tawkerbot4, Christian75, Chrislk02, Blackjack48, Thijs!bot, Epbr123, Inner Earth, Messageman3, Sturm55, AntiVandalBot, Postlewaight, Badocter, Lfstevens, JAnDbot, MER-C, Greensburger, LittleOldMe, Magioladitis, Bongwarrior, VoABot II, Fusionmix, JamesBWatson, Balloonguy, Gtman, Causesobad, NatureA16, Hdt83, MartinBot, Darklord517, Ariel., JCraw, Sven87, Nono64, Bgold4, Dglosser, C.R.Selvakumar, J.delanoy, Pharaoh of the Wizards, Bombhead, Michael Daly, It Is Me Here, BaseballDetective, D.M.N., MartinBotIII, Squids and Chips, CardinalDan, Deor, VolkovBot, Indubitably, Chris Dybala, AlnoktaBOT, 8thstar, TXiKiBoT, Jomasecu, Vipinhari, A4bot, Someguy1221, James.Spudeman, Anna Lincoln, Clarince63, Camoflauge, RedAndr, Shanata, Madhero88, Dirkbb, GoTeamVenture, Dhall27, AlleborgoBot, Praefectorian, Snetter2007, SieBot, TCO, Nubiatech, PanagosTheOther, Gerakibot, Caltas, Zbvhs, Happysailor, Allmightyduck, Bsherr, Oxymoron83, HokieJC, Cosmo0, Driedshroom, Nergaal, Aadgray, Apuldram, ClueBot, The Thing That Should Not Be, Meisterkoch, Doseiai2, CounterVandalismBot, Wceom, MindstormsKid, DragonBot, Regardless143, Excirial, Peter.C, Terra Xin, Jotterbot, P1415926535, ChrisHamburg, Horselover Frost, 7, Versus22, Killua98 killer, Aaron north, Spitfire, Rror, Lily W, Noctibus, Vonlich, Addbot, Roentgenium111, Willking1979, Tcncv, Wickey-nl, Non-dropframe, Muzekal Mike, Marx01, Aboctok, Woelen, Chamal N, Favonian, LinkFA-Bot, Jasper Deng, Alchemist-hp, Erutuon, Tide rolls, Lightbot, Legobot, Luckas-bot, Yobot, Ptbotgourou, AnomieBOT, Shootbamboo, Tryptofish, Jim1138, Piano non troppo, AdjustShift, Materialscientist, Midnightblue94tg, Limideen, The High Fin Sperm Whale, Citation bot, E2eamon, Frankenpuppy, Xqbot, Piza pokr, Imphras Heltharn, Capricorn42, Elvim, Bojangleskelly, Lop242438, RibotBOT, RomanHunt, AntiAbuseBot, Jilkmarine, FrescoBot, Dogposter, Sky Attacker, Kcandnikko7, JMS Old Al, Wifione, Þjóðólfr, DrilBot, Pinethicket, Raerae123, I own in the bed, Tommo65, Calum-11, Raerae11, RECTUM INSPECTUM, Teh Ploxanater, RandomStringOfCharacters, Danielj123, Mikespedia, Jujutacular, FoxBot, Double sharp, IiizTheWiikiNerd, Dinamik-bot, Vrenator, Dominic Hardstaff, Pbrower2a, Bcai388, 4, Tssa 893, DARTH SIDIOUS 2, Dhburns, AXRL, Slon02, EmausBot, WikitanvirBot, Nuujinn, Pete Hobbs, Super48paul, Mattylak, Tommy2010, K6ka, Josve05a, StringTheory11, Wayne Slam, Harrylewis101, Setha43, Tolly4bolly, Donner60, 28bot, Whoop whoop pull up, Mjbmrbot, ClueBot NG, MelioraCogito, Jack Greenmaven, Eonsword, Gilderien, Lanthanum-138, Rezabot, Widr, Aprilade, Regulov, BG19bot, Vagobot, Rhysmcb1, ElphiBot, MusikAnimal, Klilidiplomus, Lieutenant of Melkor, BattyBot, Mrt3366, The Illusive Man, Littlelizey, Pwdent, BrightStarSky, Webclient101, Lugia2453, Frosty, ComfyKem, Gabby Merger, PinkAmpersand, Abuelita12345, Entymology, BenjHolladay, Jakec, Whirlwind927, Brainiacal, DudeWithAFeud, Scarlettail, DSCrowned, Beanstash, Aryamanarora, The Chemistry Bookworm, Christina M. Joseph, Nerpderpmerpperson, KasparBot, Scolit00 and Anonymous: 533

- **Noble gas** *Source:* https://en.wikipedia.org/wiki/Noble_gas?oldid=687136225 *Contributors:* AxelBoldt, Carey Evans, Derek Ross, LC~enwiki, Mav, Tarquin, Andre Engels, Christian List, DrBob, Fonzy, Hephaestos, Olivier, Edward, Shellreef, Ahoerstemeier, Jimfbleak, Rlandmann, Salsa Shark, Nikai, GCarty, Smack, Lenaic, Stone, Jake Nelson, Tero~enwiki, Paul-L~enwiki, Taoster, Betterworld, Fvw, Shantavira, Donarreiskoffer, Gentgeen, Robbot, Sander123, Jakohn, Romanm, Merovingian, Rfc1394, Flauto Dolce, Meelar, Mervyn, Hadal, JackofOz, Robinh, Lupo, Dina, Giftlite, DocWatson42, Herbee, Monedula, Karn, Everyking, Slyguy, Kandar, Quackor, Andycjp, Antandrus, 1297, Icairns, Sam Hocevar, Gscshoyru, Deglr6328, Adashiel, Mike Rosoft, DanielCD, Discospinster, Rich Farmbrough, Vsmith, Ponder, Paul August, NeilTarrant, Geoking66, RJHall, El C, Dnwq, Shanes, Remember, Sietse Snel, Art LaPella, Femto, Marco Polo, Shenme, Viriditas, SpeedyGonsales, Severious, Obradovic Goran, Haham hanuka, Nsaa, Eddideigel, Orangemarlin, Ranveig, Jumbuck, Alansohn, Gary, Jared81, Keenan Pepper, Plumbago, Cjthellama, InShaneee, Suruena, Vuo, Gene Nygaard, Feline1, Kay Dekker, Boothy443, Woohookitty, Cimex, TarmoK, LOL, Pol098, WadeSimMiser, Mpatel, Wayward, Shanedidona, Palica, Stevey7788, Paxsimius, Graham87, David Levy, DePiep, Canderson7, Rjwilmsi, Mfwills, Vary, HappyCamper, Matjlav, Vuong Ngan Ha, RobertG, Latka, Nihiltres, Strangnet, RexNL, Mjp797, DevastatorIIC, Goudzovski, Scerri, Kri, King of Hearts, Chobot, DVdm, Bgwhite, EamonnPKeane, YurikBot, Chaser, Ollie holton, Stephenb, Yyy, NawlinWiki, Bachrach44, Jaxl, Adamn, Semperf, Zirland, Bota47, T-rex, Thetoaster3, Wknight94, Tetracube, Leptictidium, Phgao, Zzuuzz, Adilch, Scoutersig, Keepiru, HereToHelp, Katieh5584, NeilN, GrinBot~enwiki, DVD R W, Tom Morris, That Guy, From That Show!, Itub, Anthony Duff, SmackBot, FocalPoint, Tarret, KnowledgeOfSelf, Shoy, Kilo-Lima, Edgar181, Gaff, Aksi great, Gilliam, Isaac Dupree, Pslawinski, Durova, Bluebot, Bduke, SchfiftyThree, Moshe Constantine Hassan Al-Silverburg, CSWarren, Robth, DHN-bot~enwiki, Darth Panda, Gyrobo, Tsca.bot, Eric Olson, MJCdetroit, Rrburke, Aldaron, Nakon, Eganev, Pwjb, Smokefoot, DMacks, Wizardman, The undertow, SashatoBot, Lambiam, Krashlandon, Titus III, John, Zaphraud, Jaganath, Breno, Anoop.m, IronGargoyle, JHunterJ, Slakr, Beetstra, Noah Salzman, Optimale, Kpengboy, SandyGeorgia, AdultSwim, MTSbot~enwiki, BranStark, Iridescent, StephenBuxton, Jaksmata, Tawkerbot2, Swampgas, JForget, Irwangatot, Ruslik0, CompRhetoric, Dorothybaez, Bentleymrk, Gogo Dodo, A Softer Answer, Hibou8, SteveMcCluskey, Mattisse, Thijs!bot, Epbr123, Kablammo, Headbomb, JaimeAnnaMoore, Straussian, Werdnanoslen, Dantheman531, Mentifisto, AntiVandalBot, Luna Santin, Opelio, Cinnamon42, Scepia, LibLord, Xnuiem, Gökhan, IanOsgood, Nicholas Tan, Easchiff, Animaly2k2, Magioladitis, Bongwarrior, VoABot II, Hasek is the best, Ling.Nut, LorenzoB, Thibbs, Vssun, DerHexer, JaGa, Awolnetdiva, Mattinbgn, Hdt83, ChemNerd, Polartsang, CommonsDelinker, AlexiusHoratius, Leyo, J.delanoy, Pharaoh of the Wizards, Phillip.northfield, Hans Dunkelberg, Dhruv17singhal, Uncle Dick, Jeri Aulurtve, Extransit, Cpiral, Bombhead, Wandering Ghost, Shay Guy, Coppertwig, Plasticup, Andraaide, Belovedfreak, Acey365, NewEnglandYankee, Najlepszy, Numerjeden, Matthardingu, KChiu7, WinterSpw, Brvman, Wilhelm meis, Squids and Chips, Idioma-bot, Wikieditor06, Deor, VolkovBot, Eakka, JGHowes, Jeff G., Chris Dybala, AlnoktaBOT, Nousernamesleft, Philip Trueman, Photonikonman, TXiKiBoT, Tavix, GimmeBot, Muro de Aguas, Rei-bot, Slysplace, Enigmaman, Davidmwhite, CephasE, Sylent, TinribsAndy, Owainbut, AlleborgoBot, Surfrat60793, SieBot, Calliopejen1, OTAVIO1981, Graham Beards, BotMultichill, ToePeu.bot, Jauerback, Nathan, Triwbe, Agesworth, Keilana, Flyer22 Reborn, The Evil Spartan, Arbor to SJ, Sohelpme, Scorpion451, Enok Walker, Lightmouse, OKBot, Nielg, Nimbusania, Nergaal, Escape Orbit, PerpetualSX, Runtishpaladin, UKe-CH, Martarius, ClueBot, Artichoker, The Thing That Should Not Be, Cygnis insignis, Manbearpig4, Franamax, Blanchardb, Piledhigheranddeeper, ChandlerMapBot, Puchiko, DragonBot, Excirial, Sidias300, GngstrMNKY, Jusdafax, Finch-HIMself, Estirabot, Poigol5043, Cenarium, Zomno, Jotterbot, Bellax22, Chaser (away), Werson, Boatcolour, SeanFarris, Thingg, Aitias, Dank, Versus22, RexxS, Boleyn, Neuralwarp, Feinoha, Little Mountain 5, Skarebo, Frood, Freestyle-69, CalumH93, Addbot, Mr0t1633, Roentgenium111, DOI bot, Theleftorium, Popopee, Ronhjones, Jncraton, Moosehadley, CanadianLinuxUser, AnnaFrance, Jasper Deng, Alchemist-hp, Numbo3-bot, Tide rolls, Zorrobot, Angrysockhop, Arimareiji, Legobot, Seresin.public, Luckas-bot, ZX81, Yobot, IsFari, TaBOT-zerem, Rsquire3, Bloody Mary (folklore), KamikazeBot, Widey, Synchronism, Andme2, AnomieBOT, Lolcopter666, Jcsdude, Navneethmohan, Jim1138, IRP, Law, Materialscientist, Citation bot, E2eamon, Maxis ftw, ArthurBot, Xqbot, Capricorn42, Nickkid5, Tad Lincoln, Turk oğlan, NocturneNoir, Lop242438, Pmlineditor, GrouchoBot, Doulos Christos, Antonjad, Jilkmarine, Smot94, Robo37, OgreBot, Citation bot 1, AstaBOTh15, Pinethicket, HRoestBot, Calmer Waters, I own in the bed, Marine79, Double sharp, TobeBot, Yopure, 777sms, Navy101, Reach Out to the Truth, Minimac, DARTH SIDIOUS 2, AXRL, Mean as custard, RjwilmsiBot, Japheth the Warlock, Ripchip Bot, Salvio giuliano, Deagle AP, EmausBot, WikitanvirBot, RA0808, Jordan776, Wikipelli, P. S. F. Freitas, AvicBot, ZéroBot, Fingerginger1, Maxviwe, StringTheory11, H3llBot, Makecat, Wagino 20100516, L Kensington, Donner60, Whoop whoop pull up, JohnMCrain, Mjbmrbot, Special Cases, Washington Irving Esquire, ClueBot NG, Rich Smith, Jack Greenmaven, Hon-3s-T, Skoot13, Ethanpiot, Lanthanum-138, Widr, Lolm8, Bibcode Bot, Swamphlosion, Lowercase sigmabot, Gluonman, TCN7JM, Iankhou, Sandbh, MusikAnimal, Altaïr, WikisucksKNOBlegasses, VictorParker, Jimbo2440, Tycho Magnetic Anomaly-1, Softballbaby984, ThomasRules, BattyBot, Justincheng12345-bot, Abilanin, ChrisGualtieri, EuroCarGT, Dexbot, Webclient101, TwoTwoHello, King jakob c, RandomLittleHelper, Reatlas, Cteung, DavidLeighEllis, Ugog Nizdast, Ginsuloft, Noyster, DudeWithAFeud, Skr15081997, Matthewweber12, HotHabenero, Hotta stuffu, Sony Vark XIII, Monkbot, HiYahhFriend, ZYjacklin, Narky Blert, Jodihe93, Selimozd20, Anbgsm07, UZawMoeNaing, SandKitty256, Supdiop, KasparBot, Sat cheat, Lolhappyface and Anonymous: 594

- **Dmitri Mendeleev** *Source:* https://en.wikipedia.org/wiki/Dmitri_Mendeleev?oldid=686512212 *Contributors:* Magnus Manske, Mav, Tarquin, Koyaanis Qatsi, Malcolm Farmer, Css, Fnielsen, Danny, XJaM, William Avery, Drbug, Heron, Fonzy, Ewen, Erik Zachte, Ezra Wax, Wapcaplet, Ixfd64, Dcljr, Nine Tail Fox, Ahoerstemeier, Suisui, Angela, Александър, Lupinoid, Error, Kwekubo, Rob Hooft, Hashar, Reddi, Malcohol, DJ Clayworth, Tpbradbury, Maximus Rex, Taxman, Paul-L~enwiki, Rnbc, Jose Ramos, Bevo, Xyb, Jerzy, Gentgeen, Robbot, Vardion, Hankwang, Fredrik, Baldhur, Altenmann, Henrygb, Flauto Dolce, Auric, Wikibot, JackofOz, NeoThe1, Guy Peters, Jooler, Xyzzyva, Centrx, Giftlite, Pmerriam, Sj, Nunh-huh, Tom harrison, Fastfission, Monedula, No Guru, Curps, Alison, Pashute, Mellum, Solipsist, Bobblewik, Explendido Rocha, Chowbok, Alexf, Toytoy, Chirlu, Blankfaze, MarkSweep, Piotrus, Untifler, Gene s, Sebbe, Icairns, Lumidek, Joyous!, Picapica, Adashiel, Grstain, D6, Atrian, Freakofnurture, Discospinster, Rich Farmbrough, Rhobite, Vsmith, Slipstream, Zazou, SpookyMulder, Kenb215, DcoetzeeBot~enwiki, Tgies, Janderk, Mashford, Eric Forste, MyNameIsNotBob, RJHall, Miraceti, RoyBoy, Adambro, Bobo192, Smalljim, John Vandenberg, BrokenSegue, Enric Naval, Shenme, Larsie, Juzeris, SpeedyGonsales, Pschemp, Sam Korn, Nsaa, Lysdexia, Ranveig, Jumbuck, Alansohn, Ben davison, Jeltz, Cjthellama, Riana, Lectonar, Ayeroxor, SidP, Helixblue, Docboat, RainbowOfLight, Cmaprn, Ghirlandajo, Redvers, KTC, Nuno Tavares, Benji2~enwiki, Jeffrey O. Gustafson, OwenX, Woohookitty, Camw, Benbest, Scjessey, Lenar, Dionyziz, Karmosin, Kralizec!, MarcoTolo, Marudubshinki, Emerson7, Mandarax, Graham87, Deltabeignet, FreplySpang, DePiep, Edison, Canderson7, Ketiltrout, Sjö, Rjwilmsi, Lordkinbote, Ligulem, Peripatetic, Olessi, Matt Deres, Ecelan, Titoxd, Jwkpiano1, FlaBot, SchuminWeb, RobertG, Nihiltres, Itinerant1, Kmorozov, Anzelm, JYOuyang, RexNL, Gurch, Wars, RasputinAXP, GreyCat, Russavia, Physchim62, Mallocks, Imnotminkus, Introvert, Chobot, Bgwhite, Cactus.man, Gwernol, EamonnPKeane, The Rambling Man, YurikBot, Wavelength, Vuvar1, Brandmeister (old), Pip2andahalf, Rylz, Cyberherbalist, NAveryW, Conscious, Splash, Akamad, Alex Bakharev, Wimt, Lusanaherandraton, Ornilnas, NawlinWiki, Wiki alf, Bachrach44, Aeusoes1, LaszloWalrus, Howcheng, Arima, Tomburbine, Ragesoss, Anetode, Ravedave, Raven4x4x, Sfnhltb, Zwobot, Aaron Schulz, BOT-Superzerocool, DeadEyeArrow, Private Butcher, 2over0, Lt-wiki-bot, Closedmouth, Donald Albury, GraemeL, Dr U, JoanneB, Bandurist, Ordinary Person, T. Anthony, Mikus, Nixer, Kungfuadam, Thomas Blomberg, GrinBot~enwiki, Lunch, Itub, Attilios, SmackBot, Monkeyblue, Haverpopper, Bobet, Brianyoumans, InverseHypercube, Royalguard11, Kitchka, Delldot, Eskimbot, Kintetsubuffalo, Gilliam, Ohnoitsjamie, Master Jay, Bluebot, Keegan, F382d56d7a18630cf764a5b576ea1b4810467238, MalafayaBot, BrendelSignature, H i-c h-a M~enwiki, DHN-bot~enwiki, Colonies Chris, Darth Panda, Marblefluss, DTR, Can't sleep, clown will eat me, Scott3, Eschbaumer, Writtenright, Chlewbot, Rrburke, GeorgeMoney, Addshore, Nuklear, Khoikhoi, Jmlk17, Emact, BostonMA,

Dirk gently~enwiki, Khukri, Nakon, Coolag12345, Derek R Bullamore, Badgerpatrol, Dantadd, DMacks, Er Komandante, Jóna Þórunn, Springnuts, Vina-iwbot~enwiki, RossF18, Drmaik, Ohconfucius, Pinktulip, SashatoBot, Swatjester, JzG, John, Kipala, Breno, This user has left wikipedia, Barry Kent~enwiki, Thraxas, IronGargoyle, Frokor, Anatopism, Ryulong, Manifestation, Dr.K., MTSbot~enwiki, Fromeout11, Angryxpeh, Hu12, Joseph Solis in Australia, Suresh K. Sheth, Domitori, Blehfu, Murf661, Esn, Tawkerbot2, Chelydra, Lbr123, Emote, JForget, Mellery, Meisam.fa, Unionhawk, Scohoust, KyraVixen, Nczempin, Orayzio, Kylu, Moreschi, Chicheley, ElPoojmar, Cydebot, Kanags, Galassi, Steel, Scottiscool, Doomed Rasher, Gogo Dodo, JFreeman, Corpx, Booty3535, Tkynerd, Christian75, Codetiger, DumbBOT, Omicronpersei8, Danielil, Lordhatrus, Epbr123, Divyangmithaiwala, CopperKettle, Pampas Cat, Marek69, John254, NorwegianBlue, James086, Yettie0711, Jonny-mt, Zachary, Lithpiperpilot, SusanLesch, Natalie Erin, Hempfel, Escarbot, Ilion2, Thadius856, AntiVandalBot, RobotG, Fedayee, Eamezaga, Richiel101, Jj137, NSH001, Modernist, Malcolm, Spartaz, Phanerozoic, Caper13, Ioeth, JAnDbot, Husond, MER-C, Matthew Fennell, Plm209, Andonic, Connormah, VoABot II, JNW, Kajasudhakarababu, Think outside the box, Caroldermoid, CTF83!, Waacstats, Eldumpo, Dirac66, 28421u2232nfenfcenc, Rebecca777, Glen, DerHexer, JaGa, Edward321, Lelkesa, Hbent, MartinBot, BetBot~enwiki, Arjun01, Jaystar11, Rettetast, Kateshortforbob, CommonsDelinker, LittleOldMe old, John Duncan, Tgeairn, J.delanoy, Greenflower89, Nev1, Filll, DrKay, Trusilver, Skeptic2, Hans Dunkelberg, Maurice Carbonaro, Yonidebot, Derwig, Smeira, Petersec, Ephebi, Mikhail Dvorkin, NewEnglandYankee, Mufka, ObseloV, Entropy, Burzmali, Spitacular, MishaPan, CardinalDan, Pqwo, Matty202033, To my arse, Sumo su, Wikieditor06, Lights, Timotab, VolkovBot, Mp3boy3239, Jeff G., AlnoktaBOT, Philip Trueman, Muut21, TXiKiBoT, Slvrstn, Dale134, Kovbasa, Rei-bot, Miguel Chong, Qxz, Minol, Anna Lincoln, Lradrama, Seungfire, Broadbot, Farever, LeaveSleaves, Xelda, Maroonedsorrow, Luuva, Duncan.Hull, Maxim, Plazak, Ashnard, David Marjanović, Rhopkins8, T0nyM0ntana 420, Jamiemouse90, BrianY, Jjdon, AlleborgoBot, Symane, Carnelain, Tvinh, Pediaknowledge, Billytrousers, Demmy, Sergwiki, Rozmysl, SieBot, Ttonyb1, PlanetStar, Tiddly Tom, Scarian, WereSpielChequers, BotMultichill, Gerakibot, Seeyardee, LeadSongDog, GrooveDog, Arbor to SJ, Dwiakigle, Monegasque, Oxymoron83, Ddxc, KoshVorlon, Seth Whales, Vojvodaen, MadmanBot, Gb-oh6, Mygerardromance, Denisarona, Kanonkas, Samcristiano7, SallyForth123, Mr. Granger, Cjc15153, Atif.t2, Loren.wilton, Martarius, ClueBot, Snigbrook, The Thing That Should Not Be, All Hallow's Wraith, Techdawg667, FileMaster, Tomas e, Drmies, Uncle Milty, Polyamorph, CounterVandalismBot, Megam~enwiki, Polker03, Kgandrews, Mindmelter, DragonBot, Ktr101, Excirial, Christine1107, Rohbat, Vivio Testarossa, Eemyaj, Jotterbot, Njardarlogar, Diehard4.0, Lenary, SchreiberBike, Thehelpfulone, Thingg, G3421hi, Aitias, 05bysstern, DumZiBoT, XLinkBot, Fastily, Nathan Johnson, BodhisattvaBot, Dthomsen8, Nicki.The.Random.Chick, Little Mountain 5, Kwjbot, Good Olfactory, Airplaneman, Kbdankbot, HexaChord, Decanc, Addbot, Sarah jamieson, Jojhutton, Ronhjones, KorinoChikara, CanadianLinuxUser, NjardarBot, Cst17, Mentisock, CarsracBot, Tangoed1whiskey, Cheesepieman, Wurk a kurk, Debresser, Favonian, Tubesidiom, Numbo3-bot, Tide rolls, Tressor, David0811, Greyhood, Legobot, Luckasbot, Yobot, Veraladeramanera, TaBOT-zerem, Heidas, II MusLiM HyBRiD II, Tavy08, Tilnakk, Widey, Jobroluver98, IW.HG, Tempodivalse, DiverDave, AnomieBOT, Rubinbot, Götz, Jim1138, IRP, AdjustShift, Kingpin13, Seanaloisi, Flewis, Citation bot, Geregen2, Weirdskateguy, Neurolysis, ArthurBot, PhilAnG, Stewart96, Parthian Scribe, Xqbot, Mick10793, I have a cool name, Agleeson, Capricorn42, Drilnoth, Renaissancee, Notmax, Maxximus514, HUZZAH123, Omnipaedista, Ducki17, RibotBOT, SassoBot, Chris.urs-o, Nedim Ardoğa, GhalyBot, Ninja Scaley, The Sceptical Chymist, Thermokarst, FrescoBot, Wikipe-tan, Lagelspeil, VS6507, Happykg, BenzolBot, Jamesooders, GrayScaleRainbow, Vlp92, Louperibot, Citation bot 1, Eightofnine, Ntse, RMN1390, Bryant james, Traleo, Abductive, Rameshngbot, Plucas58, Tomcat7, SpaceFlight89, Fixer88, Tlhslobus, PrinceRegentLuitpold, Merlion444, FoxBot, Double sharp, TobeBot, نوری, ئاراس, Lotje, Javierito92, Dinamik-bot, Vrenator, Actoreng1, Earthandmoon, Brambleclawx, Kidkaos707, DARTH SIDIOUS 2, RjwilmsiBot, Ripchip Bot, DASHBot, EmausBot, GregZak, Wikipelli, HiW-Bot, ZéroBot, StringTheory11, Plotfeat, Battoe19, Zloyvolsheb, Masteratc, Kafern, Fanofnaruto2, Alfio66, JeanneMish, Mayur, Orange Suede Sofa, Negovori, ChuispastonBot, Xero200, UDHAUEIFD, Xanchester, ClueBot NG, Jnorton7558, Goose friend, Mpaa, ساجد امجد ساجد, Helpful Pixie Bot, Sceptic1954, ಬಷ್ಟ ಕಾಂಡ್ಲ್ಯ, Athkalany, Wald, Calabe1992, Vagobot, Sergeispb-10, OttawaAC, The Almightey Drill, TheEditor12345, Writ Keeper, Snow Blizzard, Uikmi, Lizzyah, Unknownkarma, Wrath X, Vassto, Ninmacer20, JYBot, Ukrained2012, Jamesx12345, Passengerpigeon, HobbyGD, Monkbot, Acagastya, FalconJackson, Jonarnold1985, Ghiutun, Josjos69, KasparBot and Anonymous: 805

- **Radionuclide** *Source:* https://en.wikipedia.org/wiki/Radionuclide?oldid=684853797 *Contributors:* AxelBoldt, Kpjas, Trelvis, Mav, Fnielsen, Ubiquity, Lir, Shellreef, Jketola, Tannin, Looxix~enwiki, Ahoerstemeier, Aarchiba, Andres, Jordi Burguet Castell, Smack, Stone, Taxman, Donarreiskoffer, Gentgeen, Robbot, Kristof vt, Ojigiri~enwiki, Drstuey, DocWatson42, Herbee, Andycjp, Antandrus, Icairns, Sam Hocevar, CALR, Discospinster, Bender235, Joanjoc~enwiki, Bobo192, 018, Arcadian, Alansohn, Keenan Pepper, AjAldous, Mlessard, Uffish, Vuo, Zereshk, Voxadam, Mindmatrix, WadeSimMiser, Eleassar777, Isnow, Graham87, Tangotango, Ems57fcva, Ayla, Kolbasz, Butros, Essaregee, Chobot, DaGizza, DVdm, YurikBot, Wavelength, Borgx, Wimt, Zwobot, Tweeq, Kkmurray, LeonardoRob0t, SmackBot, FocalPoint, Jclerman, Unyoyega, Chris the speller, Postoak, Sbharris, TheGerm, V1adis1av, OrphanBot, Addshore, Drphilharmonic, Kukini, John, H.sand01, Gobonobo, Dicklyon, Citicat, PetaRZ, Gungasdindin, CmdrObot, Harej bot, Nilfanion, A876, ST47, Raomap, Optimist on the run, JLD, Epbr123, Pjvpjv, Gierszep, AntiVandalBot, Deberle, JAnDbot, Arch dude, Albany NY, Sophie means wisdom, I80and, VoABot II, AdamWalker, Animum, Cgingold, Dirac66, Squidonius, MartinBot, Rogerrluo, Maurice Carbonaro, Stan J Klimas, Actarux, Philip Trueman, TXiKiBoT, Seb az86556, PGWG, Breakyunit, Keilana, Yerpo, Rhenning007, Simonbayly, Verytas, KathrynLybarger, Jessiehawkes, Nergaal, Jons63, ClueBot, The Thing That Should Not Be, Wysprgr2005, Niceguyedc, Nuclearmedzors, PixelBot, Jotterbot, Dekisugi, Johndoe616, Panos84, Vegetator, Jonverve, Jeremycenus, TNTM64, XLinkBot, Avoided, WikHead, Teslaton, Dfoxvog, Addbot, Gregisfat2, Haruth, Leszek Jańczuk, CUSENZA Mario, Tide rolls, Ben Ben, Luckas-bot, AnomieBOT, Zhieaanm, Ciphers, Sz-iwbot, Materialscientist, Citation bot, ArthurBot, Xqbot, Braindamagehurts, Biggieyankfan, Hamburgers1212, Dan6hell66, LucienBOT, Slastic, Citation bot 1, Minivip, FoxBot, Double sharp, Lotje, DARTH SIDIOUS 2, Ripchip Bot, EmausBot, WikitanvirBot, Look2See1, Demomoer, Steve123963, Donner60, Research new, ChuispastonBot, Cgt, Xrayburst1, ClueBot NG, Ecubar, Smm201`0, Reify-tech, Maripaz21, Bibcode Bot, MusikAnimal, Lazord00d, Zedshort, BattyBot, Pratyya Ghosh, Sssciencce, Dexbot, Burzuchius, CensoredScribe, Tinyscooter, Karl Hess666, NewEnglandDr, Teddyktchan, Trackteur, SStewartGallus, IiKkEe, DSCrowned, Pinnate foliage, KasparBot and Anonymous: 174

- **Neutronium** *Source:* https://en.wikipedia.org/wiki/Neutronium?oldid=675084412 *Contributors:* Trelvis, Bryan Derksen, The Anome, Manning Bartlett, PierreAbbat, Alan Peakall, Looxix~enwiki, Mark Foskey, Stone, David Latapie, Omegatron, Pakaran, Daran, Lumos3, ChrisO~enwiki,Netizen, Rursus, Bkell, Cyberia23, Xanzzibar, David Gerard, Eequor, Urhixidur, MementoVivere, Aaryna, Rich Farmbrough, Cacycle,SocratesJedi, MarkS, Chairboy, Shenme, Evgeny, Rbj, Arcadian, I9Q79oL78KiL0QTFHgyc, Pearle, Dillee1, Anthony Appleyard, Nik42,Keenan Pepper, Sligocki, Mac Davis, Caesura, Ayeroxor, Sobolewski, KapilTagore, RJFJR, Joriki, TomTheHand , BillC, Christopher Thomas,Seminumerical, Ilyak, Rjwilmsi, Nightscream, Vegaswikian, Brighterorange, Yamamoto Ichiro, FlaBot, TiagoTiago, Gurch, Iggy Koopa,Spacepotato, Bhny, Hydrargyrum, Gaius Cornelius, Salsb, Hyuri, David R. Ingham, Complainer, Dhollm, Kkmurray, SamuelRiv, SFH, Nikkimaria, Jwissick, Arthur Rubin, Reyk, KenoSarawa, Tobyk777, That Guy, From That Show!, AndrewWTaylor, SmackBot, Incnis Mrsi, Melchoir, Edgar181, Betacommand, Thumperward, Sbharris, VMS Mosaic, DMacks, Leon..., Soumyasch, A. Parrot, JoeBot, OS2Warp, Rwflammang, Ruslik0, Jsmaye, Vorlon19, Osssua, Ebyabe, Nonagonal Spider, Headbomb, Marek69, Chris goulet, JustAGal, JHFTC, J. Langton, Spartaz, CosineKitty, Acroterion, Klaxton, Magioladitis, Dirac66, Uncle Dick, Maurice Carbonaro, Happy8, Anchovee, Dorftrottel, John Darrow, Jeff G., Jmrowland, Themel, Philip Trueman, Pjstewart, Hqb, Someguy1221, Oracle7168, ^demonBot2, Gavinmcq, PlanetStar, Caltas, The Great Attractor, JohnnyMrNinja, Copeland3300, Ljofa, Prottos007, Gordon Ecker, Alexbot, Tlesinski, Jtle515, Plasmic Physics, DoctorEric, CaptainVideo890, Neuralwarp, XLinkBot, Sittnick, Ziggy Sawdust, Dsimic, Addbot, DOI bot, Toyokuni3, Dsmith77, Lightbot, OlEnglish, Zorrobot, Yobot, AnomieBOT, Götz, 3Juno3, Materialscientist, Citation bot, Geregen2, 78.26, FrescoBot, LucienBOT, Robo37, Citation bot 1, Jonesey95, Spidey104, Tom.Reding, Double sharp, 4, Optiguy54, GoingBatty, We hope, ZéroBot, Quondum, AManWithNoPlan, Donner60, Bigbigreader, Whoop whoop pull up, ClueBot NG, Lanthanum-138, Braincricket, Cabazap, Helpful Pixie Bot, Bibcode Bot, Saltof, TheMan4000, Allthingsmadscience, Robert the Devil, Meszz12321, EuroCarGT, Singularitydnb, Redactor1802, ToFeignClef, Tokan one, Crbeals, Rolf h nelson, Owllord97, Lostmeow, Deepeshshrestha, Cn145912 and Anonymous: 150

- **Chemical Galaxy** *Source:* https://en.wikipedia.org/wiki/Chemical_Galaxy?oldid=648451856 *Contributors:* Mxn, David Gerard, Codepoet, Rich Farmbrough, Supercoop, RoyBoy, Maurreen, Eaolson, Rjwilmsi, Nightscream, Tardis, Reyk, SmackBot, Quarty~enwiki, DJBullfish, Headbomb, DadaNeem, Pjstewart, Polyamorph, Addbot, DOI bot, Francisco Leandro, Materialscientist, Blackguard SF, Citation bot 1, Double sharp, Wikielwikingo, Bibcode Bot, Monkbot and Anonymous: 12

- **Systematic element name** *Source:* https://en.wikipedia.org/wiki/Systematic_element_name?oldid=681147318 *Contributors:* Bryan Derksen, Cyrek, Heron, Michael Hardy, Karada, Eric119, SebastianHelm, Ahoerstemeier, David Latapie, Dysprosia, Furrykef, Gentgeen, Robbot, RedWolf, Nyh, Rursus, JesseW, Smjg, Graeme Bartlett, Herbee, Wwoods, Filceolaire, DragonflySixtyseven, Icairns, Ponder, Nickj, Func, Ardric47, Anthony Appleyard, MatthewWilcox, Typhlosion, DLJessup, Melaen, Nightstallion, Kenyon, Isnow, Waldir, Rogo, Graham87, DePiep, Cuenca, Makaristos, FlaBot, Glenn L, Physchim62, Chobot, Rewster, YurikBot, Borgx, RussBot, Icarus3, Shawn81, Anetode, Ospalh, BazookaJoe, PDD, Kurykh, TdanTce, AxG, Pancake4lyfe, WeggeBot, Mycroft.Holmes, Barticus88, Headbomb, AntiVandalBot, C.A.T.S. CEO, TomS TDotO, DJPohly, Trvsdrlng, DorganBot, Ratfox, Funandtrvl, Dave Andrew, Jeff G., Seattle Skier, TXiKiBoT, Legoktm, RSStockdale, OKBot, Nergaal, Bentu, Egmontaz, Addbot, Roentgenium111, Lightbot, Vroo, Xqbot, Funky junk9, RedBot, Double sharp, ZéroBot, StringTheory11, Humorahead01, ClueBot NG, IluvatarBot, Trans1000, 3.14159265358pi, 🔲🔲🔲, BanunterX and Anonymous: 67

- **Primordial nuclide** *Source:* https://en.wikipedia.org/wiki/Primordial_nuclide?oldid=682536394 *Contributors:* TakuyaMurata, SebastianHelm, Cherkash, Quickbeam, Stone, Jni, Bkell, Kwamikagami, Keenan Pepper, DePiep, Rjwilmsi, Kolbasz, Samuel Curtis, Itub, SmackBot, Incnis Mrsi, Michbich, Sbharris, Vladis1av, AStext, Headbomb, Magioladitis, WolfmanSF, BatteryIncluded, Dirac66, Robin S, SireSpanky, R'n'B, Hans Dunkelberg, Maurice Carbonaro, Dawright12, UnitedStatesian, Dufo, Johntobey, Alexis Brooke M, Addbot, Roentgenium111, CanadianLinuxUser, OlEnglish, Jarble, Luckas-bot, Robert Treat, 4th-otaku, AnomieBOT, LilHelpa, Citation bot 1, Coekon, Achim1999, Double sharp, EmausBot, Wikipelli, RockMagnetist, Teaktl17, Colapeninsula, Fauzan, Snotbot, Frietjes, Bibcode Bot, BG19bot, Danjirokatsujima, Khazar2, Burzuchius, Poppy Appletree, Reatlas, Teddyktchan and Anonymous: 26

- **Electron shell** *Source:* https://en.wikipedia.org/wiki/Electron_shell?oldid=681782312 *Contributors:* Bryan Derksen, The Anome, LA2, Rmhermen, Heron, Bth, Tim Starling, Kku, Julesd, Glenn, Smack, Wereon, Stirling Newberry, Giftlite, Eequor, Alexf, AliveFreeHappy, Discospinster, FT2, Hidaspal, Bender235, Chairboy, Bobo192, Sasquatch, Kjkolb, Pearle, Alansohn, Snowolf, HenkvD, Benbest, Riumplus, Slocombe, Isnow, Ashmoo, Graham87, V8rik, DePiep, Saperaud~enwiki, Rjwilmsi, Strait, Tintazul, Pumeleon, DyluckTRocket, Fresheneesz, Goudzovski, CiaPan, JWB, SpuriousQ, Stephenb, Giro720, Nsmith 84, Beanyk, Wknight94, Leptictidium, 2over0, Sbyrnes321, Luk, SmackBot, Zs, Incnis Mrsi, KnowledgeOfSelf, Eskimbot, Provelt, Canthusus, Yamaguchi🔲🔲, Gilliam, Skizzik, Rmosler2100, Chris the speller, Kurykh, Jprg1966, Famspear, Khoikhoi, Weirdy, PieRRoMaN, Gokmop, Inositol, DMacks, Sadi Carnot, Lambiam, Titus III, Quasispace, AB, Hemmingsen, A. Parrot, Beetstra, Aeluwas, Kyoko, Supaman89, Sam Li, Tawkerbot2, Gigaslav, Sohum, FlyingToaster, No1lakersfan, MC10, Gogo Dodo, DumbBOT, Epbr123, Baron162, Pcbene, AntiVandalBot, Astavats, Res2216firestar, JAnDbot, Bongwarrior, VoABot II, Jespinos, 28421u2232nfenfcenc, Error792, Cpl Syx, Arjun01, Keith D, EdBever, Stlava, Melamed katz, Hans Dunkelberg, Uncle Dick, Coppertwig, Ybk33, Imaginary Pi Slicer, NewEnglandYankee, Deor, Iosef, TXiKiBoT, The Original Wildbear, Anonymous Dissident, BotKung, Monty845, Sue Rangell, SieBot, Dongoose34, Oxymoron83, Onesspite, Dposte46, Nandobike, Dear Reader, Escape Orbit, Razrsharp67, ClueBot, Umptious, Polyamorph, Michał Sobkowski, LizardJr8, Dupdater, Officer781, Darkveilis5, Excirial, Tassos Kan., Danmichaelo, R1255, ZJB3, Blueagle77, JKeck, Skarebo, Sgpsaros, Sami Lab, Addbot, Blechnic, Ronhjones, TutterMouse, Fieldday-sunday, Moosehadley, Ashanda, WikiUserPedia, 84user, Tide rolls, Function95, Steak, Gail, आशीष भटनागर, Luckas-bot, Yobot, 2D, TaBOT-zerem, THEN WHO WAS PHONE?, Reindra, Tempodivalse, Jim1138, Materialscientist, Xqbot, DSisyphBot, Raiyyan123, Grim23, AuricBlofeld, Ryryrules100, Fragma08, Spacecadet262, Kmecholsky, Redrose64, Pinethicket, I dream of horses, Hamtechperson, Crchambers89, Merlion444, December21st2012Freak, Double sharp, DixonDBot, Heavyweight Gamer, Vrenator, Lmp883, JV Smithy, Minimac, Onel5969, DASHBot, EmausBot, Hhhippo, JSquish, Dariusjjack, L0ngpar1sh, GeorgeBarnick, Jwollbold, Gottlob Gödel, ClueBot NG, Movyn, Lanthanum-138, Widr, MerllwBot, Helpful Pixie Bot, Zach mielhausen, Lowercase sigmabot, Fenntil, Medicalfizzicist, 🔲🔲🔲🔲🔲🔲, Hallows AG, Anbu121, Pratyya Ghosh, Elyuimnim, Agregory898, Badtouch5, HamishMorgan, Ugog Nizdast, Tictsc654, Cautious Chemist, Aspenzimmerman, The urgg and Anonymous: 360

- **Group 4 element** *Source:* https://en.wikipedia.org/wiki/Group_4_element?oldid=686448999 *Contributors:* Eric119, Ahoerstemeier, Stone, Denelson83, Gentgeen, Sander123, Rfc1394, David Edgar, Wwoods, Rpyle731, Yekrats, 1297, Icairns, Vsmith, Joanjoc~enwiki, Ranveig, Jumbuck, Gary, Arakin, Feline1, LukeSurl, DePiep, Rjwilmsi, Koavf, Scerri, Chobot, Bgwhite, YurikBot, Jimp, Wikispork, Gaius Cornelius, Bota47, Ray Chason, Bomac, Hugo-cs, DHN-bot~enwiki, Steve Pucci, Titus III, Courcelles, Cryptic C62, A7x, Nick Number, JAnDbot, Magioladitis, STBot, Skier Dude, VolkovBot, ABF, TXiKiBoT, Lamro, SieBot, BotMultichill, KoshVorlon, Nergaal, Polyamorph, Doprendek, Addbot, Queenmomcat, CanadianLinuxUser, LaaknorBot, CarsracBot, LinkFA-Bot, Alchemist-hp, Luckas-bot, Yobot, Amirobot, AnomieBOT, Materialscientist, Citation bot, LilHelpa, Brane.Blokar, Xqbot, GrouchoBot, Jezhotwells, RibotBOT, Aashaa, R8R Gtrs, LucienBOT, Firq, Citation bot 1, Double sharp, TobeBot, Seahorseruler, AXRL, Sverigekillen, WikitanvirBot, GoingBatty, ZéroBot, StringTheory11, Chemicalinterest, H3llBot, ChuispastonBot, Lanthanum-138, Parcly Taxel, Helpful Pixie Bot, Bibcode Bot, 2009ict043, Szczureq, Cyberbot II, Monkbot, KasparBot and Anonymous: 29

- **Carbon group** *Source:* https://en.wikipedia.org/wiki/Carbon_group?oldid=683163884 *Contributors:* Fonzy, Michael Hardy, Eric119, Ahoerstemeier, Stone, Robbot, Seglea, Rfc1394, Ojigiri~enwiki, Bkell, Inter, Xerxes314, Dratman, Yekrats, CryptoDerk, 1297, Icairns, AliveFreeHappy, Vsmith, Femto, Jag123, Eddideigel, Alansohn, SteinbDJ, Gene Nygaard, Gurkha711, Feline1, LukeSurl, Boothy443, Woohookitty,

Isnow, Eras-mus, DePiep, Rjwilmsi, Scerri, YurikBot, Hairy Dude, RussBot, Stephenb, E2mb0t~enwiki, Reyk, Sbyrnes321, Dandelions, SmackBot, Edgar181, HalfShadow, Hmains, Hugo-cs, DHN-bot~enwiki, French user, SashatoBot, Lambiam, Titus III, IronGargoyle, Margoz, FunPika, Christian75, West Brom 4ever, Mentifisto, AntiVandalBot, Clamster5, Res2216firestar, VoABot II, Obliterator, R'n'B, Adavidb, Silverxxx, HighKing, Izno, Squids and Chips, VolkovBot, Chris Dybala, Al.locke, Philip Trueman, Claidheamohmor, Dpvwia, Burntsauce, Seano12345, SieBot, PlanetStar, Scorpion451, Nergaal, ClueBot, PipepBot, KevinXcore, Puchiko, Jusdafax, Estirabot, NuclearWarfare, SchreiberBike, Versus22, Addbot, Roentgenium111, CanadianLinuxUser, LaaknorBot, CarsracBot, DFS454, LinkFA-Bot, Jasper Deng, Bbernet13, Legobot, Luckas-bot, Fraggle81, Amirobot, Tonyrex, AnomieBOT, Daniele Pugliesi, Materialscientist, MidnightBlueMan, ArthurBot, Brane.Blokar, Xqbot, GrouchoBot, Sapna3654, Jilkmarine, FrescoBot, LucienBOT, Gerard1894, Pinethicket, 10metreh, Getheren, Double sharp, AXRL, TjBot, Darnir redhat, EmausBot, John of Reading, ZéroBot, StringTheory11, Donner60, Whoop whoop pull up, ClueBot NG, TechnoCat, Lanthanum-138, Widr, Captain dave86, ChrisGualtieri, GoShow, JYBot, Mogism, Burzuchius, King jakob c, Dzongha, Jakec, Ginsuloft, Shahsohan7, KasparBot and Anonymous: 101

- **Group 10 element** *Source:* https://en.wikipedia.org/wiki/Group_10_element?oldid=637384238 *Contributors:* Heron, Dcljr, Ahoerstemeier, Stone, Donarreiskoffer, Gentgeen, Rfc1394, Wayland, Wwoods, Yekrats, Brockert, Icairns, Grm wnr, Vsmith, Joanjoc~enwiki, Femto, Yono, Nk, DePiep, Scerri, Chobot, Midgley, Moe Epsilon, SmackBot, Unyoyega, Hugo-cs, DHN-bot~enwiki, Steve Pucci, SashatoBot, Titus III, Anoop.m, JorisvS, Headbomb, Btards, Sss333, JAnDbot, TARBOT, NatureA16, Sir Intellegence, R'n'B, LordAnubisBOT, Chelitaannal, Blood Oath Bot, VolkovBot, Philip Trueman, EmxBot, Scorpion451, Nergaal, Chem-awb, Alexbot, Plasmic Physics, Aaron north, Addbot, LaaknorBot, Luckas-bot, Yobot, Amirobot, KamikazeBot, AnomieBOT, Materialscientist, RibotBOT, Erik9bot, A.amitkumar, LucienBOT, Firq, Getheren, Double sharp, AXRL, TjBot, Rlholden, WikitanvirBot, ZéroBot, StringTheory11, Chemicalinterest, ChuispastonBot, Lanthanum-138, BG19bot, TomeHale, Szczureq, BattyBot, Justincheng12345-bot, ChrisGualtieri, JYBot and Anonymous: 18

- **Group 3 element** *Source:* https://en.wikipedia.org/wiki/Group_3_element?oldid=686821842 *Contributors:* Malcolm Farmer, Heron, RTC, Michael Hardy, Paul A, Eric119, Ahoerstemeier, Jimfbleak, Angela, EdH, Fvincent, Gentgeen, Rfc1394, Auric, Wwoods, Yekrats, Icairns, Zfr, Vsmith, Joanjoc~enwiki, Femto, Eddideigel, Jumbuck, Arakin, Feline1, Duplode, DePiep, Rjwilmsi, Scerri, King of Hearts, Flying Jazz, Yurik-Bot, RobotE, Gadget850, Bota47, Allens, Bomac, The Ronin, Edgar181, Freddy S., Hugo-cs, Chris the speller, Bluebot, DHN-bot~enwiki, Raistuumum, Archimerged, Rifleman 82, Epbr123, Headbomb, Escarbot, 1of3, JAnDbot, Magioladitis, Jackfirst, Mschel, J.delanoy, STBotD, AlnoktaBOT, Petergans, SieBot, The way, the truth, and the light, Nergaal, Polyamorph, Alexbot, Estirabot, Istilidion, Addbot, Laaknor-Bot, LinkFA-Bot, Alchemist-hp, Lightbot, Luckas-bot, Yobot, Amirobot, Materialscientist, Citation bot, ArthurBot, LilHelpa, Brane.Blokar, Xqbot, Aashaa, R8R Gtrs, LucienBOT, Firq, Intelligentsium, RedBot, Double sharp, Trappist the monk, RjwilmsiBot, EmausBot, GA bot, Fishing Chimp, H3llBot, Music Sorter, SBaker43, ChuispastonBot, ClueBot NG, KLindblom, Lanthanum-138, Helpful Pixie Bot, Bibcode Bot, BattyBot, Khazar2, Jim Carter, Berrenerreb and Anonymous: 49

- **Group 11 element** *Source:* https://en.wikipedia.org/wiki/Group_11_element?oldid=686515990 *Contributors:* The Anome, Rgamble, Robert Foley, Heron, Octothorn, Edward, Rambot, Eric119, Ahoerstemeier, Den fjättrade ankan~enwiki, Hick ninja, Malbi, PS4FA, Stone, Ike9898, Zoicon5, Pakaran, Zandperl, Chris 73, Rfc1394, Alexwcovington, Robin Patterson, Karn, Everyking, Yekrats, Jason Quinn, Brockert, Pne, Bobblewik, 1297, Pmanderson, Icairns, Neutrality, Fg2, Chmod007, Vsmith, Femto, Bobo192, Enric Naval, Ranveig, Caesura, Polyparadigm, Chochopk, Graham87, BD2412, DePiep, Oxydo~enwiki, FlaBot, Scerri, Elfguy, Eweisser, YurikBot, Nirvana2013, IslandGyrl, Searchme, SmackBot, Unyoyega, Mom2jandk, Hugo-cs, Kurykh, DHN-bot~enwiki, KingAlanI, Steve Pucci, Ixnayonthetimmay, Titus III, Anoop.m, Cowbert, Thefamouseccles, Christian75, Epbr123, AntiVandalBot, JAnDbot, Hut 8.5, VoABot II, Watashiwabakayo, R'n'B, PrestonH, Lord-dAnubisBOT, SJP, VolkovBot, TXiKiBoT, SieBot, ToePeu.bot, Prestonmag, Avidallred, Alex.muller, Dabomb87, Nergaal, ClueBot, Foxj, Mild Bill Hiccup, Estirabot, Esbboston, 7, Djavko, DumZiBoT, Addbot, LaaknorBot, LinkFA-Bot, Alchemist-hp, Zorrobot, Luckas-bot, Yobot, Amirobot, Washerboy, AnomieBOT, Materialscientist, Brane.Blokar, Xqbot, Sophus Bie, Erik9bot, LucienBOT, Jschnur, Mikespedia, Getheren, Double sharp, Cthomas sysplan, Minimac, AXRL, TjBot, EmausBot, Thomasmathewsj, ZéroBot, StringTheory11, Donner60, ChuispastonBot, Whoop whoop pull up, ClueBot NG, Aswn, Lanthanum-138, Widr, MusikAnimal, JYBot, TwoTwoHello, Jwoodward48wiki, St170e, GimMorh and Anonymous: 89

- **Metalloid** *Source:* https://en.wikipedia.org/wiki/Metalloid?oldid=686916116 *Contributors:* WojPob, Mav, Tarquin, Andre Engels, Heron, Ahoerstemeier, Kimiko, Smack, Charles Matthews, Timwi, Zoicon5, IceKarma, GPHemsley, Gentgeen, Robbot, Sander123, Dirgela, Romanm, Rfc1394, Hippietrail, Radagast, Giftlite, Karn, Yekrats, Eregli bob, Icairns, Gscshoyru, Iantresman, Neutrality, Churchilljrhigh, Discospinster, Rich Farmbrough, ESkog, El C, Gilgamesh he, C1k3, Remember, Army1987, Reuben, Giraffedata, Kjkolb, Rkuchta, Jumbuck, Alansohn, Anthony Appleyard, Neonumbers, Keenan Pepper, Snowolf, Wtshymanski, Feline1, Benbest, Tabletop, Eaolson, SDC, DePiep, Rjwilmsi, Ctdunstan, Yamamoto Ichiro, FlaBot, Margosbot~enwiki, Nihiltres, TimSE, Czar, Chobot, Jaraalbe, DVdm, Bgwhite, YurikBot, Bhny, Shawn81, Deskana, Dureo, Pyrotec, Scs, Gadget850, DeadEyeArrow, Bota47, Wknight94, FF2010, Lt-wiki-bot, Nikkimaria, Cobblet, JoanneB, TLSuda, Attilios, SmackBot, KocjoBot~enwiki, Eskimbot, Jab843, Srnec, Gilliam, Brianski, Hugo-cs, Chris the speller, Jfsamper, DHN-bot~enwiki, Suicidalhamster, DR04, Mr.Z-man, Mwtoews, DMacks, N Shar, Curly Turkey, SashatoBot, Lambiam, Khazar, John, JorisvS, CaptainVindaloo, Munita Prasad, Mr Stephen, Waggers, Quarty~enwiki, Poonu, Keitei, KlaudiuMihaila, Plankton 22, R~enwiki, Tawkerbot2, Dlohcierekim, Binks, Cryptic C62, DKqwerty, Lahiru k, J Milburn, Dgw, Christian75, DumbBOT, Dylant07, Smeazel, Thijs!bot, Epbr123, Pajz, Headbomb, John254, Jayron32, Darklilac, Ioeth, JAnDbot, MER-C, Db099221, Andonic, Rothorpe, Bencherlite, Magioladitis, Bongwarrior, Sodabottle, Dirac66, Lethaniol, Cpl Syx, Pasketti, UnfriendlyFire, Leyo, J.delanoy, MaximumPC, Johnbod, Yromemtnatsisrep, Wikimac007, Chinagirl5566, 83d40m, DorganBot, Squids and Chips, VolkovBot, CWii, AlnoktaBOT, Philip Trueman, TXiKiBoT, Xx8zaxo8xx, Pumapo, Martin451, Axiosaurus, Broadbot, Bcharles, Lamro, Petergans, SieBot, StAnselm, Mikemoral, Scarian, ToePeu.bot, Lord British, Techman224, Moletrouser, Dillard421, Joelster, Nergaal, Mr. Granger, ClueBot, Mild Bill Hiccup, Piledhigheranddeeper, Ishiho555, ChandlerMapBot, Redchasteen, Clanton2, Eeekster, Lartoven, Sun Creator, Esbboston, Arjayay, Polly, Indopug, Rror, Dthomsen8, SilvonenBot, Feministo, On the other side, Addbot, Paper Luigi, Roentgenium111, Laurinavicius, CarsracBot, LinkFA-Bot, Fryedpeach, Legobot, Luckas-bot, Yobot, Fraggle81, Ayrton Prost, AnomieBOT, BlackRaspberry, Sixxgun11, JackieBot, Klimenok, Law, Materialscientist, Citation bot, LilHelpa, Xqbot, R0pe-196, Capricorn42, Eru Ilúvatar, GrouchoBot, RibotBOT, Mirroredlens, Thehelpfulbot, Green Cardamom, R8R Gtrs, FrescoBot, Deutsch Fetisch, VI, HJ Mitchell, DivineAlpha, Citation bot 1, I dream of horses, Achim1999, Lightlowemon, Orenburg1, Double sharp, J4V4, 777sms, FrozenPencil, Onel5969, Smartiger, EmausBot, Dead Horsey, Faolin42, Natgel, NotAnonymous0, Tommy2010, P. S. F. Freitas, 15turnsm, Josve05a, Makecat, Natalefarrell, Donner60, TYelliot, DASHBotAV, NeonGas, Petrb, Mikhail Ryazanov, ClueBot NG, Lanthanum-138, Delusion23, Braincricket, Parcly Taxel, MerlIwBot, Helpful Pixie Bot, Bibcode Bot, BG19bot, Sandbh, Megakacktus, FutureTrillionaire, Glacialfox, Hockeyman7157, Sandman212, StarryGrandma, GoShow, Khazar2, Dexbot,

Webclient101, TwoTwoHello, Frosty, Jamesx12345, Dictatoracharyya123, Corinne, Thechaka135, Forgot to put name, AmericanLemming, Sterlinggreeson, Tentinator, SmartyPants1999, Jakec, Atongichk, Mahusha, Jonas Vinther, Games610, Bhavyaalok, Macofe, Christy LRC, Bammie73, Crystallizedcarbon, Peeyush.tripathi97, ASchnauzerCalledFred, RannyThePhilosopher, Divyanshi jain, Ccrivass, Swastik2001 and Anonymous: 304

- **Transition metal** *Source:* https://en.wikipedia.org/wiki/Transition_metal?oldid=686736827 *Contributors:* Derek Ross, Sodium, Lee Daniel Crocker, Mav, Bryan Derksen, Koyaanis Qatsi, Rjstott, Karen Johnson, William Avery, FvdP, Xoder, Michael Hardy, JakeVortex, Wapcaplet, Eric119, Looxix~enwiki, Ellywa, Ahoerstemeier, Andrewa, Kimiko, Hashar, Lfh, Kaal, Echidna, Jusjih, Chuunen Baka, Gentgeen, Robbot, Chris 73, Arkuat, Rfc1394, Elysdir, Timemutt, Sushi~enwiki, Centrx, Christopher Parham, Dratman, Darrien, Patteroast, Alexf, Yath, Icairns, Zfr, Erc, Felix Wan, Discospinster, Vsmith, Remember, RoyBoy, Femto, Smalljim, Reinyday, Cmdrjameson, Dungodung, Sam Korn, Lysdexia, Passw0rd, Jumbuck, Alansohn, Gerweck, Keenan Pepper, Ronline, Benjah-bmm27, Ahruman, Axl, Walkerma, Bart133, Trampled, Reaverdrop, Gurkha711, Stemonitis, Benbest, Polyparadigm, Jeff3000, MONGO, Macaddct1984, SDC, Prashanthns, Graham87, DePiep, Josh Parris, Rjwilmsi, Commander, HappyCamper, Yamamoto Ichiro, Yakiea, FlaBot, Latka, Tijuana Brass, Physchim62, King of Hearts, Chobot, Flying Jazz, DVdm, Gwernol, Tone, Roboto de Ajvol, YurikBot, Wavelength, RobotE, NTBot~enwiki, Jimp, Wambo, Wiki alf, Excession, Bota47, Wknight94, Tetracube, Gulliveig, CWenger, QmunkE, Katieh5584, Nippoo, SmackBot, Unyoyega, KocjoBot~enwiki, Eskimbot, Edgar181, Aksi great, Gilliam, Fogster, Hugo-cs, DHN-bot~enwiki, Pandora Xero, Jumping cheese, Nakon, Smokefoot, DMacks, Kukini, SashatoBot, Titus III, JorisvS, Mgiganteus1, Olin, A. Parrot, Mtodorov 69, Rkmlai, George The Dragon, Atraxendeluge, Waggers, Spiel496, Epitaf, Hetar, Iridescent, Twas Now, Courcelles, Irwangatot, Chrumps, WeggeBot, Gwdr500, LouisBB, Gogo Dodo, Optimist on the run, Coldphoenix182, Thijs!bot, Epbr123, Lanky, Pjvpjv, CharlotteWebb, Woppit, AntiVandalBot, Joan-of-arc, Micro.pw, Lfstevens, MER-C, WmRowan, Andonic, Tarif Ezaz, Bongwarrior, VoABot II, Fphilx, Dirac66, 28421u2232nfenfcenc, Jackfirst, Thibbs, Schmloof, Vigyani, ChemNerd, Ravichandar84, J.delanoy, Bogey97, Uncle Dick, StonedChipmunk, Eivindgh, Mangwanani, Touch Of Light, Cometstyles, Treisijs, Xiahou, Squids and Chips, CardinalDan, VolkovBot, ABF, Philip Trueman, Rei-bot, Kumorifox, Melsaran, Axiosaurus, Cremepuff222, Kingpin4646, Wikiisawesome, Kryan5, Synthebot, Spinningspark, AlleborgoBot, Petergans, SieBot, OMCV, Scarian, WereSpielChequers, Monkeyskate1, BotMultichill, Dawn Bard, Caltas, Yintan, Wiff&Hoos, Ayudante, Gunmetal Angel, Orcoteuthis, 🔲🔲🔲🔲~enwiki, Nergaal, Denisarona, ClueBot, Fyyer, The Thing That Should Not Be, Polyamorph, CounterVandalismBot, Runnerboy4444, Measles92793, Excirial, Rockincon1, Legalchemist, Jotterbot, Thingg, BlueDevil, DumZiBoT, SilvonenBot, NellieBly, Ziggy Sawdust, MystBot, Addbot, Denali134, Willking1979, Fieldday-sunday, LaaknorBot, LinkFA-Bot, Jasper Deng, Frozenguild, Ehrenkater, Erutuon, Tressor, Teles, Gail, Frehley, Legobot, Luckas-bot, Yobot, Tohd8BohaithuGh1, Newportm, MarcoAurelio, THEN WHO WAS PHONE?, Espeon, Heart of a Lion, Eric-Wester, AnomieBOT, Shootbamboo, Piano non troppo, Flewis, Materialscientist, Hunnjazal, ImperatorExercitus, Citation bot, Orienteer05, Xqbot, Capricorn42, Eru Ilúvatar, Nickkid5, GrouchoBot, RibotBOT, Mathonius, ScatheMote100, AntiAbuseBot, Adrignola, R8R Gtrs, Lucien-BOT, Haosys, VS6507, Cannolis, Ahmer Jamil Khan, Pinethicket, MastiBot, Double sharp, Sintau.tayua, TobeBot, Sumone10154, 777sms, Ismathsadhir, Weedwhacker128, RjwilmsiBot, TjBot, EmausBot, John of Reading, Jullfiqur, IncognitoErgoSum, Zubyrhassan1, Wikipelli, Thecheesykid, Fæ, StringTheory11, Azuris, Dax max, Wikfr, Bandn, Fairweva, ChuispastonBot, Staticd, DASHBotAV, ClueBot NG, Satellizer, Lanthanum-138, Bibcode Bot, Sandbh, Mark Arsten, Cncmaster, Benzband, Tony Tan, Joehill11, Adsurikar, Adi.akbartauhidin, Mogism, Frosty, Timothy, Ekips39, 14bemery, Revolution1221, Deepak harshal nagle, TrollyPolly4256, Adeoluwakiisi, Bezbeliiii, Yx007yx007, KasparBot and Anonymous: 435

- **Metal** *Source:* https://en.wikipedia.org/wiki/Metal?oldid=676815256 *Contributors:* Wesley, Bryan Derksen, Koyaanis Qatsi, AstroNomer~enwiki, Christian List, Ortolan88, William Avery, Roadrunner, Robert Foley, Heron, Jaknouse, Fransvannes, Ixfd64, Alfio, Ahoerstemeier, Theresaknott, TUF-KAT, Александър, Julesd, Poor Yorick, Andres, EdH, Smack, Hashar, Emperorbma, Tantalate, Vanished user 5zariu3jisj 0j4irj, Stone, Xanthine, Zoicon5, Selket, SEWilco, Phoebe, Lypheklub, Shizhao, Jusjih, Johnleemk, Frazzydee, Gromlakh, Gentgeen, Robbot, Jred-mond, Lowellian, Rursus, Hadal, HaeB, Dina, Alan Liefting, Centrx, Giftlite, Graeme Bartlett, Everyking, Anville, Curps, Bensaccount, Erdal Ronahi, Jackol, Pne, Gadfium, Utcursch, SoWhy, Gazibara, Slowking Man, Antandrus, Beland, Latitude0116, Icairns, ELApro, MikeRosoft, Discospinster, Rich Farmbrough, Cacycle, Vsmith, Pavel Vozenilek, Paul August, SpookyMulder, ESkog, RJHall, El C, Shanes, Re-member, EurekaLott, Triona, Femto, Adambro, Bobo192, Fir0002, Harald Hansen, Smalljim, Malafaya, Maurreen, ParticleMan, Man vyi, La goutte de pluie, Jojit fb, David Gale, Haham hanuka, Passw0rd, Jumbuck, Stephen G. Brown, Danski14, Alansohn, Anthony Appleyard, Andrewpmk, Iothiania, Riana, Lectonar, MarkGallagher, Walkerma, Mysdaao, SidP, L33th4x0rguy, Tycho, Docboat, Tc 191, Amorymeltzer, RainbowOfLight, Vuo, OwenX, Nuggetboy, Scjessey, Pol098, MONGO, Mouvement, Eaolson, SDC, 🔲🔲🔲🔲🔲🔲, Crucis, CPES, Prashan-thns, Palica, Joe Roe, Dysepsion, Mandarax, Tslocum, Graham87, Magister Mathematicae, FreplySpang, DePiep, Nanite, Canderson7, Rjwilmsi, Tangotango, MZMcBride, HappyCamper, Boccobrock, Graibeard, DoubleBlue, Cyclometh, Joe056, FlaBot, Latka, Nihiltres, Nivix, RexNL, Gurch, Ayla, TeaDrinker, BradBeattie, Chobot, Flying Jazz, Bornhj, Gwernol, Elfguy, YurikBot, Wavelength, Mahahahaneap-neap, Pip2andahalf, RussBot, Stephenb, Shell Kinney, CambridgeBayWeather, Rsrikanth05, Wimt, NawlinWiki, Nirvana2013, Trovatore, Nick, E rulez, RL0919, DeadEyeArrow, Bota47, User27091, Searchme, FF2010, Wikilackey, Zzuuzz, Closedmouth, Ketsuekigata, Жованв6, Dspradau, CWenger, Fram, Ordinary Person, Kevin, Willtron, GinaDana, Jaranda, ArielGold, Garion96, Smurfy, Katieh5584, Kungfuadam, JDspeeder1, GrinBot~enwiki, Serendipodous, Mejor Los Indios, CIreland, AndrewWTaylor, Luk, SpLoT, SmackBot, Ashill, Reedy, Knowl-edgeOfSelf, Kilo-Lima, Thenickdude, Onebravemonkey, Edgar181, Srnec, Ekilfeather, Gilliam, Talinus, ERcheck, Andy M. Wang, Squiddy, Anwar saadat, Chris the speller, Improbcat, Bluebot, Persian Poet Gal, Ian13, Papa November, SchfiftyThree, Nbarth, DHN-bot~enwiki, Colonies Chris, Hallenrm, 56, Royboycrashfan, Can't sleep, clown will eat me, Shalom Yechiel, Tim Pierce, Yidisheryid, Rrburke, LouSchef-fer, SundarBot, Nakon, Blake-, Smokefoot, EdGl, ILike2BeAnonymous, Acdx, Ck lostsword, Pilotguy, Kukini, The undertow, SashatoBot, Lambiam, Swatjester, Brysonlochte, AlanD, Molerat, Vanished user 9i39j3, John, Scientizzle, Jan.Smolik, JoshuaZ, BladeHamilton, Ste-lio, Ckatz, Slakr, Werdan7, Stwalkerster, Beetstra, Mr Stephen, Ace Frahm, Ahering@cogeco. ca, MTSbot~enwiki, Jose77, Dik~enwiki, BranStark, Wizard191, Iridescent, Muéro, StephenBuxton, Twas Now, Blehfu, Thesexualityofbereavement, Mcwatson, JayHenry, Tawkerbot2, Billde, Ghaly, Joostvandeputte~enwiki, JForget, Sleeping123, Ale jrb, Megaboz, Juliantwd, Bill K, Dgw, WLior, NickW557, FlyingToaster, MarsRover, INVERTED, The Enslaver, Cydebot, Ryan, Gundampilotspaz, MC10, Rifleman 82, Nick Wilson, Gogo Dodo, A Softer Answer, Guitardemon666, ObeyMe, Q43, HitroMilanese, Christian75, Inhumer, Englishnerd, Thijs!bot, VoABot, Epbr123, Barticus88, N5iln, Trevyn, Marek69, A3RO, A b, James086, Dfrg.msc, CharlotteWebb, Big Bird, Escarbot, Menti fisto, Danceangle90, AntiVandalBot, Luna Santin, Dark-Audit, Modernist, Farosdaughter, Spencer, Alphachimpbot, Astavats, Elaragirl, Myanw, Gökhan, MikeLynch, Mikenorton, JAnDbot, Barek, MER-C, CosineKitty, Plantsurfer, NE2, Hamsterlopithecus, Andonic, Roleplayer, Beaumont, Bongwarrior, VoABot II, AndriusG, JamesBWat-son, Avicennasis, Cgingold, Mtd2006, Allstarecho, Spellmaster, Darkrangerj, Glen, DerHexer, JaGa, Edward321, Esanchez7587, Mattinbgn,

Nevit, InvertRect, Supahfreekeh, Oroso, Gwern, Xiru, Jtir, MartinBot, Schmloof, Mermaid from the Baltic Sea, Axlq, Polartsang, Rettetast, Anaxial, RWyn, Redshoe2, J-t-m, Jargon777, J.delanoy, Pharaoh of the Wizards, Trusilver, Bogey97, Numbo3, Hans Dunkelberg, TheHurt-Procces, 12dstring, NerdyNSK, Reedy Bot, Icseaturtles, Katalaveno, James A. Stewart, Roni2204, Smeira, P4k, Jeepday, Samtheboy, Joey-rockyhorror, AntiSpamBot, Jcwf, Darrendeng, NewEnglandYankee, In Transit, Nwbeeson, RayChang, Cobi, Seamarine, Zeo197, Cmichael, MatTticus, Juliancolton, Cometstyles, Jamesontai, Jamesofur, Ken g6, Bonadea, IceDragon64, Idioma-bot, Wikieditor06, ACSE, Lights, Deor, Macedonian, Jeff G., Indubitably, Nburden, Philip Trueman, Fran Rogers, TXiKiBoT, Cosmo1976, Xenophrenic, Katoa, Moogwrench, Not-mattepp, A Red Pirate, Anonymous Dissident, Arnon Chaffin, Qxz, Una Smith, Jaenop, Dlae, AtaruMoroboshi, Cremepuff222, Luuva, Maxim, ARUNKUMAR P.R, Wolfrock, Skogen11341, Synthebot, Sweetnshort, Rhopkins8, Falcon8765, GlassFET, George bennett, Monty845, An-thony197427, Fddf, Headsets, Nagy, Face-2-face, Symane, Hilts99, Rock2e, Borne nocker, I am grate, EmxBot, Hmwith, The Random Editor, SieBot, JoeGucciardi, PlanetStar, Tiddly Tom, Scarian, Caffm8, Dawn Bard, Rufus Turner, Nology, Yintan, Mangostar, Leeman056, Bentogoa, Tiptoety, Radon210, Belinrahs, Wilson44691, Oiws, Taemyr, Yerpo, Oxymoron83, Avnjay, KoshVorlon, Steven Crossin, Iain99, Techman224, Alex.muller, Voltron, Fratrep, Rosiestep, Latics, Jacob.jose, Ascidian, Superbeecat, ShexRox, MansonScar, Nergaal, Denisarona, Stained Illu-sion, TheCatalyst31, Metalbul, WikipedianMarlith, Faithlessthewonderboy, BlnLiCr, Loren.wilton, ClueBot, Badger Drink, Foxj, The Thing That Should Not Be, Ndenison, Buxbaum666, Meekywiki, The Incredible Editor, Fossiliferous, Regibox, Back and Forth, Xenon54, Van-dalCruncher, LizardJr8, Neo001~enwiki, Riverwire, Alexostamp, Franksbnetwork, Puchiko, Say2anniyan, Gamunday, Graysen98, Awickert, Excirial, Alexbot, Jusdafax, CrazyChemGuy, Anon lynx, PixelBot, Limmy11frog, Ludwigs2, Abrech, Lartoven, EBY3221, Paramaya, TER-ANADON, Alisterg, Iohannes Animosus, Yran01, Kaiba, Razorflame, Mikaey, Ottawa4ever, JakeHawley, Thingg, 07andy07, Gazza BWFC, Aitias, Versus22, LieAfterLie, Limonns, SoxBot III, Averizi, XLinkBot, Roxy the dog, AshidoX, Rror, Avoided, WikHead, Ziggy Sawdust, Badgernet, Alexius08, WikiDao, Bird747, Carlmalone, Press-yes, Felix Folio Secundus, Addbot, Denali134, Narayansg, W4410ck, Element16, Landon1980, Hellboy2hell, Yobmod, TutterMouse, Njaelkies Lea, Fieldday-sunday, KorinoChikara, CanadianLinuxUser, Juliemfreeman1, Divyansh.ip15, Ashanda, Cst17, CarsracBot, Glane23, CUSENZA Mario, Deamon138, Elen of the Roads, Im anoob68, Alchemist-hp, Gem-iniey, Pomlikespee, Dr. Beaster, Tide rolls, Bfigura's puppy, Hardness, NHJG2, WikiDreamer Bot, 123theman123, Bubonic plague, Albert galiza, EJF (huggle), Z897623, TehTomeh, Frehley, Luckas-bot, Yobot, 2D, Pink!Teen, Chipthief00, Ptbotgourou, Mmxx, عالم بوبحم, The Flying Spaghetti Monster, Monorailx, AnomieBOT, Hello108, Rubinbot, Daniele Pugliesi, Jim1138, Galoubet, JackieBot, Piano non troppo, Kingpin13, Ulric1313, Flewis, Materialscientist, ImperatorExercitus, OllieFury, Bob Burkhardt, Frankenpuppy, Serveux, LovesMacs, Dark Master87, Pxlnight13, Xqbot, Sketchmoose, Sionus, JimVC3, Capricorn42, PrometheusDesmotes, 4twenty42o, Nickkid5, Hardymahi, YBG, Nayvik, Callumpuffy, Omnipaedista, Logger9, PeterWiki09, Supermake, Shadowjams, Klin06, Middle 8, Haploidavey, Thehelpful-bot, K Fuchs 45784578, Chair Noble, FrescoBot, Ziyaad97, Tommyknee, Nub22, Pepper, Wikipe-tan, HANZBLIX, L7tr4h, Sky Attacker, Recognizance, Alaphent, Recon Unit, EmoCon, Citation bot 1, Redrose64, Longstead, Pinethicket, I dream of horses, Pokemonned, Bob-bychong, ThePillock, MastiBot, SirMoo, TheGreatAwesomeness, Aetylus, SpaceFlight89, Σ, Footwarrior, Jujutacular, Robvanvee, FoxBot, Double sharp, Sweet xx, Ticklewickleukulele, Animalparty, Zaz3494, Vrenator, Jonathanr668, TBloemink, Pbrower2a, Starfawcks, Dian-naa, Navy101, Sammetsfan, Tbhotch, Marie Poise, DARTH SIDIOUS 2, Cromulant, Mean as custard, TjBot, Curt5876, EmausBot, John of Reading, WikitanvirBot, Katherine, Rajkiandris, Slightsmile, Wikipelli, JSquish, ZéroBot, Fæ, StringTheory11, Chemicalinterest, Hazard-SJ, Monterey Bay, Imaprincezz, FinalRapture, Hephaestus III, Wayne Slam, Shrigley, Fanyavizuri, Puffin, RockMagnetist, Peter Karlsen, Pka977, PookBB, Eg-T2g, Waxcaptain45, DASHBotAV, ResearchRave, Petrb, ClueBot NG, Lanthanum-138, Monsoon Waves, Imhungryyy, Rezabot, Widr, Nijanand, Soccerkid97, ImminentFate, Helpful Pixie Bot, Nightenbelle, Calabe1992, BZTMPS, Lowercase sigmabot, Vagobot, Sci-enceSage, Author1979, Sandbh, Wiki13, Happfeetoneohone, Mark Arsten, Dipankan001, Franspinale, Colton11901, 63boom, MrBean4160, Stevelot, Birmingham99, Justincheng12345-bot, Some old guy over 9000, Chivesud, ChrisGualtieri, GoShow, CarrieVS, EuroCarGT, Duck-nish, Dexbot, Webclient101, Lfcwill97, Jbaugher13, Carl00876, Ryan Smedley, Frosty, K. Böhme, Lewisbrown2000, Likewhatyousee, Reatlas, Mikemike8787mad, Epicgenius, Nicolalouise2012, Light Peak, Backendgaming, Babitaarora, Ugog Nizdast, My name is not dave, Ginsuloft, My-2-bits, Gokul.gk7, Liz, Iliyan110, JaconaFrere, Hussain bin farrukh, Razmir1337, Mahusha, William3576, Shivu goyal, Mat2001, Tristan-Wix, IiKkEe, Amortias, Galax.z, Flyingchocolate, Crystallizedcarbon, Vikas121212, Vikas kumar yadav1, Aishwaryauv1999, Provingmatt, Dmartinez 208, Julian.losekoot, KasparBot and Anonymous: 1215

- **Nonmetal** *Source:* https://en.wikipedia.org/wiki/Nonmetal?oldid=686603180 *Contributors:* Andre Engels, Heron, Ewen, Ubiquity, Shell-reef, Eric119, Looxix~enwiki, Ahoerstemeier, Smack, Paul-L~enwiki, Geraki, Jusjih, Bearcat, Branddobbe, Gentgeen, Romanm, Rfc1394, Carnildo, Alan Liefting, Graeme Bartlett, Kim Bruning, Michael Devore, Semprini, Beland, Icairns, Jcw69, D6, Vsmith, Remember, Bobo192, Orbst, Dungodung, Jumbuck, Keenan Pepper, Benjah-bmm27, Ahruman, Gene Nygaard, Feline1, Nuggetboy, GregorB, DePiep, Mendaliv, Saperaud~enwiki, Rjwilmsi, Vary, HappyCamper, Margosbot~enwiki, Nihiltres, Nivix, Chobot, Bgwhite, Roboto de Ajvol, YurikBot, Wimt, Gadget850, ☐☐☐☐ robot, KnowledgeOfSelf, Bomac, Eskimbot, Edgar181, Srnec, Gilliam, Hugo-cs, 56, Mr.Z-man, DMacks, Just plain Bill, Anoop.m, JorisvS, Mgiganteus1, Aleenf1, IronGargoyle, Munita Prasad, BranStark, KlaudiuMihaila, Wjejskenewr, Paul Foxworthy, Cour-celles, CmdrObot, Leujohn, Funnyfarmofdoom, Nebular110, Gogo Dodo, Kw0~enwiki, Epbr123, Barticus88, Eb.eric, Mentifisto, AntiVan-dalBot, Indian Chronicles, JAnDbot, How to save a life, Connormah, VoABot II, Smihael, Kayau, MartinBot, BetBot~enwiki, ChemNerd, Kateshortforbob, Leyo, Tgeairn, J.delanoy, Trusilver, Cocoaguy, TomS TDotO, Genius695, Monkeysack420, Gallador, KylieTastic, Squids and Chips, CardinalDan, VolkovBot, AlnoktaBOT, Fluffybun, Kostaki mou, TXiKiBoT, A4bot, Rei-bot, GcSwRhIc, Cosmium, Broadbot, LeaveSleaves, Natg 19, Lova Falk, Unused0030, SieBot, Nubiatech, RJaguar3, Sunny910910, Adabow, SethBoldt, Nergaal, Denisarona, Im-ageRemovalBot, Sfan00 IMG, ClueBot, Ficbot, Otolemur crassicaudatus, Jotterbot, SoxBot III, Wnt, Life of Riley, XLinkBot, BodhisattvaBot, TheOner, Frood, SkyLined, Acaeton, Addbot, Roentgenium111, Element16, Ronhjones, Leszek Jańczuk, CarsracBot, LinkFA-Bot, Thewon-deridiot, Numbo3-bot, Tide rolls, Legobot, Luckas-bot, Yobot, Fraggle81, Amirobot, AnomieBOT, Piano non troppo, Sz-iwbot, Flewis, Mate-rialscientist, ArthurBot, Xqbot, Sketchmoose, Capricorn42, Eru Ilúvatar, Nickkid5, DSisyphBot, YBG, GrouchoBot, Nasirkhan, Chris.urs-o, Mathonius, 14albeev, Sahehco, FrescoBot, Wikipe-tan, HamburgerRadio, Redrose64, Pinethicket, I dream of horses, Shiva Khanal, Shade-ofTime09, Double sharp, Sumone10154, DARTH SIDIOUS 2, Mean as custard, RjwilmsiBot, EmausBot, Acather96, GA bot, Immunize, Razor2988, Liamgales, TuHan-Bot, P. S. F. Freitas, ZéroBot, StringTheory11, Mikhail Ryazanov, ClueBot NG, This lousy T-shirt, Lanthanum-138, Widr, Helpful Pixie Bot, Bibcode Bot, Krishnaprasaths, Sandbh, Rjlanc, Jazzlw, DPL bot, Mrt3366, Tenisben, Khazar2, Faizan, Epic-genius, Ugog Nizdast, ThePokemon850, Qwertyxp2000, Sarahjrae, Vikas121212, Sonukumar2014, Hoodihood, KasparBot, Abhiuo0007 and Anonymous: 229

- **Refractory metals** *Source:* https://en.wikipedia.org/wiki/Refractory_metals?oldid=673646797 *Contributors:* Ugen64, Stone, Drxenocide, Meelar, Seth Ilys, DocWatson42, Madoka, Deus Ex, Pjabbott, Chadernook, Burschik, Smalljim, DePiep, Ketiltrout, Rjwilmsi, Srleffler, Mrbehemoth, Jengelh, Gaius Cornelius, Shaddack, Gadget850, Lion Roller, Groyolo, That Guy, From That Show!, SmackBot, Commander Keane bot, Hongooi, Gracenotes, Frap, Makyen, Beetstra, NinjaCharlie, Wizard191, Rambam rashi, TheHerbalGerbil, Cybernetic, Headbomb, Adamelk, TonyGerillo, Funandtrvl, Barneca, Rocketmagnet, OlavN, Xresonance, Andy Dingley, Gerakibot, Flyer22 Reborn, TechTube, Nergaal, Antediluvian67, LordJesseD, Addbot, Lightbot, Luckas-bot, Materialscientist, Citation bot, GrouchoBot, Prari, Citation bot 1, Trappist the monk, AJBryhan, EmausBot, GoingBatty, ZéroBot, SporkBot, Whoop whoop pull up, Pokbot, Snotbot, Lanthanum-138, Helpful Pixie Bot, Bibcode Bot, Pop-up casket, JeffGw, Monkbot and Anonymous: 53

- **Noble metal** *Source:* https://en.wikipedia.org/wiki/Noble_metal?oldid=678876296 *Contributors:* Kpjas, The Anome, DopefishJustin, Ike9898, Furrykef, Robbot, Rursus, Enochlau, Everyking, Icairns, Abdull, Rich Farmbrough, Vsmith, Shad0, Art LaPella, Whosyourjudas, Viriditas, Keenan Pepper, Sligocki, Wtshymanski, Benbest, Polyparadigm, Knuckles, Bluemoose, Mandarax, Enz, Margosbot~enwiki, Rune.welsh, Common Man, Physchim62, Chobot, Roboto de Ajvol, YurikBot, Wavelength, Borgx, Jimp, Hede2000, Bhny, Voyevoda, Mccready, Rayc, Skittle, Itub, SmackBot, Edgar181, Bluebot, Cybercobra, RandomP, Smokefoot, JorisvS, Wizard191, Metre01, Patrickwooldridge, N2e, Rifleman 82, Bsdaemon, Alaibot, Satori Son, Thijs!bot, Muaddeeb, Escarbot, Gioto, Wayiran, Amberroom, PhilKnight, McDoobAU93, STBotD, Idiomabot, Christophenstein, TXiKiBoT, JhsBot, UnitedStatesian, Bcharles, Cmjayakumar, SieBot, ToePeu.bot, KoshVorlon, Sanya3, Chem-awb, Fangjian, Twinsday, Michał Sobkowski, Piledhigheranddeeper, Tablemajorrt5, Scyldscefing, EgraS, DumZiBoT, BodhisattvaBot, Heeero60, Addbot, DOI bot, Numbo3-bot, Donfbreed, Smackeldorf, AnomieBOT, Magog the Ogre 2, Materialscientist, Citation bot, Xqbot, GrouchoBot, Asfarer, DrilBot, Pinethicket, Achim1999, Bgpaulus, Double sharp, LilyKitty, Mtz1010, WikitanvirBot, Rami radwan, Chemicalinterest, Scientific29, Yugo312, ClueBot NG, Wikiphysicsgr, Hari Eswar SM, Sasakubo1717, Bibcode Bot, BG19bot, Softballbaby984, Nedgreiner, Wieldthespade, T.J.S.1 and Anonymous: 65

- **Periodic trends** *Source:* https://en.wikipedia.org/wiki/Periodic_trends?oldid=685385321 *Contributors:* William Avery, Stone, Robbot, Xezbeth, Paul August, Sol~enwiki, RoyBoy, Bart133, Somody, Feline1, Jak86, TheIguana, Yamamoto Ichiro, Shultzc, DVdm, Bgwhite, Nolookingca, Malcolma, Tetracube, Brammers, Canthusus, Chris the speller, Jprg1966, Thatoldaccount, Flyguy649, T-borg, Tfl, Beetstra, Gegnome, Eastlaw, JForget, Vanished user vjhsduheuiui4t5hjri, Kaldosh, Rifleman 82, Christian75, Azztech, JamesAM, Epbr123, Frozenport, Omegakent, Sluzzelin, VoABot II, Rich257, DerHexer, Mirek2, Rettetast, Nescalona, M-le-mot-dit, Juliancolton, Reelrt, DarkNiGHTs, Truthanado, Yintan, Nerdygeek101, ClueBot, Cptmurdok, Ulao, Lartoven, Grey Matter, Tnxman307, Noosentaal, DumZiBoT, Stickee, Nicoguaro, Addbot, Musicyea, Jamesbond12345, Aunva6, Tayzhian, Tide rolls, Yobot, Ninjalemming, AnomieBOT, Jim1138, Materialscientist, Matttoothman, Puppydog12345, Erik9bot, A.amitkumar, Wikipedian0791, Dotarulez2, Pinethicket, Double sharp, Reaper Eternal, Diannaa, Chipmunkdavis, Super48paul, Slightsmile, Maxim Gavrilyuk, Onyxqk, Rocketrod1960, ClueBot NG, Antiowner, Lanthanum-138, Salehkn, Widr, Metaknowledge, Pluma, Zontras Gry, Pratyya Ghosh, Tirtha bose12, Padenton, Mogism, Th232, KeymasterCZ, Lugia2453, Isarra (HG), Andrebolle, Ribhavgupta, UndoMeister, DavidLeighEllis, Sciencegeekemc, Al.Ilseman, JaconaFrere, Vaishvi16, Dimitar G Slavov, Jpskycak, Jmbsouth and Anonymous: 203

- **Aufbau principle** *Source:* https://en.wikipedia.org/wiki/Aufbau_principle?oldid=685957495 *Contributors:* SimonP, Andres, Smack, DocWatson42, Pietz, OwenBlacker, DragonflySixtyseven, Zaheen, Rgdboer, Gershwinrb, Spoon!, Jon the Geek, Bobo192, Andrew Gray, SidP, Unconventional, Ringbang, Killing Vector, Firsfron, Prashanthns, Graham87, DePiep, Wragge, John Baez, Physchim62, DVdm, Bgwhite, YurikBot, Postglock, Rsrikanth05, Arthur Rubin, Tomj, Guillom, Itub, SmackBot, Jrockley, Meatmanek, PJTraill, Pieter Kuiper, Colonies Chris, Melkhior, BWDuncan, DMacks, Stefano85, Robofish, Rundquist, Mgiganteus1, Neelix, Cydebot, Rifleman 82, Btharper1221, TAnthony, Boleslaw, Mitosh mora, Dirac66, User A1, Joker99352, Rettetast, T.vanschaik, Hans Dunkelberg, CardinalDan, SimDarthMaul, Julia Neumann, Philip Trueman, Pjstewart, Shureg, Moonraker12, Tesi1700, Bschaeffer~enwiki, Atif.t2, ClueBot, Divye, Niceguyedc, Versus22, ErkangZhu, Addbot, Some jerk on the Internet, DOI bot, Escherichia coli, Laurinavicius, Wang lvan, Drova, Ettrig, II MusLiM HyBRiD II, Spaghettipoop, Citation bot, Obersachsebot, قل‌ی زادگان, Rice.brendan, Chemmix, A. di M., R8R Gtrs, Citation bot 1, Arvindsrm2894, Amit kumar roy, Double sharp, Ianprime0509, Jdlawlis, Aaleo20, DexDor, EmausBot, Ibot, ZéroBot, ClueBot NG, Lanthanum-138, Pluma, Helpful Pixie Bot, Bibcode Bot, DuranFam6, MusikAnimal, Pozitron969, MeanMotherJr, Mrt3366, Bourgeo, Webclient101, TwoTwoHello, MaatyBot, Antrocent, HMSLavender, Tymon.r, Dharmik148, Mnster98 and Anonymous: 87

- **History of the periodic table** *Source:* https://en.wikipedia.org/wiki/History_of_the_periodic_table?oldid=687172027 *Contributors:* AxelBoldt, The Anome, William Avery, Ewlloyd, D, Dcljr, Eric119, Ellywa, Ahoerstemeier, Stone, Jerzy, Gentgeen, Fredrik, Korath, Chris 73, Altenmann, Romanm, Arkuat, SoLando, Diberri, Everyking, Tweenk, Pne, Rrw, Alexf, Goog, Fredcondo, Ruzulo, DragonflySixtyseven, RetiredUser2, PFHLai, Icairns, Deglr6328, Mike Rosoft, Discospinster, Vsmith, Bishonen, SpookyMulder, Rubicon, Bletch, Bobo192, Fir0002, Smalljim, Honeycake, Wtmitchell, Helixblue, RainbowOfLight, Dominic, Eztli, Feline1, Kazvorpal, Angr, Linas, DonPMitchell, Camw, MarcoTolo, Allen3, RuM, DePiep, -DjD-, Drbogdan, HappyCamper, Ems57fcva, Bubba73, The wub, Bhadani, Dracontes, Titoxd, TheMidnighters, Gurch, Scerri, Physchim62, DVdm, Cactus.man, Jimp, Kordas, Conscious, Wimt, NawlinWiki, Shreshth91, Grafen, Darkmeerkat, DGJM, Lockesdonkey, Kkmurray, Fabiob~enwiki, Wknight94, Zzuuzz, Closedmouth, Dspradau, Petri Krohn, Kevin, Itub, SmackBot, MattieTK, Unschool, Reedy, KnowledgeOfSelf, TestPilot, VigilancePrime, Hydrogen Iodide, Davewild, Cool3, Gilliam, Skizzik, Persian Poet Gal, Bduke, Jprg1966, SchfiftyThree, Darth Panda, Jahiegel, Eliyahu S, Neo139, Pieter1, Konczewski, RedHillian, AiOlorWile, Edwtie, RJN, DMacks, Kuru, John, Microchip08, Jaganath, Shlomke, JoshuaZ, Noah Salzman, Arkrishna, DabMachine, Mikehelms, Tawkerbot2, Dgw, Black and White, Glenn4pr, Equendil, Krauss, Christian75, DumbBOT, Danogo, Thijs!bot, Mojo Hand, Purple Paint, Headbomb, Marek69, Kathovo, Majorly, Luna Santin, RapidR, Shift6, Wan nni, Gökhan, TuvicBot, JAnDbot, MER-C, Kerotan, .anacondabot, Bongwarrior, VoABot II, Romtobbi, Dirac66, Thibbs, DerHexer, Davidbws23, Amitchell125, SquidSK, Rikpotts, Rettetast, J.delanoy, Rhinestone K, Uncle Dick, Tdadamemd, DoubleParadox, Ephebi, Juliancolton, IceDragon64, DarkNiGHTs, X!, Jeff G., JohnBlackburne, Philip Trueman, Gigo12, Anna Lincoln, Gekritzl, Leafyplant, One half 3544, BigDunc, Vector Potential, Insanity Incarnate, HeirloomGardener, AlleborgoBot, Logan, LuigiManiac, Jehorn, Graham Beards, Caltas, Matthew Brandon Yeager, Keilana, Atilema, Happysailor, Cat1993127, Oxymoron83, Aflumpire, AnonGuy, OKBot, Anchor Link Bot, Nergaal, Pmigdal, Elassint, ClueBot, The Thing That Should Not Be, S Levchenkov, Pi zero, CounterVandalismBot, Delta1989, Pointillist, Djr32, Excirial, Pumpmeup, Attheweaneiffer, PixelBot, Drsrisenthil, Razorflame, Schreiber-Bike, BOTarate, Versus22, Epiclolz, Romaine, PCHS-NJROTC, SoxBot III, Jean-claude perez, Dark Mage, Avoided, Airplaneman, Kinokaru, Mootros, Laurinavicius, Vishnava, Fluffernutter, Download, Favonian, Drova, Ehrenkater, Tide rolls, Narayan, MissAlyx, Finbob83, Cflm001, Cottonshirt, Paul Siebert, Eric-Wester, Synchronism, AnomieBOT, Rubinbot, Ginyild, Piano non troppo, Aditya, Materialscientist, The High Fin Sperm Whale, Srinivas, Alexphudson, LilHelpa, Xqbot, Zad68, S h i v a (Visnu), Mattstead93, 78.26, Ignoranteconomist, Clementisbad, Amrosabra, A.amitkumar, Dougofborg, Eyea0012, Liamhaha, Paine Ellsworth, VS6507, Pinethicket, I dream of horses, Serols, Merlion444, Utility Monster, Cnwilliams, Tim1357, Double sharp, Tubby23, Vrenator, TBloemink, Oenrhysbiggs, Extra999, Allen4names,

Aoidh, Laurielaurielaurie, Hornlitz, DARTH SIDIOUS 2, Onel5969, NerdyScienceDude, DASHBot, Orphan Wiki, Acather96, Immunize, Sponk, RenamedUser01302013, Ilikefod, Tommy2010, Wikipelli, K6ka, Anirudh Emani, PS., Dfern22, Fæ, StringTheory11, Kieran Nash, Openstrings, Flightx52, MonoAV, Donner60, Negovori, DASHBotAV, Petrb, ClueBot NG, Satellizer, Alex Nico, Widr, Antiqueight, TOR-NELLcello, BG19bot, Betty Noire, Wiki13, Soerfm, Paolo Raneses, Sni56996, Klilidiplomus, Razzat99, Simeondahl, Hghyux, Mrt3366, BlaBlaBaberBabe, Mediran, MadGuy7023, TBBT Chase, Dexbot, Sam 365, Lugia2453, SFK2, Little green rosetta, Kevin12xd, Bangladesh News, Epicgenius, Sarsarpow, I am One of Many, Tentinator, George8211, Ameer Hasan Khan, JaconaFrere, BillyTanjung, Polymathica, Filedelinkerbot, Ash.flowers.palad, ArdentWhiteraven, Dude128123, Jacobo95, Zondaj, ChemWarfare, KSFT, 420BlazeIt69Sex, Thefalse-historybuff and Anonymous: 586

- **Alternative periodic tables** *Source:* https://en.wikipedia.org/wiki/Alternative_periodic_tables?oldid=680295645 *Contributors:* Jagged, Ixfd64, Johnmarks, Tyler McHenry, Freakofnurture, Vsmith, RJHall, Kwamikagami, Femto, Bawolff, RHaworth, V8rik, DePiep, HappyCamper, Glenn L, Shanel, Tetracube, Huangcjz, SmackBot, Bduke, DMacks, Garudabd, Cbuckley, CapitalR, Eassin, GiantSnowman, Thermochap, Woudloper, Headbomb, MichaelMaggs, Widefox, Mohdabubakr, Sikory, Baccyak4H, Dirac66, Johnmarks9, ChemNerd, Osquar F, Richard777, Coppertwig, BUztanAtor, Cowplopmorris, JohnBlackburne, Chris Dybala, Pjstewart, OlavN, Falcon8765, Ajrocke, Karmaxul, Thomas andreas kies, Derdoctor~enwiki, Martarius, ClueBot, Polyamorph, Vasco.bonifacio, BirgerH, Jean-claude perez, XLinkBot, Dark Mage, Parejkoj, Ad-dbot, Blethering Scot, WFPM, Drova, Latenightwithjimmyfallon, Luckas-bot, Yobot, Jgmoxness, Mardeg, Daniele Pugliesi, Materialscien-tist, Mononomic, Armbrust, Aashaa, I dream of horses, Efficiency1101e, Serols, Double sharp, RjwilmsiBot, J. Garai, EmausBot, Andres Rojas, ZéroBot, Fæ, StringTheory11, Capaccio, Housewatcher, Sven nestle2, JuTa, ClueBot NG, Lanthanum-138, Robertharrison95, Addi-hockey10 (automated), Curb Chain, ChrisGualtieri, Bc239, Ericwm6, Carpelogos, Piazzalunga, Richard Yin, ShenHuiZhang, MetazoanMarek and Anonymous: 66

- **Extended periodic table** *Source:* https://en.wikipedia.org/wiki/Extended_periodic_table?oldid=684533395 *Contributors:* Mav, Bryan Derk-sen, Tarquin, Fonzy, Dcljr, Ahoerstemeier, Darkwind, LittleDan, Smack, Tpbradbury, Donarreiskoffer, Bearcat, Robbot, Psychonaut, Rfc1394, Bkell, DocWatson42, Pne, Phe, Icairns, Jayjg, Sfeldman, Rich Farmbrough, Kwamikagami, Gershwinrb, WhiteTimberwolf, Femto, Eddideigel, Jumbuck, Keenan Pepper, Alinor, Pauli133, Feline1, Czolgolz, Vartan84, Richard Arthur Norton (1958-), Georgia guy, Waldir, Mandarax, DePiep, Rjwilmsi, Koavf, AED, Gurch, Glenn L, Chobot, YurikBot, Ozabluda, Ospalh, Tetracube, Cffrost, Sbyrnes321, SmackBot, Incnis Mrsi, Chris the speller, Bduke, Flyguy649, DMacks, John, Mgiganteus1, Valoem, INVERTED, Meodipt, Thijs!bot, Keraunos, Mojo Hand, Headbomb, Luna Santin, Guy Macon, Seaphoto, JAnDbot, Kruckenberg.1, Gavia immer, VoABot II, Swpb, R'n'B, RockMFR, Captain panda, Choihei, MNRoss, Cowplopmorris, Jim Swenson, RingtailedFox, Philip Trueman, TXiKiBoT, McM.bot, UnitedStatesian, Petergans, SieBot, Rihanij, PlanetStar, Droog Andrey, Oxymoron83, Dixongrove, Nergaal, Jkj115, ClueBot, IceUnshattered, DanielDeibler, LizardJr8, Excirial, LarryMorseDCOhio, Psinu, DumZiBoT, Iv0202, Addbot, Roentgenium111, LaaknorBot, Ehrenkater, OlEnglish, Teles, MTM, Yinweichen, Luckas-bot, Yobot, Alakasam, AnomieBOT, AndrooUK, Williamb4, Citation bot, Vuerqex, ArthurBot, Brane.Blokar, Xqbot, Nippashish, Shadowjams, FrescoBot, Ionutzmovie, TachyonJack, ינון גלעדי, Humanrace378, Finalius, Robo37, Redrose64, Pinethicket, I dream of horses, Efficiency1101e, Mikespedia, Double sharp, Vrenator, 4, 09curranm, Unbitwise, Patriot8790, EGroup, Solomonfromfinland, AvicBot, Dia-mond0304E, StringTheory11, Bamyers99, AManWithNoPlan, Kolo-Dearney, L Kensington, Mythic219, Whoop whoop pull up, Nookone77, ClueBot NG, LKS Corea, Ypnypn, Hackne, Lanthanum-138, Moneya, Parcly Taxel, Widr, Iulius C, Helpful Pixie Bot, Bibcode Bot, Ewan1999, BG19bot, 3.14159265358pi, BattyBot, Kc kennylau, APerson, Dexbot, LightandDark2000, Numbermaniac, Titanic225, Rush Revisionz, Ruby Murray, Eyesnore, Kiran Khatri Chhetri, Ben373119, Mfb, Monkbot, Nerd in Texas, Amortias, TerryAlex, Whizzkiddi and Anonymous: 133

- **Rare earth element** *Source:* https://en.wikipedia.org/wiki/Rare_earth_element?oldid=682524705 *Contributors:* Mav, Heron, Hephaestos, Fred Bauder, Kku, Ixfd64, Pcb21, Looxix~enwiki, Cherkash, Janko, Stone, Carlossuarez46, Owain, Fredrik, Chris 73, Altenmann, Merovin-gian, Alan Liefting, Graeme Bartlett, DocWatson42, Yekrats, Pne, Edcolins, Jurema Oliveira, Andycjp, Yath, Beland, AndrewKeenanRichard-son, Icairns, Neutrality, ELApro, Thorwald, Woolstar, Pjacobi, Vsmith, Adambro, Gunark, Smalljim, DougBTX, Kjkolb, RPaschotta, Jared81, Eric Kvaalen, Geo Swan, Wtmitchell, Isaac, Vuo, Skatebiker, Axeman89, Feline1, Blaxthos, Kay Dekker, Joriki, Karnesky, Mindmatrix, Benbest, Polyparadigm, Tabletop, SDC, Teemu Leisti, Stevey7788, Ryoung122, BD2412, DePiep, Vegaswikian, FlaBot, AlexCovarrubias, Vonspringer, Srleffler, Chobot, Benlisquare, Banaticus, YurikBot, Wavelength, Gaius Cornelius, Moe Epsilon, Elkman, Wknight94, 2over0, Cyrus Grisham, Arthur Rubin, Moogsi, CharlesHBennett, Allens, Erudy, Itub, SmackBot, FocalPoint, Incnis Mrsi, Originalbigj, C.Fred, KVDP, Underwater, Edgar181, Gilliam, BertholdD, Chris the speller, RootsLINUX, Hibernian, Sbharris, Bruce Marlin, Egsan Bacon, Shalom Yechiel, Frap, Valich, Mr.Z-man, TheMaster42, Vina-iwbot~enwiki, Krashlandon, Takamaxa, John, Rigadoun, Saluton~enwiki, Slogby, Hrmanu, JorisvS, Peterlewis, IronGargoyle, JHunterJ, Billybob1yay!, CmdrObot, Scohoust, Jsmaye, Jac16888, Gogo Dodo, Stuston, Chachilongbow, Asterphage, Headbomb, Hcobb, Heroeswithmetaphors, Guy Macon, Paul from Michigan, Fayenatic london, Storkk, Gcm, Txomin, Dream Focus, Bongwarrior, VoABot II, Nikevich, ZackTheJack, Animum, Jackfirst, Roricka, ChemNerd, Parveson, AlexiusHoratius, Machegav, TSullivan, Mohawk82, Eivindgh, Gurchzilla, GilHamiltonTheArm, Hokieengr, Bonadea, MkClark, VolkovBot, Gmoose1, Timmonsgray, Nug, SergeyKurdakov, Alxhotel, Grammarmonger, Riedemann, TXiKiBoT, Lradrama, Manchurian candidate, Raymondwinn, UnitedState-sian, BotKung, Telecineguy, Tri400, Plazak, Enkyo2, SieBot, Swliv, Mahmoodyaqub, BotMultichill, Cavarooni, The way, the truth, and the light, Orthorhombic, Jimthing, Paolo.dL, Maelgwnbot, Kaiserkuo, Mr. Stradivarius, Nergaal, CStack3, Arugia, ClueBot, Devki123, Ciu-dadanoGlobal, John.D.Ward, ImperfectlyInformed, Anigif, Drmies, Polyamorph, Jwihbey, Shjacks45, Lessogg, Excirial, Chka1, Muhan-des, Ark25, ChrisHodgesUK, DJ Sturm, AlanM1, Dthomsen8, Rreagan007, MrN9000, Whitetail 31, Deineka, Addbot, Roentgenium111, DOI bot, Nohomers48, Kmacwi, Jim10701, Download, Sillyfolkboy, SamatBot, 5 albert square, Raminagrobis fr, Frozendairy, VASANTH S.N., Angrense, Tide rolls, Jarble, Rojypala, Legobot, Faustbanana, Luckas-bot, Yobot, Librsh, Ytterby, AnomieBOT, Wikieditoroftoday, Ciphers, Czabaczaba, Ulric1313, Naiduhemant, Materialscientist, Citation bot, Kbamfield, Eumolpo, Quebec99, Geologyrocksmyworld, Lil-Helpa, Xqbot, Imphras Heltharn, Srich32977, Coretheapple, Almabot, RibotBOT, Gbruin, Bellerophon, Amaury, Johnnie Rico, Nolimits5017, FrescoBot, Kkj11210, Citation bot 1, Redrose64, Biker Biker, Pinethicket, Half price, RedBot, Phoenix7777, Mikespedia, Fartherred, Dac04, Spidey71, Kgrad, Double sharp, Hansschwerin, Bloodbottler, Grammarxxx, Vrenator, RjwilmsiBot, TGCP, EmausBot, AmigoCgn, Ebe123, RenamedUser01302013, Pierrejcd, Mitartep, Sanescience, Jkurutz, StringTheory11, RaymondSutanto, Iceymike, H3llBot, Michaelenandry, QEDK, Wingman4l7, Alfbar1, Hang Li Po, Ego White Tray, Orange Suede Sofa, Bill william compton, Jcaraballo, Strand Academy, Rmash-hadi, San9663, RLivergood, Woyingle, Kidela, Acanizales1, ClueBot NG, Rareearthelements, Raremetalmining, RareMetalDragon, Uzma Gamal, MelbourneStar, Jenova20, Beorn59, Lanthanum-138, Widr, Helpful Pixie Bot, Bibcode Bot, Abushell, Therareearths, Splavendr, Fes-termunk, Krenair, Bths83Cu87Aiu06, Northamerica1000, MusikAnimal, Cold Season, Hillmf6, Rlivergood, Gsc18, Bbippy, BattyBot, Zu-luKane, Mrt3366, Brendangomes, Jimdotcom, IjonTichyIjonTichy, BrightStarSky, Suradnik50, Cwalters2012, Calvinatlan, Lugia2453, Reg-molar, Rfassbind, Myall blues, Brielmannh, Everymorning, Marusworld, Tomevans23, DarkestElephant, Fattychicken6, Elekid147, Monkbot, Tigercompanion25, Teddyktchan, Annil Singh, Julietdeltalima, Gameblitzer, Unknowk123456789, Tornado241, KasparBot and Anonymous: 257

- **Abundance of the chemical elements** *Source:* https://en.wikipedia.org/wiki/Abundance_of_the_chemical_elements?oldid=686758960 *Contributors:* Bryan Derksen, Timo Honkasalo, The Anome, Dcljr, Cherkash, Smack, SEWilco, Pakaran, Jni, Donarreiskoffer, Arkuat, Rursus, Al-khowarizmi, MisfitToys, Karol Langner, FT2, Pjacobi, Vsmith, Chad okere, Paul August, Blade Hirato~enwiki, ESkog, Art LaPella, Kjkolb, Pearle, Andrewpmk, Flying fish, Rodii, Woohookitty, Oliphaunt, Rend~enwiki, TotoBaggins, CharlesC, Magister Mathematicae, BD2412, Eteq, DePiep, Drbogdan, Saperaud~enwiki, Rjwilmsi, R.e.b., AndyKali, Alphachimp, Flcelloguy, Roboto de Ajvol, Wavelength, Russoc4, Rsrikanth05, Joel7687, Syrthiss, Brainwad, ASmartKid, NielsenGW, Ordinary Person, Cmglee, Itub, SmackBot, Hydrogen Iodide, Elminster Aumar, Onebravemonkey, Michbich, Bluebot, Sbharris, Darth Panda, Tamfang, Polonium, DMacks, Nishkid64, Titus III, JorisvS, JHunterJ, Hypnosifl, Geologyguy, Mdanziger, Dan Gluck, Wizard191, Iridescent, Courcelles, PavelCurtis, Vaughan Pratt, Thermochap, Runningonbrains, Korandder, Sopoforic, Xminivann, Pcu123456789, Headbomb, Marek69, Porqin, Paul from Michigan, Gdo01, Frankie816, Drollere, Talon Artaine, Mabuhelwa, PaulTaylor, SBarnes, ChemNerd, J.delanoy, GoatGuy, Ian.thomson, AntiSpamBot, Warut, Nwbeeson, Rex07, Atropos235, VoidLurker, Joeinwap, 28bytes, Holme053, Philip Trueman, Zidonuke, McM.bot, Wingedsubmariner, Billinghurst, Falcon8765, Insanity Incarnate, Junkinbomb, Czmtzc, Minion87, Jauerback, Orthorhombic, Scorpion451, R0uge, Lightmouse, Tombomp, Nn123645, Nergaal, Denisarona, ClueBot, PipepBot, The Thing That Should Not Be, CounterVandalismBot, DragonBot, Sun Creator, LarryMorseDCOhio, SchreiberBike, Count Truthstein, DumZiBoT, Addbot, Foggynight, Roentgenium111, Substar, Marx01, CanadianLinuxUser, Favonian, LinkFA-Bot, Tide rolls, 仮, Arbitrarily0, Luckas-bot, Yobot, AnomieBOT, Floozybackloves, Materialscientist, Citation bot, Frankie0607, Riventree, Citation bot 1, Pinethicket, I dream of horses, Tom.Reding, RedBot, Mikespedia, Tim1357, Double sharp, Trappist the monk, Vrenator, Pbrower2a, DARTH SIDIOUS 2, EmausBot, Katherine, ScottyBerg, Mz7, ZéroBot, MacHyver, Wayne Slam, ClueBot NG, KlappCK, Widr, Reify-tech, Helpful Pixie Bot, Bibcode Bot, Churchgoer251, Glacialfox, Wastednow, Dexbot, Dissident93, EvergreenFir, Abitslow, Trackteur, IiKkEe, Firetraner, Sleepneeder, Stepbang and Anonymous: 200

- **List of elements** *Source:* https://en.wikipedia.org/wiki/List_of_elements?oldid=686692760 *Contributors:* Ktsquare, Ubiquity, BenRG, Wilberth, Icairns, Histrion, Mike Rosoft, D6, Discospinster, Gilward Kukel, Spangineer, Tintin1107, Woohookitty, Mandarax, DePiep, Astronaut, Ryk, TheDJ, Glenn L, Mercury McKinnon, Wavelength, RussBot, Janke, Moe Epsilon, Wap, EEMIV, Dspradau, Anthony717, Jagged 85, Tam-fang, Leob, Lambiam, Gobonobo, JorisvS, Zwart, Mihaiparparita, Reywas92, Kaldosh, Difluoroethene, Christian75, Seaphoto, QuiteUnusual, JAnDbot, Tedickey, Klausok, Dirac66, Adrian J. Hunter, R'n'B, 2help, Treisijs, 28bytes, AlphaPyro, Yintan, Jdaloner, Nergaal, Denisarona, Drmies, Niceguyedc, Copyeditor42, Markgriz, LarryMorseDCOhio, Tullywinters, WikHead, Addbot, Roentgenium111, Yoenit, Ronhjones, Cfajohnson, Jasper Deng, Ehrenkater, Arbitrarily0, Angrysockhop, Yobot, DemocraticLuntz, Quangbao, Materialscientist, Xqbot, Jayarathina, Koektrommel, YBG, Kyng, Deni42, R8R Gtrs, DrilBot, Pinethicket, Cyanos, MastiBot, Ilvon, FoxBot, Double sharp, BlogTVer, EmausBot, WikitanvirBot, Slightsmile, P. S. F. Freitas, StringTheory11, Markinvancouver, Italienmoose, SBaker43, Chris857, ClueBot NG, Wd930, Lokata~enwiki, Snotbot, Lanthanum-138, Braincricket, Parcly Taxel, Widr, MerlIwBot, Darlough, DBigXray, ۱۶۱رضا, Lefty7788, The Il-lusive Man, Ducknish, Alvin Chris Antony, TwoTwoHello, Henryleonard, Stewwie, Jnargus, Reatlas, InfoManPerson, Ashishben, Eyesnore, Sw02, Jemee012, J.meija, Sam Sailor, Usama8800, Vieque, BethNaught, Erbium pls, KH-1, Nazriyanazim, Asher4544, GeneralizationsAre-Bad, Janmesh patel, Kordinatum and Anonymous: 190

- **Table of nuclides** *Source:* https://en.wikipedia.org/wiki/Table_of_nuclides?oldid=686937777 *Contributors:* David spector, Heron, Ewen, Edward, Timwi, Rorro, Halibutt, Giftlite, Rich Farmbrough, Kwamikagami, Alansohn, Eric Kvaalen, MattWade, Strait, KaiMartin, Wavelength, JWB, Limulus, Ntouran, Itub, Sbharris, Mikhajist, My Flatley, Thijs!bot, Greg L, Escarbot, Colinsweet, D-rew, User A1, Rod57, TraceyR, Jmocenigo, Quilbert, Kristhof, NukeMan, Billinghurst, Bob Bryan, ImageRemovalBot, Shinkolobwe, DenverRedhead, ZooFari, Addbot, Nevermore4ever, WFPM, CarsracBot, WuBot, Yobot, Cflm001, LastRanger, Donko XI, RibotBOT, Locobot, Mnmngb, Achim1999, Efficiency1101e, Minivip, Abc518, Tjlafave, Saintonge235, 777sms, EmausBot, Frietjes, MerlIwBot, ChrisGualtieri, Khazar2, Burzuchius, Jemee012, Tantalum180m, Tomdickson and Anonymous: 28

- **Timeline of chemical element discoveries** *Source:* https://en.wikipedia.org/wiki/Timeline_of_chemical_element_discoveries?oldid=686495166 *Contributors:* Tobias Hoevekamp, Magnus Manske, Derek Ross, Lee Daniel Crocker, Mav, Bryan Derksen, Olof, Tarquin, Malcolm Farmer, Eob, Rgamble, Rmhermen, Toby Bartels, William Avery, DavidLevinson, Zoe, Heron, Fonzy, Ewen, Tucci528, Edward, Michael Hardy, Axl-rosen, TakuyaMurata, Ahoerstemeier, G~enwiki, Sugarfish, Jiang, 212, Gh, Raven in Orbit, Malbi, Timwi, Reddi, Stone, Fibonacci, Phoebe, Lord Emsworth, Bcorr, Gentgeen, Sappe, Zandperl, Chris 73, Psychonaut, Arkuat, Rursus, Centrx, Fastfission, Wwoods, Chameleon, ManuelAnastácio, Zeimusu, Yath, Ctachme, Icairns, Creidieki, Tsemii, Guanabot, FT2, Vsmith, Florian Blaschke, Mani1, Pavel Vozenilek, Uppland, Blade Hirato~enwiki, SpookyMulder, JoeSmack, Reinyday, Ctrl build, Como, Nsaa, Jakew, Anthony Appleyard, Keenan Pepper, Benjah-bmm27, Cloud Strife~enwiki, Bootstoots, Dave.Dunford, BDD, Nightstallion, Dismas, Marianika~enwiki, Zntrip, Megan1967, Linas, Geor-gia guy, Carcharoth, Benbest, Fbriere, Graham87, DePiep, Drbogdan, Rjwilmsi, Eoghanacht, Koavf, JanSuchy, Naraht, Mariocki, Nihiltres, RexNL, Bogdzovski, Benlisquare, Gdrbot, YurikBot, Wavelength, Kwarizmi, Gaius Cornelius, DeadEyeArrow, Silverhill, Wknight94, Samuel-Riv, TheMadBaron, Closedmouth, Abune, Whobot, Meegs, Itub, SpLoT, SmackBot, Mira, Tarret, Ccalvin, Jagged 85, Renesis, Gilliam, Kurykh, WikiFlier, TheGeck0, Gracenotes, Can't sleep, clown will eat me, Skydiver, Runefurb, MrPMonday, RandomP, Koepsell, SadiCarnot, Pilotguy, Lambiam, Perfectblue97, Smartyllama, Mgiganteus1, Olin, IronGargoyle, Alatius, Novangelis, JeffW, Newone, IvanLanin, Tawkerbot2, JRSpriggs, CmdrObot, Glenn4pr, Rifleman 82, Christian75, Headbomb, SGGH, Escarbot, Oreo Priest, AntiVandalBot, Nisselua, Tpth, Dylan Lake, Shift6, Dougher, Figma, Plantsurfer, Briancollins, Dricherby, LittleOldMe, Quantockgoblin, Hbent, Patstuart, Schmieder, R'n'B, EmleyMoor, Celephicus, DRKS, Collegebookworm, Moon Ranger, Warut, Bob, Kraniel, Sstrebel, JavierMC, Squids and Chips, PhilipTrueman, Rei-bot, Kv75, CloakedHorror, Inx272, Rwell3471, Petergans, PlanetStar, Rfts, Nergaal, The sunder king, Mario Žamić, Clue-Bot, Surfeited, S Levchenkov, Jan1nad, Mild Bill Hiccup, J8079s, Excirial, Jusdafax, Eeekster, Estirabot, ZrikiSvargla, DumZiBoT, Dr-jezza, XLinkBot, Stickee, Avoided, Skarebo, Nsim, Sami Lab, Jamieb561, Roentgenium111, Lancshero, DOI bot, Guoguo12, Bezuidenhout, Ashanda, Icantouchmytoes, Tassedethe, Sanchitblazer, Legobot, Dor Cohen, Kilom691, Azylber, AnomieBOT, Piano non troppo, Materi-alscientist, Citation bot, Herr Mlinka, Coretheapple, Trongphu, Chris.urs-o, Spesh531, SD5, Riventree, Msary80, Trewal, Robo37, Trdsf, Citation bot 1, Intelligentsium, Redrose64, Wdcf, Double sharp, UTrunn, Armando-Martin, RjwilmsiBot, Skamecrazy123, EmausBot, Johnof Reading, Syncategoremata, GoingBatty, XinaNicole, Peterindelft, ZéroBot, BAICAN XXX, Josve05a, StringTheory11, H3llBot, Make-cat, Kevjonesin, RockMagnetist, ClueBot NG, Gareth Griffith-Jones, Matt5595, Lanthanum-138, O.Koslowski, Rezabot, Helpful Pixie Bot,

Bibcode Bot, Bths83Cu87Aiu06, MusikAnimal, Soerfm, Cengime, AlanPalgut, Siuenti, Dexbot, Cwobeel, Burzuchius, XXN, Jc86035, Limitderivative, Kevin12xd, Makecat (public), Monkbot, Hashimmmm, Hockey100050, Fattbutts and Anonymous: 221

37.9.2 Images

- **File:2005metal_import.PNG** *Source:* https://upload.wikimedia.org/wikipedia/commons/3/3f/2005metal_import.PNG *License:* CC BY-SA 3.0 *Contributors:* Transferred from en.wikipedia to Commons by Stefan4 using CommonsHelper. *Original artist:* Anwar saadat at English Wikipedia

- **File:3192M-fluorite1.jpg** *Source:* https://upload.wikimedia.org/wikipedia/commons/0/0d/3192M-fluorite1.jpg *License:* CC BY-SA 3.0 *Contributors:* Own work *Original artist:* CarlesMillan

- **File:ADOMAH_periodic_table_-_electron_orbitals_(polyatomic).svg** *Source:* https://upload.wikimedia.org/wikipedia/commons/f/f1_periodic_table_-_electron_orbitals_%28polyatomic%29.svg *License:* CC BY-SA 3.0 *Contributors:* Own work *Original artist:* DePiep

- **File:Aluminium-4.jpg** *Source:* https://upload.wikimedia.org/wikipedia/commons/5/5d/Aluminium-4.jpg *License:* CC BY 3.0 *Contributors:* http://images-of-elements.com/ *Original artist:* Unknown

- **File:Ambox_important.svg** *Source:* https://upload.wikimedia.org/wikipedia/commons/b/b4/Ambox_important.svg *License:* Public domain *Contributors:* Own work, based off of Image:Ambox scales.svg *Original artist:* Dsmurat (talk · contribs)

- **File:Americium-241.jpg** *Source:* https://upload.wikimedia.org/wikipedia/commons/1/16/Americium-241.jpg *License:* Public domain *Contributors:* Own work (Original text: *I (Whitepaw (talk)) created this work entirely by myself.*) *Original artist:* Whitepaw (talk)

- **File:Americium-241_Sample_from_Smoke_Detector.JPG** *Source:* https://upload.wikimedia.org/wikipedia/commons/9/93/Americium Sample_from_Smoke_Detector.JPG *License:* CC BY-SA 3.0 *Contributors:* Own photo *Original artist:* MedicalReference

- **File:Antimony-4.jpg** *Source:* https://upload.wikimedia.org/wikipedia/commons/5/5c/Antimony-4.jpg *License:* CC BY 3.0 *Contributors:* http://images-of-elements.com/ *Original artist:* Unknown

- **File:Apollo_CSM_lunar_orbit.jpg** *Source:* https://upload.wikimedia.org/wikipedia/commons/c/c0/Apollo_CSM_lunar_orbit.jpg *License:* Public domain *Contributors:* http://www.hq.nasa.gov/office/pao/History/alsj/a15/as15-88-11963.jpg *Original artist:* NASA

- **File:ArTube.jpg** *Source:* https://upload.wikimedia.org/wikipedia/commons/2/2f/ArTube.jpg *License:* CC BY-SA 2.5 *Contributors:* usermade *Original artist:* User:Pslawinski

- **File:Argon-glow.jpg** *Source:* https://upload.wikimedia.org/wikipedia/commons/5/53/Argon-glow.jpg *License:* CC BY 3.0 *Contributors:* http://images-of-elements.com/argon.php *Original artist:* Jurii

- **File:Argon_Spectrum.png** *Source:* https://upload.wikimedia.org/wikipedia/commons/3/37/Argon_Spectrum.png *License:* CC BY-SA 3.0 *Contributors:* Own work http://goiphone5.com/ *Original artist:* Abilanin

- **File:Argon_discharge_tube.jpg** *Source:* https://upload.wikimedia.org/wikipedia/commons/8/87/Argon_discharge_tube.jpg *License:* GFDL 1.2 *Contributors:* Own work *Original artist:* Alchemist-hp (<a href='//commons.wikimedia.org/wiki/User_talk:Alchemist-hp' title='User talk: Alchemist-hp'>talk</a>) (www.pse-mendelejew.de)

- **File:Argon_ice_1.jpg** *Source:* https://upload.wikimedia.org/wikipedia/commons/0/0d/Argon_ice_1.jpg *License:* CC-BY-SA-3.0 *Contributors:* No machine-readable source provided. Own work assumed (based on copyright claims). *Original artist:* No machine-readable author provided. Deglr6328~commonswiki assumed (based on copyright claims).

- **File:Arsen_1a.jpg** *Source:* https://upload.wikimedia.org/wikipedia/commons/7/7b/Arsen_1a.jpg *License:* CC-BY-SA-3.0 *Contributors:*

- Arsen_1.jpg *Original artist:* Arsen_1.jpg: Original uploader was Tomihahndorf at de.wikipedia

- **File:Arsenic_trioxide.jpg** *Source:* https://upload.wikimedia.org/wikipedia/commons/a/a7/Arsenic_trioxide.jpg *License:* Public domain *Contributors:* ? *Original artist:* ?

- **File:Asterisks_one.svg** *Source:* https://upload.wikimedia.org/wikipedia/commons/4/49/Asterisks_one.svg *License:* CC BY-SA 3.0 *Contributors:* Own work *Original artist:* DePiep

- **File:Asterisks_one_(right).svg** *Source:* https://upload.wikimedia.org/wikipedia/commons/1/1c/Asterisks_one_%28right%29.svg *License:* CC BY-SA 3.0 *Contributors:* Own work *Original artist:* DePiep

- **File:Asterisks_two.svg** *Source:* https://upload.wikimedia.org/wikipedia/commons/3/3f/Asterisks_two.svg *License:* CC BY-SA 3.0 *Contributors:* Own work *Original artist:* DePiep

- **File:Atomic_number_depiction.jpg** *Source:* https://upload.wikimedia.org/wikipedia/commons/1/18/Atomic_number_depiction.jpg *License:* CC0 *Contributors:* Atomic Number Depiction.jpg *Original artist:* en:User:Materialscientist

- **File:Baiyunebo_ast_2006181.jpg** *Source:* https://upload.wikimedia.org/wikipedia/commons/b/be/Baiyunebo_ast_2006181.jpg *License:* Public domain *Contributors:* http://earthobservatory.nasa.gov/IOTD/view.php?id=77723&src=eoa-iotd *Original artist:* NASA Earth Observatory

- **File:Band_filling_diagram.svg** *Source:* https://upload.wikimedia.org/wikipedia/commons/9/9d/Band_filling_diagram.svg *License:* CC0 *Contributors:* Own work *Original artist:* Nanite

- **File:Blue_Light.JPG** *Source:* https://upload.wikimedia.org/wikipedia/commons/c/cc/Blue_Light.JPG *License:* CC BY-SA 3.0 *Contributors:* Own work *Original artist:* Hunleybluelight

- **File:Bohr-atom-PAR.svg** *Source:* https://upload.wikimedia.org/wikipedia/commons/5/55/Bohr-atom-PAR.svg *License:* CC-BY-SA-3.0 *Contributors:* Transferred from en.wikipedia to Commons. *Original artist:* Original uplo:JabberWok]] at en.wikipedia

- **File:Boron_R105.jpg** *Source:* https://upload.wikimedia.org/wikipedia/commons/1/19/Boron_R105.jpg *License:* CC BY 3.0 *Contributors:* "Walking Tour of the Elements" CD and "Rediscovery of the Elements" DVD, ISBN 978-0-615-30795 *Original artist:* James L Marshall

- **File:Bromine-3D-vdW.png** *Source:* https://upload.wikimedia.org/wikipedia/commons/b/bd/Bromine-3D-vdW.png *License:* Public domain *Contributors:* ? *Original artist:* ?

- **File:Bromine_25ml.jpg** *Source:* https://upload.wikimedia.org/wikipedia/commons/7/7c/Bromine_25ml.jpg *License:* CC BY-SA 3.0 *Contributors:* http://woelen.homescience.net/science/index.html *Original artist:* W. Oelen

- **File:Bromine_vial_in_acrylic_cube.jpg** *Source:* https://upload.wikimedia.org/wikipedia/commons/3/35/Bromine_vial_in_acrylic_cube.jpg *License:* CC BY-SA 3.0 de *Contributors:* Own work *Original artist:* Alchemist-hp (pse-mendelejew.de)

- **File:ChemicalGalaxy_Longman_1951.jpg** *Source:* https://upload.wikimedia.org/wikipedia/en/5/54/ChemicalGalaxy_Longman_1951.jpg *License:* Fair use *Contributors:*
  From website.
  Original artist: ?

- **File:ChemicalGalaxy_Stewart_2004.jpg** *Source:* https://upload.wikimedia.org/wikipedia/en/a/a3/ChemicalGalaxy_Stewart_2004.jpg *License:* Fair use *Contributors:*
  From website.
  Original artist: ?

- **File:Chemical_Galaxy_II.jpg** *Source:* https://upload.wikimedia.org/wikipedia/en/d/d9/Chemical_Galaxy_II.jpg *License:* Fair use *Contributors:*
  Philip Stewart
  Original artist: ?

- **File:Chlorine-3D-vdW.png** *Source:* https://upload.wikimedia.org/wikipedia/commons/8/87/Chlorine-3D-vdW.png *License:* Public domain *Contributors:* ? *Original artist:* ?

- **File:Chlorine_ampoule.jpg** *Source:* https://upload.wikimedia.org/wikipedia/commons/f/f4/Chlorine_ampoule.jpg *License:* CC BY-SA 3.0 *Contributors:* http://woelen.homescience.net/science/index.html *Original artist:* W. Oelen

- **File:Circuit_Breaker_115_kV.jpg** *Source:* https://upload.wikimedia.org/wikipedia/commons/2/21/Circuit_Breaker_115_kV.jpg *License:* Public domain *Contributors:* Own work *Original artist:* Wtshymanski at en.wikipedia

- **File:Close_packing_box.svg** *Source:* https://upload.wikimedia.org/wikipedia/commons/a/a7/Close_packing_box.svg *License:* Public domain *Contributors:* Own work *Original artist:* en:User:Twisp

- **File:Coloured-transition-metal-solutions.jpg** *Source:* https://upload.wikimedia.org/wikipedia/commons/5/57/Coloured-transition-metal-.jpg *License:* Public domain *Contributors:* No machine-readable source provided. Own work assumed (based on copyright claims). *Original artist:* No machine-readable author provided. Benjah-bmm27 assumed (based on copyright claims).

- **File:Commons-logo.svg** *Source:* https://upload.wikimedia.org/wikipedia/en/4/4a/Commons-logo.svg *License:* ? *Contributors:* ? *Original artist:* ?

- **File:Compact_Isotope_Table,_Right_Lobe.jpg** *Source:* https://upload.wikimedia.org/wikipedia/en/f/fb/Compact_Isotope_Table%2C_Right_Lobe.jpg *License:* CC-BY-SA-3.0 *Contributors:*
  I created this work entirely by myself, based on Wikipedia Isotope Table in Chen format.
  Original artist:
  DenverRedhead (talk)Aran David Stubbs

- **File:Copper_germanium.jpg** *Source:* https://upload.wikimedia.org/wikipedia/commons/7/74/Copper_germanium.jpg *License:* CC BY-SA 3.0 *Contributors:* http://commons.wikimedia.org/wiki/File:Metals_for_jewellery.jpg *Original artist:* Mauro Cateb

- **File:Cyclooctasulfur-above-3D-balls.png** *Source:* https://upload.wikimedia.org/wikipedia/commons/7/7e/Cyclooctasulfur-above-3D-balls.png *License:* Public domain *Contributors:* ? *Original artist:* ?

- **File:DIMendeleevCab.jpg** *Source:* https://upload.wikimedia.org/wikipedia/commons/c/c8/DIMendeleevCab.jpg *License:* Public domain *Contributors:* Transferred from ru.wikipedia
  Original artist: —. Original uploader was Serge Lachinov at ru.wikipedia

- **File:Diamond-and-graphite-with-scale.jpg** *Source:* https://upload.wikimedia.org/wikipedia/commons/4/4b/Diamond-and-graphite-with.jpg *License:* CC BY-SA3.0 *Contributors:* File:Diamond-39513.jpg and File:Graphite-tn19a.jpg *Original artist:* Rob Lavinsky/iRocks.com

- **File:Dibromine-2D-dimensions.png** *Source:* https://upload.wikimedia.org/wikipedia/commons/6/66/Dibromine-2D-dimensions.png *License:* Public domain *Contributors:* ? *Original artist:* ?

- **File:Dichlorine-2D-dimensions.png** *Source:* https://upload.wikimedia.org/wikipedia/commons/6/63/Dichlorine-2D-dimensions.png *License:* Public domain *Contributors:* ? *Original artist:* ?

- **File:Difluorine-2D-dimensions.png** *Source:* https://upload.wikimedia.org/wikipedia/commons/5/53/Difluorine-2D-dimensions.png *License:* Public domain *Contributors:* ? *Original artist:* ?

- **File:Diiodine-2D-dimensions.png** *Source:* https://upload.wikimedia.org/wikipedia/commons/4/44/Diiodine-2D-dimensions.png *License:* Public domain *Contributors:* ? *Original artist:* ?

- **File:Dinitrogen-2D-dimensions.png** *Source:* https://upload.wikimedia.org/wikipedia/commons/4/4e/Dinitrogen-2D-dimensions.png *License:* Public domain *Contributors:* ? *Original artist:* ?

- **File:Discovery_of_chemical_elements.svg** *Source:* https://upload.wikimedia.org/wikipedia/commons/3/3d/Discovery_of_chemical_elements.svg *License:* CC BY-SA 3.0 *Contributors:* Wikimedia Commons. *Original artist:* Sandbh

- **File:Dmitri_Ivanowitsh_Mendeleev.jpg** *Source:* https://upload.wikimedia.org/wikipedia/commons/b/b3/Dmitri_Ivanowitsh_Mendeleev.jpg *License:* Public domain *Contributors:* New York Public Library Archives *Original artist:* Historical and Public Figures Collection

- **File:Dmitry_Mendeleyev_Osnovy_Khimii_1869-1871_first_periodic_table.jpg** *Source:* https://upload.wikimedia.org/wikipedia/commons/1/1d/Dmitry_Mendeleyev_Osnovy_Khimii_1869-1871_first_periodic_table.jpg *License:* Public domain *Contributors:* Chemical Heritage Foundation *Original artist:* Dmitry Ivanovich Mendeleyev, 1834-1907

- **File:Edelmetalle.jpg** *Source:* https://upload.wikimedia.org/wikipedia/commons/9/90/Edelmetalle.jpg *License:* CC-BY-SA-3.0 *Contributors:* de:Image:Edelmetalle.jpg. *Original artist:* de:User:Tomihahndorf.

- **File:Edit-clear.svg** *Source:* https://upload.wikimedia.org/wikipedia/en/f/f2/Edit-clear.svg *License:* Public domain *Contributors:* The *Tango! Desktop Project. Original artist:*

  The people from the Tango! project. And according to the meta-data in the file, specifically: "Andreas Nilsson, and Jakub Steiner (although minimally)."

- **File:Eight_Allotropes_of_Carbon.png** *Source:* https://upload.wikimedia.org/wikipedia/commons/f/f8/Eight_Allotropes_of_Carbon.png *License:* CC-BY-SA-3.0 *Contributors:* Created by Michael Ströck (mstroeck) *Original artist:* Created by Michael Ströck (mstroeck)

- **File:Electron_affinity_of_the_elements.svg** *Source:* https://upload.wikimedia.org/wikipedia/commons/6/6c/Electron_affinity_of_the_.svg *License:* CC BY-SA 3.0 *Contributors:* Based on Electron affinities of the elements 2.png by Sandbh. *Original artist:* DePiep

- **File:Electron_orbitals.svg** *Source:* https://upload.wikimedia.org/wikipedia/commons/1/11/Electron_orbitals.svg *License:* Public domain *Contributors:* own work by Patricia.fidi and Lt Paul - Originally from pl:Grafika:Orbitale.png, author pl:Wikipedysta:Chemmix. *Original artist:* Patricia.fidi

- **File:Electron_shell_003_Lithium_-_no_label.svg** *Source:* https://upload.wikimedia.org/wikipedia/commons/a/ae/Electron_shell_003_-_no_label.svg *License:* CC BY-SA2.0uk *Contributors:* File:Electron shell003Lithium.svg *Original artist:* Pumbaa(original work byGregRobson)

- **File:Electron_shell_010_Neon_-_no_label.svg** *Source:* https://upload.wikimedia.org/wikipedia/commons/3/3e/Electron_shell_010_Neon_-_no_label.svg *License:* CC BY-SA 2.0 uk *Contributors:* http://commons.wikimedia.org/wiki/Category:Electron_shell_diagrams (corresponding labeled version) *Original artist:* commons:User:Pumbaa (original work by commons:User:Greg Robson)

- **File:Elemental_abundances.svg** *Source:* https://upload.wikimedia.org/wikipedia/commons/0/09/Elemental_abundances.svg *License:* Public domain *Contributors:* http://pubs.usgs.gov/fs/2002/fs087-02/ *Original artist:* Gordon B. Haxel, Sara Boore, and Susan Mayfield from USGS; vectorized by User:michbich

- **File:Elementspiral_(polyatomic).svg** *Source:* https://upload.wikimedia.org/wikipedia/commons/c/ce/Elementspiral_%28polyatomic%29.svg *License:* CC BY-SA 3.0 *Contributors:* Own work *Original artist:* DePiep

- **File:Empirical_atomic_radius_trends.png** *Source:* https://upload.wikimedia.org/wikipedia/commons/b/bc/Empirical_atomic_radius_trends.png *License:* GFDL *Contributors:* Own work *Original artist:* StringTheory11

- **File:Endohedral_fullerene.png** *Source:* https://upload.wikimedia.org/wikipedia/commons/e/e1/Endohedral_fullerene.png *License:* GFDL *Contributors:* Own work *Original artist:* Hajv01

- **File:Ernest_Orlando_Lawrence.jpg** *Source:* https://upload.wikimedia.org/wikipedia/commons/8/88/Ernest_Orlando_Lawrence.jpg *License:* Public domain *Contributors:* ? *Original artist:* ?

- **File:Fibreoptic4.jpg** *Source:* https://upload.wikimedia.org/wikipedia/commons/4/42/Fibreoptic4.jpg *License:* CC BY-SA 3.0 *Contributors:* http://en.wikipedia.org/wiki/File:Fibreoptic.jpg *Original artist:* BigRiz

- **File:Filament.jpg** *Source:* https://upload.wikimedia.org/wikipedia/commons/0/08/Filament.jpg *License:* Public domain *Contributors:* https://en.wikipedia.org/w/index.php?title=File:Filament.jpg&action=edit *Original artist:* Lander777

- **File:First_Ionization_Energy.svg** *Source:* https://upload.wikimedia.org/wikipedia/commons/1/1d/First_Ionization_Energy.svg *License:* CC BY-SA 3.0 *Contributors:* http://commons.wikimedia.org/wiki/File:Erste_Ionisierungsenergie_PSE_color_coded.png *Original artist:* User:Sponk

- **File:Flag_of_Russia.svg** *Source:* https://upload.wikimedia.org/wikipedia/en/f/f3/Flag_of_Russia.svg *License:* PD *Contributors:* ? *Original artist:* ?

- **File:Fluorine-3D-vdW.png** *Source:* https://upload.wikimedia.org/wikipedia/commons/3/30/Fluorine-3D-vdW.png *License:* Public domain *Contributors:* ? *Original artist:* ?

- **File:Folder_Hexagonal_Icon.svg** *Source:* https://upload.wikimedia.org/wikipedia/en/4/48/Folder_Hexagonal_Icon.svg *License:* Cc-by-sa-3.0 *Contributors:* ? *Original artist:* ?

- **File:Gallium1_640x480.jpg** *Source:* https://upload.wikimedia.org/wikipedia/commons/9/92/Gallium_crystals.jpg *License:* CC-BY-SA-3.0 *Contributors:* Own work *Original artist:* en:user:foobar

- **File:Glenn_Seaborg_-_1964.jpg** *Source:* https://upload.wikimedia.org/wikipedia/commons/4/47/Glenn_Seaborg_-_1964.jpg *License:* Public domain *Contributors:* NAIL Control Number: NWDNS-326-COM-12 NARA (enter "Glenn Seaborg" in search form under Digital Copies tab) *Original artist:* Atomic Energy Commission. (1946 - 01/19/1975)

- **File:Gold-crystals.jpg** *Source:* https://upload.wikimedia.org/wikipedia/commons/d/d7/Gold-crystals.jpg *License:* CC BY-SA 3.0 de *Contributors:* Own work *Original artist:* Alchemist-hp (<a href='//commons.wikimedia.org/wiki/User_talk:Alchemist-hp' title='User talk:)www.pse-mendelejew.de

- **File:GoldNuggetUSGOV.jpg** *Source:* https://upload.wikimedia.org/wikipedia/commons/5/5b/GoldNuggetUSGOV.jpg *License:* Public domain *Contributors:* http://resourcescommittee.house.gov/subcommittees/emr/usgsweb/photogallery/ *Original artist:* Unknown

- **File:Goodyear-blimp.jpg** *Source:* https://upload.wikimedia.org/wikipedia/commons/2/2a/Goodyear-blimp.jpg *License:* Public domain *Contributors:* user-made *Original artist:* Derek Jensen (Tysto)
- **File:Graphite2.jpg** *Source:* https://upload.wikimedia.org/wikipedia/commons/4/48/Graphite2.jpg *License:* Public domain *Contributors:* [1] [2] *Original artist:* U.S. Geological Survey
- **File:Half-logarithm_graph.jpg** *Source:* https://upload.wikimedia.org/wikipedia/commons/0/0b/Half-logarithm_graph.jpg *License:* CC BY-SA 4.0 *Contributors:* Own work *Original artist:* Alexey Alekseenko
- **File:Halogens.jpg** *Source:* https://upload.wikimedia.org/wikipedia/commons/c/c6/Halogens.jpg *License:* CC BY-SA 3.0 *Contributors:* Science Made Alive: Chemistry/Elem - Halogens *Original artist:* W. Oelen
- **File:HeTube.jpg** *Source:* https://upload.wikimedia.org/wikipedia/commons/1/1f/HeTube.jpg *License:* CC BY-SA 2.5 *Contributors:* user-made *Original artist:* User:Pslawinski
- **File:HeavyMineralsBeachSand.jpg** *Source:* https://upload.wikimedia.org/wikipedia/commons/8/81/HeavyMineralsBeachSand.jpg *License:* Public domain *Contributors:* Original photograph *Original artist:* Photograph taken by Mark A. Wilson (Department of Geology, The College of Wooster). [1]
- **File:Helium-glow.jpg** *Source:* https://upload.wikimedia.org/wikipedia/commons/0/00/Helium-glow.jpg *License:* CC BY 3.0 *Contributors:* http://images-of-elements.com/helium.php *Original artist:* Jurii
- **File:Helium_discharge_tube.jpg** *Source:* https://upload.wikimedia.org/wikipedia/commons/8/82/Helium_discharge_tube.jpg *License:* GFDL 1.2 *Contributors:* Own work *Original artist:* Alchemist-hp(<a href='//commons.wikimedia.org/wiki/User_talk:Alchemist-hp' title='User talk:Alchemist-hp'>talk</a>) (www.pse-mendelejew.de)
- **File:Helium_spectra.jpg** *Source:* https://upload.wikimedia.org/wikipedia/commons/c/c3/Helium_spectra.jpg *License:* Public domain *Contributors:* Transferred from en.wikipedia; transferred to Commons by User:João Sousa using CommonsHelper. *Original artist:* (teravolt (talk)). Original uploader was Teravolt at en.wikipedia
- **File:Helium_spectrum.jpg** *Source:* https://upload.wikimedia.org/wikipedia/commons/8/80/Helium_spectrum.jpg *License:* Public domain *Contributors:* http://imagine.gsfc.nasa.gov/docs/teachers/lessons/xray_spectra/worksheet-specgraph2-sol.html *Original artist:* NASA
- **File:Henning_brand.jpg** *Source:* https://upload.wikimedia.org/wikipedia/en/7/79/Henning_brand.jpg *License:* Public domain *Contributors:* ? *Original artist:* ?
- **File:Henry_Moseley.jpg** *Source:* https://upload.wikimedia.org/wikipedia/en/d/dd/Henry_Moseley.jpg *License:* PD-US *Contributors:* ? *Original artist:* ?
- **File:Hf-crystal_bar.jpg** *Source:* https://upload.wikimedia.org/wikipedia/commons/3/38/Hf-crystal_bar.jpg *License:* CC BY-SA 3.0 de *Contributors:* Own work *Original artist:* Alchemist-hp (<a href='//commons.wikimedia.org/wiki/User_talk:Alchemist-hp' title='User talk')(www.pse-mendelejew.de)
- **File:Hot_metalwork.jpg** *Source:* https://upload.wikimedia.org/wikipedia/commons/a/a9/Hot_metalwork.jpg *License:* GFDL 1.2 *Contributors:* Own work *Original artist:* **fir0002 | flagstaffotos.com.au**
- **File:Ilmenit_-_Miask,_Ural.jpg** *Source:* https://upload.wikimedia.org/wikipedia/commons/e/eb/Ilmenit_-_Miask%2C_Ural.jpg *License:* CC BY-SA 3.0 *Contributors:* Own work *Original artist:* Ra'ike (see also: de:Benutzer:Ra'ike)
- **File:Iod_kristall.jpg** *Source:* https://upload.wikimedia.org/wikipedia/commons/7/7c/Iod_kristall.jpg *License:* Public domain *Contributors:* de:Image:Iod_kristall.jpg, selbst gemachtes Foto aus meiner Sammlung chemischer Elemente. *Original artist:* de:user:Tomihahndorf
- **File:Iodine-3D-vdW.png** *Source:* https://upload.wikimedia.org/wikipedia/commons/0/07/Iodine-3D-vdW.png *License:* Public domain *Contributors:* ? *Original artist:* ?
- **File:Iodinecrystals.JPG** *Source:* https://upload.wikimedia.org/wikipedia/commons/a/ae/Iodinecrystals.JPG *License:* Public domain *Contributors:* Own work *Original artist:* Greenhorn1
- **File:Ionization_energies.png** *Source:* https://upload.wikimedia.org/wikipedia/commons/2/27/Ionization_energies.png *License:* Public domain *Contributors:* Self-made; based on data from: Martin, W. C.; Wiese, W. L. (1996) *Atomic, Molecular, & Optical Physics Handbook*, American Institute of Physics ISBN 156396242X *Original artist:* RJHall
- **File:Klechkovski_rule.svg** *Source:* https://upload.wikimedia.org/wikipedia/commons/9/95/Klechkovski_rule.svg *License:* CC-BY-SA-3.0 *Contributors:* No machine-readable source provided. Own work assumed (based on copyright claims). *Original artist:* No machine-readable author provided. Bono~commonswiki assumed (based on copyright claims).
- **File:Klechkowski_rule_2.svg** *Source:* https://upload.wikimedia.org/wikipedia/commons/3/30/Klechkowski_rule_2.svg *License:* CC-BY-SA-3.0 *Contributors:* Own work *Original artist:* Sharayanan
- **File:Known-elements-1700-2000.png** *Source:* https://upload.wikimedia.org/wikipedia/commons/d/dc/Known-elements-1700-2000.png *License:* CC BY-SA 3.0 *Contributors:* Own work *Original artist:* Soerfm
- **File:KrTube.jpg** *Source:* https://upload.wikimedia.org/wikipedia/commons/e/e7/KrTube.jpg *License:* CC BY-SA 2.5 *Contributors:* user-made *Original artist:* User:Pslawinski
- **File:Krypton-glow.jpg** *Source:* https://upload.wikimedia.org/wikipedia/commons/9/9c/Krypton-glow.jpg *License:* CC BY 3.0 *Contributors:* http://images-of-elements.com/krypton.php *Original artist:* Jurii
- **File:Krypton_Spectrum.jpg** *Source:* https://upload.wikimedia.org/wikipedia/commons/a/a6/Krypton_Spectrum.jpg *License:* Public domain *Contributors:* en wikipedia ([1]) *Original artist:* Mrgoogfan

- **File:Krypton_discharge_tube.jpg**_Source:_https://upload.wikimedia.org/wikipedia/commons/5/50/Krypton_discharge_tube.jpg_License:_1.2 _Contributors:_Own work_Original artist:_Alchemist-hp(<a href='//commons.wikimedia.org/wiki/User_talk:Alchemist-hp'title='User talk:Alchemist-hp'>talk</a>) (www.pse-mendelejew.de)

- **File:Lavoisier.jpg** _Source:_ https://upload.wikimedia.org/wikipedia/commons/7/7e/Lavoisier.jpg _License:_ Public domain _Contributors:_ University of Texas , Austin _Original artist:_ David

- **File:Lead_electrolytic_and_1cm3_cube.jpg** _Source:_ https://upload.wikimedia.org/wikipedia/commons/e/e6/Lead_electrolytic_and_1cm3_cube.jpg _License:_ FAL _Contributors:_ Own work _Original artist:_ **Alchemist-hp (<a href='//commons.wikimedia.org/wiki/User_talk: 'title='User talk:Alchemist-hp'>talk</a>) (www.pse-mendelejew.de)**

- **File:Liquid_fluorine_tighter_crop.jpg** _Source:_ https://upload.wikimedia.org/wikipedia/commons/9/91/Liquid_fluorine_tighter_crop.jpg _License:_ CC BY-SA 3.0 _Contributors:_

- Liquid_fluorine.jpg _Original artist:_ Liquid_fluorine.jpg: Prof B. G. Mueller

- **File:Lutetium_sublimed_dendritic_and_1cm3_cube.jpg** _Source:_ https://upload.wikimedia.org/wikipedia/commons/7/74/Lutetium_ _dendritic_and_1cm3_cube.jpg_License:_FAL_Contributors:_Own work_Original artist:_Alchemist-hp(<a href='//commons.wikimedia.org/wiki/User_ talk:Alchemist-hp'title='User talk:Alchemist-hp'>talk</a>) (www.pse-mendelejew.de)

- **File:Madelung_rule.svg** _Source:_ https://upload.wikimedia.org/wikipedia/commons/0/02/Madelung_rule.svg _License:_ GFDL _Contributors:_ Own work _Original artist:_ Chemmix

- **File:Medeleeff_by_repin.jpg** _Source:_ https://upload.wikimedia.org/wikipedia/commons/b/b3/Medeleeff_by_repin.jpg _License:_ Public domain _Contributors:_ http://www.picture.art-catalog.ru/picture.php?id_picture=4318 _Original artist:_ Ilya Repin

- **File:Mendeleev'{}s_periodic_table_(1869).svg** _Source:_ https://upload.wikimedia.org/wikipedia/commons/4/46/Mendeleev%27s_periodic_table_%281869%29.svg _License:_ Public domain _Contributors:_ https://archive.org/stream/zeitschriftfrch12unkngoog#page/n414/mode/2up _Original artist:_ Dimitri Mendeleev (in Zeitschrift für Chemie (1869))

- **File:Mendelejevs_periodiska_system_1871.png** _Source:_ https://upload.wikimedia.org/wikipedia/commons/5/55/Mendelejevs_periodiska_system_1871.png _License:_ Public domain _Contributors:_ Källa:Dmitrij Ivanovitj Mendelejev (1834 - 1907). Originally from sv.wikipedia; description page is/was here. _Original artist:_ Original uploader was Den fjättrade ankan at sv.wikipedia

- **File:Mendelejew_signature.jpg** _Source:_ https://upload.wikimedia.org/wikipedia/commons/8/8a/Mendelejew_signature.jpg _License:_ Public domain _Contributors:_ http://www.prometeus.nsc.ru/archives/exhibit2/prmendel.ssi _Original artist:_ ?

- **File:Mendeleyev_gold_Barry_Kent.JPG** _Source:_ https://upload.wikimedia.org/wikipedia/commons/3/31/Mendeleyev_gold_Barry_Kent.JPG _License:_ CC-BY-SA-3.0 _Contributors:_ Own work _Original artist:_ Robert Wielgórski (Barry Kent)

- **File:Modern_3T_MRI.JPG** _Source:_ https://upload.wikimedia.org/wikipedia/commons/b/bd/Modern_3T_MRI.JPG _License:_ CC-BY-SA-3.0 _Contributors:_ Photographed by User:KasugaHuang on Mar 27, 2006 at Tri-Service General Hospital, Taiwan. _Original artist:_ User:KasugaHuang

- **File:Monazit_-_Mosambik,_O-Afrika.jpg**_Source:_https://upload.wikimedia.org/wikipedia/commons/4/4b/Monazit_-_Mosambik%2C.jpg _License:_ CC BY-SA 3.0 _Contributors:_ Own work _Original artist:_ Ra'ike (see also: de:Benutzer:Ra'ike)

- **File:NatCopper.jpg** _Source:_ https://upload.wikimedia.org/wikipedia/commons/f/f0/NatCopper.jpg _License:_ CC-BY-SA-3.0 _Contributors:_

- Native_Copper_Macro_Digon3.jpg _Original artist:_ Native_Copper_Macro_Digon3.jpg: "Jonathan Zander (Digon3)"

- **File:Ndslivechart.png** _Source:_ https://upload.wikimedia.org/wikipedia/commons/b/b0/Ndslivechart.png _License:_ Public domain _Contributors:_ Own work _Original artist:_ Minivip

- **File:NeTube.jpg** _Source:_ https://upload.wikimedia.org/wikipedia/commons/8/88/NeTube.jpg _License:_ CC BY-SA 2.5 _Contributors:_ usermade _Original artist:_ User:Pslawinski

- **File:Neon-glow.jpg** _Source:_ https://upload.wikimedia.org/wikipedia/commons/f/f8/Neon-glow.jpg _License:_ CC BY 3.0 _Contributors:_ http://images-of-elements.com/neon.php _Original artist:_ Jurii

- **File:Neon_discharge_tube.jpg** _Source:_ https://upload.wikimedia.org/wikipedia/commons/4/46/Neon_discharge_tube.jpg _License:_ GFDL 1.2 _Contributors:_ Own work _Original artist:_ Alchemist-hp (<a href='//commons.wikimedia.org/wiki/User_talk:Alchemist-hp' title='User talk: Alchemist-hp'>talk</a>) (www.pse-mendelejew.de)

- **File:Neon_spectra.jpg** _Source:_ https://upload.wikimedia.org/wikipedia/commons/9/99/Neon_spectra.jpg _License:_ Public domain _Contributors:_ Transferred from en.wikipedia
Original artist: Teravolt at en.wikipedia

- **File:Newlands_periodiska_system_1866.png**_Source:_https://upload.wikimedia.org/wikipedia/commons/e/e5/Newlands_periodiska_system_1866.png_License:_Public domain_Contributors:_John Alexander Reina Newlands(1838–1898)_Original artist:_John Alexander Reina Newlands

- **File:Nickel_electrolytic_and_1cm3_cube.jpg**_Source:_https://upload.wikimedia.org/wikipedia/commons/6/6a/Nickel_electrolytic_and_1cm3_cube.jpg_License:_FAL_Contributors:_Own work_Original artist:_Alchemist-hp(<a href='//commons.wikimedia.org/wiki/User_talk:Alchemist-hp'title='User talk:Alchemist-hp'>talk</a>) (www.pse-mendelejew.de)

- **File:Niels_Bohr.jpg** _Source:_ https://upload.wikimedia.org/wikipedia/commons/6/6d/Niels_Bohr.jpg _License:_ Public domain _Contributors:_ Niels Bohr's Nobel Prize biography, from 1922 _Original artist:_ The American Institute of Physics credits the photo [1] to AB Lagrelius & Westphal, which is the Swedish company used by the Nobel Foundation for most photos of its book series _Les Prix Nobel._

- **File:NuclearReaction.png** _Source:_ https://upload.wikimedia.org/wikipedia/commons/7/7d/NuclearReaction.png _License:_ CC BY-SA 3.0 _Contributors:_ Own work _Original artist:_ Michalsmid

- **File:Nucleosynthesis_periodic_table.svg** _Source:_ https://upload.wikimedia.org/wikipedia/commons/3/31/Nucleosynthesis_periodic_table.svg _License:_ CC BY-SA 3.0 _Contributors:_ Own work _Original artist:_ Cmglee

- **File:NuclideMap_small_preview.jpg** *Source:* https://upload.wikimedia.org/wikipedia/commons/a/a5/NuclideMap_small_preview.jpg *License:* Public domain *Contributors:* National Nuclear Data Center, information extracted from the NuDat 2 database, http://www.nndc.bnl.gov/nudat2/ *Original artist:* The viewer

- **File:NuclideMap_stitched_small_preview.png** *Source:* https://upload.wikimedia.org/wikipedia/commons/7/79/NuclideMap_stitched_preview.png *License:* Public domain *Contributors:* National Nuclear Data Center, information extracted from the NuDat2 database, http://www.nndc.bnl.gov/nudat2/ *Original artist:* The viewer

- **File:Nuvola_apps_edu_science.svg** *Source:* https://upload.wikimedia.org/wikipedia/commons/5/59/Nuvola_apps_edu_science.svg *License:* LGPL *Contributors:* http://ftp.gnome.org/pub/GNOME/sources/gnome-themes-extras/0.9/gnome-themes-extras-0.9.0.tar.gz *Original artist:* David Vignoni / ICON KING

- **File:Office-book.svg** *Source:* https://upload.wikimedia.org/wikipedia/commons/a/a8/Office-book.svg *License:* Public domain *Contributors:* This and myself. *Original artist:* Chris Down/Tango project

- **File:P2M-1.png** *Source:* https://upload.wikimedia.org/wikipedia/commons/8/89/P2M-1.png *License:* Public domain *Contributors:* Own work *Original artist:* Dhatfield

- **File:P2M0.png** *Source:* https://upload.wikimedia.org/wikipedia/commons/d/d4/P2M0.png *License:* Public domain *Contributors:* Own work *Original artist:* Dhatfield

- **File:P2M1.png** *Source:* https://upload.wikimedia.org/wikipedia/commons/5/5a/P2M1.png *License:* Public domain *Contributors:* Own work *Original artist:* Dhatfield

- **File:Palladium.jpg** *Source:* https://upload.wikimedia.org/wikipedia/commons/2/23/Palladium.jpg *License:* CC BY 3.0 *Contributors:* http://images-of-elements.com/palladium.php *Original artist:* Jurii

- **File:People_icon.svg** *Source:* https://upload.wikimedia.org/wikipedia/commons/3/37/People_icon.svg *License:* CC0 *Contributors:* OpenClipart *Original artist:* OpenClipart

- **File:Periodic_Table_of_Elements_showing_Electron_Shells.svg** *Source:* https://upload.wikimedia.org/wikipedia/commons/a/a8/Periodic_Table_of_Elements_showing_Electron_Shells.svg *License:* CC BY-SA 2.5 *Contributors:* ? *Original artist:* ?

- **File:Periodic_Table_overview_(standard).svg** *Source:* https://upload.wikimedia.org/wikipedia/commons/d/d4/Periodic_Table_overview_%28standard%29.svg *License:* CC BY-SA 3.0 *Contributors:* Own work *Original artist:* DePiep

- **File:Periodic_Table_overview_(wide).svg** *Source:* https://upload.wikimedia.org/wikipedia/commons/f/f3/Periodic_Table_overview_%2829.svg *License:* CC BY-SA 3.0 *Contributors:* Own work *Original artist:* DePiep

- **File:Periodic_table_(metals–metalloids–nonmetals,_32_columns).png** *Source:* https://upload.wikimedia.org/wikipedia/commons/7/78/Periodic_table_%28metals%E2%80%93metalloids%E2%80%93nonmetals%2C_32_columns%29.png *License:* CC BY-SA 4.0 *Contributors:* Own work *Original artist:* DePiep

- **File:Periodic_table_(polyatomic).svg** *Source:* https://upload.wikimedia.org/wikipedia/commons/9/98/Periodic_table_%28polyatomic%29.svg *License:* CC BY-SA 3.0 *Contributors:* Own work -- Actually "inspired by"/forked from earlier free versions on Wikipedia/Commons like this, but there is no option to note this in Upload. *Original artist:* DePiep

- **File:Periodic_table_blocks_spdf_(32_column).svg** *Source:* https://upload.wikimedia.org/wikipedia/commons/f/f2/Periodic_table_blocks_spdf_%2832_column%29.svg *License:* CC BY-SA 3.0 *Contributors:* https://commons.wikimedia.org/wiki/File:Periodic_Table_2.svg *Original artist:* User:DePiep

- **File:Periodic_table_by_Mendeleev,_1869.svg** *Source:* https://upload.wikimedia.org/wikipedia/commons/c/ce/Periodic_table_by_Mende2C_1869.svg *License:* Public domain *Contributors:* Own work *Original artist:* NikNaks

- **File:Periodic_table_by_Mendeleev,_1871.svg** *Source:* https://upload.wikimedia.org/wikipedia/commons/a/aa/Periodic_table_by_Mende2C_1871.svg *License:* Public domain *Contributors:* Own work *Original artist:* NikNaks

- **File:Periodic_table_compilation.svg** *Source:* https://upload.wikimedia.org/wikipedia/commons/0/04/Periodic_table_compilation.svg *License:* CC BY 3.0 *Contributors:* a compilation of: files: Telluric screw of De Chancourtois.gif; A New System of Chemical Philosophy fp.jpg; Mendeleev law.jpg; Lavoisier Traité élémentaire de chimie p192.jpg; Periodic table large.svg *Original artist:* authors described in individual files

- **File:Periodic_table_monument.jpg** *Source:* https://upload.wikimedia.org/wikipedia/commons/0/00/Periodic_table_monument.jpg *License:* CC BY-SA 2.0 *Contributors:* http://www.flickr.com/photos/mmmdirt/279349599 *Original artist:* http://www.flickr.com/people/mmmdirt/

- **File:Periodic_trends.svg** *Source:* https://upload.wikimedia.org/wikipedia/commons/f/fe/Periodic_trends.svg *License:* CC0 *Contributors:* Own work *Original artist:* Mirek2

- **File:Periodic_variation_of_Pauling_electronegativities.png** *Source:* https://upload.wikimedia.org/wikipedia/commons/b/b4/Periodic_of_Pauling_electronegativities.png *License:* CC-BY-SA-3.0 *Contributors:* Own work *Original artist:* Physchim62

- **File:Phosphorus2.jpg** *Source:* https://upload.wikimedia.org/wikipedia/commons/5/55/Phosphorus2.jpg *License:* CC BY-SA 3.0 *Contributors:* http://woelen.homescience.net/science/index.html *Original artist:* W. Oelen

- **File:Platinum-nugget.jpg** *Source:* https://upload.wikimedia.org/wikipedia/commons/6/6a/Platinum-nugget.jpg *License:* FAL *Contributors:* Own work *Original artist:* Alchemist-hp (<a href='//commons.wikimedia.org/wiki/User_talk:Alchemist-hp' title='User talk:Alchemist-hp'>tal)(www.pse-mendelejew.de)

- **File:Polycrystalline-germanium.jpg** *Source:* https://upload.wikimedia.org/wikipedia/commons/0/08/Polycrystalline-germanium.jpg *License:* CC BY 3.0 *Contributors:* http://images-of-elements.com/germanium.php *Original artist:* Jurii

- **File:Portal-puzzle.svg** *Source:* https://upload.wikimedia.org/wikipedia/en/f/fd/Portal-puzzle.svg *License:* Public domain *Contributors:* ? *Original artist:* ?

- **File:Question_book-new.svg** *Source:* https://upload.wikimedia.org/wikipedia/en/9/99/Question_book-new.svg *License:* Cc-by-sa-3.0 *Contributors:*
  Created from scratch in Adobe Illustrator. Based on Image:Question book.png created by User:Equazcion *Original artist:*
  Tkgd2007

- **File:RareEarthOreUSGOV.jpg** *Source:* https://upload.wikimedia.org/wikipedia/commons/1/10/RareEarthOreUSGOV.jpg *License:* Public domain *Contributors:* ? *Original artist:* ?

- **File:Rareearth_production.svg** *Source:* https://upload.wikimedia.org/wikipedia/commons/9/93/Rareearth_production.svg *License:* Public domain *Contributors:* This is a vector conversion of <a href='//commons.wikimedia.org/wiki/File:Rare-earths.gif' class='image'><img alt='Rare-earths.gif' src='https://upload.wikimedia.org/wikipedia/commons/thumb/c/c1/Rare-earths.gif/200px-Rare-earths.gif' width='200' height='121' srcset='https://upload.wikimedia.org/wikipedia/commons/thumb/c/c1/Rare-earths.gif/300px-Rare-earths.gif 1.5x, https://upload.wikimedia.org/wikipedia/commons/thumb/c/c1/Rare-earths.gif/400px-Rare-earths.gif 2x' data-file-width='442' data-file-height='268' /></a>. The SVG version is nearly identical to the original, though Wikimedia's renderer shows differences. *Original artist:* BMacZero

- **File:Rareearthoxides.jpg** *Source:* https://upload.wikimedia.org/wikipedia/commons/5/55/Rareearthoxides.jpg *License:* Public domain *Contributors:* http://www.ars.usda.gov/is/graphics/photos/jun05/d115-1.htm *Original artist:* Peggy Greb, US department of agriculture

- **File:S1M0.png** *Source:* https://upload.wikimedia.org/wikipedia/commons/1/1d/S1M0.png *License:* Public domain *Contributors:* Own work *Original artist:* Dhatfield

- **File:S2M0.png** *Source:* https://upload.wikimedia.org/wikipedia/commons/8/8f/S2M0.png *License:* Public domain *Contributors:* Own work *Original artist:* Dhatfield

- **File:Scandium_sublimed_dendritic_and_1cm3_cube.jpg** *Source:* https://upload.wikimedia.org/wikipedia/commons/e/e6/Scandium_sub dendritic_and_1cm3_cube.jpg*License:*FAL*Contributors:*Own work*Original artist:*Alchemist-hp(<a href='//commons.wikimedia.org/wiki/User_talk:Alchemist-hp'title='User talk:Alchemist-hp'>talk</a>) (www.pse-mendelejew.de)

- **File:Selenium_black_(cropped).jpg** *Source:* https://upload.wikimedia.org/wikipedia/commons/d/d3/Selenium_black_%28cropped%29.jpg *License:* CC BY-SA 3.0 *Contributors:* File:Selenium black.jpg *Original artist:* W. Oelen

- **File:Semiconductor-1.jpg** *Source:* https://upload.wikimedia.org/wikipedia/commons/8/8d/Semiconductor-1.jpg *License:* CC BY-SA 3.0 *Contributors:* Own work *Original artist:* Lejla peace

- **File:ShortPT20b.png** *Source:* https://upload.wikimedia.org/wikipedia/commons/c/c9/ShortPT20b.png *License:* CC BY-SA 3.0 *Contributors:* Own work *Original artist:* Sandbh

- **File:SiliconCroda.jpg** *Source:* https://upload.wikimedia.org/wikipedia/commons/e/e9/SiliconCroda.jpg *License:* Public domain *Contributors:* Transferred from en.wikipedia *Original artist:* Original uploader was Enricoros at en.wikipedia

- **File:Silver_crystal.jpg** *Source:* https://upload.wikimedia.org/wikipedia/commons/5/55/Silver_crystal.jpg *License:* CC BY-SA 3.0 *Contributors:* Own work (additional processed by Waugsberg) *Original artist:* Alchemist-hp (<a href='//commons.wikimedia.org/wiki/User_talk:Alchemist-hp' title='User talk:Alchemist-hp'>talk</a>) (www.pse-mendelejew.de)

- **File:Sn-Alpha-Beta.jpg** *Source:* https://upload.wikimedia.org/wikipedia/commons/2/2b/Sn-Alpha-Beta.jpg *License:* CC BY-SA 3.0 de *Contributors:* Own work *Original artist:* Alchemist-hp (<a href='//commons.wikimedia.org/wiki/User_talk:Alchemist-hp' title='User talk:Alchemist-hp'>talk</a>)(www.pse-mendelejew.de)

- **File:SolarSystemAbundances.png** *Source:* https://upload.wikimedia.org/wikipedia/commons/e/e6/SolarSystemAbundances.png *License:* CC BY-SA 3.0 *Contributors:* Transferred from en.wikipedia to Commons. *Original artist:* The original uploader was 28bytes at English Wikipedia

- **File:Sommerfeld_ellipses.svg** *Source:* https://upload.wikimedia.org/wikipedia/commons/7/75/Sommerfeld_ellipses.svg *License:* Public domain *Contributors:* Own work *Original artist:* Pieter Kuiper

- **File:Spacer.gif** *Source:* https://upload.wikimedia.org/wikipedia/commons/5/52/Spacer.gif *License:* Public domain *Contributors:* ? *Original artist:* ?

- **File:Speaker_Icon.svg** *Source:* https://upload.wikimedia.org/wikipedia/commons/2/21/Speaker_Icon.svg *License:* Public domain *Contributors:* ? *Original artist:* ?

- **File:Symbol_book_class2.svg** *Source:* https://upload.wikimedia.org/wikipedia/commons/8/89/Symbol_book_class2.svg *License:* CC BY-SA 2.5 *Contributors:* Mad by Lokal_Profil by combining: *Original artist:* Lokal_Profil

- **File:Tellurium2.jpg** *Source:* https://upload.wikimedia.org/wikipedia/commons/c/c1/Tellurium2.jpg *License:* CC BY 3.0 *Contributors:* http://images-of-elements.com/tellurium.php *Original artist:* Unknown

- **File:Text_document_with_red_question_mark.svg** *Source:* https://upload.wikimedia.org/wikipedia/commons/a/a4/Text_document_with_red_question_mark.svg *License:* Public domain *Contributors:* Created by bdesham with Inkscape; based upon Text-x-generic.svg from the Tango project. *Original artist:* Benjamin D. Esham (bdesham)

- **File:Titan-crystal_bar.JPG** *Source:* https://upload.wikimedia.org/wikipedia/commons/d/db/Titan-crystal_bar.JPG *License:* CC BY-SA 3.0 *Contributors:* Own work *Original artist:* Alchemist-hp (pse-mendelejew.de)

- **File:Transparent.gif** *Source:* https://upload.wikimedia.org/wikipedia/commons/c/ce/Transparent.gif *License:* Public domain *Contributors:* Own work *Original artist:* Edokter

- **File:Unbalanced_scales.svg** *Source:* https://upload.wikimedia.org/wikipedia/commons/f/fe/Unbalanced_scales.svg *License:* Public domain *Contributors:* ? *Original artist:* ?

- **File:Universe_content_pie_chart.jpg** *Source:* https://upload.wikimedia.org/wikipedia/commons/3/30/Universe_content_pie_chart.jpg *License:* Public domain *Contributors:* http://map.gsfc.nasa.gov/media/080998/index.html *Original artist:* Credit: NASA / WMAP Science Team

- **File:Warszawa_Centrum_Nowych_Technologii_UW-4.jpg***Source:*https://upload.wikimedia.org/wikipedia/commons/9/9a/Warszawa_ Nowych_Technologii_UW-4.jpg *License:* CC BY-SA 3.0 *Contributors:* Own work *Original artist:* Halibutt
- **File:WikiProject_Geology.svg** *Source:* https://upload.wikimedia.org/wikipedia/commons/e/e7/WikiProject_Geology.svg *License:* CC BY-SA 2.5 *Contributors:* ? *Original artist:* ?
- **File:Wiki_letter_w_cropped.svg** *Source:* https://upload.wikimedia.org/wikipedia/commons/1/1c/Wiki_letter_w_cropped.svg *License:* CC-BY-SA-3.0 *Contributors:*
- Wiki_letter_w.svg *Original artist:* Wiki_letter_w.svg: Jarkko Piiroinen
- **File:Wikibooks-logo-en-noslogan.svg***Source:* https://upload.wikimedia.org/wikipedia/commons/d/df/Wikibooks-logo-en-noslogan.svg *License:* CC BY-SA 3.0 *Contributors:* Own work *Original artist:* User:Bastique, User:Ramac et al.
- **File:Wikibooks-logo.svg** *Source:* https://upload.wikimedia.org/wikipedia/commons/f/fa/Wikibooks-logo.svg *License:* CC BY-SA 3.0 *Contributors:* Own work *Original artist:* User:Bastique, User:Ramac et al.
- **File:Wikinews-logo.svg** *Source:* https://upload.wikimedia.org/wikipedia/commons/2/24/Wikinews-logo.svg *License:* CC BY-SA 3.0 *Contributors:* This is a cropped version of Image:Wikinews-logo-en.png. *Original artist:* Vectorized by Simon 01:05, 2 August 2006 (UTC) Updated by Time3000 17 April 2007 to use official Wikinews colours and appear correctly on dark backgrounds. Originally uploaded by Simon.
- **File:Wikiquote-logo.svg** *Source:* https://upload.wikimedia.org/wikipedia/commons/f/fa/Wikiquote-logo.svg *License:* Public domain *Contributors:* ? *Original artist:* ?
- **File:Wikisource-logo.svg** *Source:* https://upload.wikimedia.org/wikipedia/commons/4/4c/Wikisource-logo.svg *License:* CC BY-SA 3.0 *Contributors:* Rei-artur *Original artist:* Nicholas Moreau
- **File:Wikiversity-logo-Snorky.svg** *Source:* https://upload.wikimedia.org/wikipedia/commons/1/1b/Wikiversity-logo-en.svg *License:* CC BY-SA 3.0 *Contributors:* Own work *Original artist:* Snorky
- **File:Wikiversity-logo.svg** *Source:* https://upload.wikimedia.org/wikipedia/commons/9/91/Wikiversity-logo.svg *License:* CC BY-SA 3.0 *Contributors:* Snorky (optimized and cleaned up by verdy_p) *Original artist:* Snorky (optimized and cleaned up by verdy_p)
- **File:Wiktionary-logo-en.svg** *Source:* https://upload.wikimedia.org/wikipedia/commons/f/f8/Wiktionary-logo-en.svg *License:* Public domain *Contributors:* Vector version of Image:Wiktionary-logo-en.png. *Original artist:* Vectorized by Fvasconcellos (talk · contribs), based on original logo tossed together by Brion Vibber
- **File:XeF2.png** *Source:* https://upload.wikimedia.org/wikipedia/commons/2/27/XeF2.png *License:* Public domain *Contributors:* user-made *Original artist:* User:Smokefoot
- **File:XeTube.jpg** *Source:* https://upload.wikimedia.org/wikipedia/commons/5/59/XeTube.jpg *License:* CC BY-SA 2.5 *Contributors:* user-made *Original artist:* User:Pslawinski
- **File:Xenon-glow.jpg** *Source:* https://upload.wikimedia.org/wikipedia/commons/5/5d/Xenon-glow.jpg *License:* CC BY 3.0 *Contributors:* http://images-of-elements.com/xenon.php *Original artist:* Jurii
- **File:Xenon-tetrafluoride-3D-vdW.png** *Source:* https://upload.wikimedia.org/wikipedia/commons/e/e3/Xenon-tetrafluoride-3D-vdW.png *License:* Public domain *Contributors:* user-made *Original artist:* User:Benjah-bmm27
- **File:Xenon_Spectrum.jpg** *Source:* https://upload.wikimedia.org/wikipedia/commons/6/67/Xenon_Spectrum.jpg *License:* Public domain *Contributors:* Transferred from en.wikipedia; transferred to Commons by User:Homer Landskirty using CommonsHelper. *Original artist:* Teravolt (talk). Original uploader was Teravolt at en.wikipedia
- **File:Xenon_discharge_tube.jpg** *Source:* https://upload.wikimedia.org/wikipedia/commons/d/d7/Xenon_discharge_tube.jpg *License:* FAL *Contributors:* Own work *Original artist:* Alchemist-hp (<a href='//commons.wikimedia.org/wiki/User_talk:Alchemist-hp' title='User talk: Alchemist-hp'>talk</a>) (www.pse-mendelejew.de)
- **File:Xenon_short_arc_1.jpg** *Source:* https://upload.wikimedia.org/wikipedia/commons/9/9e/Xenon_short_arc_1.jpg *License:* CC BY 2.5 *Contributors:* user-made *Original artist:* provided that proper attribution of my copyright is made. - Atlant 19:15, 26 August 2005 (UTC)
- **File:Yttrium_sublimed_dendritic_and_1cm3_cube.jpg***Source:*https://upload.wikimedia.org/wikipedia/commons/1/19/Yttrium_sublimed dendritic_and_1cm3_cube.jpg *License:* FAL *Contributors:* Own work *Original artist:* Alchemist-hp (www.pse-mendelejew.de)
- **File:Zirconium_crystal_bar_and_1cm3_cube.jpg** *Source:* https://upload.wikimedia.org/wikipedia/commons/9/92/Zirconium_crystal_bar_ and_1cm3_cube.jpg *License:* FAL *Contributors:* Own work *Original artist:* Alchemist-hp (pse-mendelejew.de).
- **File:Zn_reaction_with_HCl.JPG** *Source:* https://upload.wikimedia.org/wikipedia/commons/d/d4/Zn_reaction_with_HCl.JPG *License:* Public domain *Contributors:* Own work *Original artist:* Chemicalinterest
- **File:Дмитрий_Иванович_Менделеев_4.gif** *Source:* https://upload.wikimedia.org/wikipedia/commons/2/26/%D0%94%D0%BC%D0% B8%D1%82%D1%80%D0%B8%D0%B9_%D0%98%D0%B2%D0%B0%D0%BD%D0%BE%D0%B2%D0%B8%D1%87_%D0%9C%D0 %B5%D0%BD%D0%B4%D0%B5%D0%BB%D0%B5%D0%B5%D0%B2_4.gif*License:*Public domain*Contributors:*Weeks,Mary Elvira(1933)*The Discovery of the Elements*,Easton,PA:Journal of Chemical Education,p.208ISBN:0766138720.*Original artist:*Unknown

37.9.3 Content license

- Creative Commons Attribution-Share Alike 3.0

49121055R00210

Made in the USA
Lexington, KY
26 January 2016